Computational Fluid Dynamics for Engineers and Scientists

Sreenivas Jayanti

Computational Fluid Dynamics for Engineers and Scientists

Sreenivas Jayanti
Department of Chemical Engineering
Indian Institute of Technology Madras
Chennai
India

ISBN 978-94-024-1511-7 ISBN 978-94-024-1217-8 (eBook)
https://doi.org/10.1007/978-94-024-1217-8

Softcover re-print of the Hardcover 1st edition 2018

Printed on acid-free paper

This Springer imprint is published by Springer Nature
The registered company is Springer Science+Business Media B.V.
The registered company address is: Van Godewijckstraat 30, 3311 GX Dordrecht, The Netherlands

Preface

Computational fluid dynamics (CFD) has its origins in fluid mechanics and mathematics but finds applications in diverse fields of engineering and science. The rapid spread of CFD from aerospace applications to almost every branch of engineering—especially over the past two decades, and especially in the form of easy-to-use CFD codes with attractive graphical user interfaces and with promises of being capable of simulating "anything that flows"—has created a large pool of new students of CFD. Many of these would have had an elementary course in fluid mechanics and perhaps one in numerical methods, but would otherwise have had no grounding in CFD. This book is written for two types of such students: for those who are taking up CFD for the first time in the form of a semester-long taught course of about 40 to 50 lectures, and for those practising engineers and scientists who are already using CFD as an analysis tool in their professions but would like to deepen and broaden their understanding of the subject. Presuming no more knowledge than college-level understanding of the core subjects, the book puts together all the necessary topics to give the reader a comprehensive introduction to CFD. This is done in the following way.

We begin with a rather unusual foray into CFD by taking up two simple but seemingly tough fluid flow problems that we work out in the CFD way. We show how the partial differential equation (PDE) governing the fluid flow in each case can be converted into a set of easily solvable linear algebraic equations by making some approximations which are at the core of CFD. In the second chapter, we derive, from basics, all the necessary fluid flow equations that we would be solving in CFD. We discuss the nature of these equations and introduce the concept of wellposedness of the fluid flow problem in a mathematical sense. In the next chapter, we discuss the basic concepts that underlie CFD with reference to the numerical solution of a typical PDE encountered in fluid flow. We discuss the finite difference approximation of a derivative to desired accuracy; the consistency and stability of a finite difference approximation of a governing equation; the diffusive and dispersive errors induced therein; and finally the measures that have been devised to control these errors. In the next chapter, we use the lessons learnt from the numerical solution of a single PDE to tackle the real problem of solving fluid flow equations, which are non-linear, coupled PDEs. Distinction is made here between issues associated with compressible subsonic flows, supersonic flows with shocks, and incompressible flows. We discuss the difficulties associated with each and see how these have been overcome. The next chapter is an acknowledgement of the deep contribution made to CFD by mathematicians who have developed special methods to solve the set of linear algebraic equations. In order to give the reader proper perspective of the subject, we discuss not only the basic methods but also a number of advanced methods that have made CFD a very efficient number cruncher. The final two chapters of the book are intended for the more advanced user. We discuss in Chap. 6 the special difficulties that arise when we wish to solve practical problems. We make a distinction between complications arising out of geometrical complexity and those arising out of the complexity of physics (and chemistry) of the problem. We discuss what the problems in each case are and how these can be overcome. Given the still-evolving nature of these issues, we keep the details to a minimum and focus on the broad ideas. The last chapter contains a brief discussion of what can be considered as the Holy Grail of CFD, namely, finding the optimal design of a fluid flow component. We briefly review the conventional routes to optimization, and see, through a case study, how CFD can be embedded in a bigger scheme of things to find the optimal shape of the fluid flow component.

The book contains many sections that are rather easy to follow but some sections require considerable effort from the reader. These have been included to give the reader an overview of the developments in CFD. A few key references are given in such cases to serve the reader as a lead for further follow-up study. Each chapter contains a problem set which is aimed at reinforcing some of the concepts discussed in the chapter. The problems also serve to identify what a student is expected to have learnt or become proficient at. Some problems require knowledge which is not covered in this book; in such cases, a suitable reference is given for the interested reader to follow-up on.

I expect that a semester-long introductory course in CFD can be put together using the material presented in Chaps. 1–3 (the first three sections), and 4 (sects. 4.2 and 4.4; sect. 4.3 requires knowledge of gas dynamics and is therefore for a specialized audience), the first four sections of Chap. 5, and a few relevant subsections of Chap. 6. A more advanced course—one that emphasizes on assignment projects involving coding—can also be put together using the material presented in sect. 3.5, all sections of Chap. 4, sect. 5.5, and up to and including sect. 6.4.4 of Chap. 6. A quick run-through of the earlier chapters can serve as an introduction to the advanced course. The end of the chapter problem sets can be used as assignments to be worked out on paper or through computer programming, as the case may be.

In preparing this book, I have benefitted from years of experience of using CFD as an analysis tool to study numerous fluid flow problems involving flow, heat transfer, chemical reactions, and multiphase flow. During this time, I had the opportunity to learn from excellent books on CFD and from insightful research articles published in reputed journals. I have also been teaching a course on CFD at IIT Madras for a number of years, and the structure of the book has definitely been influenced by this experience. I owe a debt of gratitude to Springer and their staff, especially, to Mark de Jongh, Cindy Zitter, and Tom Spicer, for their encouragement and advice. Finally, I wish to thank my wife, Lalitha Kalpana, son, Pranava Chaitanya, and daughter, Prajnana Dipika, for putting up with many inconveniences over the past few years. I dedicate this book to the memory of my father—an experimental nuclear physicist—whose path of teaching and research I subconsciously followed, and to the three formidable women of my life: my mother, my wife, and my mother-in-law.

Chennai, India Sreenivas Jayanti

Contents

Chapter 1
Introduction

Fluid flow occurs in a number of natural and industrial processes. Examples of the former are the wind flow pattern that affects our weather, and the flow of blood through arteries and veins which maintains oxygen supply to the organs and cells in our bodies. A common example of industrial processes where fluid flow is of importance is the air flow over a car which influences not only the fuel consumption but also the stability of the car when it is driven at high speeds. Another such common example is the air flow over the blades of a wind turbine where one seeks to convert the kinetic energy of the wind to shaft power and eventually to electrical power. There are many other cases where fluid flow has a strong influence in the way an allied phenomenon occurs. For example, fluid flow has an important role in the dispersion of atmospheric pollutants, in the distribution and delivery of drugs injected into the blood stream, and in the dispersion of the fuel and the subsequent ignition, combustion and heat release inside the engine of a car. In all these cases, the scientist or the engineer seeks a detailed knowledge of the flow and other parameters associated with the process. Such analysis can be very useful. The depiction of the wind velocities, temperature contours and isobars to illustrate changes in local weather and climatic conditions (such as tracking of cyclones) has now become the norm in the weather bulletins on our television screens. The reader may also have heard of the extensive use of wind tunnel testing and computational fluid dynamics (CFD) simulations that are used in the field of Formula One car racing. It is quite common for teams to run CFD simulations in-between races to fine tune their car settings to suit a particular circuit or to suit the driving style of a particular driver. Another well-publicized illustration of the deployment of CFD was the case of King's Cross Underground Station Fire which occurred in November 1987 in London killing 31 people and injuring about 100 others. A baffling aspect of the fire, namely, its rapid rise and flashover, was resolved first through CFD simulations (Simcox et al. 1992) carried out in 1988. The "trench effect" discovered through CFD simulations was initially discounted by many but had to be accepted eventually when it was confirmed later through scale-model experiments! Those were the early days of CFD applications in non-aerospace,

S. Jayanti, *Computational Fluid Dynamics for Engineers and Scientists*,
https://doi.org/10.1007/978-94-024-1217-8_1

non-mechanical engineering domains. CFD has now become an essential analytical tool in almost every branch of engineering and in many scientific disciplines.

The secret behind the success of CFD is its ability to simulate flows in close to practical conditions–in terms of tackling real, three-dimensional, irregular flow geometries and phenomena involving complex physics. This is made possible by resorting to numerical solution of the equations governing fluid flow rather than seeking an analytical solution. As we will see in Chap. 2, the equations describing the flow of common fluids, such as air and water, consist of mathematical statements of conservation of such fundamental quantities as mass, momentum and energy during fluid flow, and equations of state of the fluid defining its thermodynamic properties. The variables in these equations are three velocity components, pressure and temperature (or a related quantity such as the internal energy or the enthalpy) of the fluid. In the general case, each of these varies with location (x, y, z) and time within the flow domain. Their variation is governed and determined by the conservation equations which take the form of non-linear partial differential equations. Analytical solutions exist only in some highly idealized conditions, and are therefore not very useful in dealing with practical problems. Thus the conventional approach to fluid mechanics has limited applicability; taking the numerical solution approach widens considerably the scope of problems that can be tackled.

Computational fluid dynamics deals with the numerical solution of equations that describe fluid flow and allied phenomena. Although some of the basic numerical methods have been known for a long time and pre-date the formulation of the fluid flow equations that we use in CFD, the specialized set of algorithms and techniques that constitute CFD have been developed only over the last century. A computational fluid dynamics solution, as we shall see below, requires large number of arithmetic computations on real numbers; hence its rise coincided with the advent of computers and the rapid expansion of computer power that ensued in the subsequent decades. Many of the methods and algorithms that we use today have had their origins in the 1960s–1980s. In this chapter, we begin the exploration of the subject by seeking a CFD solution to two fairly simple flow problems which nevertheless illustrate the main ideas in this approach.

1.1 The Case of Flow in a Duct of Rectangular Cross-Section

Consider the case of steady, fully-developed, laminar flow of a Newtonian fluid through a duct of rectangular cross-section in the x-y plane as shown schematically in Fig. 1.1a. For this case, the pressure gradient in the axial direction is constant and the velocity in the duct axial direction, i.e., w, the velocity component in the z-direction, is the only non-zero velocity component. Its variation within the rectangle is governed by the following partial differential equation:

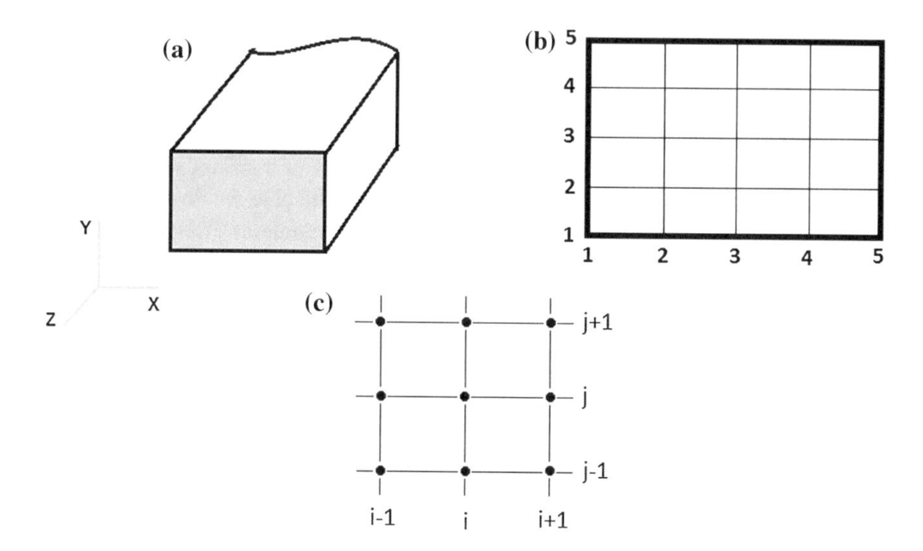

Fig. 1.1 Fully developed, steady flow through a rectangular duct: **a** computational domain, **b** discretization into 16 cells, and **c** nomenclature for finite difference approximations

$$\partial^2 \mathrm{w}/\partial x^2 + \partial^2 \mathrm{w}/\partial y^2 = (1/\mu)\, \partial p/\partial z = \mathrm{C} = \text{constant} \tag{1.1.1a}$$

Here, μ is the dynamic viscosity of the fluid and p is the pressure. The mathematical problem is fully specified when the value of the constant is specified together with the no-slip boundary condition, namely,

$$\mathrm{w} = 0 \text{ on the walls of the duct} \tag{1.1.1b}$$

By solving Eq. (1.1.1), one would be able to find w(x, y), and upon integrating it over the entire cross-section, one would be able to obtain the volumetric flow rate through the duct for a given pressure gradient, which is uniquely determined by the value of the constant and the viscosity of the fluid.

An analytical solution to Eq. (1.1.1) is possible but this may require advanced level mathematical skills. Using the CFD approach, a numerical solution can be obtained with no more than entry-level engineering mathematical skills. However, this is at the cost of some simplification, as explained below. A CFD solution involves three major steps. In the first step, we discretize the flow domain and identify the grid nodes at which we seek the solution. These points are closely spaced and are spread throughout the flow domain. In the second step, we use finite difference approximations for the derivatives and reduce the governing partial differential Eq. (1.1.1) to an algebraic equation involving the velocities at the neighbouring grid points. We do this for all the grid points at which the solution needs to be determined and arrive at a system of algebraic equations to be solved. In the third step, we solve the system of linear algebraic equations to obtain the velocities at the grid nodes. These three steps are illustrated below for the case of flow through a rectangular duct:

In step I, we seek to find the solution at pre-specified grid points (x_i, y_j) spread throughout the rectangular flow domain shown in Fig. 1.1a. Given the simple geometry, we locate these grid points at the intersections of constant x and constant y lines placed at regular intervals. Figure 1.1b shows the case for five vertical and five horizontal grid lines including those defining the boundaries. Thus, we redefine the problem given by Eq. (1.1.1) as one of finding w (x_i, y_j) where i = 1–5 and j = 1–5 and where x_i and y_j define the x- and y-coordinates of the grid points.

In step II, we find the equations that need to be solved to determine the velocities. Since the velocity on the boundary is given (Eq. 1.1.1b), there are nine interior points at which w needs to be determined. At each such grid point, we write a finite difference approximation of Eq. (1.1.1a). Several choices exist; for this example, we choose to approximate the equation as follows at grid point (i, j):

$$\left(w_{i-1,j} - 2w_{i,j} + w_{i+1,j}\right)/\Delta x^2 + \left(w_{i,j-1} - 2w_{i,j} + w_{i,j+1}\right)/\Delta y^2 = C \qquad (1.1.2)$$

where Δx and Δy are the grid spacings, i.e., the distance between successive grid points in the x- and y-directions (see Fig. 1.1c). Equation (1.1.2) can be rewritten as

$$\begin{aligned}(1/\Delta x^2)w_{i-1,j} - (2/\Delta x^2 + 2/\Delta y^2)w_{i,j} \\ + (1/\Delta x^2)w_{i+1,j} + (1/\Delta y^2)w_{i,j-1} + (1/\Delta y^2)w_{i,j+1} = C\end{aligned} \qquad (1.1.3)$$

For the specific case of $C = -100\ \mathrm{m^{-1}s^{-1}}$, $\Delta x = 0.1$ m and $\Delta y = 0.05$ m, and for the grid point i = 2, j = 2, Eq. (1.1.3) becomes (writing $w_{i,j}$ as w_{ij} as there is no confusion in this case)

$$100w_{12} - 1000w_{22} + 100w_{32} + 400w_{21} + 400w_{23} = -100 \qquad (1.1.4)$$

Substituting the boundary values that $w_{12} = 0$ and $w_{21} = 0$, we finally have

$$-1000w_{22} + 100\,w_{32} + 400\,w_{23} = -100.0 \qquad (1.1.5a)$$

We can apply the template given by Eq. (1.1.2) to grid points (3,2), (4,2), (2,3), (3,3), (4,3), (2,4), (3,4) and (4,4) sequentially, and obtain the following algebraic equations:

$$100w_{22} - 1000w_{32} + 100w_{42} + 400w_{33} = -100 \qquad (1.1.5b)$$

$$100w_{32} - 1000w_{42} + 400w_{43} = -100 \qquad (1.1.5c)$$

$$-1000w_{23} + 100w_{33} + 400w_{22} + 400w_{24} = -100 \quad (1.1.5d)$$

$$100w_{23} - 1000w_{33} + 100w_{43} + 400w_{32} + 400w_{34} = -100 \quad (1.1.5e)$$

$$100w_{33} - 1000w_{43} + 400w_{42} + 400w_{44} = -100 \quad (1.1.5f)$$

$$-1000w_{24} + 100w_{34} + 400w_{23} = -100 \quad (1.1.5g)$$

$$100w_{24} - 1000w_{34} + 100w_{44} + 400w_{33} = -100 \quad (1.1.5h)$$

$$100w_{34} - 1000w_{44} + 400w_{43} = -100 \quad (1.1.5i)$$

Equations (1.1.5a–i) are then the equations that need to be solved to find the velocity at the chosen nodes.

In step III, we solve the equations and obtain the solution. One can see that there are nine algebraic equations that need to be solved to obtain the nine values of $w_{i,j}$. These are coupled in the sense that none of them can be solved in isolation and one thus needs to solve a system of algebraic equations given by

$$Aw = b \quad (1.1.6)$$

where A is the coefficient matrix given by

$$A = \begin{bmatrix} -1000 & 100 & 0 & 400 & 0 & 0 & 0 & 0 & 0 \\ 100 & -1000 & 100 & 0 & 400 & 0 & 0 & 0 & 0 \\ 0 & 100 & -1000 & 0 & 0 & 400 & 0 & 0 & 0 \\ 400 & 0 & 0 & -1000 & 100 & 0 & 400 & 0 & 0 \\ 0 & 400 & 0 & 100 & -1000 & 100 & 0 & 400 & 0 \\ 0 & 0 & 400 & 0 & 100 & -1000 & 0 & 0 & 400 \\ 0 & 0 & 0 & 400 & 0 & 0 & -1000 & 100 & 0 \\ 0 & 0 & 0 & 0 & 400 & 0 & 100 & -1000 & 100 \\ 0 & 0 & 0 & 0 & 0 & 400 & 0 & 100 & -1000 \end{bmatrix} \quad (1.1.7a)$$

$$w = \begin{bmatrix} w_{22} & w_{32} & w_{42} & w_{23} & w_{33} & w_{43} & w_{24} & w_{34} & w_{44} \end{bmatrix}^T \quad (1.1.7b)$$

$$b = \begin{bmatrix} -100 & -100 & -100 & -100 & -100 & -100 & -100 & -100 & -100 \end{bmatrix}^T \quad (1.1.7c)$$

We need to solve Eq. (1.1.6) to get the velocities at the grid nodes. For this small system of equations, we can use an elimination technique. However, here we use the Gauss-Seidel iterative method, which is often used in CFD computations, to solve for the velocities. Accordingly, using index k to indicate the iteration number we rewrite Eqs. (1.1.5a–i) as follows:

$$w_{22}^{k+1} = \left(100 + 100w_{32}^{k} + 400w_{23}^{k}\right)/1000 \quad (1.1.8a)$$

$$w_{32}^{k+1} = \left(100 + 100w_{22}^{k+1} + 100w_{42}^{k} + 400w_{33}^{k}\right)/1000 \quad (1.1.8b)$$

$$w_{42}^{k+1} = \left(100 + 100w_{32}^{k+1} + 400w_{43}^{k}\right)/1000 \quad (1.1.8c)$$

$$w_{23}^{k+1} = \left(100 + 400w_{22}^{k+1} + 100w_{33}^{k} + 400w_{24}^{k}\right)/1000 \quad (1.1.8d)$$

$$w_{33}^{k+1} = \left(100 + 100w_{23}^{k+1} + 400w_{32}^{k+1} + 100w_{43}^{k} + 400w_{44}^{k}\right)/1000 \quad (1.1.8e)$$

$$w_{43}^{k+1} = \left(100 + 100w_{33}^{k+1} + 400w_{42}^{k+1} + 400w_{44}^{k}\right)/1000 \quad (1.1.8f)$$

$$w_{24}^{k+1} = \left(100 + 400w_{23}^{k+1} + 100w_{34}^{k}\right)/1000 \quad (1.1.8g)$$

$$w_{34}^{k+1} = \left(100 + 400w_{33}^{k+1} + 100w_{24}^{k+1} + 100w_{44}^{k}\right)/1000 \quad (1.1.8h)$$

$$w_{44}^{k+1} = \left(100 + 400w_{43}^{k+1} + 100w_{34}^{k+1}\right)/1000 \quad (1.1.8i)$$

Equations (1.1.8a)–(1.1.8i) will now be solved iteratively and sequentially with the initial guess of $\{w\}^0 = \{0\}$. The values for the first fourteen iterations are shown in Table 1.1. One can see that the successive solutions are converging as iteration progresses. The last row contains the solution to a four significant digit accuracy.

Table 1.1 Solution of Eq. (1.1.8) using Gauss Seidel method with a starting value of $\{w\}^0 = 0$

Iter #	w_{22}	w_{32}	w_{42}	w_{23}	w_{33}	w_{43}	w_{24}	w_{34}	w_{44}
0	0.0000	0.0000	0.0000	0.0000	0.0000	0.0000	0.0000	0.0000	0.0000
1	0.1000	0.1100	0.1110	0.1400	0.1580	0.1602	0.1560	0.1788	0.1820
2	0.1670	0.1910	0.1832	0.2450	0.2897	0.2750	0.2159	0.2557	0.2356
3	0.2171	0.2559	0.2356	0.3022	0.3543	0.3239	0.2464	0.2899	0.2586
4	0.2465	0.2899	0.2586	0.3326	0.3850	0.3453	0.2620	0.3061	0.2687
5	0.2620	0.3061	0.2687	0.3481	0.3993	0.3549	0.2699	0.3136	0.2733
6	0.2699	0.3136	0.2733	0.3558	0.4058	0.3592	0.2737	0.3170	0.2754
7	0.2737	0.3170	0.2754	0.3595	0.4089	0.3612	0.2755	0.3186	0.2763
8	0.2755	0.3186	0.2763	0.3613	0.4102	0.3621	0.2764	0.3194	0.2768
9	0.2764	0.3194	0.2768	0.3621	0.4109	0.3625	0.2768	0.3197	0.2770
10	0.2768	0.3197	0.2770	0.3625	0.4112	0.3627	0.2770	0.3199	0.2771
11	0.2770	0.3199	0.2771	0.3627	0.4113	0.3628	0.2771	0.3199	0.2771
12	0.2771	0.3199	0.2771	0.3628	0.4114	0.3628	0.2771	0.3200	0.2771
13	0.2771	0.3200	0.2771	0.3628	0.4114	0.3628	0.2771	0.3200	0.2771
14	0.2771	0.3200	0.2771	0.3628	0.4114	0.3628	0.2771	0.3200	0.2771

Thus, in the CFD approach, we reduce the governing partial differential equation to the more familiar problem of a system of linear algebraic equations involving the variable values at grid points as the unknowns. The algebraic equations are then solved to obtain the velocities. Other quantities of interest, such as the velocity at a particular location or the wall shear stress or the overall drag force, are derived from these velocities through interpolation or extrapolation, as required.

1.2 The Case of Flow in a Duct of Triangular Cross-Section

Consider now the case of steady, fully-developed, laminar flow of a Newtonian fluid through a duct of *triangular* cross-section shown schematically in Fig. 1.2a. The flow inside the duct is governed by the same equation as in the above case (Eq. 1.1.1a) with the same boundary condition (Eq. 1.1.1b). The difference in this case is that the boundaries do not coincide with lines of constant x and constant y. We therefore use a different approach, one based on the finite volume method, to reduce the governing partial differential equation to a system of algebraic equations. We once again use three steps to solve the problem:

In step I, we discretize the flow domain into triangular cells, which, when put together, will make up the entire flow domain. Figure 1.2b shows such discretization containing nine equilateral triangles. We seek to determine the value of the velocity at the centroid of each triangle, i.e., at nine points within the triangular domain, as shown in Fig. 1.2b. The coordinates of the centroid of the triangle can be determined from the coordinates of its three vertices using the following simple formula:

$$x_c = (x_1 + x_2 + x_3)/3; \; y_c = (y_1 + y_2 + y_3)/3 \tag{1.2.1}$$

where the subscript c denotes the centroid and the numerical subscripts denote the vertices. The cells are numbered 1–9 as shown in Fig. 1.2b.

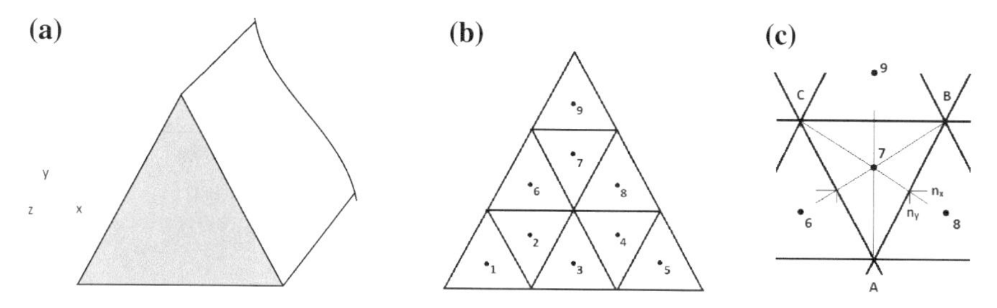

Fig. 1.2 Fully developed, steady flow through a triangular duct: **a** computational domain, **b** discretization into 9 cells, and **c** nomenclature for finite volume approximation

In step II, we reduce the governing partial differential equation to a system of algebraic equations using the finite volume method. We start by rewriting Eq. (1.1.1a) as

$$\nabla.(\nabla w) = C = -0.001 \quad (1.2.2)$$

and integrate over cell i:

$$\int_i \{\nabla.(\nabla w)\}\, dA = \int_i C dA \quad (1.2.3)$$

Using Gauss' divergence theorem, the integral over the closed surface area on the left hand side (LHS) of Eq. (1.2.3) can be converted into a line integral over the closed periphery of the cell i. Since C is constant, the integral on the right hand side (RHS) becomes equal to the product of C and the area of the cell. Thus,

$$\int_i \{\nabla.(\nabla w)\}\, dA = \int_i \{\mathbf{n}.(\nabla w)\}.dl = C\, A_i \quad (1.2.4)$$

where **n** is the outward normal vector of the perimeter. With reference to triangle ABC in Fig. 1.2c, the line integral in Eq. (1.2.4) can be evaluated as

$$\int_i \{\mathbf{n}.(\nabla w)\}.dl = \sum\nolimits_j (\mathbf{n}.\nabla w)_{j,i} l_{j,i} = C A_i \quad (1.2.5)$$

where j indicates the side of the triangle. Using the direction cosines (n_x and n_y) of the outward normal for each side, Eq. (1.2.5) can be rewritten as

$$\begin{aligned}&\left(n_x \partial w/\partial x + n_y \partial w/\partial y\right)_{AB} L_{AB} + \left(n_x \partial w/\partial x + n_y \partial w/\partial y\right)_{BC} L_{BC} \\ &+ \left(n_x \partial w/\partial x + n_y \partial w/\partial y\right)_{CA} L_{CA} = C\, A_{ABC}\end{aligned} \quad (1.2.6)$$

Here, the direction cosines are obtained from the geometry of the each side while the gradients are evaluated at mid-points of each side using values of the cell and those of the neighbouring cells or boundary conditions. For example, for cell 7 shown in Fig. 1.2c, n_x is equal to ½, 0 and −½ for sides AB, BC and CA, respectively. Similarly, n_y is equal to $-\sqrt{3}/2$, 1 and $-\sqrt{3}/2$. The length of all these sides is the same. Noting that $(\partial w/\partial x)_{BC}$ need not be evaluated as the corresponding $n_x = 0$, the gradients in Eq. (1.2.6) can be evaluated as follows:

$$\begin{aligned}&\partial w/\partial x|_{AB} = (w_7 - w_6)/(x_{c7} - x_{c6});\ \partial w/\partial y|_{AB} = (w_7 - w_6)/(y_{c7} - y_{c6})\\ &\partial w/\partial x|_{BC} = \text{not needed};\quad \partial w/\partial y|_{BC} = (w_9 - w_7)/(y_{c9} - y_{c7})\\ &\partial w/\partial x|_{CA} = (w_8 - w_7)/(x_{c9} - x_{c7});\ \partial w/\partial y|_{CA} = (w_8 - w_7)/(y_{c8} - y_{c7})\end{aligned}$$

Noting that $L_{AB} = L_{BC} = L_{CA} = l$ and the area of the triangle is equal to $(\sqrt{3}/4)l^2$, Eq. (1.2.6) can be rewritten as

$$(1/2)((w_8 - w_7)/(x_{c9} - x_{c7}))l + (-\sqrt{3}/2)((w_8 - w_7)/(y_{c8} - y_{c7}))l + (-1)((w_9 - w_7)/(y_{c9} - y_{c7}))l + (-1/2)((w_7 - w_6)/(x_{c7} - x_{c6}))l + (-\sqrt{3}/2)((w_7 - w_6)/(y_{c7} - y_{c6}))l = (\sqrt{3}/4)(l^2C) \quad (1.2.7)$$

We note that Eq. (1.2.7) is an algebraic equation containing the velocities at the cell centroids as the unknowns. The coefficients can be obtained from the geometry of each cell. Table 1.2 summarizes the evaluation of the direction cosines and the gradients for all the nine cells. Here a common nomenclature is used for the three sides; side a is the side which is horizontal (e.g., side BC of cell 7), side b is the side that makes an angle of 120° (or 300°) with the horizontal (e.g., side AB of cell 7) and side c is the one that an makes an angle of 210° (or 60°) with the horizontal (e.g. side CA of cell 7). For a side which is part of the boundary, the gradient is evaluated using the value of the cell at its centroid and the value of the boundary closest to it, i.e., at the mid-point of the side. For example, for the left side boundary (side c) of cell l, the mid-point has the coordinates of ($l/4$, $\sqrt{3}l/4$) while the centroid has the coordinates of ($l/2$, $l/(2\sqrt{3})$). The gradients on this face can thus be approximated as

$$\partial w/\partial x = (w_1 - 0)/(l/2 - l/4); \quad \partial w/\partial y = (w_1 - 0)/[l/(2\sqrt{3}) - \sqrt{3}l/4] \quad (1.2.8)$$

Using the direction cosines and the gradients for each face of cell and using the template given in Eq. (1.2.6), one can arrive at the following system of linear algebraic equations which contain the velocities at the centroid of each cell as the unknowns:

$$-138.5461\, w_1 + 34.6410\, w_2 = -4.3301 \quad (1.2.9a)$$

$$34.6410\, w_1 - 86.6025 w_2 + 34.6410 w_3 + 17.3205 w_6 = -4.3301 \quad (1.2.9b)$$

$$34.6410 w_2 - 103.7230 w_3 + 34.6410 w_4 = -4.3301 \quad (1.2.9c)$$

$$34.6410 w_3 - 86.6025 w_4 + 34.6410 w_5 + 17.3205 w_8 = -4.3301 \quad (1.2.9d)$$

$$34.6410 w_4 - 138.5640 w_5 = -4.3301 \quad (1.2.9e)$$

$$17.3205 w_2 - 121.2473 w_6 + 34.6410 w_7 = -4.3301 \quad (1.2.9f)$$

$$34.6410 w_6 - 86.6025 w_7 + 34.6410 w_8 + 34.6410 w_9 = -4.3301 \quad (1.2.9g)$$

$$17.3205 w_4 + 34.6410 w_7 - 121.2473 w_8 = -4.3301 \quad (1.2.9h)$$

$$17.3205 w_7 - 155.0846 w_9 = -4.3301 \quad (1.2.9i)$$

In step III, we solve the nine simultaneous equations using the Gauss-Seidel method. We arrange the unknowns in the cell-wise manner, i.e.,

Table 1.2 Evaluation of the geometric parameters and fluxes associated with the nine triangular cells shown in Fig. 1.2b. The first side is the horizontal side of each triangle and successive sides are numbered in anti-clockwise direction

Node	1	2	3	4	5	6	7	8	9
x_c	0.050000	0.100000	0.150000	0.200000	0.250000	0.100000	0.150000	0.200000	0.150000
y_c	0.028868	0.057735	0.028868	0.057735	0.028868	0.115470	0.144338	0.115470	0.202073
x_{m1}	0.050000	0.100000	0.150000	0.200000	0.250000	0.100000	0.150000	0.200000	0.150000
y_{m1}	0.000000	0.086603	0.000000	0.086603	0.000000	0.086603	0.173205	0.086603	0.173205
n_x	0	0	0	0	0	0	0	0	0
n_y	−1	1	−1	1	−1	−1	1	−1	−1
F_x	0	0	0	0	0	0	0	0	0
F_y	$(w_1-0)/(yc_1-ym_1)$	$(w_2-w_6)/(yc_2-yc_6)$	$(w_3-0)/(yc_3-ym)$	$(w_4-w_8)/(yc_4-yc_8)$	$(w_5-0)/(yc_5-ym)$	$(w_6-w_2)/(yc_6-yc_2)$	$(w_7-w_9)/(yc_7-yc_9)$	$(w_8-w_4)/(yc_8-yc_4)$	$(w_9-w_7)/(yc_9-yc_7)$
x_{m2}	0.07500	0.07500	0.17500	0.17500	0.27500	0.12500	0.12500	0.22500	0.17500
y_{m2}	0.04330	0.04330	0.04330	0.04330	0.04330	0.12990	0.12990	0.12990	0.21651
n_x	0.86603	−0.86603	0.86603	−0.86603	0.86603	0.86603	−0.86603	0.86603	0.86603
n_y	0.50000	−0.50000	0.50000	−0.50000	0.50000	0.50000	−0.50000	0.50000	0.50000
F_x	$(w_1-w_2)/(xc_1-xc_2)$	$(w_2-w_1)/(xc_2-xc_1)$	$(w_3-w_4)/(xc_3-xc_4)$	$(w_4-w_3)/(xc_4-xc_3)$	$(w_5-0)/(xc_5-xm_2)$	$(w_6-w_7)/(xc_6-xc_7)$	$(w_7-w_6)/(xc_7-xc_6)$	$(w_8-0)/(xc_8-xm)$	$(w_9-0)/((xc_9-xm)$
F_y	$(w_1-w_2)/(yc_1-yc_2)$	$(w_2-w_1)/(yc_2-yc_1)$	$(w_3-w_4)/(yc_3-yc_4)$	$(w_4-w_3)/(yc_4-yc_3)$	$(w_5-0)/(yc_5-ym_2)$	$(w_6-w =)/(yc_6-yc_7)$	$(w_7-w_6)/(yc_7-yc_6)$	$(w_8-0)/(yc_8-ym)$	$(w_9-0)/(yc_9-ym)$
x_{m3}	0.02500	0.12500	0.12500	0.22500	0.22500	0.07500	0.17500	0.17500	0.12500
y_{m3}	0.04330	0.04330	0.04330	0.04330	0.04330	0.12990	0.12990	0.12990	0.21651
n_x	−0.86603	0.86603	−0.86603	0.86603	−0.86603	−0.86603	0.86603	−0.86603	−0.86603
n_y	0.50000	−0.50000	0.50000	−0.50000	0.50000	0.50000	−0.50000	0.50000	0.50000
F_x	$(w_1-0)/(xc_1-xm_3)$	$(w_2-w_3)/(xc_2-xc_3)$	$(w_3-w_2)/(xc_3-xc_2)$	$(w_4-w_5)/(xc_4-xc_5)$	$(w_5-w_4)/(xc_5-xc_4)$	$(w_6-0)/(xc_6-xm)$	$(w_7-w_8)/(xc_7-xc_8)$	$(w_8-w_7)/(xc_8-xc_7)$	$(w_9-0)/(xc_9-xm)$
F_y	$(w_1-0)/(yc_1-ym_3)$	$(w_2-w_3)/(yc_2-yc_3)$	$(w_3-w_2)/(yc_3-yc_2)$	$(w_4-w_5)/(yc_4-yc_5)$	$(w_5-w_4)/(yc_5-yc_4)$	$(w_6-0)/(yc_6-ym)$	$(w_7-w_8)/(yc_7-yc_8)$	$(w_8-w_7)/(yc_8-yc_7)$	$(w_9-0)/(yc_9-ym)$

$$w = \{ w_1 \quad w_2 \quad w_3 \quad w_4 \quad w_5 \quad w_6 \quad w_7 \quad w_8 \quad w_9 \}^T \tag{1.2.10}$$

and write Eqs. (1.2.9a–i) in the form of

$$Aw = b \tag{1.2.11}$$

where the element of A are tabulated below.

$$A = \begin{bmatrix} -138.5641 & 34.6410 & 0.0000 & 0.0000 & 0.0000 & 0.0000 & 0.0000 & 0.0000 & 0.0000 \\ 34.6410 & -86.6025 & 34.6410 & 0.0000 & 0.0000 & 17.3205 & 0.0000 & 0.0000 & 0.0000 \\ 0.0000 & 34.6410 & -103.9230 & 34.6410 & 0.0000 & 0.0000 & 0.0000 & 0.0000 & 0.0000 \\ 0.0000 & 0.0000 & 34.6410 & -86.6025 & 86.6025 & 0.0000 & 0.0000 & 17.3205 & 0.0000 \\ 0.0000 & 0.0000 & 0.0000 & 34.6410 & -138.5641 & 0.0000 & 0.0000 & 0.0000 & 0.0000 \\ 0.0000 & 17.3205 & 0.0000 & 0.0000 & 0.0000 & -121.2436 & 34.6410 & 0.0000 & 0.0000 \\ 0.0000 & 0.0000 & 0.0000 & 0.0000 & 0.0000 & 34.6410 & -86.6025 & 34.6410 & 17.3205 \\ 0.0000 & 0.0000 & 0.0000 & 17.3205 & 0.0000 & 0.0000 & 34.6410 & -121.2436 & 0.0000 \\ 0.0000 & 0.0000 & 0.0000 & 0.0000 & 0.0000 & 0.0000 & 17.3205 & 0.0000 & -155.8846 \end{bmatrix}$$

Equation (1.2.11) can be solved using Gauss-Seidel method in the following way:

$$138.5461 w_1^{k+1} = 4.3301 + 34.6410 w_2^k \tag{1.2.12a}$$

$$86.6025 w_2^{k+1} = 4.3301 + 34.6410 w_1^{k+1} + 34.6410 w_3^k + 17.3205 w_6^k \tag{1.2.12b}$$

$$103.7230 w_3^{k+1} = 4.3301 + 34.6410 w_2^{k+1} + 34.6410 w_4^k \tag{1.2.12c}$$

$$86.6025 w_4^{k+1} = 4.3301 + 34.6410 w_3^{k+1} + 34.6410 w_5^k + 17.3205 w_8^k \tag{1.2.12d}$$

$$138.5640 w_5^{k+1} = 4.3301 + 34.6410 w_4^{k+1} \tag{1.2.12e}$$

$$121.2473 w_6^{k+1} = 4.3301 + 17.3205 w_2^{k+1} + 34.6410 w_7^k \tag{1.2.12f}$$

$$86.6025\, w_7^{k+1} = 4.3301 + 34.6410\, w_6^{k+1} + 34.6410\, w_8^k + 34.6410\, w_9^k \tag{1.2.12g}$$

$$121.2473\, w_8^{k+1} = 4.3301 + 17.3205\, w_4^{k+1} + 34.6410\, w_7^{k+1} \tag{1.2.12h}$$

$$155.0846\, w_9^{k+1} = 4.3301 + 17.3205\, w_7^{k+1} \tag{1.2.12i}$$

The solution obtained up to 15 iterations starting with an initial guess of $\{w\}^0 = \{0\}$ is given in the last row of Table 1.3.

Table 1.3 Solution of Eq. (1.2.12) using Gauss Seidel method with a starting value of $\{w\}^0 = 0$

Iter #	w_1	w_2	w_3	w_4	w_5	w_6	w_7	w_8	w_9
0	0	0	0	0	0	0	0	0	0
1	0.031250	0.062500	0.062500	0.075000	0.050000	0.044643	0.067857	0.065816	0.035317
2	0.046875	0.102679	0.100893	0.123520	0.062130	0.069770	0.111298	0.085160	0.040144
3	0.056920	0.127079	0.125200	0.141964	0.066741	0.085668	0.126360	0.092098	0.041818
4	0.063020	0.142421	0.136462	0.149701	0.068675	0.092163	0.132068	0.094834	0.042452
5	0.066855	0.149759	0.141487	0.153031	0.069508	0.094842	0.134361	0.095965	0.042707
6	0.068690	0.153039	0.143690	0.154472	0.069868	0.095966	0.135314	0.096443	0.042813
7	0.069510	0.154473	0.144648	0.155095	0.070024	0.096443	0.135717	0.096647	0.042857
8	0.069868	0.155095	0.145063	0.155364	0.070091	0.096647	0.135889	0.096735	0.042877
9	0.070024	0.155364	0.145243	0.155481	0.070120	0.096735	0.135963	0.096772	0.042885
10	0.070091	0.155481	0.145320	0.155531	0.070133	0.096772	0.135995	0.096789	0.042888
11	0.070120	0.155531	0.145354	0.155552	0.070138	0.096789	0.136009	0.096796	0.042890
12	0.070133	0.155552	0.145368	0.155562	0.070140	0.096796	0.136014	0.096799	0.042890
13	0.070138	0.155562	0.145374	0.155566	0.070141	0.096799	0.136017	0.096800	0.042891
14	0.070140	0.155566	0.145377	0.155567	0.070142	0.096800	0.136018	0.096801	0.042891
15	0.070141	0.155567	0.145378	0.155568	0.070142	0.096801	0.136019	0.096801	0.042891

1.3 CFD for the More Generic Case of Fluid Flow

The above two case studies illustrate the CFD approach conceptually. We obtain a solution to the partial differential equation in a rather simple way but by making compromises on the accuracy of the solution at three levels. The first compromise occurs when we re-state the problem as one of finding the solution at discrete grid points of our choice; that is, the problem is reduced to one of finding $w(x_i, y_j)$, rather than one of finding $w(x, y)$. If the velocity is required at any other point, then it has to be obtained by interpolation from neighbouring locations at which the velocity is determined. This introduces an interpolation error into the solution. The second compromise that we make is that we solve only an approximate form of the governing equation while calculating $w(x_i, y_j)$. Equations (1.1.2) and (1.2.7) are examples of the approximation of the governing partial differential Eq. (1.1.1) at grid point (i, j) and cell 7, respectively. This introduces an additional error truncation error in the numerical solution. Finally, all numerical solutions involve repetitive arithmetic operations which are carried out with finite precision arithmetic. There is thus a possibility of accumulation round-off errors. Also, one often uses iterative methods for the solution of the linear equations. Thus, even the approximate equations at discrete points in the flow domain are not solved exactly but only approximately. This introduces additional error. However, these errors can be controlled to a large extent. In a well-formulated CFD problem, the errors arising out of the first two approximations can be reduced by increasing the number of grid points. This is illustrated in Table 1.4 where the computations described Sect. 1.2 have been repeated for increasing number of cells; one can see that the total volumetric flow rate increases slightly with increasing number of cells but tends towards an asymptotic value in both cases. Figures 1.3 and 1.4 show the velocity contours in the two cases for increasing number of cells. As the latter increases, the iso-velocity lines becomes more smooth giving rise to a smoothly-varying velocity field that is expected in these cases.

Table 1.4 Computed relative volumetric flow rate with increasing number of grid points or cells for the same domain and the same value of pressure gradient for the case of fully developed flow through a duct of rectangular cross-section and triangular cross-section. We find that as the number of grid points increases, the computed volumetric flow rate through the cross-section approaches a constant value

Rectangular duct case		Triangular duct case	
No. of cells	Volumetric flow rate (relative)	No. of cells	Volumetric flow rate (relative)
5 × 5	18.0050	100	181.958
10 × 10	21.1094	400	175.399
20 × 20	21.7689	900	174.174
30 × 30	21.8806	1600	173.744
40 × 40	21.9183	2500	173.545
60 × 60	21.9447	3600	173.436

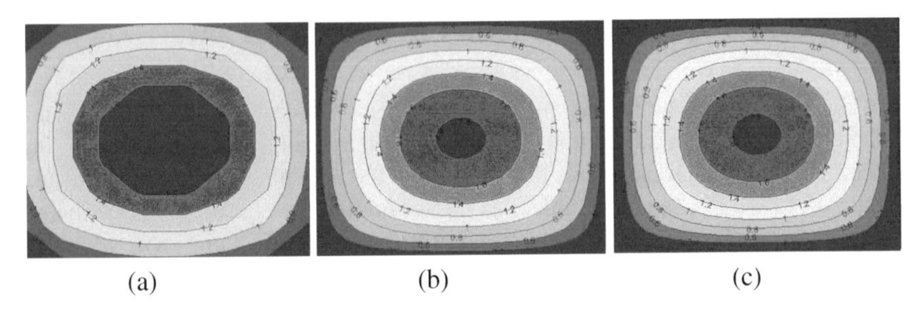

(a) (b) (c)

Fig. 1.3 Velocity contours obtained for a grid of **a** 10×10, **b** 30×30 and **c** 60×60 cells in each direction in the rectangular duct case. The contour values are relative

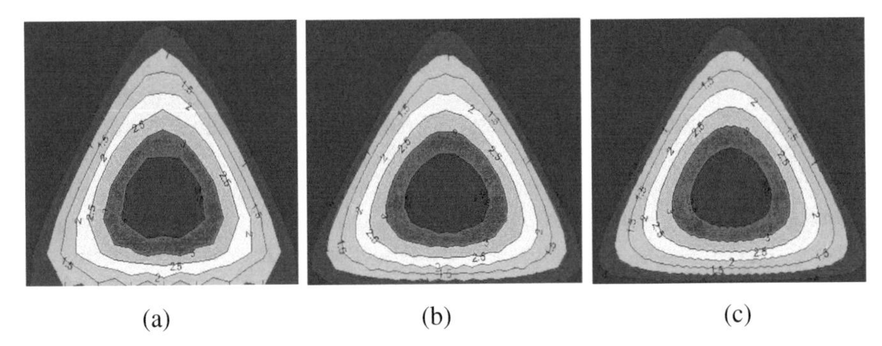

(a) (b) (c)

Fig. 1.4 Velocity contours obtained for **a** 100, **b** 400 and **c** 1600 cells for the triangular duct case. The contour values are relative

Thus, the CFD approach, despite the three-fold set of approximations made in obtaining a solution, is capable of delivering an accurate solution provided the numerical solution approach is well-formulated. It is relatively straightforward to do this for an equation such as the Poisson equation (Eq. 1.1.1). In the more generic case, the governing equations are more complicated and contain a *set* of coupled partial differential equations. Finding appropriate formulae for finite difference approximations and gradients is not easy. Special considerations come in when a system of non-linear, coupled partial differential equations has to be solved. Even the splitting of the flow domain into cells or filling the flow domain into well-spaced grid points becomes non-trivial. The solution of the linear algebraic equations too poses several challenges. These issues are discussed in depth in the rest of the book with focus on important developments that have made CFD the versatile tool that it is today.

Problems

1. Discuss a couple of examples where CFD can be used to solve a design problem and a couple of other cases where it cannot be used.
2. Discuss a couple of examples where CFD can be used to investigate a what-if kind of scenario (for example, the King's Cross Underground Fire).

3. Repeat the solution of Eq. (1.1.6) with different initial guesses and verify that the final answer is the same.
4. Repeat the solution of Eq. (1.1.2) for different values of C and verify that the average velocity varies linearly with C.
5. Repeat the rectangular duct problem for the grid shown in Fig. Q1.1. Assemble the algebraic equations and solve them using Gauss-Seidel method until a four-decimal place accuracy of solution is obtained.
6. Repeat the triangular duct problem for the grid shown in Fig. Q1.2. Assemble the algebraic equations and solve them using Gauss-Seidel method until a four-decimal place accuracy of solution is obtained.
7. Repeat the triangular duct problem for the grid shown in Fig. Q1.3. Assemble the algebraic equations and solve them using Gauss-Seidel method until a four-decimal place accuracy of solution is obtained.

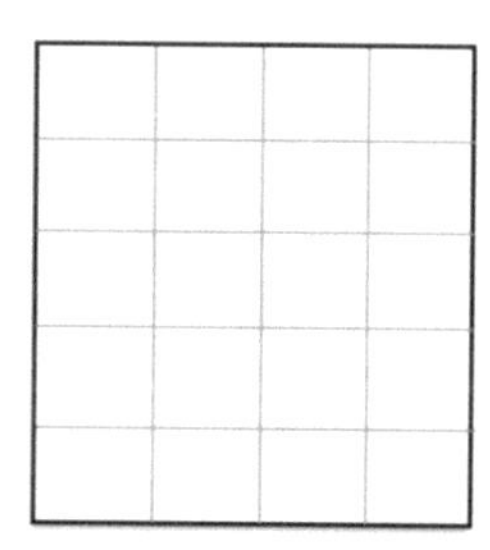

Fig. Q1.1 Computational domain and grid for Problem #5. Set up the 12 equations that need to be solved to find the velocity values at the interior grid points

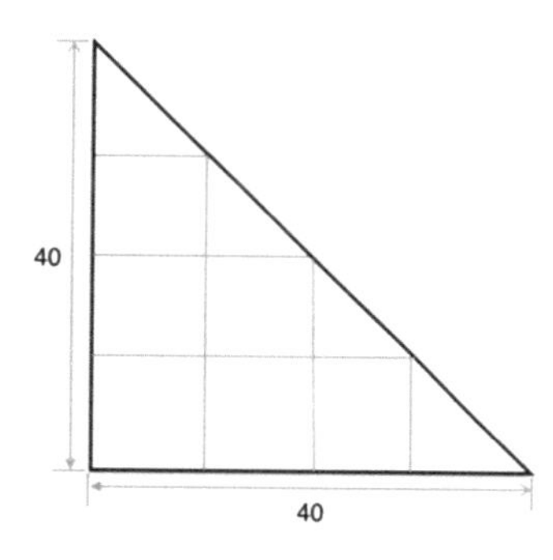

Fig. Q1.2 Computational domain and cell layout for Problem #6. Each side of the right-angled triangle is split into four equal parts. All dimensions are in millimeters

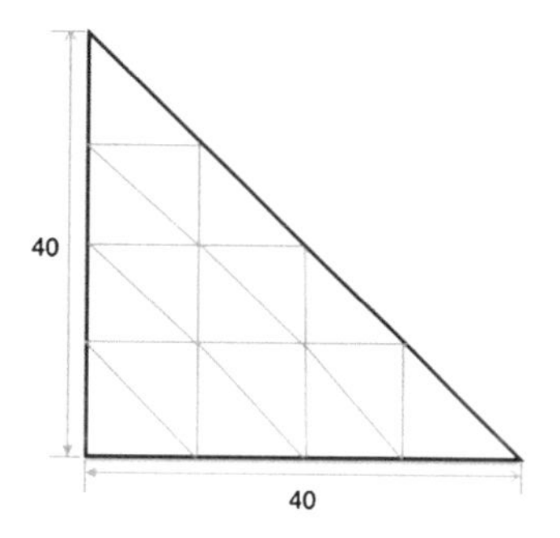

Fig. Q1.3 Computational domain and cell layout for Problem #7. All dimensions are in millimeters

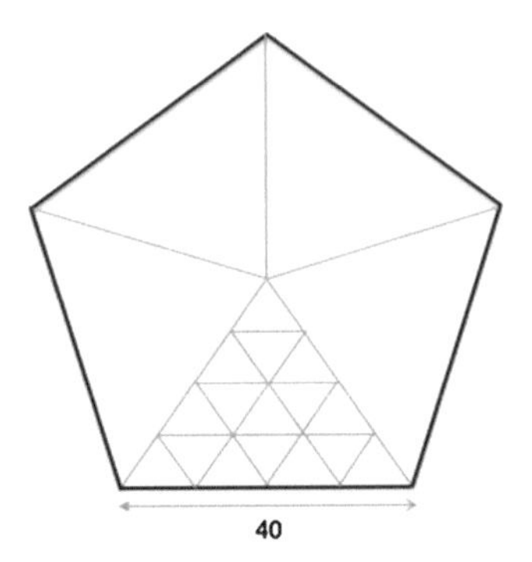

Fig. Q1.4 Computational domain and cell layout for Problem #12. All dimensions are in millimeters

8. Write a computer program to solve for the fully developed flow through a rectangular duct for an arbitrary number of grid divisions in the x- and y-directions. Run it for increasing number of grid points. In each case, find the average velocity and check that it approaches a constant value as the number of grid points increases.
9. What happens to the number of iterations needed to reach a certain level of convergence as the number of grid points increases?
10. Repeat problems 3 and 4 for the case of fully developed flow through a triangular duct.
11. Write a computer program to solve for the fully developed flow through a triangular duct (Case 2) for an arbitrary number of divisions into triangles. Run it for increasing number of triangular cells. In each case, find the average velocity and check that it approaches a constant value as the number of grid points increases.
12. Use the triangular duct case as an example to find the velocity field and average velocity for flow through a regular pentagon-shaped duct shown in Fig. Q1.4.

Chapter 2
Equations Governing Fluid Motion

In this chapter, we derive the basic equations that govern the flow of fluids. We examine the nature of these equations and discuss appropriate boundary and initial conditions that are needed to complete the mathematical description of fluid flow.

2.1 Basic Concepts of Fluid Flow

A fluid is distinguished from a solid in its response to the application of a shear stress. A fluid deforms continuously—leading to a continuous change of shape or size or both—under the action of a shear stress. While a solid also deforms upon the application of a shear stress, a fluid continues to deform as long as the shear stress exists. In contrast, in the case of a solid, equilibrium is established between the applied force which causes deformation and an internal resistive force which is proportional (within the elastic limit) to the extent of deformation suffered. At this point, there is no further change in the deformation even though the force is still acting on it. In a fluid too, an internal resistive force is set up; however, as we will see, it is proportional to the *rate* of deformation. When equilibrium is established, then the rate of deformation remains constant; the fluid continues to deform resulting in fluid flow. Thus, making a fluid flow requires the application of a shear stress, and when there is no shear stress, the fluid has no relative motion. Finally, when the applied shear stress is removed, solids tend to regain their original shape. In fluids, there is no such tendency to return to the original shape.

Fluids which exhibit the above characteristics are called simple fluids and comprise most liquids and gases. The most common fluids that we encounter in every day life, namely, air and water, are examples of such simple fluids. Some complex fluids, such as polymeric solutions and concentrated sugar solutions, exhibit a more complicated behaviour wherein the response of the fluid to a shear stress consists of a combination of fluid-like and solid-like behaviour. Such fluids

S. Jayanti, *Computational Fluid Dynamics for Engineers and Scientists*,
https://doi.org/10.1007/978-94-024-1217-8_2

are called viscoelastic fluids or fluids with memory and are discussed comprehensively by Bird et al. (1987).

At a molecular level, the movement of individual molecules is slight in a solid and the molecules do not readily move relative to one another. This makes a solid rigid. In a liquid, the movement of molecules is greater and the molecules can move past one another and thus the rigidity is lost although the force of attraction between molecules is still sufficient to keep the liquid together in a definite volume with a well-defined interface. In a gas, the force of attraction is much less than in liquids and the molecules are free to go anywhere until they are stopped by a solid or liquid boundary. Thus, a gas expands to fill the available volume in a confined space. The molecular-level movement in gases and liquids gives rise to two important properties found in fluids: pressure and viscosity. Pressure can be seen as being a result of innumerable molecular collisions on an imaginary plane in the fluid at rest. The resulting loss of momentum by the colliding molecules can be expressed as a compressive, pressure force acting on the imaginary plane. When there is relative motion within the fluid, the movement of molecules across two adjacent layers having different velocities leads to a net transport of momentum from the higher momentum layer to the lower momentum layer. This can be construed as a diffusion of momentum along the gradient such that a momentum flux exists from the layer with higher momentum to the one with lower momentum. Viscosity is a measure of this momentum diffusivity and is visualized in terms of a force which opposes relative motion within the fluid. Thus, a shear force, equivalent to the viscous force, has to be applied in order to maintain a given velocity gradient in a fluid. Kinetic theory works well for gases and is able to predict the increase of viscosity of gases with temperature. Liquids usually exhibit a more complicated behaviour and their viscosity decreases rapidly with increasing temperature. Another property which is specific to liquids is surface tension which is attributed to unequal forces of attraction on the molecules constituting the interface between the liquid and its vapour or gas. At a solid–liquid interface, the additional complexity of wetting of the surface arises. It is usual to neglect the effect of surface tension in fluid flows (except in some special cases) and the practice is followed here.

A fluid consists of a large number of molecules, each of which has a certain position, velocity and energy which vary as a result of collision with other molecules. Usually we are not interested in the motion of individual molecules but only in their average behaviour, i.e., in the variation of macroscopic quantities such as pressure, density, enthalpy etc. as functions of position and time. The fluid is therefore treated as a continuous distribution of matter, even though it consists of discrete particles. (This description of fluid as a continuum may break down in some extreme cases such as in the flow of rarified gases in the outer atmosphere or in flows through extremely small channels the diameter of which is of the order of the mean free path (White 1991). Such cases are not considered in this book.) The random motion of molecules of the fluid, also known as Brownian motion, causes, over a period of time (which however is very short compared to typical time scales of flows of practical interest), exchange of mass, momentum and heat. These phenomena cannot be treated by the continuum assumption. However, they appear as coefficients of diffusivity, viscosity and heat conduction of the fluid continuum.

The continuum assumption allows a field representation of the properties of a fluid particle, which can be defined as the smallest lump of fluid having sufficient number of molecules to permit continuum interpretation of its properties on a statistical basis. Since this fluid particle is very small, the average properties of the fluid particle are, in the limit, assigned to a point. For example, the field of a property φ can be described by an equation of the form

$$\varphi = \varphi(r, t) \quad \text{or} \quad \varphi = \varphi(x, y, z, t) \tag{2.1.1}$$

In fluid dynamics, we deal with scalar fields (e.g. density), vector fields (e.g. velocity) and tensor fields (e.g. stress). While the first two can be readily visualized, the stress tensor needs elaboration. If we construct a small cube around any point P in the fluid, there can be surface stresses acting on each of the six surfaces. Decomposing the stress on a surface into a normal component and two tangential components acting in the plane (see Fig. 2.1a), we have nine Cartesian components of the stress tensor. For an infinitesimal cube at a point, the stress can be represented as a tensor with components σ_{ij} which denotes stress acting in the jth direction on a plane normal to the ith direction.

We can associate with the cube the concept of positive faces and negative faces. Positive faces are those for which the outward normal vector is in the positive direction of a coordinate line. Negative faces are those for which the outward normal vector is aligned in the negative direction of the coordinate line. With reference to the 2-dimensional case shown in Fig. 2.1b, the (left) x-face located at $x - \Delta x/2$ is a negative face while the (right) x-face located at $x + \Delta x/2$ is a positive face. The stresses are positive when they are directed as shown in Fig. 2.1b. Notice that the directions of the stresses are reversed on the negative faces. For example, on the negative x-face, a normal stress of 100 Pa will be acting in the negative

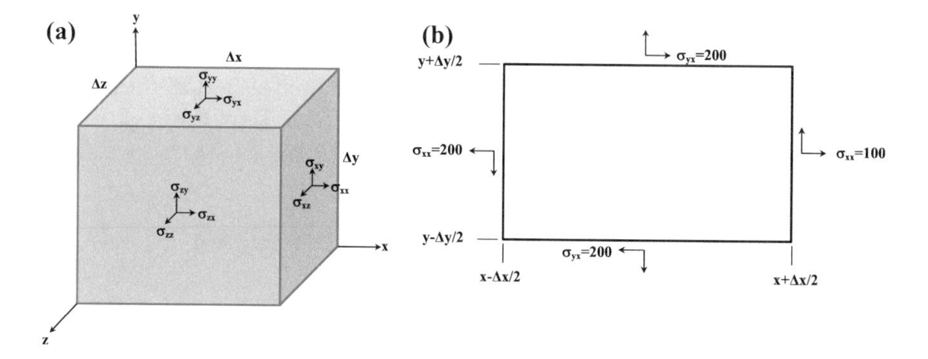

Fig. 2.1 **a** Nine components of stresses acting on a body. **b** Stresses in a two-dimensional shape showing the direction of stresses at the positive and the negative faces

x-direction. A normal stress of 100 Pa acting on the positive x-face (at $x + \Delta x/2$) would be acting in the positive x-direction. The direction of action of shear stresses is also reversed on the negative faces. Thus, σ_{yx} of 200 Pa would be acting in the +ve x-direction when it is acting on the +ve y-face (top) and in the negative x-direction when it is acting on the −ve y-face (bottom). It can be shown from the principle of conservation of angular momentum that, in the absence of externally imposed angular momentum on the fluid, the stress tensor is symmetric, i.e., $\sigma_{ij} - \sigma_{ji}$. Thus six, rather than nine, quantities suffice to determine the state of stress at a point.

In describing fluid motion, one often uses rates of change of flow variables, for example, application of force causing a rate of change of momentum. It is necessary to distinguish between two complementary view points that can be adopted in describing these derivatives. These are the *Lagrangian* view point in which the focus is always on the same set of fluid particles and fluid motion is described in terms of change of properties of these particles; and the *Eulerian* view point in which the focus is on what happens in a certain region of the flow domain and fluid motion is described in terms of the variation of flow properties as a function of location and time within this region. The Lagrangian approach can thus be seen as dealing with a system consisting of identified set of fluid elements while the Eulerian approach deals with a specified control volume. The former is generally more suitable to the description of problems in solid mechanics where the identification of the system is relatively easy. The control volume approach is more suitable to fluid mechanics in which the identification of the system may become difficult as time progresses as, by definition, the fluid keeps on deforming as long as a stress is maintained on it. Thus a fluid system will soon get completely out of shape whereas a solid system will more or less retain its shape. The interest in fluid mechanics is usually on what happens in a confined space. For example, the fluid flow around a car and the drag and lift forces that are thereby created as the car travels at a given speed is more elegantly studied by examining the region around the car. Similarly, the flow inside a chemical reactor where the reactants mix and form products while liberating or consuming heat is studied more simply by looking at the fluid mixture inside the reactor rather by following a reactant particle. Such cases are more easily described from the Eulerian perspective.

Obviously, the two approaches are mathematically equivalent. Consider a fluid property φ which is in general a function of space and time. Thus, $\varphi = \varphi(x, y, z, t)$. An observer who changes his position by a distance of $\Delta x, \Delta y, \Delta z$ over a time Δt will find a differential change in φ which amounts to

$$d\varphi = \partial\varphi/\partial x \Delta x + \partial\varphi/\partial y \Delta y + \partial\varphi/\partial z \Delta z + \partial\varphi/\partial t \Delta t \qquad (2.1.2)$$

Dividing throughout by Δt and taking the limit as Δt tends to zero and noting that $\Delta x/\Delta t = u_0$, $\Delta y/\Delta t = v_0$ and $\Delta z/\Delta t = w_0$ are the three velocity components of the *observer* in the limiting case of Δt tending to zero, we can write the total derivative as

$$\frac{d\emptyset}{dt} = \frac{\partial\emptyset}{\partial t} + u_o\frac{\partial\emptyset}{\partial x} + v_o\frac{\partial\emptyset}{\partial y} + w_o\frac{\partial\emptyset}{\partial z} \tag{2.1.3}$$

This will be the rate of change of $\emptyset$ at the point (x_0, y_0, z_0) recorded by an observer travelling at an arbitrary velocity. Consider the special case when the observer is moving with the fluid velocity, i.e., $u_0 = u$, $v_0 = v$, $w_0 = w$. In this case, the observer will be noting the changes occurring in the same set of fluid particles. The total derivative is then the rate of change in the system property from a Lagrangian point of view. To differentiate this special case, this derivative is called the substantial or material or convective derivative and is often denoted by $D\varphi/Dt$. For example, DT/Dt is the rate of increase of temperature of a fluid element consisting of the same set of fluid particles. It expresses the Lagrangian or system derivative in terms of local derivatives at a point, which is an Eulerian description of change. Thus the two are equivalent and can be written as

$$\frac{D\emptyset}{Dt} = \frac{\partial\emptyset}{\partial t} + u\frac{\partial\emptyset}{\partial x} + v\frac{\partial\emptyset}{\partial y} + w\frac{\partial\emptyset}{\partial z} \quad \text{or} \quad \frac{D\emptyset}{Dt} = \frac{\partial\emptyset}{\partial t} + u\cdot\nabla\emptyset \tag{2.1.4}$$

The first term on the right hand side is called the local derivative. The second term represents the change in φ due to the convection of the particle to the current location from another location where the value of φ is different.

The meaning of these derivatives will become clear when we consider the change in the property φ of the fluid particles contained in the control volume Ω in Fig. 2.2a. After a short time Δt, the fluid particles will be displaced slightly from their position at t. A majority of the particles will remain in Ω, but some will come

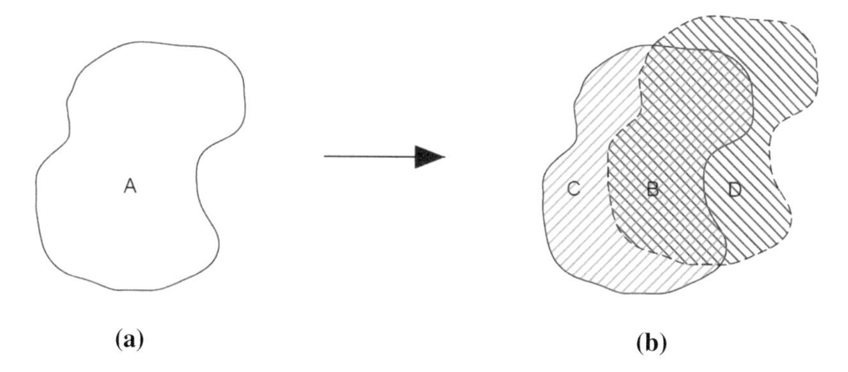

Fig. 2.2 Variation of system property Φ over a short time Δt: **a** volume A contains all the particles which belong to the system at time t, and **b** these have moved to volume B after a time of Δt. At time t, the boundaries of the system and the control volume coincide. At time $t + \Delta t$, the system boundary in (b) is shown as a dashed line and control volume boundary is shown as a solid line

out and some new particles will come in. As a result, the new system boundary containing the same set of particles is given by Ω' as denoted in Fig. 2.2b. Now let us evaluate the rate of change of a property Φ of the system. Suppose φ is the property per unit mass, i.e., intensive property, of the system. For example, if H, the enthalpy of the system, is the property under consideration, then φ will be equal to h, the specific enthalpy. Thus, $\Phi = \int \rho\varphi d\Omega$.

The rate of change of Φ of the system can be written as

$$D\Phi/Dt = \lim_{\Delta t \to 0} (\Phi(t+\Delta t) - \Phi(t))/\Delta t = \lim_{\Delta t \to 0} 1/\Delta t \left(\int_{\Omega} \rho\varphi d\Omega - \int_{\Omega'} \rho\varphi d\Omega' \right) \tag{2.1.5}$$

Now, the volume Ω' can be divided into three parts: volume B which is common to both the control volume and the system at t + Δt; volume C which is that part of the control volume which is no longer in the system; and volume D which is that part of the system which is no longer in the control volume. With a bit of manipulation, the rate of change of Φ of the system can be written as

$$D\Phi/Dt = \lim_{\Delta t \to 0} 1/\Delta t \left(\int_{B+C} \rho\varphi(t+\Delta t) d\Omega - \int_{A} \rho\varphi(t) d\Omega - \int_{C} \rho\varphi(t+\Delta t) d\Omega + \int_{D} \rho\varphi(t+\Delta t) d\Omega \right) \tag{2.1.6}$$

The first two integrals on the right hand side, which are over the same control volume Ω, lead to the temporal derivative $\partial\varphi/\partial t$. Volume C appearing in the second term can be viewed as the volume swept in time Δt by the particles sitting on the surface PQR at time t and can be written as $\int_{PQR} (-\rho \mathbf{u} \cdot dA\Delta t)$, the minus representing the fact that the velocity vector $\mathbf{u}$ and the outward normal vector of PQR are in opposite directions. Similarly, volume D can be viewed as the volume swept in time Δt by the particles sitting on the surface PSR at time t and can be written as $\int_{PSR} (\rho \mathbf{u} \cdot dA\Delta t)$. Substituting these in the above expression, and taking the limit as $\Delta t \to 0$, we get

$$\begin{aligned} D\Phi/Dt &= \lim_{\Delta t \to 0} 1/\Delta t \left(\int_{V} \partial(\rho\varphi)/\partial t d\Omega + \int_{PQR} \rho\varphi \mathbf{u} \cdot dA + \int_{PSR} \rho\varphi \mathbf{u} \cdot dA \right) \\ &= \left(\int_{CV} \partial(\rho\varphi)/\partial t d\Omega + \int_{CS} \rho\varphi \cdot dS \right) \end{aligned} \tag{2.1.7}$$

where CV and CS refer to the control volume and the closed surface enveloping the control volume, respectively. Using Gauss' divergence theorem, and dividing all the terms by the volume, the equation can be written, in terms of the intensive property φ as

$$D/Dt \int_{\Omega(t)} \rho\varphi d\Omega = \int_{\Omega(t)} \{\partial(\rho\varphi)/\partial t + \mathbf{u} \cdot \nabla(\rho\varphi)\} d\Omega \tag{2.1.8}$$

which is of the same form as Eq. (2.1.4). The left hand side indicates change in the Lagrangian frame of reference while the right hand side gives the equivalent change over a volume, and thus, the equivalent change in an Eulerian frame of reference. The first term on the right hand side represents the temporal variation of φ while the second term represents the net efflux of the property due to flow. The above equation expresses the rate of change of a variable of a system in terms of derivatives of a control volume and is called the Reynolds transport theorem. It is useful in deriving the equations of motion of a fluid in the Eulerian frame of reference, as explained below.

2.2 Laws Governing Fluid Motion

In the absence of electromagnetic effects (these can be readily included if necessary), the fundamental laws that fluid motion must obey are (i) the conservation of mass, (ii) the conservation of linear momentum, (iii) the conservation of angular momentum, (iv) the conservation of energy or the first law of thermodynamics, and (v) the second law of thermodynamics. For a system consisting of a collection of material of fixed identity contained in a volume $\Omega(t)$, these laws can be stated as follows:

Conservation of mass: In the absence of nuclear reactions, the mass of a fluid is conserved and the rate of change of mass of a given system of particles is zero. Mathematically this can be expressed as

$$D/Dt \int_{\Omega(t)} \rho d\Omega = 0 \tag{2.2.1}$$

Conservation of linear momentum: The rate of change of the linear momentum of the system is equal to the sum of external forces acting on it, i.e.,

$$D/Dt \int_{\Omega(t)} \rho \mathbf{u} d\Omega = \sum_{\Omega(t)} \mathbf{F}_{ext} \tag{2.2.2}$$

where the summation on the right hand side includes all external forces acting on the system.

Conservation of angular momentum: The rate of change of the angular momentum of the system of particles about a fixed axis is equal to the sum of the moments of the external forces about the axis, i.e.

$$D/Dt \int_{\Omega(t)} \rho(\mathbf{r} \times \mathbf{u})d\Omega = \sum_{\Omega(t)} (\mathbf{r} \times \mathbf{F}_{ext}) \tag{2.2.3}$$

Conservation of energy: The rate of change of the total energy, i.e., the sum of internal energy, E_i, kinetic energy, E_k, and potential energy, E_p, of a system of fluid particles is equal to the rate of work done on the system plus the rate of heat supplied to the system. Here, the internal energy is due to the random motion of the molecules relative to the averaged translatory motion of the molecules; the kinetic energy is due to macro-level motion, given by the fluid velocity components u, v and w, of the fluid elements; and the potential energy is due to the existence of a potential field such as the gravitational field. Thus, the conservation of energy can be written as

$$D/Dt \int_{\Omega(t)} \rho\left(e_i + {}^1\!/_2\mathbf{u}\cdot\mathbf{u} + e_p\right)d\Omega = \dot{Q} - \dot{W} \tag{2.2.4}$$

where $\mathbf{u} \cdot \mathbf{u} = u^2 + v^2 + w^2$, $\dot{Q}$ and $\dot{W}$ are the rate of heat addition to the system and the rate of work done by the system, and e_i and e_p are the internal energy and potential energy per unit mass and are thus the respective "specific" quantities.

Second law of thermodynamics: During a thermodynamic process the change of entropy, *s*, of the system plus that of the surrounding is zero or positive, i.e.,

$$D/Dt \int_{\Omega(t)} \rho s d\Omega \geq \sum_{\Omega(t)} \dot{Q}/T \tag{2.2.5}$$

The equality applies in a reversible process and the inequality applies in all irreversible processes. It should be noted that entropy is not a conserved quantity. An equation for the entropy of the flow can be derived but it is not independent from the energy equation, as will be shown later.

The above equations are written for a Lagrangian system and hence the notation of the substantial derivative is used on the left hand side terms. In addition to the above laws, there are subsidiary laws or constitutive relations which apply to specific types of fluid e.g. equation of state for a perfect gas and the relationship between stress and strain rate in a fluid. The presence of chemical reactions, turbulence and multiphase flow phenomena may necessitate a large number of auxiliary equations. This will be discussed in Chap. 6.

The conservation laws stated in Eqs. (2.2.1)–(2.2.5) refer to a system of particles and thus represent a Lagrangian point of view. As mentioned earlier, for many fluid mechanics problems, it is convenient to think in terms of a control volume fixed in space through which the fluid flows. The governing equations have to be converted therefore to this control volume or Eulerian approach. This can be done systematically using the Reynolds transport theorem derived above.

2.2.1 *Conservation of Mass*

Let us consider a fixed closed "control" volume CV having an enveloping surface CS, and apply the law of conservation of mass to the particles contained in it at time t. With reference to the Reynolds transport theorem, in the case of conservation of mass, $\phi = m$, the total mass of the system. Thus, the corresponding intensive property, φ, is equal to $d\phi/dm = 1$. Substituting this in the Reynolds transport theorem (Eq. 2.1.7) above, we get from the mass conservation Eq. (2.2.1)

$$D/Dt\left(\int_{CV} \rho d\Omega\right) = \left(\int_{CV} \partial(\rho)/\partial t d\Omega + \int_{CS} \rho \mathbf{u} \cdot dS\right) = 0 \quad (2.2.6)$$

Using Gauss divergence theorem, the surface integral can be written as a volume integral giving:

$$\int_{CV} (\partial\rho/\partial t + \mathbf{\nabla}\cdot(\rho\mathbf{u}))d\Omega = 0 \quad (2.2.7)$$

Since this is true for any arbitrary control volume, the integrand must be identically zero. Thus, we have the equation for conservation of mass as

$$\partial\rho/\partial t + \mathbf{\nabla} \cdot (\rho\mathbf{u}) = 0 \quad (2.2.8)$$

Equation 2.2.8 is also known as the continuity equation and has a particularly simple form for incompressible flows where density is constant:

$$\mathbf{\nabla} \cdot \mathbf{u} = \partial u/\partial x + \partial v/\partial y + \partial w/\partial z = 0 \quad (2.2.9)$$

Incompressibility strictly requires constant density; the density can be assumed to be more or less constant for cases where the pressure changes in the flow introduce little change in the density, which is the case for many cases involving flow of liquids. For the flow of gases, the Mach number, defined as $M = u/c$, where u is the characteristic flow velocity and c is the speed of sound in that medium, is a good indicator of incompressibility. Flows having Mach number less than 0.3 are usually treated as being incompressible. In the case of flows of mixtures, density may change, not because of pressure but because of changes in the composition of mixtures. Even here, the flow can be considered to be a flow of incompressible fluid (for example, air is a mixture of gases) if the Mach number is not too high. However, density changes from composition changes must be included and Eq. (2.2.8) should be used in this case even though the flow is incompressible in the Mach number sense.

2.2.2 *Conservation of Linear Momentum*

With reference to the Reynolds transport theorem, in the case of conservation of linear momentum, $\phi = m\mathbf{u}$. Thus, the corresponding intensive property, φ, is equal to $d\phi/dm = \mathbf{u}$, the velocity vector itself. Substituting this in the Reynolds transport theorem (Eq. 2.1.7) above, we get from the conservation of momentum Eq. (2.2.2)

$$D/Dt\left(\int_{CV} \rho\mathbf{u}d\Omega\right) = \left(\int_{CV} \partial(\rho\mathbf{u})/\partial t d\Omega + \int_{CS} \rho\mathbf{u}\mathbf{u}\cdot d\mathbf{S}\right) = \sum \mathbf{F}_{ext} \quad (2.2.10)$$

Here, the quantity $\mathbf{uu}$ is a dyadic product and is a tensor having nine components, namely, u_iu_j, where i, j = 1, 2, 3. Making use of the Gauss divergence theorem, the surface integral in Eq. (2.2.10), and the momentum conservation equation can be written as

$$\int_{CV} \{\partial(\rho\mathbf{u})/\partial t + \nabla\cdot(\rho\mathbf{u}\mathbf{u})\}d\Omega = \sum \mathbf{F}_{ext} \quad (2.2.11)$$

The external forces on the control volume can be of two types: body forces such as weight, and surface forces such as stresses. For a body force, one can define a body force per unit mass, f_b, such that $\int_{CV} \rho\mathbf{f}_b d\Omega = \mathbf{F}_b$. In the specific case of weight, $\mathbf{f_b}$ will be equal to $\mathbf{g}$, the acceleration due to gravity. Surface stresses can be represented by the stress tensor σ with the components as shown in Fig. 2.1. Each component has to be multiplied by the appropriate area of the face on which it is acting to get the equivalent force. Since the stresses acting on the positive and the negative faces act in opposite directions (Fig. 2.1b), the resulting force on the body will be due to the difference between the stress components acting on the +ve and −ve faces. Figure 2.3 shows the forces due to stress acting in the x-direction on a parallellopiped of sides Δx, Δy and Δz. The resultant stress-induced force in the x-direction can therefore be written as

$$\mathbf{F}_{ext,\sigma,x} = \left(\sigma_{xx}|_{x+\Delta x} - \sigma_{xx}|_x\right)\Delta y\Delta z + \left(\sigma_{yx}|_{y+\Delta y} - \sigma_{yx}|_y\right)\Delta z\Delta x + \left(\sigma_{zx}|_{z+\Delta z} - \sigma_{zx}|_z\right)\Delta x\Delta y \quad (2.2.12)$$

Writing $\sigma_{xx}|_{x+\Delta x} = \sigma_{xx}|_x + \Delta x\,\partial\sigma_{xx}/\partial x|_x$ etc. and substituting in Eq. (2.2.12), dividing by $\Delta x\Delta y\Delta z$ and taking the limit as Δx, Δy and Δz tend to zero, the external force in the x-direction per unit volume resulting from the stresses can be written as

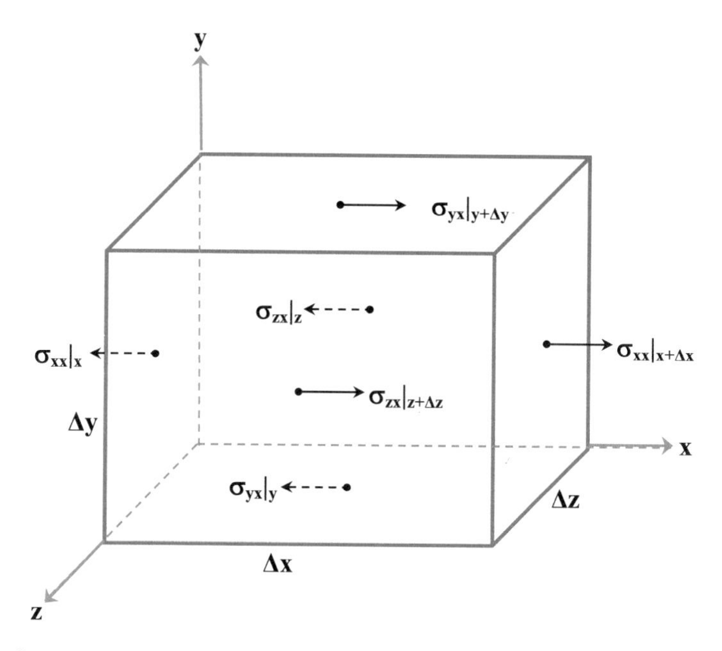

Fig. 2.3 Stresses acting in the x-direction on a rectangular parallelepiped

$$f_{ext,\sigma,x} = \partial\sigma_{xx}/\partial x + \partial\sigma_{yx}/\partial y + \partial\sigma_{zx}/\partial z = [\nabla \cdot \sigma]_x \tag{2.2.13}$$

Here $[\nabla \cdot \sigma]_x$ refers to the x-component of the vector product of the vector and the tensors, i.e., $[\nabla \cdot \sigma]$. The resultant force per unit volume due to stresses acting on the control volume can therefore be represented as $[\nabla \cdot \sigma]$.

We can thus write the momentum balance Eq. (2.2.11) as

$$\int_{CV} (\partial(\rho\mathbf{u})/\partial t + \nabla \cdot (\rho\mathbf{uu}) - \rho\mathbf{f}_b - \nabla.\sigma)d\Omega = 0$$

Since this is written for any arbitrary control volume, the integrand must be identically zero and hence the conservation of linear momentum takes the form

$$\partial(\rho\mathbf{u})/\partial t + \nabla \cdot (\rho\mathbf{uu}) = \rho\mathbf{f}_b + \nabla \cdot \sigma \tag{2.2.14}$$

At this stage, the meaning of the second term on the left hand side is not entirely clear. To investigate this, we write down the expanded form of Eq. (2.2.14) in each coordinate direction:

$$\begin{aligned}&\partial(\rho u)/\partial t + \partial(\rho u^2)/\partial x + \partial(\rho vu)/\partial y + \partial(\rho wu)/\partial z\\&\quad = \rho f_{bx} + \partial\sigma_{xx}/\partial x + \partial\sigma_{yx}/\partial y + \partial\sigma_{zx}/\partial z\end{aligned} \tag{2.2.15a}$$

$$\partial(\rho v)/\partial t + \partial(\rho uv)/\partial x + \partial(\rho v^2)/\partial y + \partial(\rho wv)/\partial z \\ = \rho f_{by} + \partial\sigma_{xy}/\partial x + \partial\sigma_{yy}/\partial y + \partial\sigma_{zy}/\partial z \quad (2.2.15b)$$

$$\partial(\rho w)/\partial t + \partial(\rho uw)/\partial x + \partial(\rho vw)/\partial y + \partial(\rho w^2)/\partial z \\ = \rho f_{bz} + \partial\sigma_{xz}/\partial x + \partial\sigma_{yz}/\partial y + \partial\sigma_{zz}/\partial z \quad (2.2.15c)$$

It is clear from the above that the term $[\mathbf{\nabla} \cdot \rho\mathbf{uu}]_i$ indicates the net convective *out*flux, (i.e., the net momentum flux *leaving* the control volume with the flow) of momentum in the ith direction through all the faces of the control volume (or through the control surface in the general case).

The stress tensor needs to be specified. Keeping in mind the two special properties of fluids that we discussed earlier, namely, pressure and viscosity, we decompose the stress tensor into two components: an isotropic, compressive pressure p which is present even when the fluid is not in motion; and a (viscous) stress, τ, which arises only when there is relative motion within the fluid. Thus, $\boldsymbol{\sigma}$ is written as

$$\boldsymbol{\sigma} = -p\mathbf{I} + \boldsymbol{\tau} \quad (2.2.16)$$

where $\mathbf{I}$ is the unit tensor and the negative sign before the first term on the right hand side indicates that it is compressive. With this decomposition, and considering gravitational force as the only external body force, the momentum conservation equation can be written as

$$\partial(\rho\mathbf{u})/\partial t + \mathbf{\nabla} \cdot (\rho\mathbf{uu}) = -\mathbf{\nabla} p + \mathbf{\nabla} \cdot \boldsymbol{\tau} + \rho\mathbf{g} \quad (2.2.17)$$

The above form of the momentum conservation equation is not particularly useful because the stresses resulting from fluid motion, namely, τ, are not known. For the incompressible flow of an inviscid fluid, i.e., for a fluid with zero viscosity, $\tau = 0$ and we have a closed system of equations involving four variables, viz., u, v, w and p and four equations, viz., the continuity equation Eq. (2.2.8) and the three momentum Eq. (2.2.17). The solution of these gives the flow field for isothermal flows. However, real fluids have non-zero viscosity and τ is therefore not zero. This introduces nine extra unknowns into the model and additional relations are needed to close the system of equations. For a compressible flow, density also becomes a variable and one more relation is required. These additional relations cannot be obtained from the application of fundamental laws governing fluid motion (for example, the use of energy conservation introduces an additional variable in the form of enthalpy or temperature), and have to be obtained empirically. The perfect gas law, relating density, pressure and temperature of a gas, is one such empirical relation. This law (or its variants) is used for compressible flows. Similar "constitutive" relations are required to relate the stresses to the velocity field.

The simplest relation (other than there being no relation at all!) between two variables is a linear relation. In solid mechanics, a linear relation, in the form of Hooke's law, is assumed between stress and strain which represents the deformation produced by the stress. By analogy, it can be assumed that there is a correspondingly linear relationship between stress and the deformation that it produces in fluids. Since a fluid continuously deforms under the action of a stress, a linear relationship is sought between the stress and the *rate* of deformation. For one dimensional flow, this takes the form of Newton's law of viscosity, viz., $\tau_{yx} \propto du/dy$, the proportionality constant being the dynamic viscosity of the fluid. This linear relation is only an assumption but it holds good in a number of cases, including for the two most common fluids on earth, namely, air and water. There are a number of other cases where the linearity assumption is not correct; in these cases, the viscosity may depend on the amount of stress applied or the stress the fluid is already subjected to. However, these cases are not considered here.

The extension of Newton's law of viscosity to the general three-dimensional case is non-trivial and requires an understanding of the kinematics of deformation in fluid flow. As a first level of generalization, let us consider the two-dimensional case shown in Fig. 2.4. The rectangular fluid element ABCD deforms into the trapezoidal element A′B′C′D′ after a short time Δt. Associated with this transformation are four types of deformation: translation, as given by the displacement, for example, of corner A to A′; rotation, as illustrated by the counterclockwise rotation of diagonal A′C′ relative to AC; extensional strain or dilatation, as demonstrated by the element A′B′C′D′ being bigger than ABCD; and shear strain, as exemplified by the distortion of the shape of ABCD from square to rhombus.

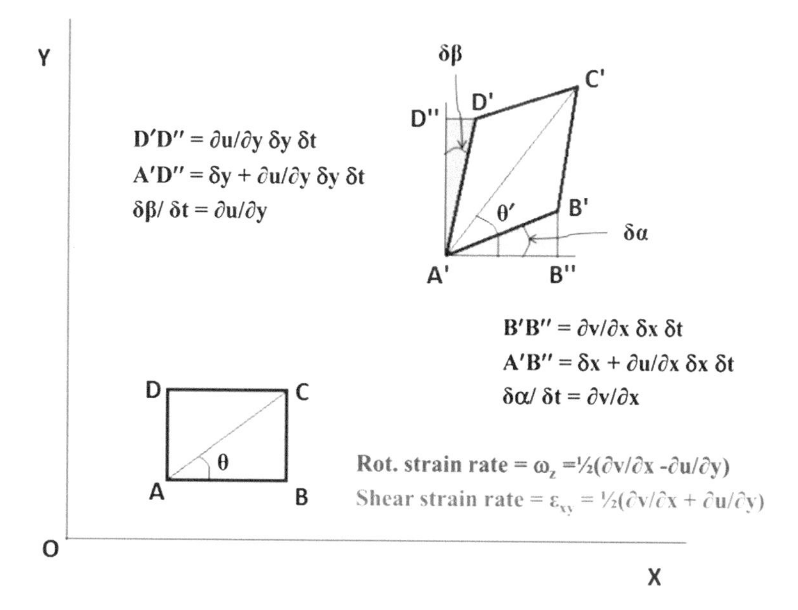

Fig. 2.4 Deformation of a rectangular fluid element ABCD to A′B′C′D′ over a short time Δt

It is necessary to describe these deformations quantitatively, and, since we are dealing with fluid flow, the focus will be on the rate of deformation rather than on deformation itself. The translation is defined by the displacement $u\Delta t$ and $v\Delta t$ of the point A in the x- and y-directions, respectively. The rate of translation is therefore u and v in the two directions. The rotation of the element ABCD over the time Δt is given by the relative angular displacement of the diagonal AC. The incremental rotation of the diagonal about the z-axis, $\delta\Omega_z$, is given by $\delta\Omega_z = \theta + \delta\alpha - 45°$, which can be expressed as $\delta\Omega_z = 1/2(\delta\alpha - \delta\beta)$ as $2\theta + \delta\alpha + \delta\beta - 90°$. Referring to Fig. 2.4, in the limiting case of $\Delta t \to 0$, we can write $\delta\alpha$ and $\delta\beta$ as

$$\delta\alpha = \lim_{\Delta t \to 0} \{\tan^{-1}[(\partial v/\partial x\, \Delta x\, \Delta t)/(\Delta x + \partial u/\partial x\, \Delta x\, \Delta t)]\} = (\partial v/\partial x)\Delta t$$
$$\delta\beta = \lim_{\Delta t \to 0} \{\tan^{-1}[(\partial u/\partial y\, \Delta y\, \Delta t)/(\Delta y + \partial v/\partial y\, \Delta y\, \Delta t)]\} = (\partial u/\partial y)\Delta t$$

Thus, the rate of rotation about the z-axis is given by

$$d/dt\ (\Omega_z) = \omega_z = 1/2(\partial v/\partial x - \partial u/\partial y)$$

The two-dimensional shear strain, which can be defined as the average decrease of the angle between two lines which are initially perpendicular, (for example, AB and AD in Fig. 2.4), can be obtained as $1/2(\delta\alpha + \delta\beta)$. Hence the rate of shear strain is given by

$$\varepsilon_{xy} = 1/2(\partial v/\partial x + \partial u/\partial y)$$

Finally, the extensional strain or dilatation is defined as the fractional increase in length. In the x-direction, the extensional strain for the side AB is thus given by

$$\{(\Delta x - \partial u/\partial x\, \Delta x\, \Delta t) - \Delta x\}/\Delta x = \partial u/\partial x\, \Delta t$$

The rate of extensional strain in the x-direction is thus given by $\varepsilon_{xx} = \partial u/\partial x$.

The above relations, obtained for a two-dimensional case can be readily extended to three dimensions. Thus,

- The rates of translation in the three directions are u, v and w.
- The rates of rotation about x-, y- and z- axis are $1/2(\partial w/\partial y - \partial v/\partial z)$, $1/2(\partial u/\partial z - \partial w/\partial x)$, and $1/2(\partial v/\partial x - \partial u/\partial y)$, respectively.
- The rates of shear strain in the xy, yz and zx planes are $\varepsilon_{xy} = 1/2(\partial v/\partial x + \partial u/\partial y) = \varepsilon_{yx}$, $\varepsilon_{yz} = 1/2(\partial w/\partial y + \partial v/\partial z) = \varepsilon_{zy}$, and $\varepsilon_{zx} = 1/2(\partial u/\partial z + \partial w/\partial x) = \varepsilon_{xz}$.
- The rates of extensional strain in the x-, y- and z-directions are $\varepsilon_{xx} = \partial u/\partial x$, $\varepsilon_{yy} = \partial v/\partial y$, and $\varepsilon_{zz} = \partial w/\partial z$.

Taken together, the extensional and shear strain rates constitute a symmetric second-order tensor which can be represented as the array

$$\varepsilon_{ij} = \begin{bmatrix} \varepsilon_{xx} & \varepsilon_{xy} & \varepsilon_{xz} \\ \varepsilon_{yx} & \varepsilon_{yy} & \varepsilon_{yz} \\ \varepsilon_{zx} & \varepsilon_{zy} & \varepsilon_{zz} \end{bmatrix} \quad (2.2.18)$$

The tensor ε_{ij} has two properties which are important: it has three invariants which are independent of direction or choice of axes:

$$\begin{aligned} I_1 &= \varepsilon_{xx} + \varepsilon_{yy} + \varepsilon_{zz} \\ I_2 &= \varepsilon_{xx}\varepsilon_{yy} + \varepsilon_{yy}\varepsilon_{zz} + \varepsilon_{zz}\varepsilon_{xx} - \varepsilon_{xy}^2 - \varepsilon_{yz}^2 - \varepsilon_{zx}^2 \\ I_3 &= \mathrm{Det}\,(\varepsilon_{ij}) \end{aligned} \quad (2.2.19)$$

Further, due to its symmetry, it has one and only one set of "principal" axes for which the off-diagonal terms vanish. In a Cartesian coordinate system with these principal axes as the coordinate axes, the strain rate tensor has only three non-zero components, namely, those along the diagonal.

Now, the velocity gradient tensor, $\partial u_i/\partial x_j$, can be decomposed into the sum of a symmetric tensor and an antisymmetric tensor:

$$\partial u_i/\partial x_j = 1/2\left(\partial u_i/\partial x_j + \partial u_j/\partial x_i\right) + 1/2\left(\partial u_i/\partial x_j - \partial u_j/\partial x_i\right) = \varepsilon_{ij} + \omega_k$$

Given that extensional and shear strain lead to deformation of the fluid element, ε_{ij} can be seen as the deformation rate tensor. That is, the velocity gradient tensor can be resolved into the sum of the deformation strain rate and the angular velocity of a fluid element. Noting that the latter does not distort the fluid element, we explore a linear relation between viscous stress (τ_{ij}) and the deformation strain rate (ε_{ij}). For the general three-dimensional case, we can write

$$\tau_{ij} \propto \varepsilon_{kl} \quad \text{or} \quad \tau_{ij} = A_{ijkl}\varepsilon_{kl} \quad (2.2.20)$$

Since each tensor has nine components, a general linear relationship involves 9×9 or 81 constants. However, both the tensors are symmetric and therefore have only six independent components The deformation rate tensor, given by $1/2(\partial u_i/\partial x_j + \partial u_j/\partial x_i)$, is obviously symmetric. The stress tensor is symmetric because of conservation of angular momentum. Therefore only 6×6 or 36 independent constants are needed to describe a general linear relationship between the two. We can further require that the deformation law be independent of the coordinate axes in which it is expressed. This enables us to seek the relation in the principal axes of each which contain only three non-zero elements. For a linear relationship to be valid, the principal directions (denoted here as x_1, x_2 and x_3) of the two tensors must coincide, and the relation requires a total of nine independent constants, as illustrated below:

$$\tau_{11} = a_1\varepsilon_{11} + a_2\varepsilon_{22} + a_3\varepsilon_{33} \tag{2.2.21a}$$

$$\tau_{22} = b_1\varepsilon_{11} + b_2\varepsilon_{22} + b_3\varepsilon_{33} \tag{2.2.21b}$$

$$\tau_{33} = c_1\varepsilon_{11} + c_2\varepsilon_{22} + c_3\varepsilon_{33} \tag{2.2.21c}$$

We make the assumption that the fluid medium is isotropic, that is, its properties are independent of direction. The concept of isotropy can be understood by looking at the non-isotropic condition. A newspaper hot-rolled in a specific direction during its preparation is a common example of a non-isotropic medium; it will tear neatly and easily in the direction of rolling but will tear with more difficulty and in a zig-zag way in the transverse direction. Polymeric fluids containing long chain molecules aligned in a preferential way can exhibit non-isotropy and their stress-deformation relation can depend on the direction in which the stress is applied. For simple fluids, this relation can be expected to be independent of direction. With this concept of isotropy in mind, we can interpret Eq. (2.2.21a) in the following way. Consider the case where we have stress only in one direction. Application of a normal stress of magnitude τ_{11} in the x_1 direction produces a strain rate of $a_1\varepsilon_{11}$ in the direction of x_1, $a_2\varepsilon_{22}$ in the x_2 direction and $a_3\varepsilon_{33}$ in the x_3 direction. One may recall that other (shear) strain rate terms are zero here. If the same amount of stress is applied in the x_2 direction (while keeping the other stresses zero), one would expect an identical behaviour for an isotropic medium except for a shift from x_1 to x_2 direction. In both cases, one would expect the transverse strain rate to be the same or $a_3\varepsilon_{33} = b_3\varepsilon_{33}$. Since the strain rate would be the same (the magnitude of applied stress being the same), $a_3 = b_3$. Similarly, $a_{11}\varepsilon_{11} = b_2\varepsilon_{22}$ and $a_2 = b_2$. One can thus show that for an isotropic medium, $a_1 = b_2 = c_3 = k_1$ and $a_2 = a_3 = b_1 = b_3 = c_1 = c_2 = k_2$. Thus, there can be only two independent constants to describe the linear relation; Eq. (2.2.21) can be rewritten as

$$\begin{aligned}
\tau_{11} &= k_1\varepsilon_{11} + k_2(\varepsilon_{11} + \varepsilon_{22} + \varepsilon_{33}) \\
\tau_{22} &= k_1\varepsilon_{22} + k_2(\varepsilon_{11} + \varepsilon_{22} + \varepsilon_{33}) \\
\tau_{33} &= k_1\varepsilon_{33} + k_2(\varepsilon_{11} + \varepsilon_{22} + \varepsilon_{33})
\end{aligned} \tag{2.2.22}$$

Transforming these from principal axes to an arbitrary axes x, y, z where the shear stresses are not zero, we can arrive at the general linear deformation law between the viscous stress and the strain-rates:

$$\tau_{ij} = \mu\left(\partial u_i/\partial x_j + \partial u_j/\partial x_i\right) + \lambda\,\partial u_k/\partial x_k \quad \text{or} \quad \boldsymbol{\tau} = \mu\left(\nabla\mathbf{u} + \nabla\mathbf{u}^T\right) + \lambda\nabla\cdot\mathbf{u} \tag{2.2.23}$$

where μ and λ are the two independent constants. Of these, μ is what is usually known as the dynamic viscosity and λ is called the second coefficient of viscosity or the coefficient of bulk viscosity as it is associated with the divergence of **u**, the dilatation of the fluid element. Not much is known about the value of λ; Stokes

(1845) assumed it to be equal to $-2/3\,\mu$. However, there is no independent proof for this assumption (White 1991). One consequence of this is that the mean pressure, defined as the mean compressive stress on a fluid element, differs from the thermodynamic pressure, p, by an amount equal to $-(\lambda + 2/3\mu)\nabla \cdot \mathbf{u}$, which is usually assumed to be negligible. Assuming $\lambda = -2/3\mu$ (as Stokes did) would bring in a neat equality of the two pressures but there is no basis for this assumption.

Substitution of the general linear deformation law (Eq. 2.2.23) into Eq. (2.2.17) gives a closed momentum conservation equation for a *Newtonian* fluid flow involving two empirical constants, viz., μ and λ:

$$\partial(\rho\mathbf{u})/\partial t + \nabla \cdot (\rho\mathbf{u}\mathbf{u}) = \rho\mathbf{g} - \nabla p + \nabla \cdot \{\mu(\nabla\mathbf{u} + \nabla\mathbf{u}^{T}) + \lambda\nabla \cdot \mathbf{u}\} \qquad (2.2.24)$$

Here u is a vector and Eq. (2.2.24) therefore constitutes a set of three equations which are called the Navier-Stokes equations (sometimes the mass conservation equation is also added to this set) in honour of Navier (1823) and Stokes (1845). Using Cartesian index notation, in which a repeated index in a term indicates summation over that index, Eq. (2.2.24) can be written as:

$$\partial(\rho u_i)/\partial t + \partial(\rho u_i u_j)/\partial x_j = \rho g_i - \partial p/\partial x_i + \partial/\partial x_j\{\mu(\partial u_i/\partial x_j + \partial u_j/\partial x_i) + \lambda\partial u_m/\partial x_m\} \qquad (2.2.25)$$

For incompressible flow, $\nabla \cdot \mathbf{u} = 0$ from continuity equation and the term involving the second coefficient of viscosity vanishes leaving the dynamic viscosity, μ, as the sole constant describing the general linear relation between shear stress and the strain-rate tensor, subject to the conditions of isotropy etc. discussed above. When the viscosity itself, which may vary with pressure and temperature, is assumed to be constant, the Navier-Stokes equations reduce to the following simple form:

$$\partial u_i/\partial t + \partial(u_i u_j)/\partial x_j = g_i - 1/\rho\, \partial p/\partial x_i + \nu\partial^2 u_i/\partial x_j \partial x_j \qquad (2.2.26)$$

Finally, using the continuity equation, Eq. (2.2.26) can be put in the slightly different form

$$\partial u_i/\partial t + u_j \partial u_i/\partial x_j = g_i - 1/\rho\, \partial p/\partial x_i + \nu\partial^2 u_i/\partial x_j \partial x_j \qquad (2.2.27)$$

The momentum equation in the form of Eq. (2.2.26) is said to be in conservative form and it is said to be in non-conservative form when written as in Eq. (2.2.27). Mathematically, there is no difference between the two; in numerical computations the conservative form may be better, as will be discussed later in Chap. 3.

2.2.3 Conservation of Angular Momentum

With reference to the Reynolds transport theorem, in the case of conservation of linear momentum, $\Phi = m(\mathbf{r} \times \mathbf{u})$. Thus, the corresponding intensive property, φ, is equal to $d\Phi/dm = \mathbf{r} \times \mathbf{u}$. Substituting this in the Reynolds transport theorem (Eq. 2.1.7) above, we get from the conservation of momentum Eq. (2.2.3)

$$D/Dt\left\{\int_{CV} \rho(\mathbf{r} \times \mathbf{u})d\Omega\right\} = \left\{\int_{CV} \partial/\partial t\{\rho(\mathbf{r} \times \mathbf{u})\}d\Omega + \int_{CS} \rho(\mathbf{r} \times \mathbf{u})\mathbf{u} \cdot d\mathbf{S}\right\} = \sum(\mathbf{r} \times \mathbf{F}_{ext}) \tag{2.2.28}$$

Considering as before that the external forces consist of body forces and surface forces, the volumetric density of which are denoted, respectively, by $\mathbf{f}_b$ and $\boldsymbol{\sigma}$, and following the same procedure as for the derivation of linear momentum, we get from (2.2.28)

$$\int_{CV} [\partial/\partial t\{\rho(\mathbf{r} \times \mathbf{u})\} + \nabla \cdot \rho(\mathbf{r} \times \mathbf{u})\mathbf{u} - (\mathbf{r} \times \rho\mathbf{f_b}) - \nabla \cdot (\mathbf{r} \times \boldsymbol{\sigma})]d\Omega = 0 \tag{2.2.29}$$

From this, we can write the angular momentum equation as

$$\partial/\partial t\{\rho(\mathbf{r} \times \mathbf{u})\} + \nabla \cdot \rho(\mathbf{r} \times \mathbf{u})\mathbf{u} = (\mathbf{r} \times \rho\mathbf{f_b}) + \nabla \cdot (\mathbf{r} \times \boldsymbol{\sigma}) \tag{2.2.30}$$

where $\mathbf{f}_b$ and $\boldsymbol{\sigma}$ have the same meaning as for the linear momentum equation.

It can be shown (Warsi 2006) that the angular momentum equation can be satisfied only if the stress tensor $\boldsymbol{\sigma}$ is symmetric. Thus, the symmetry of $\boldsymbol{\sigma}$ and hence of the viscous stress tensor $\boldsymbol{\tau}$ is a necessary consequence of the conservation of the angular momentum equation and is not contingent on the fluid being isotropic.

2.2.4 Conservation of Energy

As mentioned at the beginning of Sect. 2.2, the conservation of energy deals with the total energy, i.e., the sum of internal energy, kinetic energy and potential energy. Of these, internal energy is due to the random motion of the molecules about their mean position and is a function of the temperature of the fluid. The specific internal energy, e, is given by $e = c_v(T - T_{ref})$, where c_v is the specific heat at constant volume and T_{ref} is a reference temperature. The internal energy is a state function of the system in the thermodynamic sense. The kinetic energy per unit volume can be written as $\frac{1}{2}\rho\,\mathbf{u} \cdot \mathbf{u} = \frac{1}{2}\rho(u^2 + v^2 + w^2)$ where u, v and w are the components of the velocity in the Cartesian coordinate system. Assuming the gravitational field to be the only potential field acting on the system, the potential energy term will be zero since gravity is already included as a body force in the momentum equation. Thus, the total energy of the system can be written as

$$E_t = \int_{CV} \rho e_t d\Omega \int_{CV} \rho\{e + {}^1\!/\!_2(\mathbf{u}\cdot\mathbf{u})\}d\Omega = \int_{CV} \{e + {}^1\!/\!_2(\mathbf{u}\cdot\mathbf{u})\}dm \qquad (2.2.31)$$

With reference to the Reynolds transport theorem, in the case of conservation of energy, $\Phi = E_t$. Thus, the corresponding intensive property, φ, is equal to $dE_t/dm = e_t = e + {}^1\!/\!_2(\mathbf{u}\cdot\mathbf{u})$. Substituting this in the Reynolds transport theorem (Eq. 2.1.7) above, we get from the conservation of energy Eq. (2.2.4)

$$D/Dt\left[\int_{CV} \rho e_t d\Omega\right] = \int_{CV} \rho e_t d\Omega + \int_{CS} \rho e_t \mathbf{u}\cdot\mathbf{n}\, dS = \dot{Q} + \dot{W} \qquad (2.2.32)$$

Let us now evaluate the terms on the right hand side of the above equation, that is, the rates of heat addition to the system and work done on the system. Consider the system consisting of fluid particles enclosed in the control volume shown in Fig. 2.5 with sides Δx, Δy and Δz. For the sake of clarity, only the heat flux, q_x, through the two x-faces is shown. Then the net heat flow into the system through these two faces is $(-\partial q_x/\partial x)\Delta x\Delta y\Delta z$. When all the faces are considered, the total heat flowing into the control volume is $-(\partial q_x/\partial x + \partial q_y/\partial y + \partial q_z/\partial z)\,\Delta x\Delta y\Delta z$, or, in the more general case, $-(\nabla\cdot\mathbf{q})\,\Delta x\Delta y\Delta z$ where $\mathbf{q}$ is the heat flux vector, i.e., heat flow vector per unit area.

Neglecting heat transfer through radiation (which will be discussed separately in Chap. 6), we will assume that heat transfer is only through conduction, and that it is given by Fourier's law of heat conduction, viz.,

$$\mathbf{q} = -k\,\nabla T \qquad (2.2.33)$$

where k is the thermal conductivity of the fluid and T is the temperature. Thus the rate of heat added to the system can be written as

$$\dot{Q} = (-\nabla\cdot\mathbf{q})\,\Delta x\Delta y\Delta z = \int_{CV} (-\nabla\cdot q)d\Omega = \int_{CV} (k\nabla^2 T)d\Omega \qquad (2.2.34)$$

The work term consists of rate of work is done only by the external forces acting on the control volume. Of these, the effect of gravity comes in as a body force, f_b per unit volume. The remaining forces, in the absence of any other body forces, are the surface stresses. If F is the force component in the x-direction, the work done *on*

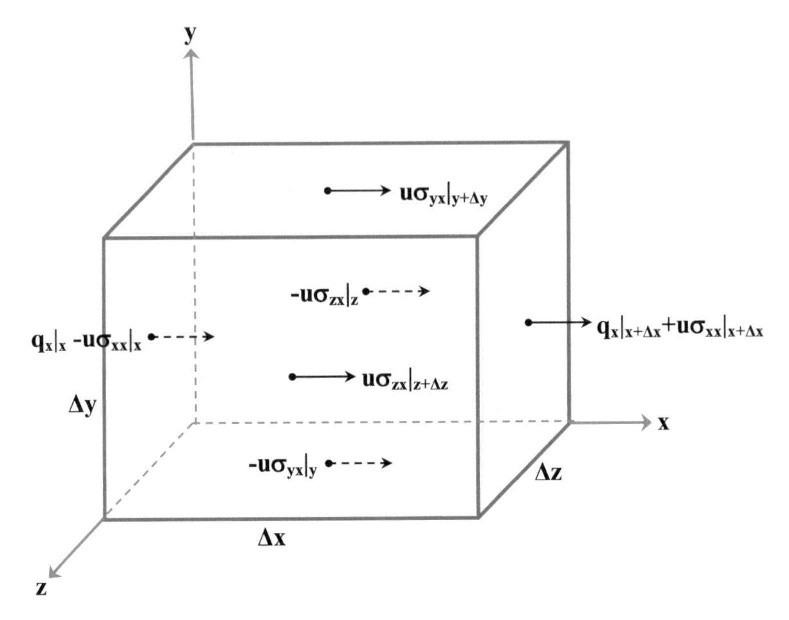

Fig. 2.5 Heat fluxes and rates of work done on the fluid in the control volume by forces acting in the x-direction

the system as it moves over a small distance dx is equal to Fdx. If this is done in a small time interval of dt, then the work done on the system is Fu, where $u = dx/dt$. For the general case where the force and the displacement are not in the same direction, the rate of work done on the system is $\mathbf{F} \cdot \mathbf{u}$. For the stresses acting in the x-direction on the system shown in Fig. 2.5, the rate of work done on the system can be evaluated, as has been done in Eqs. (2.2.12) and (2.2.13), as

$$\dot{W}_x = \Delta x \Delta y \Delta z \left[\partial(u\sigma_{xx})/\partial x + \partial(u\sigma_{yx})/\partial y + \partial(u\sigma_{zx})/\partial z - \partial(up)/\partial x + \rho u g_x \right] \tag{2.2.35}$$

Considering all the stresses acting on all the faces of the control volume, the rate of work done on the system is given by

$$\dot{W} = \int_{cv} (-\nabla \cdot \mathbf{u}p + \nabla \cdot (\mathbf{u} \cdot \tau) + \rho \mathbf{g} \cdot \mathbf{u}) dV \tag{2.2.36}$$

Substituting the expressions for rate of heat added to the system and the rate of work done on the system in Eq. (2.2.32) and using the Gauss divergence theorem and rearranging, we get

$$\int_{cv} [\partial/\partial t(\rho e_t) + \nabla \cdot \rho \mathbf{u} e_t - \nabla \cdot k\nabla T - (-\nabla \cdot \mathbf{u} p + \nabla \cdot (\mathbf{u} \cdot \tau) + \rho \mathbf{g} \cdot \mathbf{u})]dV = 0$$

or

$$\partial/\partial t(\rho e_t) + \nabla \cdot \rho \mathbf{u} e_t = \nabla \cdot k\nabla T - \nabla \cdot \mathbf{u} p + \nabla \cdot (\mathbf{u} \cdot \tau) + \rho \mathbf{g} \cdot \mathbf{u} \tag{2.2.37a}$$

The equation can be written in index notation as

$$\partial/\partial t(\rho e_t) + \partial/\partial x_j(\rho u_j e_t) = \partial/\partial x_j(k\partial T/\partial x_j) - \partial/\partial x_j(p u_j) + \partial/\partial x_j(u_i \tau_{ji}) + \rho g_j u_j \tag{2.2.37b}$$

Two other forms of energy equation, viz., for internal energy alone and for enthalpy alone are often used. These can be derived as follows. The rate of work done on the system is given as $\nabla \cdot (\mathbf{u} \cdot \boldsymbol{\tau})$. This can be decomposed into

$$\nabla \cdot (\mathbf{u} \cdot \boldsymbol{\tau}) = \mathbf{u} \cdot (\nabla \cdot \boldsymbol{\tau}) + \boldsymbol{\tau} : \nabla \mathbf{u}$$

But from the momentum conservation Eq. (2.2.24), $\nabla \cdot \boldsymbol{\tau} = \rho(D\mathbf{u}/Dt - \mathbf{g} + \nabla p)$. Therefore, $\mathbf{u} \cdot (\nabla \cdot \boldsymbol{\tau}) = \rho(\mathbf{u}\ D\mathbf{u}/Dt - \mathbf{g} \cdot \mathbf{u} + \mathbf{u} \cdot \nabla p) = \rho(D/Dt(1/2\mathbf{u}^2) - \mathbf{g} \cdot \mathbf{u} + \nabla \cdot p\mathbf{u} - p\nabla \cdot \mathbf{u})$. When substituted in this form into the energy conservation equation, all the terms except the last one will cancel out leaving an equation for internal energy of the form:

$$\partial/\partial t(\rho e) + \nabla \cdot (\rho \mathbf{u} e) = \nabla \cdot (k\nabla T) + \boldsymbol{\tau} : \nabla \mathbf{u} - p\nabla \cdot \mathbf{u} \tag{2.2.38}$$

An equation for enthalpy can be derived using the continuity equation:

$$\partial\rho/\partial t + \nabla \cdot (\rho \mathbf{u}) = \partial\rho/\partial t + \mathbf{u} \cdot \nabla\rho + \rho\nabla \cdot \mathbf{u} = D\rho/Dt + \rho\nabla \cdot \mathbf{u} = 0$$

or

$$\nabla \cdot \mathbf{u} = -1/\rho D\rho/Dt$$

Hence $p\nabla \cdot \mathbf{u} = -p/\rho D\rho/Dt = \rho D/Dt(p/\rho) - Dp/Dt$
Substituting this in the internal energy Eq. (2.2.37), we get

$$\rho D/Dt(e + p/\rho) = Dp/Dt + \nabla \cdot (k\nabla T) + \boldsymbol{\tau} : \nabla \mathbf{u} \tag{2.2.39}$$

Putting $h = e + p/\rho$ and rearranging, we get an equation for the enthalpy:

$$\partial/\partial t(\rho h) + \nabla \cdot (\rho \mathbf{u} h) = Dp/Dt + \nabla \cdot (k\nabla T) + \boldsymbol{\tau} : \nabla \mathbf{u} \tag{2.2.40a}$$

$$\partial/\partial t(\rho h) + \partial/\partial x_j(\rho u_j h) = Dp/Dt + \partial/\partial x_j(k\partial T/\partial x_j) + \tau_{ij}\partial u_i/\partial x_j \quad (2.2.40b)$$

which is another form of the energy conservation equation.

The last term in the above equation, $\tau_{ij}\partial u_i/\partial x_j$, is the only term in the energy equation involving viscosity and is called the viscous dissipation term, denoted by Φ. For a Newtonian fluid with constant properties, it can be written in Cartesian coordinates as

$$\Phi = \mu\left[2(\partial u/\partial x)^2 + 2(\partial v/\partial y)^2 + 2(\partial w/\partial z)^2 + (\partial u/\partial y + \partial v/\partial x)^2 + (\partial v/\partial z + \partial w/\partial y)^2 + (\partial u/\partial z + \partial w/\partial x)^2\right] + \lambda(\partial u/\partial x + \partial v/\partial y + \partial w/\partial z)^2 \quad (2.2.41)$$

which is always positive (provided $\lambda \geq -2/3\,\mu$ (or for incompressible flow) and $\mu > 0$). This represents the energy source due to fluid friction and results in the increase in the temperature due to pumping, for example, as frictional losses in pumping are converted into heat.

2.2.5 Entropy Equation

An equation for entropy can be obtained by manipulating the energy equation. The thermodynamic relation defining entropy, s,

$$Tds = de + p\,d(1/\rho) \quad (2.2.42)$$

can be written as

$$Tds = de - p/\rho^2 d\rho \quad \text{or} \quad de = Tds + p/\rho^2 d\rho \quad (2.2.43)$$

Writing the above in terms of rate of change of the system properties, we have

$$De/Dt = TDs/Dt + p/\rho^2 D\rho/Dt \quad (2.2.44)$$

From the continuity equation, we already have $D\rho/Dt = -\rho\nabla\cdot\mathbf{u}$. Substituting this in Eq. (2.2.44), and using Eq. (2.2.38) to evaluate the change in internal energy, we finally get

$$\rho TDs/Dt = \nabla\cdot(k\nabla T) + \tau : \nabla\mathbf{u} \quad (2.2.45)$$

which is the equation for the entropy.

In deriving this equation, heat conduction is considered as reversible heat addition. For adiabatic flow, it will be zero but the entropy will always increase because the viscous dissipation term (the second term on the right hand side) is always positive. If there is irreversible heat addition, then this should be added to the right hand side as an additional heat flux term. Also, note that no additional conservation laws are invoked in deriving the entropy equation. It is obtained by manipulating the energy equation using the definition of entropy as given by Eq. (2.2.42). Thus, the entropy equation is not independent of the energy equation and only one of the two can be used independently.

2.2.6 *Equation of State*

For a compressible flow without external heat addition or body forces, the governing equations are the continuity equation, the three momentum equations and the energy equation. These five scalar equations involve six unknown quantities: ρ, u, v, w, p and e (in addition to the transport properties μ and k). Therefore an additional equation is required to close the system of equations. This is provided by the thermodynamic relationship which exists among the fluid properties, and is known as the equation of state. According to the state principle of thermodynamics, the local thermodynamic state of the system is fixed by specifying any two independent thermodynamic state variables provided that the chemical composition of the system is not changing due to diffusion or finite-rate chemical reactions. Thus, an additional equation involving the variables is obtained, for example, if we choose the density and internal energy as the two independent variables:

$$P = p(e, \rho) \quad \text{and} \quad T = T(e, \rho) \tag{2.2.46}$$

These two would be sufficient to calculate the transport properties μ and k of the system as well.

Another example of an equation of state is the perfect gas-law which gives the relations

$$p = \rho R T; \quad e = c_v T; \quad c_v = R/(\gamma - 1) \tag{2.2.47}$$

where R is the universal gas constant and γ is the ratio of specific heats. For air at standard temperatures, R = 287 $m^2\ s^{-2}\ K^{-1}$ and $\gamma = 1.4$. Using these relations, we can get the following additional relation for pressure in terms of density and internal energy

$$p = (\gamma - 1)\rho e \tag{2.2.48}$$

which, along with the continuity equation, the three momentum equations and the internal energy equation, constitutes a set of six equations for the six unknowns (ρ, u, v, w, p, e) for the flow of a compressible perfect gas.

It should be noted that for an incompressible flow, the density becomes constant and, when specified, the pressure ceases to be a thermodynamic variable and the equation of state linking the density, internal energy and the pressure (Eq. 2.2.48) cannot be invoked. In this case, the continuity equation, the three momentum equations and the energy equation constitute a set of five governing equations for the five unknowns, viz., u, v, w, p and e.

2.2.7 Governing Equations for a Constant-Property Flow

We will end this section by listing out the equations for incompressible flow with constant thermophysical properties for the sake of future reference. Both the vectorial form and the fully expanded Cartesian coordinate form are given.

Continuity equation

$$\nabla \cdot \mathbf{u} = 0 \tag{2.2.49a}$$

$$\partial u/\partial x + \partial v/\partial y + \partial w/\partial z = 0 \tag{2.2.49b}$$

Momentum conservation equations:

$$\partial \mathbf{u}/\partial t + \nabla \cdot (\mathbf{uu}) = \mathbf{g} - 1/\rho \nabla p + \nu \nabla^2 \mathbf{u} \tag{2.2.50a}$$

$$\begin{aligned} &\partial u/\partial t + \partial(u^2)/\partial x + \partial(vu)/\partial y + \partial(wu)/\partial z \\ &\quad = -1/\varrho \partial p/\partial x + \nu\left(\partial^2 u/\partial x^2 + \partial^2 u/\partial y^2 + \partial^2 u/\partial z^2\right) + g_x \end{aligned} \tag{2.2.50b}$$

$$\begin{aligned} &\partial v/\partial t + \partial(uv)/\partial x + \partial(v^2)/\partial y + \partial(wv)/\partial z \\ &\quad = -1/\varrho \partial p/\partial y + \nu\left(\partial^2 v/\partial x^2 + \partial^2 v/\partial y^2 + \partial^2 v/\partial z^2\right) + g_y \end{aligned} \tag{2.2.50c}$$

$$\begin{aligned} &\partial w/\partial t + \partial(uw)/\partial x + \partial(vw)/\partial y + \partial\left(w^2\right)/\partial z \\ &\quad = -1/\varrho \partial p/\partial z + \nu\left(\partial^2 w/\partial x^2 + \partial^2 w/\partial y^2 + \partial^2 w/\partial z^2\right) + g_z \end{aligned} \tag{2.2.50d}$$

Energy conservation equation:

$$\frac{\partial T}{\partial t} + \nabla \cdot (\mathbf{u}T) = \left(\frac{k}{\rho c_v}\right)\nabla^2 T + \left(\frac{1}{\rho c_v}\right)\tau : \nabla \mathbf{u} \tag{2.2.51a}$$

$$\frac{\partial T}{\partial t}+\frac{\partial(uT)}{\partial x}+\frac{\partial(vT)}{\partial y}+\frac{\partial(wT)}{\partial z}=\frac{k}{\rho c_v}\left(\frac{\partial^2 T}{\partial x^2}+\frac{\partial^2 T}{\partial y^2}+\frac{\partial^2 T}{\partial z^2}\right)$$
$$+\frac{v}{c_v}\left[2\left(\frac{\partial u}{\partial x}\right)^2+2\left(\frac{\partial v}{\partial y}\right)^2+2\left(\frac{\partial w}{\partial z}\right)^2\right.$$
$$\left.+\left(\frac{\partial u}{\partial y}+\frac{\partial v}{\partial x}\right)^2+\left(\frac{\partial v}{\partial z}+\frac{\partial w}{\partial y}\right)^2+\left(\frac{\partial u}{\partial z}+\frac{\partial w}{\partial x}\right)^2\right] \tag{2.2.51b}$$

The equation of state is not necessary in this case and the fluid properties, namely, density, viscosity, specific heat (at constant volume for internal energy and at constant pressure for enthalpy) and the thermal conductivity are assumed to be known.

2.3 Boundary Conditions and Well-Posedness

The equations governing fluid flow are partial differential equations and therefore require boundary conditions to obtain a unique solution. The generic term of boundary conditions includes initial conditions for a time-dependent problem and boundary conditions which apply on the boundaries of the flow domain at any given time.

Boundary conditions can be of three types:

(a) Dirichlet type where the value of the variable is specified, e.g., zero velocity on a wall, zero normal velocity at a free surface;
(b) Neumann type, where the value of the gradient of a variable is specified, e.g., a specified heat flux on a heated surface, zero normal velocity gradient on a symmetry plane, and
(c) mixed type (also called Robin type), where both the variable and its gradient are specified, e.g., in the case of a wall with convective and/ or radiative heat transfer.

Common types of surfaces where boundary conditions are often specified are inlets, outlets, symmetry planes, walls, constant pressure planes, periodic planes, anti-symmetric planes, translating/rotating walls, open boundaries etc. Special boundary conditions are required for free surfaces as will be discussed shortly. Before we go on to the details of these, it is important to investigate the mathematical nature of the equations and the concept of well-posedness of a mathematical problem as these have a bearing on the boundary conditions to be specified for a given case.

2.3.1 Mathematical Nature of the Governing Equations

The equations governing the flow of a Newtonian fluid are quasi-linear partial differential equations of second order in the sense that the second derivatives appear as linear terms. It is known that this type of equations represent distinctly different classes of physical phenomena depending on the relative values of the coefficients. To explore this, consider the quasi-linear second-order equation of the form

$$A\partial^2\varphi/\partial x^2 + B\partial^2\varphi/\partial x\partial y + C\partial^2\varphi/\partial y^2 = D \tag{2.3.1}$$

where A, B, C and D may be non-linear functions of $x, y, \varphi, \partial\varphi/\partial x$ and $\partial\varphi/\partial y$ but not of second or higher derivatives. This equation can be written as a set of first order equations. For example, writing $\partial\varphi/\partial x = u, \partial\varphi/\partial y = v$, Eq. (2.3.1) can be written as

$$\begin{aligned} A\,\partial u/\partial x + B\,\partial u/\partial y + C\,\partial v/\partial y &= D \\ \partial v/\partial x - \partial u/\partial y &= 0 \end{aligned} \tag{2.3.2}$$

Now, a system of first order quasi-linear partial differential equations is called hyperbolic if its homogeneous part admits a wave-like solution. This means that the hyperbolic set of equations is associated with a system behaviour governed by wave-like phenomena. If the homogeneous equations admit solutions corresponding to damped waves, then it is called parabolic, and if it does not admit wave-like solutions, then the system of equations is called elliptic. The behaviour of an elliptic system is dominated by diffusion-like phenomena whereas that of a parabolic system exhibits both wave-like and diffusion-like phenomena.

This distinction of wave-like or diffusion-like behaviour is important because they are associated with the way information propagates within the system and will ultimately determine what initial and boundary conditions need to be specified. A wave-like behaviour has a certain direction and speed of propagation of information, say, of a change in temperature. A diffusion-like behaviour has no directional propagation of information, and the information of an increase in temperature propagates in all directions at a rate which is determined by the (not necessarily isotropic) diffusivity. These factors have a bearing on the solution of the equations as we will see later.

In order to find out the conditions under which Eq. (2.3.2) admits wave-like solution, we seek a simple plane wave solution propagating in the direction n:

$$u = u^* e^{j(n_x x + n_y y)}; \quad v = v^* e^{j(n_x x + n_y y)} \tag{2.3.3}$$

where n_x and n_y are the direction cosines and $j = \sqrt{-1}$. This solution is possible if the homogeneous part of Eq. (2.3.2) admits non-trivial solutions. Substituting (2.3.3), we find that this is possible if the system

$$\begin{aligned} Au^{*}n_x + B\,u^{*}n_y + Cv^{*}n_y &= 0 \\ v^{*}n_x - u^{*}n_y &= 0 \end{aligned} \tag{2.3.4}$$

has non-trivial solutions. Eliminating v* from Eq. (2.3.4), we find that for non-trivial solutions for u*, the following condition must be satisfied:

$$A(n_x/n_y)^2 + B(n_x/n_y) + C = 0$$

or

$$n_x/n_y = \left(-B \pm \sqrt{B^2 - AC}\right)/2A \tag{2.3.5}$$

The two roots of Eq. (2.3.5) are real and distinct if $B^2 - AC > 0$ indicating that two real, wave-like solutions are possible for the system. These two roots represent two characteristic lines in the x-y plane at the point (x, y). Therefore the system is hyperbolic if $B^2 - AC > 0$. If $B^2 - AC < 0$, then wave-like solutions are not possible and the system of equations will be elliptic. If $B^2 = AC$, the system is said to be parabolic.

Now let us explore the relation between the nature of the equation and the boundary conditions that are needed to find a unique solution. Consider the simple case of one-dimensional, constant property energy equation:

$$\rho c_v \partial T/\partial t + \rho u c_v \partial T/\partial x = k\partial^2 T/\partial x^2 + \mu(\partial u/\partial x)^2 \tag{2.3.6}$$

Neglecting the viscous dissipation term as it is usually small in flows with heat transfer, the above equation can be approximated as

$$\partial T/\partial x + P\partial T/\partial x = Q\partial^2 T/\partial x^2 \tag{2.3.7}$$

The first term represents the rate of increase of temperature of a fluid in a small control volume; the second and the third terms represent, respectively, the net convective and diffusive fluxes, respectively, through the control surface (in this case, the x-faces) of the control volume. In the case where diffusion is negligible, the term on the right hand side drops out and we have a first-order hyperbolic wave equation of the form

$$\partial T/\partial x + P\partial T/\partial x = 0 \tag{2.3.8}$$

which has the solution

$$T(x, t) = T(x - Pt, 0) \tag{2.3.9}$$

representing a travelling wave moving at a velocity of P in the positive x-direction. An initial profile will move with time t by a distance $P \cdot t$ as shown schematically in

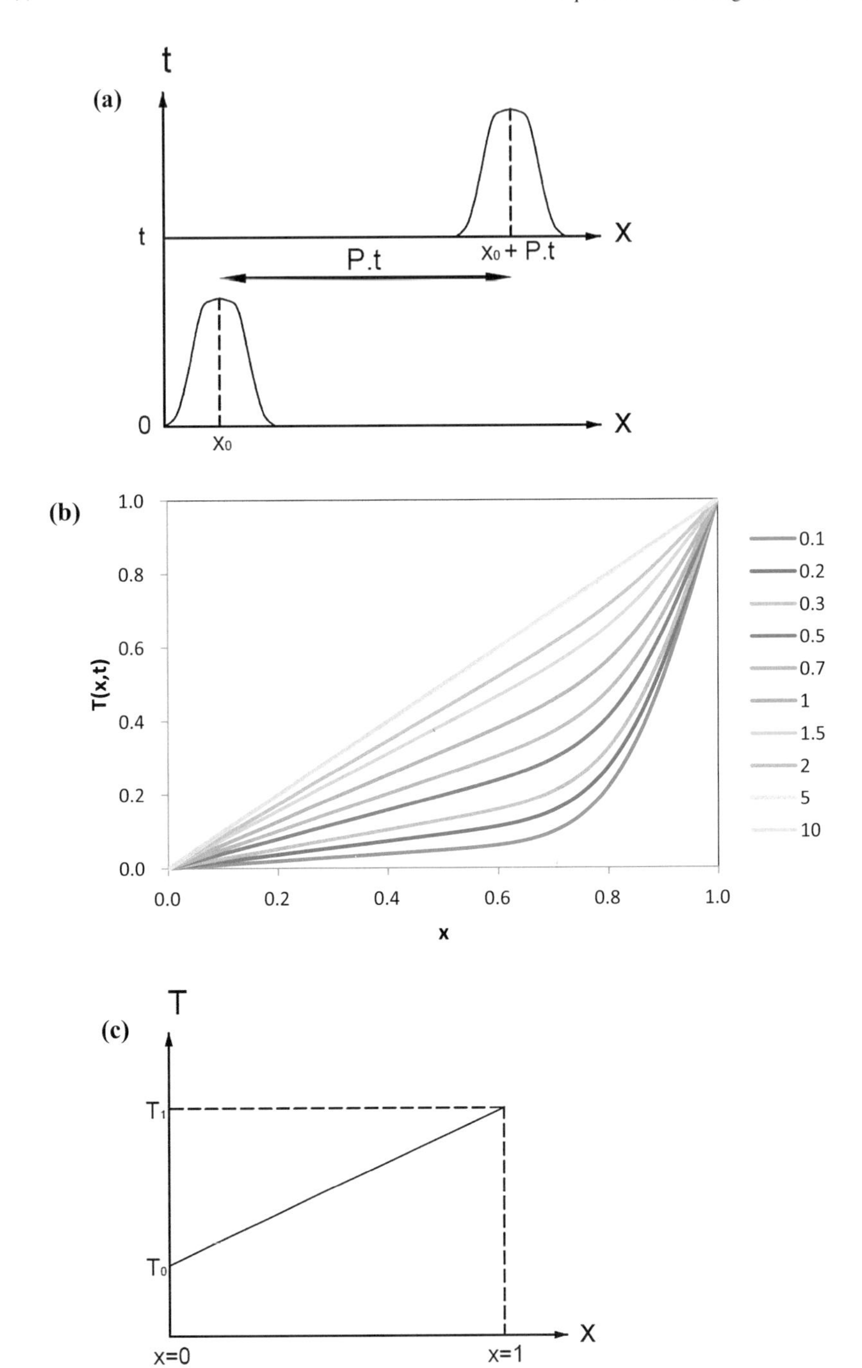

Fig. 2.6 Solution of Eq. (2.3.7) for **a** the hyperbolic case, **b** the parabolic case and **c** the elliptic case

Fig. 2.6a. It is clear that T(x, t) is influenced by the left boundary value but not by the right boundary value until the wave reaches there.

In the case where convection effects are negligible, the second term in Eq. (2.3.7) drops out and we have a parabolic equation (it can be shown to be parabolic by decomposing into a set of two first-order equations and proceeding as before to show that there is a single admissible wave-like solution):

$$\partial T/\partial x = Q\partial^2 T/\partial x^2 \qquad (2.3.10)$$

If we solve this equation subject to the boundary conditions that at x = 0, T = 0 and at $x = 1, T = T_1$ for all times and to the initial condition that at t = 0, T(x, 0) = 0, we get (see Tannehill et al. 1997) the following expression for T(x, t):

$$T(x,t) = T_1 x + \sum_{n=1}^{\infty}\{[2T_1(-1)^n]/(n\pi)\}\exp(-n^2\pi^2\alpha t)\sin(n\pi x) \qquad (2.3.11)$$

In the solution, which is shown schematically in Fig. 2.6b, the value of T(x, t) depends on both the end boundary conditions and only on past values of time.

In the case of steady state diffusion with no convection, the first two terms of Eq. (2.3.7) drop out leaving the elliptic equation

$$Q\partial^2 T/\partial x^2 = 0 \qquad (2.3.12)$$

Solving it for the same spatial boundary conditions as for Eq. (2.3.10), we get

$$T(x) = T + (T_1 - T_0)\,x \qquad (2.3.13)$$

which is shown schematically in Fig. 2.6c. Here, the solution depends on the temperature values at both the boundaries.

It is clear from the above discussion that the boundary conditions needed for the solution are different for each type of equation. This has implications for the well-posedness of a problem as discussed below.

2.3.2 *Well-Posedness of a Mathematical Problem*

Since a problem represented by a set of partial differential equations requires boundary conditions to yield a unique solution, there should be a dependence of the solution on the imposed boundary conditions. However, the solution so obtained, i.e., by solving the governing partial differential equations subject to the specified initial and boundary conditions, must be satisfactory. This condition is rigorously expressed in terms of well-posedness. A mathematical problem involving partial differential equations is said to be well-posed in the sense of Hadamard (1952) if there exists a solution; the solution is unique and it depends continuously on the boundary conditions. The last condition implies that a small change in the boundary condition at a point should produce a small change in the solution in the vicinity of the point.

To illustrate the ill-posedness of a problem, Hadamard (1952) constructed the following example involving an elliptic equation. A solution is sought for the Laplace equation

$$\partial^2\varphi/\partial x^2 + \partial^2\varphi/\partial y^2 = 0 \tag{2.3.13a}$$

in the domain $-\infty < x < \infty$ and $y > 0$ with the following boundary conditions:

$$\varphi(x,0) = 0 \quad \text{and} \quad \partial\varphi/\partial y\,(x,0) = 1/n\,\sin(nx), \quad n > 0. \tag{2.3.13b}$$

The solution of Eq. (2.3.13) can be found, using separation of variables, to be

$$\varphi(x,y) = 1/n^2 \sin(nx)\,\sinh(ny) \tag{2.3.14}$$

Although the above solution satisfies both the boundary conditions at y = 0, it is not well-behaved. For large values of n, $\partial\varphi/\partial y$ at y = 0 becomes small. However, for large values of n, φ varies as $1/n^2 \exp(ny)$ and φ becomes very large even for small values of y. Also, according to the boundary condition, $\varphi(x,0) = 0$. Thus, while the solution requires that φ must become very large within finitely small distances from y = 0, the gradient $\partial\varphi/\partial y$ must be small because it must be zero at the boundary. Thus, there is incompatibility between the solution and the boundary conditions and the problem is ill-posed. Here, the ill-posedness obviously arises from the fact that a solution to the elliptic problem is being sought in an open

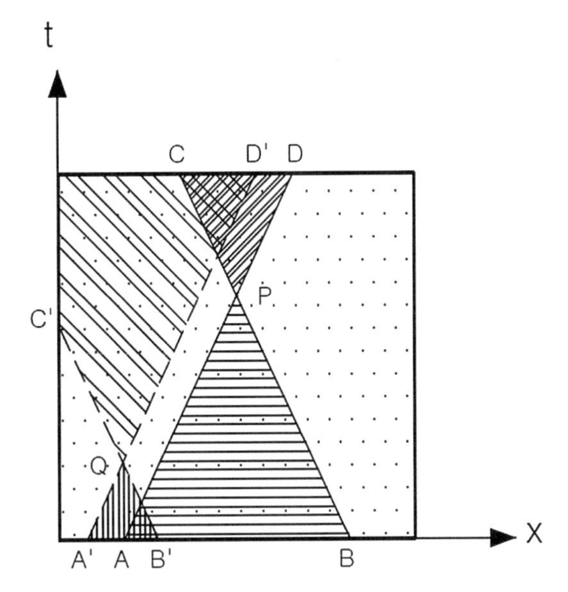

Fig. 2.7 Domains of dependence and influence for a hyperbolic problem

domain (boundary conditions being specified only for y = 0) whereas the nature of the equation requires specification of the boundary conditions over a closed domain. Further examples of ill-posed problems are discussed by Wesseling (2004).

This dependence on the boundary conditions can be discussed in terms of domains of influence and domains of dependence. For a linear hyperbolic problem, the value of the variable at point P(x, t), see Fig. 2.7, is influenced by all the points enclosed by the two characteristic lines running backward and the x-axis, i.e., the region ABPA. The initial condition (at t = 0) on the x-plane lying outside the intersection points A and B does not have any influence on the value at $P(x_P, t_P)$. Thus, the region ABPA is called the domain of dependence for point P. For a well-posed problem, a change in the initial condition in the region AB should have an effect on the value of the variable at point P. Also, a change in the initial condition outside AB should not have any effect on the value at $P(x_P, t_P)$. The value of the variable at point P will have an effect on the values enclosed by the open triangle CPD, and this zone is therefore called the domain of influence of point P. These domains of influence and dependence change with location; for example, the domain of dependence of the point Q at (x_Q, t_Q), $t_Q < t_P$ is smaller and is marked by the dashed lines A′QB′A′.

Finally, the three types of problems—hyperbolic, parabolic and elliptic—have different domains of dependence and influence, as illustrated in Fig. 2.8. For a hyperbolic problem, the domain of dependence is enclosed by the boundary and the two characteristic lines passing through P, as shown in Fig. 2.7. For a parabolic problem, there is a single characteristic line corresponding to constant t which divides the whole region into domains of dependence and influence as shown in Fig. 2.8a. For an elliptic problem, the value at point P is influenced by all the values on the enclosing boundary and the whole region is both the domain of dependence and the domain of influence. Elliptic problems require boundary conditions over a

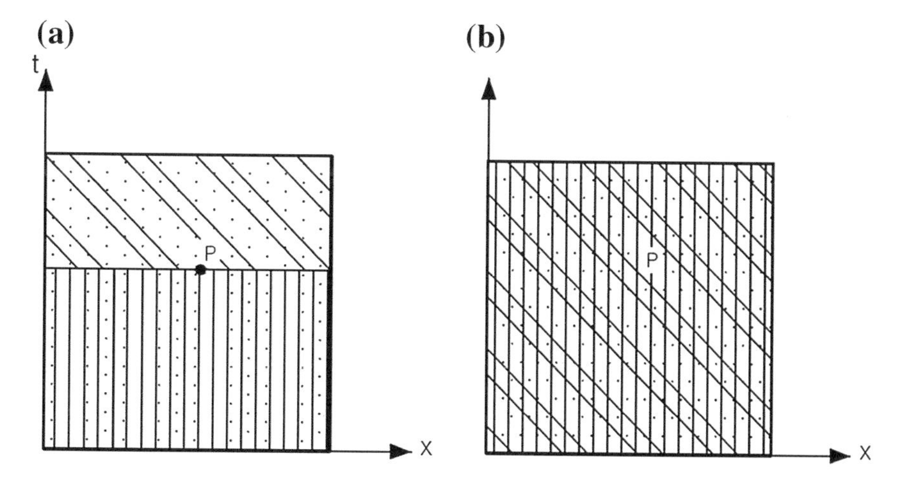

Fig. 2.8 Domains of dependence and influence for **a** a parabolic problem and **b** for an elliptic problem

closed domain and are therefore called boundary value problems while parabolic and hyperbolic problems require initial and boundary conditions over an open domain. In these cases, the solution is obtained by marching forward in time or in the direction of the time-like variable. These are therefore called initial value problems.

The equations governing the motion of a viscous fluid, namely, the continuity equation and the momentum conservation equations, are too complex to fit into the above classification of hyperbolic, parabolic or elliptic type of equations. They can be any one of them or a mixture of them depending on the specific flow and geometry. For example, unsteady inviscid compressible flow is essentially hyperbolic in nature. For steady compressible flows, the character depends on the speed of the flow. If the flow is supersonic, the equations are hyperbolic, and if the subsonic, they are elliptic. Steady transonic flows have a mixed character having both supersonic and subsonic regions, the separation between the two being usually unknown a priori. Boundary layer equations are parabolic in nature and information propagates only in the downstream direction. However, the solution of these requires pressure which is obtained by solving a potential flow equation, thus making it an elliptic-parabolic problem for subsonic conditions. When a flow has a region of recirculation, information may travel both upstream and downstream and the corresponding equations are elliptic. Since the continuity equation is of hyperbolic character, hyperbolic-parabolic problems will also arise. In flows of such mixed character, the boundary conditions should be of initial value type for the hyperbolic and parabolic components and of the boundary value type for the elliptic components. It is an important general rule that numerical methods should respect the properties of the equations that they are solving and special care should be taken in implementing the boundary conditions; see, for example, Issa and Lockwood (1977) and McGuirk and Page (1991). An extensive discussion of boundary conditions for incompressible Navier-Stokes equations is given in Gresho (1991).

2.3.3 Initial and Boundary Conditions for Fluid Flow Problems

Typical boundaries of a flow domain that one encounters in a fluid flow situation can be one of five types:

(i) a fluid-solid boundary
(ii) a free liquid surface
(iii) a liquid-vapour or liquid–liquid surface
(iv) inlet and exit sections
(v) geometry-related conditions such as symmetry, periodic planes, open boundaries

Let us examine the conditions that are applicable at these boundaries.

2.3.3.1 Conditions at a Fluid-Solid Boundary

The condition that is applied at a fluid-solid boundary for a real fluid is that there is no slip between the wall and the fluid. In the case of a liquid, the molecules are so closely packed and the mean free path is so small that the fluid particles contacting the wall should be essentially in equilibrium with the wall. Thus, for all practical purposes, the liquid sticks to the wall and will be at the wall temperature. Thus, there is no velocity slip (strictly speaking, the no-slip condition refers to the absence of tangential slip only and not to the absence of velocity jump) or temperature jump at the liquid-solid boundary:

$$U_{liq} = U_{sol} \quad T_{liq} = T_{sol} \tag{2.3.15}$$

The temperature boundary condition can also be in the form of specified heat flux (for example, zero heat flux for an insulated wall) or a combination of temperature and heat flux, e.g.,

$$-k\partial T/\partial n = h_{conv}(T - T_w) \tag{2.3.16}$$

where h_{conv} is convective heat transfer coefficient and T_w is the wall temperature. Special boundary conditions are required for radiative heat transfer and boiling heat transfer. Usually, a Neumann type or a mixed type of boundary condition is not specified for the velocity field at the wall, although special care has to be taken for turbulent flow and for flow over rough surfaces. This is discussed further in Chap. 6.

For a gas, there may be a relative slip velocity and a temperature jump if the mean free path is large (White 1991), both being of the order of Ma c_f, where Ma is the Mach number and c_f is the skin friction coefficient, relative to the characteristic velocity and the temperature driving force. As a thumb rule, no-slip conditions can be expected to prevail for Knudsen numbers greater than 0.01 where the Knudsen number, Kn, is defined as the ratio of the mean free path to the characteristic length of the flow. For inviscid flow, there can be a finite slip. Since the equations in this case would be of first order only, it would be sufficient to specify a zero normal velocity; the tangential velocity is obtained as part of the solution.

For flows with mass transfer, such as those under blowing and suction conditions, the normal velocity is fixed by the mass flow rate through the wall. The tangential velocity is set to zero for a real fluid whereas this is left floating for an inviscid flow. The temperature boundary condition usually depends on the direction of flow. If there is suction at the wall, the no-jump temperature condition, namely, $T_{fluid} = T_w$ can be taken as a fairly accurate estimate. For blowing, the temperature near the wall may depend on the temperature of the fluid being blown.

2.3.3.2 Conditions at a Free Surface

A free liquid surface is the interface between a liquid and a gas across which there is no mass transfer and at which there is no shear stress. The gas above the free surface only exerts a known pressure. One therefore will not solve for the velocity field in the gas field. The liquid phase velocity field is obtained by solving the Navier-Stokes equations with appropriate boundary conditions. Two conditions, one kinematic and one dynamic are applicable at the free surface. The kinematic condition requires that the free surface is a stream surface such that the normal component of the fluid velocity is equal to the normal component of the velocity of the free surface. For a free surface given by $\eta = \eta(x, y)$, the kinematic condition therefore requires that the upward velocity of the fluid particle, w, is equal to the motion of the free surface:

$$w(x, y, \eta) = D\eta/Dt = \partial\eta/\partial t + u\partial\eta/\partial x + v\partial\eta/\partial y \quad (2.3.16)$$

The dynamic condition for constant surface tension requires that the pressure in the liquid and the ambient pressure must be the same except for the effect of surface tension. Thus,

$$p(x, y, \eta) = p_a - \sigma(1/R_x + 1/R_y) \quad (2.3.17a)$$

where σ is the surface tension, R_x and R_y are the radii of curvature of the surface and are given by

$$1/R_x + 1/R_y = \{\partial^2\eta/\partial x^2 + \partial^2\eta/\partial y^2\}/\left\{1 + (\partial\eta/\partial x)^2 + (\partial\eta/\partial y)^2\right\}^{3/2} \quad (2.3.17b)$$

If the effect of surface tension is small and the free surface is fairly flat (so that $\partial\eta/\partial x$ and $\partial\eta/\partial y$ are small), then the above conditions reduce to

$$w = \partial\eta/\partial t \quad \text{and} \quad p = p_a$$

In the general 3-dimensional case, the free surface may be described by set of local orthogonal coordinates **n**, **t** and **s**, where **n** is the unit normal to the free surface and is directed away from it, as shown in Fig. 2.9. The unit vectors **t** and **s** lie within the plane of the free surface at a given point and are mutually orthogonal. The local curvature of the free surface, κ, is expressed in the radius of the curvature in the t and s directions:

$$\kappa = 1/R_t + 1/R_s \quad (2.3.18)$$

In this case, the kinematic condition can be expressed as

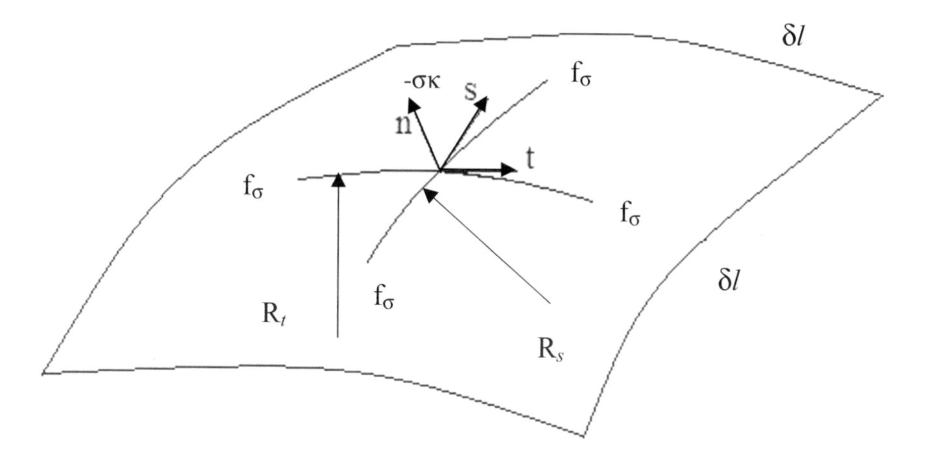

Fig. 2.9 Kinematic and dynamic boundary conditions at a gas–liquid interface. The tensile stress, $f_\sigma = \sigma\,\delta l$, created in the plane of the interface is shown. In the normal direction, it introduces a pressure-like compressive stress given by $-\sigma\kappa$

$$[(\mathbf{u} - \mathbf{u}_f) \cdot \mathbf{n}]_{fs} = \dot{m}_{li} = 0 \tag{2.3.19}$$

where $\mathbf{u}_f$ is the velocity of a fluid particle on the free surface, $\dot{m}_{li}$ is the mass transfer across the interface and the subscript fs denotes the free surface.

The dynamic condition requires that the forces acting on the free surface are in equilibrium. Denoting the liquid and the gas phases on either side of the free surface by subscripts l and g, we can write the dynamic condition for the general condition where the surface tension changes with position or time as

$$\begin{aligned} (\mathbf{n} \cdot \mathbf{T})_l \cdot \mathbf{n} &= (\mathbf{n} \cdot \mathbf{T})_g \cdot \mathbf{n} + \sigma\kappa \\ (\mathbf{n} \cdot \mathbf{T})_l \cdot \mathbf{t} &= (\mathbf{n} \cdot \mathbf{T})_g \cdot \mathbf{t} + \partial\sigma/\partial\mathbf{t} \\ (\mathbf{n} \cdot \mathbf{T})_l \cdot \mathbf{s} &= (\mathbf{n} \cdot \mathbf{T})_g \cdot \mathbf{s} + \partial\sigma/\partial\mathbf{s} \end{aligned} \tag{2.3.20}$$

Figure 2.9 shows the tensile stress, $f_\sigma = \sigma\,\delta l$, created in the plane of the interface by an interface of length δl in both t and s directions. This will act as a shear stress on the fluid. If the surface tension varies with position, for example, due to changes in temperature or concentration, then the surface tension-induced shear stress on the liquid continuum can lead to fluid motion. In the specific case of temperature variations along the surface, and given that surface tension decreases as temperature is increased, the fluid movement, known as Marangoni convection, will be from the relatively hot region to the relatively cold region. In the normal direction, surface tension introduces a pressure-like compressive stress on a convex surface shown in Fig. 2.9. Thus, it is only for a flat surface that we have the condition that $p_l = p_g$ at the interface.

2.3.3.3 Conditions at a Vapour–Liquid or Liquid–Liquid Interface

A vapour–liquid interface differs from a free surface by having a flux of mass, momentum and heat across the former. Examples of such vapour–liquid interfaces are the ocean-atmosphere interface and the vapour–liquid interface in chemical and other process equipment such as absorbers, bubble columns and steam generators. In general, one computes the velocity field in each phase by solving the phasic conservation equations subject to boundary conditions including those at the interface. At these interfaces, the equality of fluxes of momentum, heat and mass on the liquid side and the vapour side must also be ensured. Thus, we have, in addition to the kinematic and dynamic conditions described above,

$$\dot{m}_{li} = \dot{m}_{vi} \quad \tau_{li} = \tau_{vi} \quad q_{li} = q_{vi} \tag{2.3.21}$$

Of these, momentum and heat exchanges are diffusion processes, and can be evaluated in terms of the velocity or temperature gradients. Mass transfer between the two phases occurs by more complicated processes such as evaporation and condensation and case-specific relations may be necessary to evaluate the mass flux across the interface. For a liquid–liquid interface, the equality of fluxes is still applicable. The major difference in this case is in the mass transfer, which is primarily due to diffusion, and a gradient-type of boundary condition can be written for the interfacial mass flux also. A more detailed discussion of computation of multiphase flows is given in Chap. 6.

2.3.3.4 Conditions at Inlet and Outlet Sections

The inlet conditions often take the form of Dirichlet boundary conditions and the values of all the flow variables are specified in the inlet section. For incompressible flows, the inlet conditions include all the velocity components and temperature at the inlet; the pressure is not specified at the inlet (except perhaps at a point as a reference pressure) as it is the pressure gradient that drives the flow and the density is a constant. The inlet values should also be specified for additional flow variables, if any, such as concentration and turbulence quantities. The boundary conditions at the outlet are usually unknown, and often the assumption of fully developed flow is made at the outlet so that, for a variable φ, the outlet boundary condition becomes $\partial\varphi/\partial n = 0$. This would be valid (in the absence of external forces on the outlet) if the flow is essentially unidirectional, as may happen in the case of fully developed flow in a duct or if the outlet is located well away from a solid object in external flows. The flow domain is sometimes extended to enable the flow to become fully developed. It is also possible to specify constant pressure boundary conditions at the inlet and the outlet. The velocities on these planes cannot then be specified and will have to be extrapolated from within the interior. Yet another possibility is to specify a mass flux through the domain. Again, the velocities cannot be specified at either the inlet or the outlet but will have to be obtained based on a Neumann-type condition, i.e., $\partial\varphi/\partial n = 0$. A Neumann boundary condition for the velocities at the inlet and the outlet coupled with a no-slip condition on the solid surfaces could lead to the trivial solution the velocity is zero throughout. This would, of course, not satisfy the requirement of the specified mass flux and special measures have to be employed to avoid this trivial solution.

For compressible flows, it is necessary to specify both the stagnation pressure and the static pressure so that the mass flow rate can be uniquely determined from the relation

$$p_0 = p_s\left[1 + (\gamma - 1)M^2/2\right]^{\gamma/(\gamma-1)} \tag{2.3.22}$$

Similarly the stagnation enthalpy must be used as the thermal boundary condition at the inlet. The static temperature can then be calculated from the following relation:

$$T_0 = T_s\left[1 + (\gamma - 1)M^2/2\right] \tag{2.3.23}$$

If instead, the velocity is specified at the inlet along with the stagnation pressure and temperature, the static pressure and temperature are left "floating" and are obtained by extrapolation from the interior. The outflow boundary conditions require the specification of the static pressure alone for subsonic compressible flows while no outflow boundary conditions need to be specified for supersonic flows.

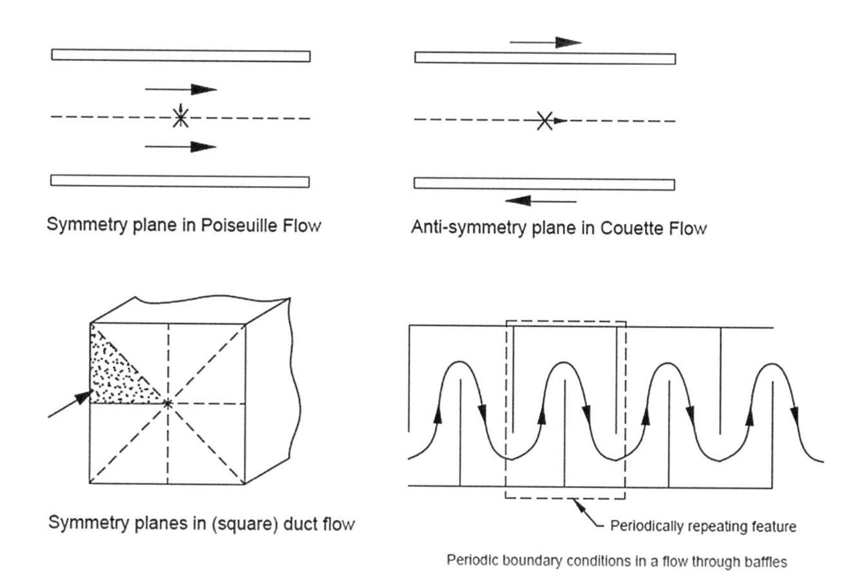

Fig. 2.10 Geometry-related boundary conditions

2.3.3.5 Geometry-Related Conditions

A flow domain may have geometry-specific boundary conditions. Principal among these types are (a) symmetry planes, (b) periodic planes and (c) anti-symmetry planes. Typical examples of these are shown in Fig. 2.10. For a symmetry plane, the normal gradient of any flow variable (including pressure) is zero. For a periodic plane, the variable values are the same at the inlet and at the outlet.

A special type of geometry-related boundary condition is the open or far-field boundary conditions. For flow over an isolated body, the computational domain is in theory infinite and needs to be very large. The specification of the approach and the downstream boundaries as well as the computational domain in the transverse direction can pose difficulties. Normally, in such cases, one knows the free stream conditions and the angle of attack. One needs to choose the approach boundary sufficiently upstream of the forward stagnation point on the body. The downstream boundary also needs to be very far as the effect of a wake can persist for quite a distance. The boundaries in the transverse direction should ideally be located 20–50 chord lengths away from the body so that free stream boundary conditions can be applied. In case of highly compressible flows, one needs to ensure that the far field boundaries do not introduce unphysical pressure reflection effects (Thomas and Salas 1986; Wesseling 2004). Special treatment is sometimes used to bring the infinite boundaries closer to the surface so as to reduce the size of the computational domain and to deal with special cases such as lifting bodies and turbomachinery flows. In such cases, free-stream conditions at a far-away boundary are replaced by artificial boundary conditions. The interested reader is referred to the works of Bayliss and Turkel (1982), Whitfield and Janus (1984), Thomas and Salas (1986), Sani and Gresho (1994) and Renardy (1997) to further explore the subject of appropriate boundary conditions for open boundaries. Implementation of boundary conditions on structured and unstructured grids has been discussed in detail by Blazek (2001).

2.3.3.6 Initial Conditions

For transient flow problems, the initial conditions of all the flow variables must be specified throughout the domain. For incompressible flow, for example, these include u, v, w and p throughout the domain. Strictly speaking, the velocity should be such that it is divergence free so that it satisfies the continuity equation. Gresho (1991) makes the point that the specified velocity field on the boundary also should be divergence-free at time = 0 and for all times. Although boundary conditions valid for all times may be specified, only part of these boundary conditions may be needed for a well-posed initial boundary value problem.

2.4 Summary

In this chapter, we have derived the fundamental equations which govern the flow of a fluid. These are the conservation of mass or the continuity equation, the conservation of momentum or the Navier-Stokes equations and the conservation of energy equation. We have also discussed the mathematical character of these equations and its influence on the necessary initial and boundary conditions for a well-posed mathematical problem. Some of the boundary conditions that are commonly used in fluid flow calculations are listed.

It is perhaps the right way to end the discussion with consequences of under- or over-specification of boundary conditions. Fletcher (1991) remarks that under-specification of boundary conditions normally leads to the failure of not being able to obtain a unique solution. Over-specification of the boundary conditions gives rise to flow solutions with unphysical boundary layers close to the boundaries where they are applied. Most of the attention in the literature has been focussed on outflow boundary conditions. For many process industry-related problems, the inlet conditions, which usually require the specification of all the variables on the inlet section, as well as the initial conditions, which require the values of all the flow variables throughout the domain, are equally, if not more, uncertain and prone to error. Usually the specification is based on some estimates of the quantities from known or imposed values of the primary variables such as the velocity, temperature and concentration. Where there is doubt about the exact boundary condition to be specified, a systematic sensitivity analysis should be carried out to verify whether or not the boundary condition is critical in determining the flow field. If it is, more care should be taken in specifying it. If more knowledge of the exact boundary condition is not available, then this ignorance factor must be considered when the results of the simulations are evaluated. The ignorance factor should reflect both the degree of uncertainty about the boundary condition and the influence of the boundary condition on the flow field. This is especially true in the context of the wide prevalence of commercial CFD codes using which one can make quick calculations of even quite complicated flow cases. The accuracy of such simulations depends both on the equations that are used to describe the flow phenomena and the choice of the computational domain and initial and boundary conditions. Validation and sensitivity analyses should be carried out to generate confidence in the solutions.

Problems

1. Identify the elements of the deformation rate tensor, $\partial u_i/\partial x_j$, for the following cases of fluid flow:

   a. fully developed flow between two parallel plates: $u = C(1 - y^2/h^2)$ where y is the distance from the mid-height and h is the half the separation distance between the two plates.

b. forced vortex: $u_r = 0, u_\theta = Cr$
c. free vortex: $u_r = 0, u_\theta = C/r$
d. potential flow past a circular cylinder of radius a with circulation:

$$u_r = U_\infty \cos\theta(1 - a^2/r^2), \quad u_\theta = -U_\infty \sin\theta(1 + a^2/r^2) + K/r$$

e. creeping flow around a sphere of radius a:

$$u_r = U_\infty \cos\theta\left\{1 - 3/2(a/r) + 1/2\,(a/r)^3\right\},$$
$$u_\theta = -U_\infty \sin\theta\left\{1 - 3/4\,(a/r) - 1/4(a/r)^3\right\}$$

For gradient operators in cylindrical and spherical coordinate systems, see Deen (1998).

2. Write down the continuity equation for (a) compressible flows, and (b) incompressible flows in vector form. Use this equation and the gradient operators to write the fully expanded form of these equations in cylindrical and spherical coordinate systems for the three types of 2-d flows.
3. Write down the 3-d continuity equation in cylindrical and spherical coordinate systems.
4. Write down the vector form of the Navier-Stokes equations. Use the gradient and divergence operator forms in cylindrical and spherical coordinates (see Deen 1998) to write the N-S equations in these coordinate systems for (a) variable and (b) constant property cases.
5. Write down the fully expanded form of the energy equation in terms of (a) internal energy and (b) enthalpy in Cartesian coordinate system for a variable property case.
6. Derive the following dimensionless form of Euler equations from Navier-Stokes equations for inviscid flow using x_r, ρ_r, u_r as the reference length, density and velocity as the non-dimensionalizing parameters:

$$\text{Continuity:} \quad \partial\rho/\partial t + \partial m/\partial x = 0 \tag{Q2.1}$$

$$\text{Momentum balance:} \quad \partial m/\partial t + \partial(um)/\partial x + 1/(\gamma M^2)\partial p/\partial x = 0 \tag{Q2.2}$$

$$\text{Energy balance:} \quad \partial p/\partial t + \gamma p \partial u/\partial x + u \partial p/\partial x = 0 \tag{Q2.3}$$

Here $m = \rho u$ and ρ, u, p and t are rendered dimensionless using x_r, ρ_r, u_r and t_r ($=x_r/u_r$). Also, γ is the ratio of specific heats, $M = u_r/c_r$ is the Mach number where $c_r = (\gamma p_r/\rho_r)^{1/2}$ = the speed of sound.

7. Show that if an inviscid flow is also adiabatic and non-heat conducting (zero thermal conductivity), the flow is also isentropic and that along a streamline the entropy is constant.

8. Derive the compressible and incompressible potential flow equations with inviscid, irrotational flow assumptions.
9. Show that potential flow with no heat addition (or removal) is also homentropic flow, i.e. entropy is constant everywhere in the flow.
10. Derive the following equation for pressure by taking divergence of the momentum conservation equations in 2-d (see Sect. 4.3):

$$\partial^2 p/\partial x^2 + \partial^2 p/\partial y^2 = 2\rho(\partial u/\partial x\, \partial v/\partial y - \partial u/\partial y\, \partial v/\partial x) \tag{Q2.4}$$

11. A system of first order partial differential equations of the form

$$\partial U/\partial t + A\partial U/\partial x + U = 0 \tag{Q2.5}$$

is said to be hyperbolic in t and x if the eigenvalues of A are real and distinct. Use this to show that the system of Euler equations given by Eqs. (Q2.1)–(Q2.3), see problem 6, is hyperbolic.

12. A system of steady state partial differential equations of the form

$$A\partial U/\partial x + B\partial U/\partial y + U = 0 \tag{Q2.6}$$

has its mathematical character defined by the roots of

$$\mathrm{Det}\,(An_x + Bn_y) = 0 \tag{Q2.7}$$

where n_x and n_y are the direction cosines of the wave-like solution propagating in the x-y plane. If n_x/n_y obtained from Eq. (Q2.7) are real and distinct, the system is hyperbolic. If all the roots are imaginary, then it is elliptic. If some are real and some imaginary, it will have a mixed hyperbolic/ elliptic character. Use these ideas to show that the following system of equations is elliptic:

$$\partial u/\partial x + \partial v/\partial y = 0$$
$$\partial v/\partial x - \partial u/\partial y = 0$$

13. Using the same reasoning as in Problem #12, show that the system of governing equations for steady, inviscid, incompressible two-dimensional flow is of mixed hyperbolic/elliptic character.
14. Using the same reasoning as in Problem #12, show that the system of governing equations for steady viscous, incompressible two-dimensional flow is of elliptic character.
15. Consider the one-dimensional form of the steady advection-diffusion equation given by

$$U d\varphi/dx + D\, d^2\varphi/dx^2 = 0 \quad 0 \le x \le 1 \tag{Q2.8}$$

with the boundary conditions

$$\varphi(0) = a \quad \text{and} \quad d\varphi(1)/dx = b \tag{Q2.9}$$

Show that the solution is given by

$$\varphi(x) = a + b/Pe\{e^{(x-1)Pe} - e^{-Pe}\} \tag{Q2.10}$$

and that the problem is well-posed for $U > 0$ but is ill-posed when $U < 0$ because if $Pe \ll -1$, $\varphi(x)$ is very sensitive to small perturbations in b (see Wesseling 2004). Thus, a Neumann boundary condition at inlet renders it ill-posed for this problem.

16. Show that the steady advection-diffusion case of Problem #15 is well-posed when the Neumann boundary condition is imposed at the outlet, i.e., when $U > 0$ in Eq. (Q2.8).
17. Show that the steady advection-diffusion equation of Problem #15 (Eq. Q2.8) is well-posed when Dirichlet boundary conditions are imposed at the inlet and the outlet.
18. Show that the Helmholtz equation

$$\partial^2\varphi/\partial x^2 + \partial^2\varphi/\partial y^2 + \lambda^2\varphi = 0 \quad 0 \le x \le \pi, \ 0 \le y \le \pi \tag{Q2.11}$$

with $\varphi = 0$ on the boundary is ill-posed because the solution given by

$$\varphi(x, y) = \sin(\lambda x)\ \sin(\lambda y), \quad \lambda = 0, 1, 2, \ldots \tag{Q2.12}$$

is not unique.

19. Consider the one-dimensional unsteady diffusion equation given by

$$\partial\varphi/\partial t = D\partial^2\varphi/\partial x^2 \quad 0 \le t \le T, 0 \le x \le 1 \tag{Q2.13}$$

with the boundary conditions

$$\varphi(t, 0) = \varphi(t, 1) = 0 \tag{Q2.14a}$$

and the “initial” condition

$$\varphi(T, x) = 1/m\ \sin(m\pi x) \tag{Q2.14b}$$

Show (see Wesseling 2004) that although a solution given by

$$\varphi(t, x) = 1/m\ \exp\{m^2\pi^2(T - t)\}\ \sin(m\pi x) \tag{Q2.15}$$

exists, the “final” solution given by

$$\varphi(0, x) = 1/m \exp(m^2\pi^2 T) \sin(m\pi x) \qquad \text{(Q2.16)}$$

is too sensitive to small changes in m for large m and that the problem is therefore ill-posed. Thus, for parabolic problems, marching backward in time is not admissible as a well-posed problem.

20. Write down the simplified form of the governing equations, identify the relevant flow domain and suggest suitable initial and boundary conditions for the following flow problems:

   a. *Stokes' first problem*: A plate of infinite width and length immersed in an initially quiescent, infinite expanse of liquid is suddenly set into motion at a uniform velocity of U_0 in the x-direction. This induces a variation of u in the normal direction. One needs to find u(t, y).
   b. *Steady, developing flow between parallel plates*: Flow enters a rectangular slot of infinite width and length L such that $L \gg H$, the separation distance between the two plates. One needs to find out the distance after which the flow becomes fully developed.
   c. *Fully-developed flow through an annulus*: We need to find the velocity profile for the case of fully developed flow between two cylinders of radii r_1 and r_2.
   d. *Plane Couette flow with a pressure gradient*: Flow takes place between two flat pates of infinite width and length with a constant pressure gradient $C_1 (C_1 < 0)$ in which one plate is stationary and the other plate moves at constant speed of U_p in the horizontal direction. Assume the flow to be laminar, steady and fully developed.
   e. *Cross-flow over a cylinder*: Flow takes place over a cylinder of diameter D and infinite length immersed in an infinite fluid medium. The approach velocity is U_∞. Note that the flow is steady and symmetric only for small Reynolds numbers ($Re = \rho U_\infty D/\mu$). Formulate the problem for the steady, symmetric case as well as for the unsteady, asymmetric case. Assume two-dimensional flow in both cases.
   f. *Creeping flow over a sphere*: A Newtonian fluid with a freestream velocity U_∞ flows over a sphere of diameter D. We need to find the velocity field. Consider the special case of creeping flow for which the Reynolds number ($Re = \rho U_\infty D/\mu \ll 1$). Non-dimenionalize the governing equations using U_∞, D and μ and apply the condition that $Re \ll 1$ to obtain a simplified set of governing equations. Set up the CFD problem to solve $u_r(r, \theta), u_\theta(r, \theta)$ and $p(r, \theta)$. Discuss how you can obtain the overall drag force on the sphere from the velocity and pressure field.
   g. Boundary layer flow over a flat plate. Consider the steady flow over a horizontal flat plate of finite length L, infinite width W and infinitesimal thickness immersed in an infinite fluid medium. Consider the case of a sufficiently high freestream velocity (U_∞) so that boundary layer type of flow occurs, and that δ, the boundary layer thickness, is much less than the length of the plate, i.e., $\delta/L \ll 1$. Non-dimensionalize the governing

equations using L and U_{∞}, and derive the "parabolized" form of the Navier-Stokes equations in which the streamwise second derivative term drops out of the diffusion term. You wish to calculate the velocity field, i.e., u(x, y) and v(x, y) for a steady two-dimensional flow. Formulate the CFD problem and discuss how you would derive the drag force acting on the plate.

h. *Lid-driven cavity problem*: Consider the two-dimensional version of the problem in which the fluid inside a rectangular cavity of length L and height H is made to circulate within the cavity by the steady horizontal movement of the lid at a velocity of U. Formulate this steady flow problem which is often used as a benchmark problem for validation of CFD codes, see Ghia et al. (1982).

i. *Flow over a backward/forward facing step*: Consider the Poiseuille flow between two parallel plates. In the case of flow over a backward facing step, there is a step (abrupt) change in the height between the two plates at a sufficiently long distance x_0 for the flow to become fully developed in the first part of the channel. Thus, for $x < x_0$, the height between the two plates is H_1 and for $x \geq x_0$, it is H_2. If the flow is from left to right, then the problem is considered as a backward facing step when $H_2 > H_1$ and a forward facing step when $H_2 < H_1$. The former is often used as a benchmark problem for incompressible flows (see Gresho et al. 1993) and the latter for compressible flows with a shock (see Woodward and Colella 1984). Formulate these steady state two-dimensional problems.

j. *Quenching of a rectangular block in water*: Consider a rectangular block of dimensions $L \times W \times H$. It is initially at a temperature of T_0 throughout. It is suddenly immersed in water at a temperature T_w. You wish to find out how the temperature inside the block varies with time. Formulate the transient heat conduction problem. Note that the heat loss to the water appears as a boundary condition and that the convective heat transfer within water is by natural convection and is therefore likely to change from wall to wall (as in bottom wall, side wall, top wall etc.) and even over a wall. Would you complicate the problem further by allowing for the water to boil? This is an example of the difficulty in problem formulation in tackling real world problems.

Chapter 3
Basic Concepts of CFD

In a numerical solution, we seek the flow field or the values of the flow variables such as the velocity components and the pressure at discrete, predetermined grid points spread throughout the flow domain. The location of these grid points is determined as part of the grid generation process which is discussed in detail in Chap. 6. The governing equations, which are often partial differential equations, are then discretized at these grid points, and are in the process converted to algebraic equations, the unknowns of which are the values of the variables at each grid point. Solution of these algebraic equations gives us the solution at these discrete points. In this chapter, we will look at the basic concepts underlying this numerical solution procedure, and will incorporate these into a template for solving a typical conservation equation that we normally encounter in fluid flow. The solution of the Navier-Stokes equations is discussed in Chap. 4.

The equations governing fluid motion have already been discussed in Chap. 2. Two principal approaches—the structured mesh and the unstructured mesh—can be used to discretize the physical space. A structured mesh or grid is one in which the grid nodes are located at intersections of families of curvilinear coordinate lines (Fig. 3.1), and a grid cell is therefore topologically rectangular in two dimensions and cubical in three dimensions. If a consistent numbering system is employed, then identification of neighbouring grid nodes is straightforward. An unstructured mesh does not have such structure and can be made up of arbitrarily shaped cells (Fig. 3.2). Meshing flow domains of irregular shapes, especially in three dimensions, is easier with an unstructured mesh. Local grid refinement and grid adaptation are also simpler as compared to that in a structured mesh. However, the solution of the discretized equations is more difficult, and the construction of higher order discretization schemes is more difficult. We will postpone the details of how to generate a grid—structured or otherwise—to a later chapter and focus in this chapter on how to solve a given differential equation on a grid.

S. Jayanti, *Computational Fluid Dynamics for Engineers and Scientists*,
https://doi.org/10.1007/978-94-024-1217-8_3

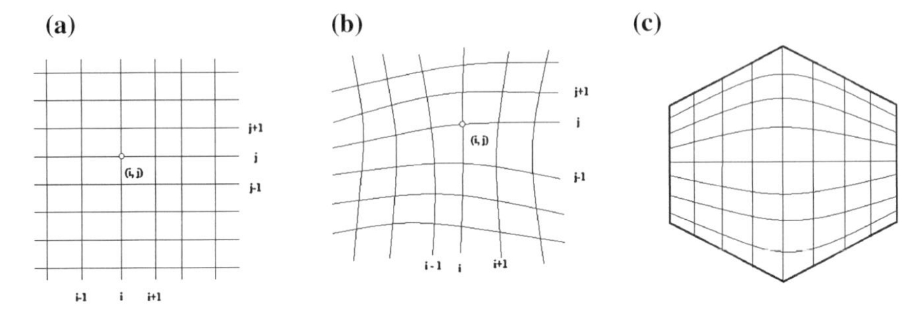

Fig. 3.1 Examples of structured grid in 2-dimensions

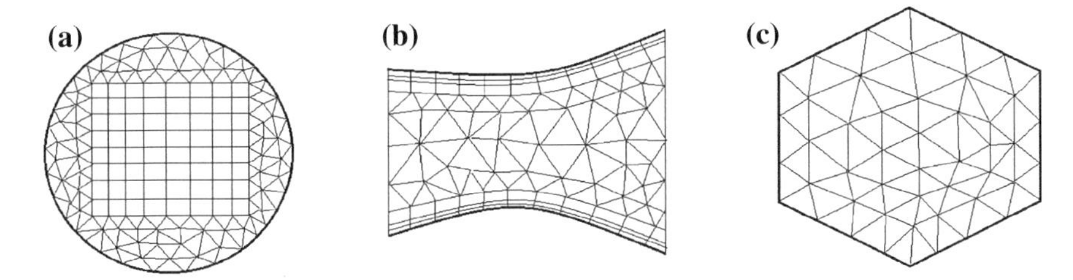

Fig. 3.2 Examples of unstructured grids in 2-dimensions for some simple computational domains

We will illustrate the concepts and issues behind this by taking a simple structured grid shown in Fig. 3.3. Here, the rectangular two-dimensional space is divided into cells of width Δx and height Δy. The grid nodes are located at intersections of constant x planes, each Δx apart, and constant y-planes, each Δy apart. Thus, the point P lying at the intersection of the ith x-plane and jth y-plane has the coordinates (x_i, y_j) where $x_i = i\ \Delta x$ and $y_j = j\ \Delta y$ and the point P can be identified by the grid node (i, j). The point Q, which is the immediate left neighbour of P, is identified as $(i - 1, j)$. Similarly, the immediate right (R), top (T) and bottom (S) neighbours can be identified as $(i + 1, j)$, $(i, j + 1)$ and $(i, j - 1)$, respectively. In the simple grid considered, the grid points are spread throughout the computational domain at regular intervals.

Once the physical discretization of the flow domain is completed, the discretization of the governing equations can be carried out. This process, as we have seen in Chap. 1, results in algebraic equations in which the values of the flow variables at these grid points are the unknowns. These equations have to be solved simultaneously (except in cases where an explicit scheme is used for a marching-forward type of problem, as we shall see later). There are several methods for the discretization of the equations, viz., finite difference methods, finite volume methods, finite element methods, spectral methods, collocation techniques, etc. While all these have been used for fluid flow problems, the first two are the most

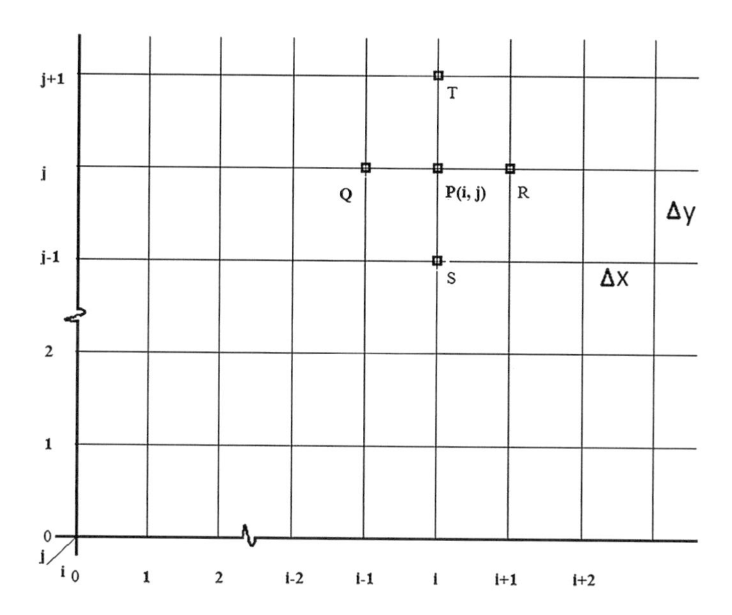

Fig. 3.3 Simple, Cartesian grid for a rectangular geometry

popular methods and can be said to be the first choice for systems of equations involving coupled non-linear partial differential equations, which may be the case, for example, with a turbulent combusting flow. Of these two methods, the finite difference method is the older and far simpler approach and is said to have started with Euler about 250 years ago. The finite volume method is of more recent origin and some of the early papers using this method have been published by McDonald (1971). The finite difference method requires a structured mesh while the finite volume method can be used also on an unstructured mesh. In the present chapter, the basic concepts of CFD are discussed in the context of the finite difference method; the finite volume method of discretization of the governing equations is discussed in Chap. 6 in the context of dealing with flow domains of irregular shape which is where the method comes into its own.

3.1 The Finite Difference Method

3.1.1 Finite Difference Approximation of a Derivative

The finite difference method for the discretization of an equation is based on the properties of Taylor series expansion of algebraic functions using which each derivative appearing in the equation is written approximately in terms of finite differences. Consider a function u(x) and its first derivative, du/dx at x in a one-dimensional space discretized using a uniform mesh spacing of Δx (Fig. 3.2a). A finite difference approximation for this can be obtained from a Taylor series expansion of $u(x + \Delta x)$ around point x:

$$u(x+\Delta x) = u(x) + du/dx|_x \Delta x + d^2u/dx^2|_x (\Delta x)^2/2! + d^3u/dx^3|_x (\Delta x)^3/3! + \cdots \quad (3.1.1)$$

We assume that u(x) is continuous so that all the higher derivatives, all of which are evaluated at point x, are defined and finite. The series converges if Δx is small. Since the nth term in the series expansion contains the product of the (n − 1)th derivative of u (which is finite for a smoothly varying function) and the (n − 1)th power of Δx, the contribution of a particular term decreases as Δx is reduced. For sufficiently small Δx and for sufficiently smooth variation of u(x), the magnitude of successive terms of the series becomes progressively smaller. Any term of the series which contains Δx^p is referred to as pth order term. A pth order term will therefore be larger in magnitude than a (p + 1)th order term. This enables one to terminate the series after a sufficiently large number of terms. The question as to how many terms to retain depends on both u(x) and Δx. For small Δx and smooth u(x), it is sufficient if a small number of terms are retained. In the trivial case of linear variation of u(x), retaining the first two terms is enough to get an exact expression for $u(x + \Delta x)$!

An approximation for $u(x + \Delta x)$ can be made involving only a finite number of terms in the series. For example, retaining only first three terms on the right hand side (RHS) of the series, we can write $u(x + \Delta x)$ approximately as:

$$u(x+\Delta x) = u(x) + du/dx|_x \Delta x + d^2u/dx^2|_x (\Delta x^2)/2! + O(\Delta x^3) \quad (3.1.2)$$

where $O(\Delta x^3)$ indicates terms of the order of Δx^3 or higher. The contribution of the neglected terms to the value of $u(x + \Delta x)$ is termed as the truncation error. Since the magnitude of successive terms is expected to become progressively smaller, the order of approximation or truncation of the series is denoted by the order of the leading term or the first (and thus the largest) term of the truncation error. The more the number of terms included in the approximation, the smaller will be the truncation error.

Taylor series expansions can be used to derive finite difference approximations for derivatives. Rearranging (3.1.2), we can get a finite difference approximation for the first derivative at x as

$$du/dx|_x = \{u(x+\Delta x) - u(x)\}/\Delta x - d^2u/dx^2|_x \Delta x/2 + O(\Delta x^2) \quad (3.1.3)$$

Retaining only the first two terms of Eq. (3.1.3) and using the notation given in Fig. 3.3, we can write an approximate formula for the first derivative at grid node i as

$$du/du|_i = (u_{i+1} - u_i)/\Delta x + O(\Delta x) \quad (3.1.4)$$

We may note that this is a gross approximation for the first derivative as only the first two terms of the Taylor series expansion are used. It will be accurate only for small grid spacing and smoothly varying u(x). Despite this, first order approximations are often used in CFD for reasons of convenience as will be discussed later. Hence the obsession of a CFD engineer with grid sensitivity of the solution.

It may be noted that in Eq. (3.1.4), the grid spacing, Δx, occurs as a first power in the leading term in the neglected part of the Taylor series, namely, $\Delta x/2!\, d^2u/dx^2$ at x. This implies that the error in the approximation for the first derivative decreases as $(\Delta x)^1$ as Δx decreases. Hence, this approximation is said to be first order-accurate, and it introduces, in theory, more error than, say, a second order accurate approximation. Expanding $u(x - \Delta x)$ around point x, another approximation can be derived for the first derivative. Thus,

$$u(x - \Delta x) = u(x) - du/dx|_x\Delta x + d^2u/dx^2|_x\Delta x^2/2! - d^3u/dx^3|_x\Delta x^3/3! + \cdots \quad (3.1.5)$$

Rearranging Eq. (3.1.5), we get

$$du/dx|_x = 1/\Delta x\{u(x) - u(x - \Delta x)\} + d^2u/dx^2|_x\Delta x/2 + O(\Delta x^2) \quad (3.1.6)$$

or

$$du/du|_i = (u_i - u_{i-1})/\Delta x + O(\Delta x) \quad (3.1.7)$$

which again gives a first-order approximation for the first derivative.

Yet another approximation for du/dx at x can be obtained by subtracting (3.1.5) from (3.1.1):

$$u(x + \Delta x) - u(x - \Delta x) = 2\Delta x\, du/dx|_x + 2(\Delta x)^3/3!\, d^3u/dx^3|_x + \cdots \quad (3.1.8)$$

Rearranging Eq. (3.1.8), we obtain

$$du/dx|_x = \{u(x + \Delta x) - u(x - \Delta x)\}/(2\Delta x) + d^3u/dx^3|_x(\Delta x)^2/3! + \cdots \quad (3.1.9a)$$

or

$$du/dx|_i = \{u_{i+1} - u_{i-1}\}/(2\Delta x) + O(\Delta x^2) \quad (3.1.9b)$$

Since the grid spacing occurs in the leading term of the truncation error as $(\Delta x)^2$, the error in the approximation decreases as $(\Delta x)^2$. Thus, if the grid spacing is decreased by a factor of two, the discretization error decreases by a factor of four provided Δx is already quite small. Thus, the approximation given by (3.1.9) is said to be second order accurate.

The three approximations for the first derivative, given by Eqs. (3.1.4), (3.1.7) and (3.1.9) are schematically represented in Fig. 3.4a, and are respectively called forward, backward and central differencing schemes. The forward difference approximation is given by the slope of the line joining point i and i + 1, the backward by the slope of the straight line joining points i − 1 and i, and the central scheme by the slope of the line joining points i − 1 and i + 1. It can be seen that the central scheme gives a better approximation to the true first derivative, given by the slope of the tangent to the curve at point i, than either the first or the second derivative for small values of grid spacing. For large grid spacing for a rapidly varying solution (see Fig. 3.4b), it is not necessary that the central scheme is always better. Thus, the benefits of higher accuracy of approximation can be realized only on a sufficiently fine grid. Generally, one-sided differences (i.e., forward and backward differences) are less accurate than central schemes for the same number of points. One-sided differences are required for approximating the derivatives at boundaries and also in compressible flows.

It is possible to construct finite difference approximations of higher accuracy but this requires inclusion of more number of adjacent points (which ultimately leads to a more complicated system of discretized equations). Let us construct a third-order difference approximation for the first derivative at i using forward differencing. As we will see later, this requires four adjacent points, namely, i, i + 1, i + 2 and i + 3. Let us seek therefore an approximation of the form

$$du/dx|_i = (a\,u_i + b\,u_{i+1} + c\,u_{i+2} + d\,u_{i+3})/\Delta x + O(\Delta x^3) \tag{3.1.11}$$

where a, b, c and d are constants to be determined. Noting that a third-order approximation for the first derivative implies elimination of the second and the third derivatives from the Taylor series approximation (just as the second order approximation of the central scheme in Eq. (3.1.9b) meant elimination of the

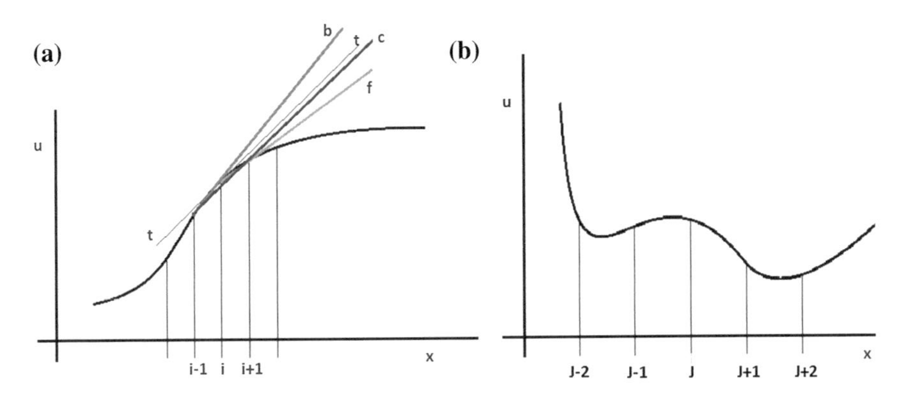

Fig. 3.4 **a** Graphical representation of finite difference approximation of the first derivative at point i using f-forward differencing, b-backward differencing, c-central differencing, and t-true slope. **b** Coarse grid discretization of a rapidly varying function; backward, central and forward difference approximations are the best at points (J − 1), J and (J + 1), respectively

second derivative in the approximation, see Eq. (3.1.9a)), Eq. (3.1.11) can also be written as

$$\begin{aligned}\Delta x\, du/dx|_i &= (au_i + bu_{i+1} + cu_{i+2} + du_{i+3}) + (0)\Delta x^2/2!\, d^2u/dx^2|_i \\ &\quad + (0)\Delta x^3/3!\, d^3u/dx^3|_i + (e)\Delta x^4/4!\, d^4u/dx^4|_i + O(\Delta x^5)\end{aligned} \tag{3.1.12}$$

Or

$$\begin{aligned}au_i + bu_{i+1} + cu_{i+2} + du_{i+3} &= (0)\, u_i + \Delta x\, du/dx|_i + (0)\Delta x^2/2!\, d^2u/dx^2|_i \\ &\quad + (0)\Delta x^3/3!\, d^3u/dx^3|_i + (e)\Delta x^4/4!\, d^4u/dx^4|_i + O(\Delta x^5)\end{aligned} \tag{3.1.13}$$

where constants a, b, c and d are to be determined, and e is a non-zero constant. Note that Eq. (3.1.13) is equivalent to Eq. (3.1.11), which is what we seek to obtain. Expanding u_{i+1}, u_{i+2} and u_{i+3} in terms of Taylor series around u_i, we have

$$\begin{aligned}u_{i+1} &= u_i + \Delta x\, du/dx|_i + \Delta x^2/2!\, d^2u/dx^2|_i + \Delta x^3/3!\, d^3u/dx^3|_i \\ &\quad + \Delta x^4/4!\, d^4u/dx^4|_i + \cdots\end{aligned} \tag{3.1.14a}$$

$$\begin{aligned}u_{i+2} &= u_i + 2\Delta x\, du/dx|_i + (2\Delta x)^2/2!\, d^2u/dx^2|_i + (2\Delta x)^3/3!\, d^3u/dx^3|_i \\ &\quad + (2\Delta x)^4/4!\, d^4u/dx^4|_i + \cdots\end{aligned} \tag{3.1.14b}$$

$$\begin{aligned}u_{i+3} &= u_i + 3\Delta x\, du/dx|_i + (3\Delta x)^2/2!\, d^2u/dx^2|_i + (3\Delta x)^3/3!\, d^3u/dx^3|_i \\ &\quad + (3\Delta x)^4/4!\, d^4u/dx^4|_i + \cdots\end{aligned} \tag{3.1.14c}$$

Evaluating (au_i + bu_{i+1} + cu_{i+2} + du_{i+3}), and rearranging the terms, we get

$$\begin{aligned}au_i + bu_{i+1} + cu_{i+2} + du_{i+3} &= (a + b + c + d)\, u_i + (b + 2c + 3d)\Delta x\, du/dx|_i \\ &\quad + (b + 4c + 9d)\, \Delta x^2/2!\, d^2u/dx^2|_i \\ &\quad + (b + 8c + 27d)\Delta x^3/3!\, d^3u/dx^3|_i \\ &\quad + (b + 16c + 81d)\Delta x^4/4!\, d^4u/dx^4|_i + \cdots.\end{aligned} \tag{3.1.15}$$

Comparing the coefficients of the like terms in Eqs. (3.1.13) and (3.1.15), we obtain the following relations to determine the coefficients:

$$\begin{aligned} &a + b + c + d = 0 \\ &b + 2c + 3d = 1 \\ &b + 4c + 9d = 0 \\ &b + 8c + 27d = 0 \\ &b + 16c + 81d = e \end{aligned} \tag{3.1.16}$$

Simultaneous solution of these yields a = −11/6, b = 3, c = −3/2, d = 1/3 and e = 6. Thus, the third order approximation for the first derivative at point i using forward differences can be written as

$$du/dx|_i = (-11u_i + 18u_{i+1} - 9u_{i+2} + 2u_{i+3})/(6\Delta x) + 6\Delta x^3/4!\, d^4u/dx^4|_i + O(\Delta x^4) \tag{3.1.17}$$

A second order accurate scheme for the first derivative at point i using forward differences can similarly be derived as

$$du/dx|_i = (-3u_i + 4u_{i+1} - u_{i+2})/(2\Delta x) + O(\Delta x^2) \tag{3.1.18}$$

and using backward differences as

$$du/dx|_i = (3u_i - 4u_{i+1} + u_{i+2})/(2\Delta x) + O(\Delta x^2) \tag{3.1.19}$$

Comparing expressions (3.1.18), (3.1.19) and (3.1.9), we find that a second order scheme for the first derivative using forward or backward scheme requires three adjacent points whereas a central scheme requires only two points. Also, comparing expressions (3.1.4), (3.1.18) and (3.1.17) for the first derivative with first, second and third order accuracy, we find that the increasing the accuracy of a scheme by one order requires the inclusion of one additional adjacent point. The result can be generalized and a general method for the derivation of an approximation of any order of accuracy has been discussed by Hildebrand (1956).

The derivation of finite difference approximations for higher derivatives is straightforward. For example, the second derivative of u with respect to x at x can be approximated as

$$\begin{aligned} d^2u/dx^2|_i = d/dx(du/dx)|_i &= (1/\Delta x)\left(du/dx|_{i+1/2} - du/dx|_{i-1/2}\right) \\ &= (1/\Delta x)\{(u_{i+1} - u_i)/\Delta x - (u_i - u_{i-1})/\Delta x\} \\ &= (u_{i+1} - 2u_i - u_{i-1})/\Delta x^2 \end{aligned} \tag{3.1.20}$$

In the above approximation, central differences have been used throughout over a mesh spacing of Δx/2 and the resulting accuracy of the scheme is second-order, as can be verified by a Taylor series expansion. Thus,

$$d^2u/dx^2|_i = (u_{i+1} - 2u_i - u_{i-1})/\Delta x^2 + O(\Delta x^2) \quad (3.1.21)$$

The second derivative can also be evaluated in terms of forward differences alone. Thus,

$$\begin{aligned} d^2u/dx^2|_i = d/dx(du/dx)|_i &= (1/\Delta x)\left(du/dx|_{i+1} - du/dx|_i\right) \\ &= (1/\Delta x)\{(u_{i+2} - u_{i+1})/\Delta x - (u_{i+1} - u_i)/\Delta x\} \\ &= (u_{i+2} - 2u_{i+1} - u_i)/\Delta x^2 \end{aligned} \quad (3.1.22)$$

Since forward differences are used throughout, this approximation will be of first order accuracy. Thus,

$$d^2u/dx^2|_i = (u_{i+2} - 2u_{i+1} - u_i)/\Delta x^2 + O(\Delta x) \quad (3.1.23)$$

We may note however that the right hand side of Eq. (3.1.23) would be a second order accurate expression for $d^2u/dx^2|_{i+1}$.

A backward difference approximation for the second derivative at i would look similar Eq. (3.1.23) but would involve function values at points i, (i − 1) and (i − 2). A combination of backward and forward differences leads to a central difference approximation. An approximation for the third derivative can also be obtained in similar way. Thus,

$$\begin{aligned} d^3u/dx^3|_i &= d/dx\left(d^2u/dx^2\right)|_i = \{1/(2\Delta x)\}\left(d^2u/dx^2|_{i+1} - d^2u/dx^2|_{i-1}\right) \\ &= \{(u_{i+2} - 2u_{i+1} + u_i)/\Delta x^2 - (u_i - 2u_{i-1} + u_{i-2})/\Delta x^2\}/(2\Delta x) \\ &= (u_{i+2} - 2u_{i+1} + 2u_{i-1} - u_{i-2})/(2\Delta x^3) + O(\Delta x^2) \end{aligned} \quad (3.1.24)$$

where central difference approximations for d^2u/dx^2 are substituted at points i + 1 and i − 1. The resulting expression is second order accurate.

It is instructive to compare the central, second order accurate approximations for the first, the second and the third derivatives given by Eqs. (3.1.9), (3.1.21) and (3.1.24), respectively. It can be seen that the even derivatives will have the value at the point i in the difference formula while the odd-derivatives do not involve the value at the point where the derivative is being evaluated. This aspect is important for the efficient solution of the difference equations, as we shall see in Chap. 5. Also, for the same order of accuracy, increasing the order of derivative by one order requires the inclusion of an additional point. Thus, a general result can be stated that if p is the order the derivative being evaluated, q is the accuracy of discretization and n is the number of adjacent points used in the discretization, then n = p + q − 1 for central differencing and n = p + q for one-sided differencing. This result is valid only for uniform mesh size where some of the terms cancel out. The case of non-uniform mesh size is discussed in Sect. 3.1.4.

Finally, the evaluation of mixed derivatives, which usually occur in fluid mechanics problems as a result of coordinate transformation to a non-orthogonal system (see Chap. 6), is also straightforward. The second derivative of u with respect to x and y, $\partial^2 u/\partial x \partial y$, at the point (i, j) can be expressed as

$$\begin{aligned}(\partial^2 u/\partial x\partial y)|_{i,j} &= \partial/\partial x(\partial u/\partial y)|_{i,j} = \left(\partial u/\partial y|_{i+1,j} - \partial u/\partial y|_{i-1,j}\right)/(2\Delta x) \\ &= [\{(u_{i+1,j+1} - u_{i+1,j-1}) - (u_{i-1,j+1} - u_{i-1,j-1})\}/(2\Delta y)]/(2\Delta x) \\ &= (u_{i+1,j+1} - u_{i+1,j-1} - u_{i-1,j+1} + u_{i-1,j-1})/(4\Delta x\Delta y) + O(\Delta x^2, \Delta y^2)\end{aligned} \quad (3.1.25)$$

Since central schemes are used throughout, the above expression is second-order accurate. Changing the order of differentiation and using backward differences, we get the following first order accurate expression for $\partial^2 u/\partial x\partial y$ at point (i, j):

$$\partial^2 u/\partial x\partial y|_{i,j} = (u_{i,j} - u_{i-1,j} - u_{i,j-1} + u_{i-1,j-1})/(\Delta x\Delta y) + O(\Delta x,\ \Delta y) \quad (3.1.26)$$

It is possible to use a combination of one-sided and central schemes, as may be required, for example, at a boundary. Thus,

$$\begin{aligned}\partial^2 u/\partial x\partial y|_{i,j} &= \partial/\partial x(\partial u/\partial y)|_{i,j} = (\partial u/\partial y|_{i+1,j} - \partial u/\partial y|_{i-1,j})/(2\Delta x) \\ &= \{(au_{i+1,j+2} + bu_{i+1,j+1} + cu_{i+1,j})/\Delta y \\ &\quad - (au_{i-1,j+2} + bu_{i-1,j+1} + cu_{i-1,j})/\Delta y\}/(2\Delta x) \\ &= (au_{i+1,j+2} + bu_{i+1,j+1} + cu_{i+1,j} - au_{i-1,j+2} \\ &\quad - bu_{i-1,j+1} - cu_{i-1,j})/(2\Delta x\Delta y) + O(\Delta x^2, \Delta y^2)\end{aligned} \quad (3.1.27)$$

where a second-order forward difference scheme (with appropriate constants a, b and c) is used to evaluate the derivatives with respect to y.

Thus, a very large number of finite difference approximations can be obtained for the derivatives of functions.

3.1.2 *Discretization of Differential Equations Using Finite Differences*

Finite difference approximations for differential (both ordinary and partial) equations can be obtained by writing the approximation for each derivative. Let us illustrate this with the case of a two-dimensional Poisson equation of the form

$$\partial^2 u/\partial x^2 + \partial^2 u/\partial y^2 = f(x, y) \tag{3.1.28}$$

subject to the boundary condition that u is given by a Dirichlet condition on all the boundaries (see Fig. 1.1). For an interior point (i, j), the first term on the left hand side of Eq. (3.1.28) can be written using central differencing as

$$\partial^2 u/\partial x^2|_{i,j} = \left(u_{i+1,j} - 2u_{i,j} + u_{i-1,j}\right)/\Delta x^2 + \partial^3 u/\partial x^3|_{i,j}\,(\Delta x)^2/3! + O(\Delta x^3)$$

Similarly, the second term on the left hand side can be written as

$$\partial^2 u/\partial y^2|_{i,j} = \left(u_{i,j+1} - 2u_{i,j} + u_{i,j-1}\right)/\Delta y^2 + \partial^3 u/\partial y^3|_{i,j}(\Delta y)^2/3! + O(\Delta y^3)$$

Denoting by $f_{i,j}$ the value of f at the point (x_i, y_j), the discretized form of the Eq. (3.1.28) can be written as

$$\begin{aligned}&\left(u_{i+1,j} - 2u_{i,j} + u_{i-1,j}\right)/\Delta x^2 + \left(u_{i,j+1} - 2u_{i,j} + u_{i,j-1}\right)/\Delta y^2\\&= f_{i,j} - \{\partial^3 u/\partial x^3|_{i,j}(\Delta x)^2/3! + O(\Delta x^3)\\&\quad + \partial^3 u/\partial y^3|_{i,j}(\Delta y)^2/3! + O(\Delta y^3)\}\end{aligned}$$

The quantity in square brackets on the right hand side is the overall truncation error in discretizing Eq. (3.1.28); the final difference equation can be written as

$$\left(u_{i+1,j} - 2u_{i,j} + u_{i-1,j}\right)/\Delta x^2 + \left(u_{i,j+1} - 2u_{i,j} + u_{i,j-1}\right)/\Delta y^2 = f_{i,j} + O(\Delta x^2, \Delta y^2) \tag{3.1.29}$$

Since Dirichlet boundary conditions are used throughout the boundary, the variable values over the boundary are specified and need not be determined. Equation (3.1.29) can therefore be used for all interior points, i.e., for $2 \le i \le N_i$ and $2 \le j \le N_j$, where N_i and N_j are the number of divisions in the ith and the jth directions, respectively. These result in $N_i \times N_j$ number of algebraic equations which have to be solved simultaneously for $u_{i,j}$. If Neumann or Robin boundary condition is given over part of the boundary, then the values of $u_{i,j}$ on this boundary should be determined by solving additional algebraic equations. For example, along the boundary AB in Fig. 3.5, the gradient is specified, viz., $\partial u/\partial y = c_1$. Approximating the gradient using forward difference, and noting that j = 1 for points on the boundary AB, we can write for these boundary points

$$\left(u_{i,j+1} - u_{i,j}\right)/\Delta y = c_1 \quad \text{or} \quad u_{i,1} = u_{i,2} - c_1\Delta y \tag{3.1.30}$$

However, the above approximation is only first order accurate. Using a second order accurate one-sided formula for the first derivative, we can write

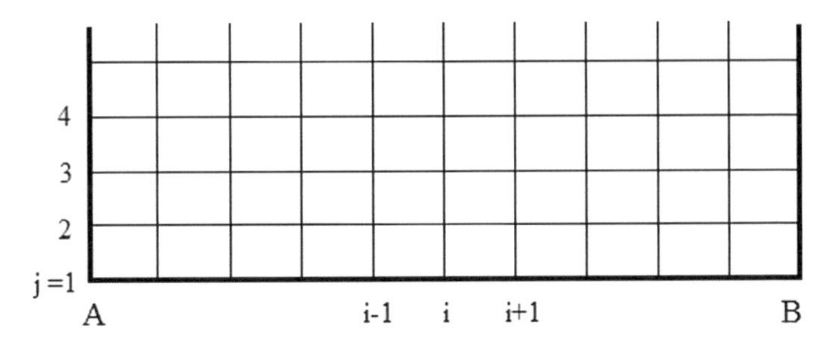

Fig. 3.5 Application of a Neumann boundary condition at wall

$$\partial u/\partial y|_{i,j} = \left(a u_{i,j+2} + b u_{i,j+1} + c u_{i,j}\right)/\Delta y = c_1$$

or

$$u_{i,1} = (c_1 \Delta y - a\, u_{i,3} - b\, u_{i,2})/c \tag{3.1.31}$$

Another boundary condition that is often encountered in heat conduction problems is the convective boundary condition, namely, that, the heat flux from the surface is given by $q_w = h(T_{inf} - T_w)$ where h is the (specified) convective heat transfer coefficient, T_{inf} is the (specified) free stream temperature of air and T_w is the local wall temperature which is the unknown quantity. This boundary condition can be implemented readily by noting that q_w is also given by $q_w = -k\partial T/\partial y|_{wall}$ within the computational domain. Using a second order accurate forward differencing formula for the derivative, we thus have

$$h\left(T_{inf} - T_{i,j}\right) = -k\, \partial T/\partial y|_{wall} = -k\left(a T_{i,j+2} + b T_{i,j+1} + c\, T_{i,j}\right)/\Delta y$$

or

$$T_{i,1} = (h\, \Delta y\, T_{inf}/k - a\, T_{i,3} - b\, T_{i,2})/(h\, \Delta y - c\, k) \tag{3.1.32}$$

The above approach can be used to determine the wall temperature from a known temperature gradient or the wall heat flux from a known (calculated) temperature distribution.

It is usual to maintain the same order of accuracy of discretization for each term. However, if the situation warrants, for example, if terms of unequal importance are present in the equation, it is sufficient to use a higher order approximation for the significant terms; using a lower order of approximation for the unimportant terms may be computationally efficient. For instance, the convective terms in creeping flow have very little effect and are usually neglected in analytical solutions. In a numerical solution, it would be sufficient to treat these with a first order scheme while treating the diffusive term with a second order accurate scheme. Another case in question is the boundary layer flows where streamwise gradients are likely to be less significant at regions well away from the stagnation points. Similarly, when a time-dependent formulation is used for a steady problem (see Chap. 4), the interest lies in the steady state solution and the time-dependent terms may therefore be treated in a simple manner.

3.1.3 Discretization of Time-Dependent Equations

The discretization of time-dependent equations introduces a peculiar dilemma. Consider the unsteady state, one-dimensional heat conduction equation:

$$\partial T/\partial t = \alpha\,\partial^2 T/\partial x^2 \tag{3.1.33}$$

where α is the thermal diffusivity. Consider a uniform discretization in space with a step of Δx and in time with a time step of Δt. Using the subscript i to denote the ith space step and the superscript n to denote the nth time step, i.e., $x_i = i\Delta x$ and $t^n = n\Delta t$, we seek a discretization of this equation of the form

$$\partial T/\partial t|_i^n = \alpha\,\partial^2 T/\partial x^2|_i^n \tag{3.1.34}$$

The left hand side of Eq. (3.1.34) can be readily discretized using, for example, forward differencing:

$$\partial T/\partial t|_i^n = \left(T_i^{n+1} - Ti^n\right)/\Delta t + O(\Delta t) \tag{3.1.35}$$

The spatial discretization of the right hand side term is straightforward. For example, using central differencing and taking α to be constant, we have

$$\alpha\partial^2 T/\partial x^2|_i = (\alpha/\Delta x^2)(T_{i+1} - 2T_i + T_{i-1}) + O(\Delta x^2) \tag{3.1.36}$$

With regard to the time step at which the RHS of Eq. (3.1.37) is to be evaluated, three choices suggest themselves: evaluate it at the nth time step; evaluate it at (n + 1)th time step; or evaluate it at n* where $n < n^* < n + 1$, e.g., $n^* = (n + n + 1)/2$. Accordingly, we can write (3.1.36) as

$$\alpha\partial^2 T/\partial x^2|_i^n = (\alpha/\Delta x^2)\left(T_{i+1}^n - 2T_i^n + T_{i-1}^n\right) + O(\Delta x^2) \tag{3.1.37a}$$

or

$$\alpha\partial^2 T/\partial x^2|_i^{n+1} = (\alpha/\Delta x^2)\left(T_{i+1}^{n+1} - 2T_i^{n+1} + T_{i-1}^{n+1}\right) + O(\Delta x^2) \tag{3.1.37b}$$

or,

$$\alpha\partial^2T/\partial x^2|_i^{n*} = (\alpha/\Delta x^2)\left(T_{i+1}^{n*} - 2T_i^{n*} + T_{i-1}^{n*}\right) + O(\Delta x^2) \tag{3.1.37c}$$

The first option (Eq. 3.1.37a) leads to the following discretization of Eq. (3.1.33):

$$\left(T_i^{n+1} - T_i^n\right)/\Delta t = (\alpha/\Delta x^2)\left(T_{i+1}^n - 2T_i^n + T_{i-1}^n\right)$$

or

$$T_i^{n+1} = T_i^n + (\alpha\Delta t/\Delta x^2)\left(T_{i+1}^n - 2T_i^n + T_{i-1}^n\right) \tag{3.1.38}$$

Equation (3.1.38) has the special feature that all the terms on the right hand side are known at the beginning of a new time step of calculation. Hence, Eq. (3.1.38) gives an explicit formulation for the calculation of T_i^{n+1} in terms of known values of variables at the nth time step. It allows one to sweep through all the spatial points systematically at a given time step, and then move on to the next time step. Using this explicit formulation, one can march forward in space and time starting from initial conditions.

The second option (Eq. 3.1.37b) leads to the following discretization of Eq. (3.1.33):

$$\left(T_i^{n+1} - T_i^n\right)/\Delta t = (\alpha/\Delta x^2)\left(T_{i+1}^{n+1} - 2T_i^{n+1} + T_{i-1}^{n+1}\right)$$

or

$$T_i^{n+1} = \left\{T_i^n + (\alpha\Delta t/\Delta x^2)\left(T_{i+1}^{n+1} + T_{i-1}^{n+1}\right)\right\}/(1 + 2\,\alpha\Delta t/\Delta x^2) \tag{3.1.39}$$

This implicit formulation does not allow one to march forward in time and space because in order to calculate T_i^{n+1}, both T_{i+1}^{n+1} and T_{i-1}^{n+1} should be known whereas only one of these two is known (T_{i-1}^{n+1} if one is marching in increasing i direction and T_{i+1}^{n+1} if one is marching in decreasing i direction). It is therefore necessary to solve equations for all the grid points simultaneously to get T_i^{n+1}. Hence the implicit method leads to a more complicated solution scheme. However, implicit methods are more stable than explicit methods and therefore can be used with larger time steps, as we will see in Sect. 3.2.

The third choice (Eq. 3.1.37c) leads to the following discretization of Eq. (3.1.33):

$$\left(T_i^{n+1} - T_i^n\right)/\Delta t = (\alpha/\Delta x^2)\left(T_{i+1}^{n*} - 2T_i^{n*} + T_{i-1}^{n*}\right)$$

or

$$T_i^{n+1} = T_i^n + (\alpha\Delta t/\Delta x^2)\left(T_{i+1}^{n*} - 2T_i^{n*} + T_{i-1}^{n*}\right) \tag{3.1.40}$$

This will have some implicit component to it and cannot therefore be evaluated in marching forward manner. Here, special mention must be made of two schemes: the Crank-Nicolson scheme and the Dufort-Frankel scheme. In the former (Crank and Nicolson 1947), each term on the right hand side of Eq. (3.1.36) is taken as the average of the values at n and n + 1. Thus, the discretized Eq. (3.1.33) becomes

$$\left(T_i^{n+1} - T_i^n\right)/\Delta t = \left(\alpha/2\Delta x^2\right)\left(T_{i+1}^n + T_{i+1}^{n+1} - 2T_i^n - 2T_i^{n+1} + T_{i-1}^n + T_i^{n+1}\right)$$

or

$$-r\,T_{i-1}^{n+1} + (2+2r)\,T_i^{n+1} - r\,T_{i+1}^{n+1} = r\,T_{i-1}^n + (2-2r)\,T_i^n + r\,T_{i+1}^n \tag{3.1.41}$$

which is implicit and second-order accurate. In the Dufort-Frankel scheme (DuFort and Frankel 1953), T_i^n is taken as the average of the T_i values at n − 1 and n + 1 time steps, i.e.,

$$T_i^n = \left(T_i^{n-1} + T_i^{n+1}\right)/2 \tag{3.1.42}$$

Substituting this into the explicit scheme (Eq. 3.1.38), we obtain

$$\left(T_i^{n+1} - T_i^{n-1}\right)/2\Delta t = \left(\alpha/\Delta x^2\right)\left(T_{i+1}^n - T_i^{n+1} - T_i^{n-1} + T_{i-1}^n\right) \tag{3.1.43}$$

which can be shown to be second order accurate in both time and space. It is also explicit and lends itself to marching forward type of solution.

Some generalization can now be made. Let us write the representative value of T at i over the time step interval n to (n + 1), $T_i^{n^*}$, be expressed as a linear combination of the values of T_i at n and n + 1, i.e.,

$$T_i^{n*} = \theta\,T_i^n + (1-\theta)\,T_i^{n+1} \tag{3.1.44}$$

Explicit and implicit schemes represent the two extreme cases where θ is taken as one and zero, respectively, while the Crank-Nicolson scheme has a θ value of ½. In a function that varies smoothly with time, the Crank-Nicolson scheme is likely to be the most accurate of the three. If the characteristic time of the system transient is very large compared to Δt, then an explicit scheme should be acceptable. If the system transient is very fast, then an implicit scheme would be more appropriate. In a typical numerical solution, other considerations, such as stability and the difficulty of solving the discretized equations, would also play a part in the choice of time discretization. These issues are discussed later.

3.1.4 Finite Difference Method on Non-uniform Meshes

Non-uniform meshes are often necessary in locations within flow domain where there is a rapid change in the properties of one or more flow variables. Common examples of such cases are the regions close to a solid wall where velocities may change rapidly in the direction normal to the wall, regions in which rapid change of flow direction takes place, regions around shocks and other discontinuities. In order to resolve the flow variables accurately in these regions, it becomes necessary to use a fine mesh. However, the use of such a fine mesh throughout the flow domain may become impossible and is often wasteful in terms of computational resources. Hence, a non-uniform mesh (the mesh spacing often reduces in a geometric progression locally, especially near the wall) is employed. Here, we examine the implications of using such non-uniform meshes.

The derivation of finite difference approximations on a non-uniform mesh is rather straightforward. Consider the non-uniform mesh in one-dimension shown in Fig. 3.6. Here, the spacing between point i and point (i + 1) is denoted by Δx_i and that between points (i + 1) and (i + 2) by Δx_{i+1}, and so on. A forward difference approximation for the first derivative at point i can be readily obtained as before by expanding $u(x_{i+1})$ about point i. Thus,

$$u(x_{i+1}) = u(x_i + \Delta x_i) = u(x_i) + \Delta x_i du/dx|_i + d^2u/dx^2_{|i}\Delta x_i^2/2! + \cdots. \tag{3.1.45}$$

Or

$$du/dx|_i = (u_{i+1} - u_i)/\Delta x_i - d^2u/dx^2|_i \Delta x_i + O(\Delta x_i^2) \tag{3.1.46}$$

Similarly, a backward difference approximation can be obtained by expanding u (x_{i-1}) about point i. Thus,

$$u(x_{i-1}) = u(x_i - \Delta x_{i-1}) = u(x_i) - \Delta x_{i-1} du/dx|_i + d^2u/dx^2_{|i}\Delta x_{i-1}^2/2! + \cdots. \tag{3.1.47}$$

Or

$$du/dx|_i = (u_i - u_{i-1})/\Delta x_{i-1} - d^2u/dx^2|_i \Delta x_{i-1} + O(\Delta x_{i-1}^2) \tag{3.1.48}$$

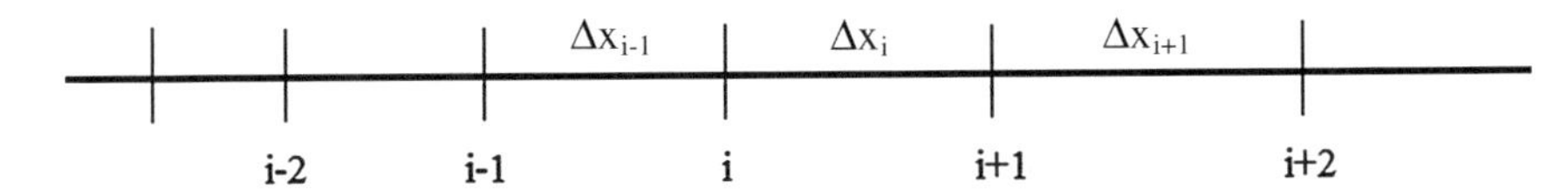

Fig. 3.6 Nomenclature for deriving finite difference approximations on a non-uniform mesh

So far, there is no difference between uniform and non-uniform meshes except for the subscript to the mesh spacing. If we subtract Eq. (3.1.47) from Eq. (3.1.45) in order to get a central difference approximation for $du/dx|_i$, we obtain, after algebraic manipulations,

$$du/dx|_i = (u_{i+1} - u_{i-1})/(\Delta x_i + \Delta x_{i-1}) - d^2u/dx^2|_i(\Delta x_i - \Delta x_{i-1})/4 \\ - d^3u/dx^3|_i(\Delta x_i^2 - \Delta x_i \Delta x_{i-1} + \Delta x_{i-1}^2)/3! + \cdots \quad (3.1.49)$$

Here we see that the second derivative term does not cancel out unless $\Delta x_{i-1} = \Delta x_i$ and a first order accuracy results if $(\Delta x_i - \Delta x_{i-1}) \sim \Delta x_i$, as may happen with a sudden change in grid spacing. However, instead of arithmetic averaging of the forward and backward difference approximations, if a weighted average, using the mesh spacing as the relative weights, is used, a second-order accurate central difference scheme may be obtained. Multiplying Eq. (3.1.45) by Δx_{i-1} and Eq. (3.1.47) by Δx_i and adding, we find that the second derivatives cancel out, and we get the following central scheme after some simplifications:

$$du/dx|_i = 1/[\Delta x_{i-1} \Delta x_i(\Delta x_{i-1} + \Delta x_i)]\ [u_{i+1}(\Delta x_{i-1})^2 - u_{i-1}(\Delta x_i)^2 + u_i(\Delta x_{i-1}^2 - \Delta x_i^2)] \\ - \Delta x_{i-1} \Delta x_i/3!\, d^3u/dx^3|_i + \cdots . \quad (3.1.50)$$

which is thus second order accurate. Similarly, the following second-order accurate forward difference formula can be obtained for the first derivative on a non-uniform mesh:

$$du/dx|_i = [(\Delta x_i + \Delta x_{i+1})/\Delta x_{i+1}][(u_{i+1} - u_i)/\Delta x_i] - [\Delta x_i/\Delta x_{i+1}] \\ [(u_{i+2} - u_i)(\Delta x_i + \Delta x_{i+1})] + \Delta x_i(\Delta x_i + \Delta x_{i+1})/6\, d^3u/dx^3 + \cdots \quad (3.1.51)$$

Similarly, evaluating the second derivative at i using the central scheme

$$d^2u/dx^2|_i = [(du/dx)|_{i+1/2} - (du/dx)|_{i-1/2}]/[½(\Delta x_i + \Delta x_{i+1})]$$

results in the following central differencing formula for the second derivative on a non-uniform mesh:

$$d^2u/dx^2|_i = [u_{i+1}\Delta x_{i-1} - u_{i-1}\Delta x_i - u_i(\Delta x_{i-1} + \Delta x_i)]/\ [1/2(\Delta x_{i-1} + \Delta x_i)\Delta x_{i-1}\Delta x_i] \\ - (\Delta x_i - \Delta x_{i-1})/3\, d^3u/dx^3|_i - (\Delta x_{i-1}^3 + \Delta x_i^3)/[12(\Delta x_{i-1} + \Delta x_i)]\ d^4u/dx^4 \quad (3.1.52)$$

In the above expression, the leading term in the truncation error is proportional to the difference between successive mesh spacings and the scheme is therefore of first order accuracy if $\Delta x_i \sim 2\Delta x_{i-1}$. Using a small geometric progression factor, say between 0.8 and 1.2, will reduce the effect of this considerably.

3.2 Analysis of Discretized Equations

We can see from the above discussion that, using finite difference methods, a given partial differential equation of continuous variables can be converted into approximate algebraic equations of discrete variables, which are the values of the variables (such as the velocity components and pressure) at pre-selected grid points. We have also seen that this discretization is by no means unique and that a number of alternative difference equations can be obtained. While some approximations are indeed of different degrees of accuracy, some alternative formulations of similar order of accuracy of approximation can also be derived. The question that we should then be concerned with is whether or not these various formulations will lead us, eventually, to an approximate solution, and preferably, a solution which approaches the true solution as we reduce the error of finite difference approximation by reducing the grid spacing. At first sight, the question appears trivial as the finite difference approximation of each derivative tends to the true derivative in the limiting case of zero grid spacing. However, a differential equation contains several terms and it is possible that the errors due to the approximation of one term may interact with those from another term destructively and lead to a build-up of the error as the solution evolves. Far-fetched though this seems, it seems all too common as the examples of Sect. 3.2.1 will show. Given that this is indeed possible, we will develop methods in Sect. 3.2.2 by which a given difference approximation of a differential equation can be analyzed a priori to determine whether or not it will give a satisfactory solution. Using these methods, we will develop, in Sect. 3.3, methods of discretization of the generic scalar transport equation which would lead to a reasonably approximate solution of the partial differential equation.

3.2.1 Need for Analysis: Simple Case Studies

Let us illustrate the difficulties that may be encountered in a naïve application of the finite differences to approximate a partial differential equation. Consider the one-dimensional linear convection (wave) equation

$$\partial u/\partial t + a\partial u/\partial x = 0 \tag{3.2.1}$$

where u is the unknown function of x and t and a is the convection speed or the wave speed. This equation is a simplification of the general transport equation and can be obtained by putting the diffusion and the source terms to zero. To complete the specification of the problem, we consider the following initial and boundary conditions:

$$\begin{array}{lll} t=0 & u(x,0)=f(x) & 0\leq x\leq L \\ x=0 & u(0,t)=g(t) & t\geq 0 \end{array} \tag{3.2.2}$$

Let us consider three simple and straightforward discretizations of Eq. (3.2.1); we will use a forward difference approximation for the time derivative and an explicit, central difference approximation for the spatial derivative. This results in a finite difference approximation of the form

$$\left(u_i^{n+1}-u_i^n\right)/\Delta t+a\left(u_{i+1}^n-u_{i-1}^n\right)/(2\Delta x)=0 \tag{3.2.3}$$

It is clear that this explicit, FTCS ("forward in time and central in space") method lends itself to an easily implementable marching-forward type of solution. We will compare this solution with that obtained using an explicit, FTFS scheme in which the spatial derivative is approximated using a forward difference and that obtained using an explicit FTBS scheme where the spatial derivative is evaluated using a backward difference approximation. It can be readily shown through a Taylor series expansion that the time derivative is first order accurate in all cases. The central differencing is second-order accurate while both FS and BS methods are first order accurate in space. Using the standard index notation that we have used earlier and assuming uniform time step Δt and grid spacing Δx, these approximations can be written as follows:

$$\text{FTBS:}\quad u_i^{n+1}=u_i^n-\sigma\left(u_i^n-u_{i-1}^n\right) \tag{3.2.4a}$$

$$\text{FTCS:}\quad u_i^{n+1}=u_i^n-\sigma\left(u_{i+1}^n-u_{i-1}^n\right)/2 \tag{3.2.4b}$$

$$\text{FTFS:}\quad u_i^{n+1}=u_i^n-\sigma\left(u_{i+1}^n-u_i^n\right) \tag{3.2.4c}$$

Here, u_i^n refers to u(x, t) = u(iΔx, nΔt), i.e., to the value of the variable u at the ith space location and nth time step and σ = aΔt/Δx where a is the wave speed, Δt and Δx are the time and space increments.

From the orders of truncation errors, we may expect that the first method will give us the best solution and that it would approach the true solution as Δt and Δx are progressively reduced. We may expect in a similar vein that the FTFS and the FTBS solutions will be less accurate, probably by the same extent in each case, but that they too will approach the true solution as Δt and Δx are progressively reduced. We may note that the exact solution of the partial differential Eq. (3.2.1), $\widetilde{u}$, is one of pure translation, i.e.,

$$\widetilde{u}(x,t)=f(x-at) \tag{3.2.5}$$

The exact solution for a convection speed, a, of 1 m/s is illustrated graphically in Fig. 3.7 for an initial solution which is composed of three non-zero segments: a triangular distribution, a rectangular pulse distribution followed by a sinusoidal distribution over 0.1 < x < 0.5 m. Over the rest of the spatial domain, u = 0 at t = 0. It is seen that as time progresses, the initial shape is convected downstream

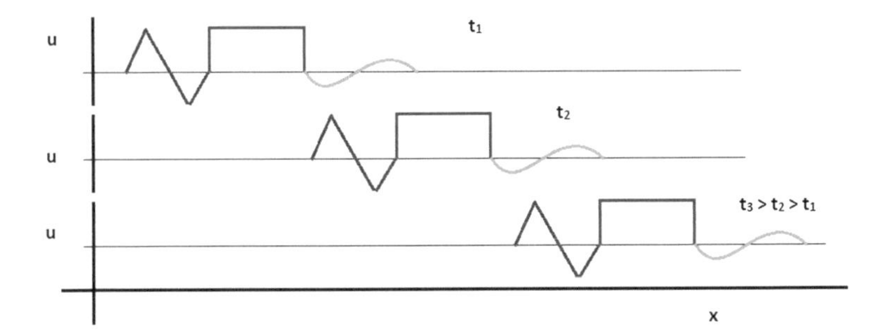

Fig. 3.7 The exact solution to the linear wave equation in 1-d. The initial u(x) is convected in the +ve x-direction at constant speed with unchanged shape

(in the positive x-direction) at a speed of 1 m/s and that the composite pulse remains unchanged in shape. If the convection speed is negative, then the initial distribution will move in the negative x-direction, i.e., to the left. We will see if this behaviour is captured by our numerical solution.

Let us investigate the behaviour of above three schemes (Eq. 3.2.4) for the initial condition in the form of a square wave given by

$$\begin{aligned} f(x) &= 0 \quad x < 0.1 \\ &= 1 \quad 0.1 \leq x \leq 0.2 \\ &= 0 \quad x > 0.2 \end{aligned} \tag{3.2.6}$$

The solution obtained by the FTBS scheme is plotted in Fig. 3.8 at 0.1, 0.2 and 0.5 s for three values of the parameter σ, namely, 0.5, 0.9 and 1.05. In each case, Δx is fixed and Δt is varied to obtain different values of σ. We see that the solution is different in each case and different from the expected true solution. While a pulse-like propagation is apparent for σ of 0.5 and 0.9, the solution is very different for the slightly higher value of 1.05. The situation with FTFS and the FTCS schemes is worse. Totally wrong values of the solution are obtained much earlier for all values of σ. Even the more accurate FTCS scheme is incorrect even for small values of t. Surprisingly, these different solutions have been obtained just by changing the value of the time step, Δt. Further, one would have expected that reducing the value of Δt would increase the accuracy; the computed solution in Fig. 3.8a appears to be worse that that in Fig. 3.8b, even though the time step in the latter is larger. Such sensitivity to the time step is clearly unacceptable.

Let us consider another simple example, this time incorporating another feature of the generic scalar transport equation, namely, unsteady diffusion. We again consider the one-dimensional, constant property case and write the governing partial differential equation as

$$\partial \varphi / \partial t = \alpha \, \partial^2 \varphi / \partial x^2 \tag{3.2.7}$$

Let us write an explicit FTCS finite difference approximation for it; this gives

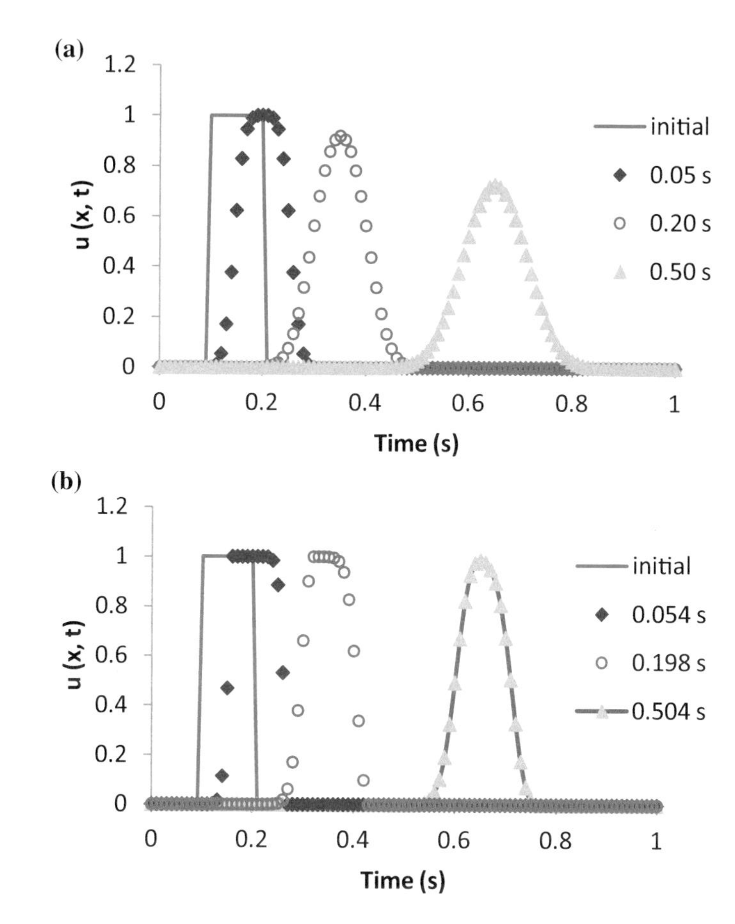

Fig. 3.8 Computed solution with FTBS for courant no. of **a** 0.5, **b** 0.9 and **c** 1.05; FTFS solution with courant no. of **d** 0.5 and **e** 0.9 and **f** CTCS solution with courant no. of 0.5

$$\left(\varphi_i^{n+1} - \varphi_i^n\right)/\Delta t = \alpha\left(\varphi_{i+1}^n - 2\,\varphi_i^n + \varphi_{i-1}^n\right)/\Delta x^2$$

or

$$\varphi_i^{n+1} = \varphi_i^n + \beta\left(\varphi_{i+1}^n - 2\,\varphi_i^n + \varphi_{i-1}^n\right) \tag{3.2.8}$$

Here, the parameter β depends on the diffusivity α and the spacing in the x- and t-coordinates. Let us try different values of Δt and Δx and compute the solution for a square pulse as the initial condition. Given that Eq. (3.2.7) represents unsteady diffusion, we expect, as time progresses, the initial square pulse to collapse slowly and become more rounded at the corners while spreading in the positive and the negative directions. We can see in Fig. 3.9 that this behaviour is obtained for some values of β but not for some other values. The FTCS scheme exhibits this behaviour for $\beta = 0.25$; for $\beta = 0.5$, some short wavelength oscillations appear, and for

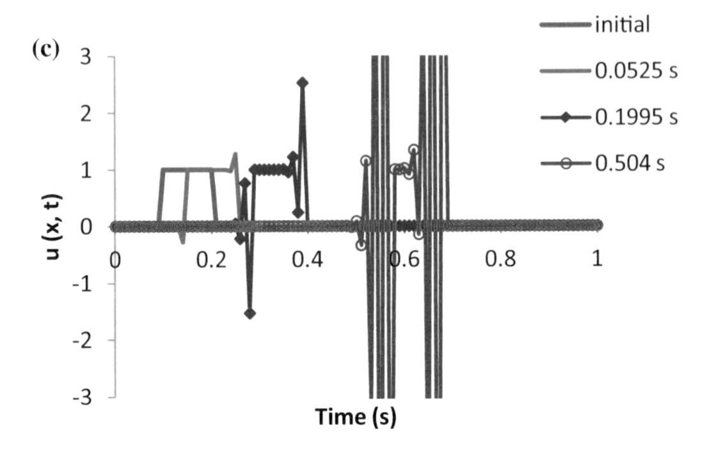

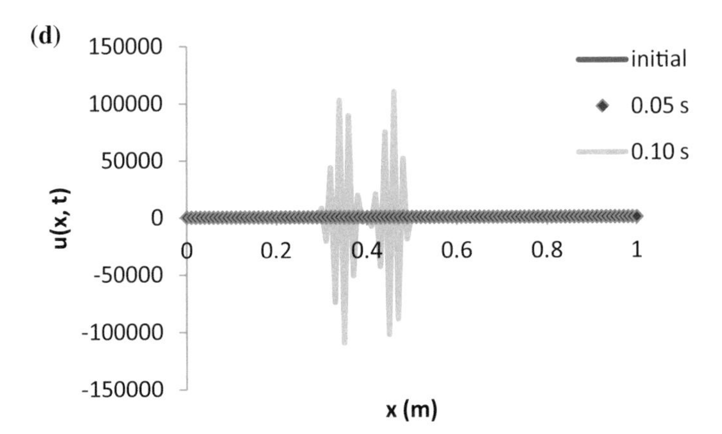

Fig. 3.8 (continued)

$\beta = 0.55$, the solution is obviously incorrect. The more accurate CTCS scheme gives incorrect solutions even for $\beta = 0.25$ at $t = 0.002$ s!

Thus, we can see that, for the simple one-dimensional unsteady convection or diffusion equation, both of which are core transport processes occurring in fluid flows, computed solutions ranging from the exact to highly unsatisfactory can be obtained using different discretization schemes. Also, for the same discretization scheme, the nature of the solution may be satisfactory or unsatisfactory depending on the choice of the parameters, in this case, the values of Δt and Δx. Such a sensitivity of the computed solution to the discretization scheme and the choice of Δt and Δx is unacceptable when dealing with a general case where the exact solution may not be known.

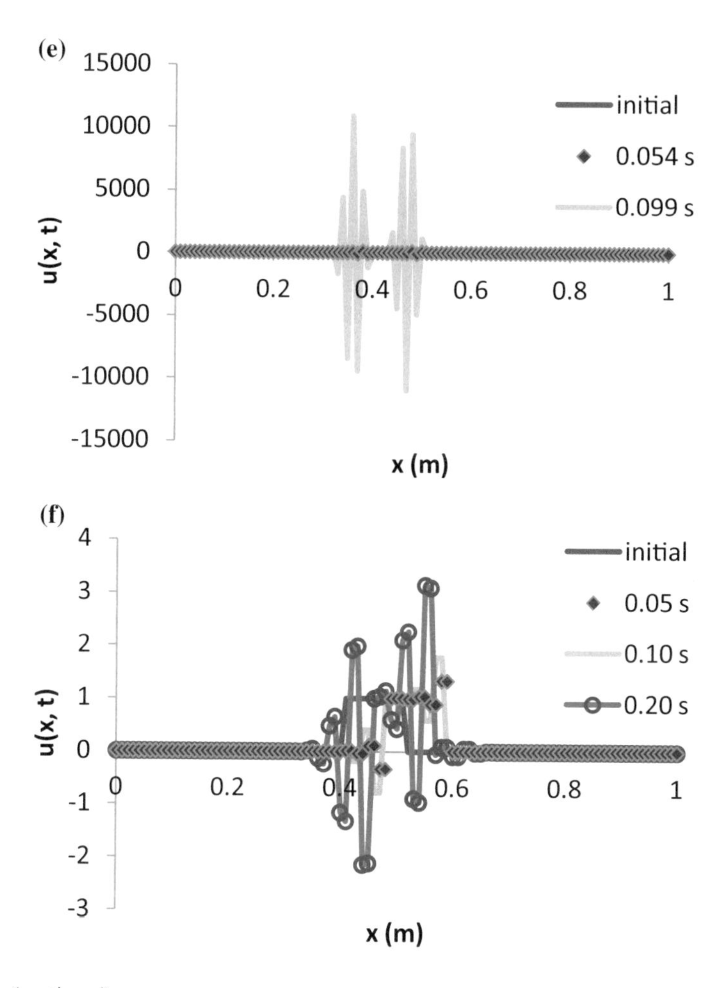

Fig. 3.8 (continued)

Another feature of these unsatisfactory solutions is the rapid build-up of errors. The error does not appear to be a result of a slow and gradual accumulation of round-off errors attributable to machine accuracy of numerical computation. It appears to be a fundamental problem with the scheme, i.e., with the way the derivatives are approximated. This can be established by constructing the solution for a case where the gradient is constant. Figure 3.10 compares the exact solution with the solution computed using FTBS and FTFS schemes with a Courant number value of 0.5 for a case where the initial condition is of a steady gradient starting at x = 0.2. The predicted solutions obtained after 10 time steps (each of 0.005 s) are compared with the exact solution at t = 0.05 s. As before, one can see that FTBS

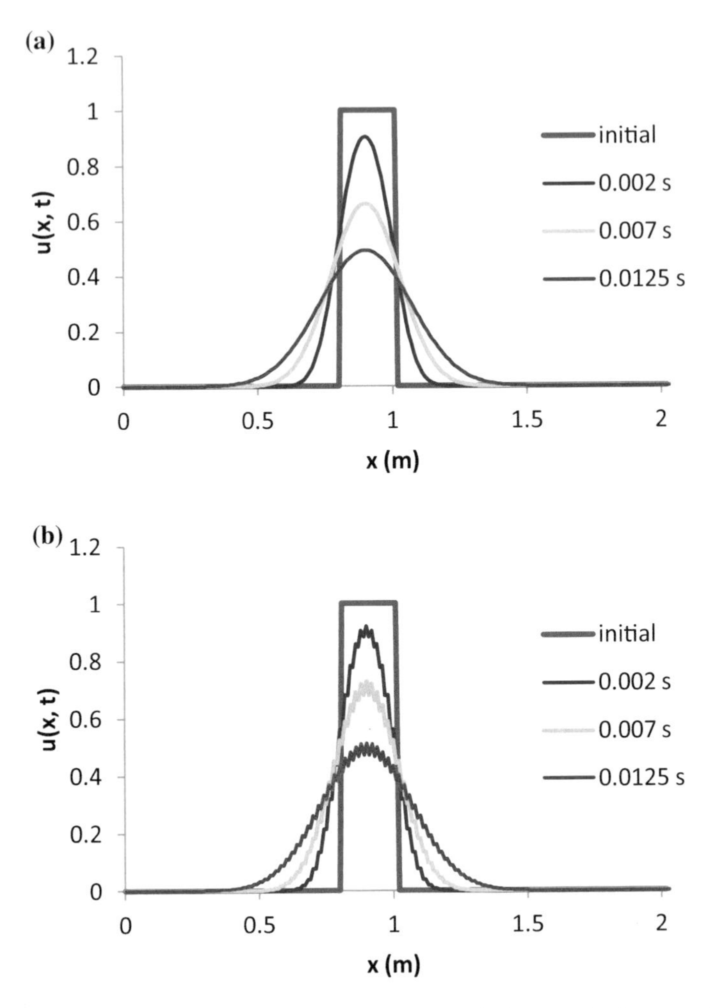

Fig. 3.9 Computed solution for the one-dimensional unsteady heat conduction equation using FTCS with **a** $\beta = 0.25$, **b** $\beta = 0.5$, and **c** $\beta = 0.55$ and **d** using CTCS with $\beta = 0.25$

solution is only slightly off the exact solution while the FTFS does not seem to be too bad for $x > 0.25$ m although wild oscillations are seen for $x < 0.2$ m. One may note from the initial condition that a change of slope of u occurs at $x = 0.2$ and that this causes immediate problems for the FTFS scheme but not for the FTBS scheme. Wrong estimation of the spatial gradient (at the transition, i.e., at $x = 0.2$ m) seems to give rise to problems immediately in the case of the FTFS scheme leading to further errors in the next time step which feed on the errors made in the previous time step, and so on. Since a wrong estimate of $\partial u/\partial x$ can lead to a wrong estimate of $\partial u/\partial t$ as per Eq. (3.2.1), the errors in u_i^{n+1}, which is computed as $u_i^{n+1} = u_i^n + \partial u/\partial t * \Delta t$, can build up very quickly.

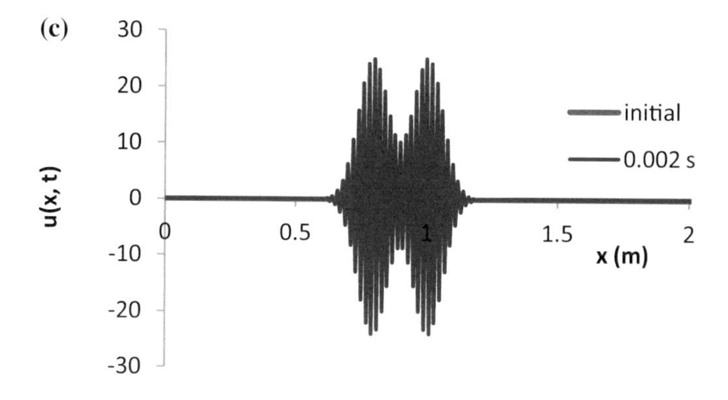

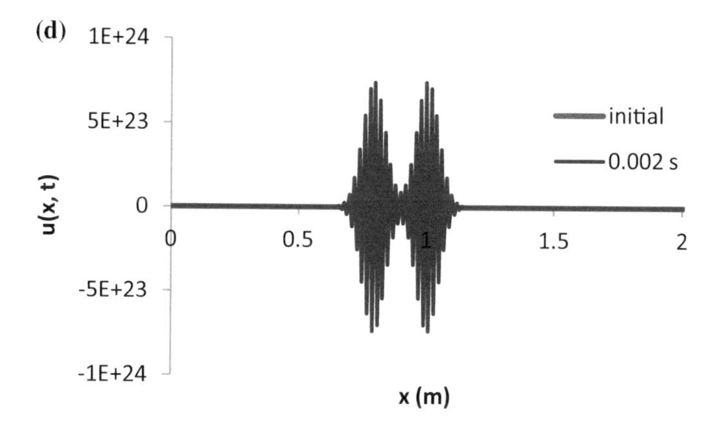

Fig. 3.9 (continued)

Clearly, there is a need to know more about the discretized equations and to understand conditions under which an accurate solution, approaching the exact solution of the partial differential equation, can be obtained. A formal analysis procedure addressing this practical need has been developed over the years; this is explained in the next section.

3.2.2 *Consistency, Stability and Convergence*

The important question of how accurate a computed solution is (or its potential to reach a required level of accuracy) compared to the exact solution of the differential equation is answered in three stages involving conditions on three important

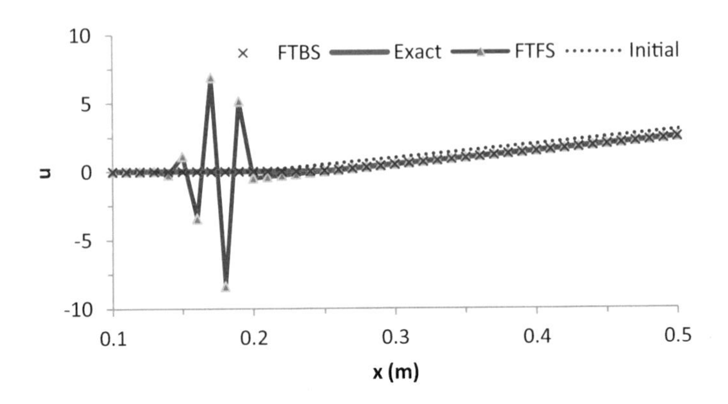

Fig. 3.10 Build-up of errors can be attributed to wrong estimation of spatial gradient. In the example shown in the figure, $\partial u/\partial x$ is constant for $x > 0.2$ m. The *change* in slope at $x = 0.2$ m is giving problems to the FTFS scheme

properties of numerical schemes, viz., consistency, stability and convergence. Essentially, they encapsulate the idea that if one wants good agreement between the computed solution and the exact solution, then one should be solving the correct equations and solve them correctly. This requires a three-level equivalence: between the discretized equation and the differential equation; between the computed solution and the exact solution of the discretized equation; and between the computed solution and the exact, analytical solution of the differential equation. The consistency condition defines the equivalence between the differential equation and its discretized counterpart. The stability condition establishes a relation between the computed solution and the exact solution of the discretized equation. The convergence condition connects the computed solution of the discretized equation to the exact solution of the differential equation. Together, they can be used to test the goodness of a solution to be obtained from a numerical scheme. These concepts are elaborated below.

A discretized equation is said to be consistent if it tends to the original differential equation when Δt, Δx etc. tend to zero. Consistency of a numerical scheme can be verified by formally expanding each term in Taylor series around the point u (x, t). This allows the determination of the truncation error involved in the discretization of each term of the differential equation. Consistency is assured if the total truncation error of the discretized equation goes to zero as Δt and Δx tend to zero, i.e.,

$$\lim_{\Delta t, \Delta x \to 0} \left| TE_i^n \right| = 0 \text{ at fixed } x_i = i\Delta x \quad \text{and} \quad t_n = n\Delta t. \tag{3.2.9}$$

Stability is associated with the build-up of errors. It can be seen from Figs. 3.8, 3.9 and 3.10 that, in several cases, the difference scheme appears to amplify errors as we progress from one time step to another. A numerical solution method is said to be stable if it does not amplify the errors (due to rounding off, truncation errors, compounding of approximations etc.) that appear in the course of a numerical solution. Thus, a condition for stability can be formulated as the requirement that any error between the computed solution, u_i^n, and the exact solution of the difference equation, $\hat{u}_i^n$, should remain uniformly bounded as $n \to \infty$ at fixed Δt. If we define the error at the point (x_i, t_n), ε_i^n, as the difference between the computed solution and the exact solution of the discretized equation at (x_i, t_n), i.e.,

$$\varepsilon_i^n = \hat{u}_i^n - u_i^n \tag{3.2.10}$$

then the stability condition can be written as

$$\lim_{n \to \infty} \left| \varepsilon_i^n \right| \le K \text{ at fixed } \Delta t \tag{3.2.11}$$

with K being independent of n. For time-dependent problems, stability guarantees that the method produces a bounded solution if the exact solution itself is bounded.

A numerical scheme is said to be convergent if the computed solution of the discretized equations tends to the exact solution of the differential equation as the grid and time spacing tend to zero. Formally, this can be defined as follows. The computed solution u_i^n should approach the exact solution $\tilde{u}(x, t)$ of the differential equation at any point $x_i = i\ \Delta x$ and $t_n = n\ \Delta t$ when Δx and Δt tend to zero while keeping x_i and t^n constant. In other words, the error

$$\tilde{\varepsilon}_i^n = \tilde{u}_i^n - u_i^n \tag{3.2.12}$$

satisfies the following convergence condition:

$$\lim_{\Delta t, \Delta x \to 0} \left| \tilde{\varepsilon}_i^n \right| = 0 \text{ at fixed } x_i = i\Delta x \quad \text{and} \quad t_n = n\Delta t. \tag{3.2.13}$$

In practical calculations Δx and Δt can never go to zero. However, what the satisfaction of the convergence condition assures us is that a grid- and time-step independent solution, i.e., a solution which does not change significantly with further decrease in grid spacing and time step, can be taken as the correct solution of the differential equation. This however may not agree with an experimental measurement if the differential equation being solved itself is a poor model of the physical phenomenon. Another possibility is that there is considerable error in the experimental data!

Clearly, the concepts of consistency, stability and convergence are inter-related. For a linear problem, this is expressed by the equivalence theorem of Lax (Richtmyer and Morton 1967) which states that for a well-posed, linear, initial value problem with a consistent discretization, stability is the necessary and sufficient condition for convergence of the numerical scheme. For non-linear problems, there is no such general theorem. Here, local stability analysis may be performed by linearizing the problem in a small solution domain. The reasonableness of the computed solution is thus assured for a well-posed linear or linearized initial value problem if the consistency and the stability of the numerical scheme can be ascertained. We will see how this can be done.

3.2.3 *Analysis for Consistency*

Analysis for consistency requires the determination of the total truncation error of the discretized equation. This can be illustrated by the following example. Consider the FTFS scheme given by Eq. (3.2.4a). The terms u_i^{n+1} and u_{i+1}^n appearing in this equation can be expanded in a Taylor series about u(x, t), i.e., about u_i^n (as discussed in Sect. 3.1). Substituting these into the differential equation and simplifying, one gets

$$\begin{aligned}&(u_i^{n+1} - u_i^n)/\Delta t + a(u_{i+1}^n - u_i^n)/\Delta x - (u_t + au_x)_i^n \\ &\quad = (\Delta t/2)(\partial^2 u/\partial t^2)_i^n + (\Delta x/2)(a\partial^2 u/\partial x^2)_i^n + O(\Delta t^2, \Delta x^2)\end{aligned} \tag{3.2.14}$$

The terms on the right hand side indicate the truncation error. It can be seen that it is of first order in both time and space and that it vanishes in the limit of Δt and Δx tending to zero. Thus, the FTFS scheme satisfies the consistency condition. The truncation error for the FTBS and the FTCS schemes can be evaluated similarly to be as follows:

$$\begin{aligned}&(u_i^{n+1} - u_i^n)/\Delta t + a(u_i^n - u_{i-1}^n)/\Delta x - (u_t + au_x)_i^n \\ &\quad = (\Delta t/2)(\partial^2 u/\partial t^2)_i^n + (\Delta x/2)(a\partial^2 u/\partial x^2)_i^n + O(\Delta t^2, \Delta x^2)\end{aligned} \tag{3.2.15}$$

$$\begin{aligned}&(u_i^{n+1} - u_i^n)/\Delta t + a(u_{i+1}^n - u_{i-1}^n)/(2\Delta x) - (u_t + au_x)_i^n \\ &\quad = (\Delta t/2)(\partial^2 u/\partial t^2)_i^n + (\Delta x^2/6)(a\partial^3 u/\partial x^3)_i^n + O(\Delta t^2, \Delta x^4)\end{aligned} \tag{3.2.16}$$

It can be seen that these two schemes are also consistent as the truncation error vanishes in each case as Δt and Δx tend to zero. The fact that, despite being consistent, they give erratic results (see Fig. 3.8) shows that consistency alone is not sufficient to guarantee a satisfactory solution.

It appears from the three examples that every discretized equation, arrived at through formal approximation of each term in the partial differential equation, must be consistent. However, this is not the case as the following example shows. Consider the Stokes problem (White 1991) of fluid motion induced in a semi-infinite expanse when the plate bounding one side is set suddenly set into motion at a uniform horizontal velocity of U. The differential equation under consideration is similar to that of unsteady heat conduction, namely, the one-dimensional parabolic equation already mentioned above:

$$\partial u/\partial t = \alpha \partial^2 u/\partial x^2 \tag{3.2.17}$$

Using forward differencing for the first derivative and central differencing for the spatial derivative, we get, in an explicit formulation,

$$\left(u_i^{n+1} - u_i^n\right)/\Delta t = \alpha\left(u_{i+1}^n - 2\,u_i^n + u_{i-1}^n\right)/\Delta x^2 + O\left(\Delta t, \Delta x^2\right) \tag{3.2.18}$$

This FTCS scheme is consistent and explicit but, as we shall see, is only conditionally stable ($\alpha\Delta t/\Delta x^2 \leq ½$) and first order accurate in time. Using central differencing for the time derivative, i.e.,

$$\left(u_i^{n+1} - u_i^{n-1}\right)/2\Delta t = \alpha\left(u_{i+1}^n - 2\,u_i^n + u_{i-1}^n\right)/\Delta x^2 + O\left(\Delta t^2, \Delta x^2\right) \tag{3.2.19}$$

would make it second-order accurate in time but the resulting scheme is unconditionally unstable. DuFort and Frankel (1953) proposed a cure for the instability by writing u_i^n as

$$u_i^n = \left(u_i^{n+1} + u_i^{n-1}\right)/2 \tag{3.2.20}$$

in the FTCS scheme. This results in the following discretized equation

$$\left(u_i^{n+1} - u_i^{n-1}\right)/2\Delta t = \alpha\left(u_{i+1}^n - u_i^{n+1} - u_i^{n-1} + u_{i-1}^n\right)/\Delta x^2 + \varepsilon_T \tag{3.2.21}$$

where ε_T is the truncation error. Equation (3.2.21) is one of a class of "leap frog" schemes in which two independent solutions develop as time progresses because the value of u_i^{n+1} does not depend on u_i^n. It can be shown to be unconditionally stable, which is rare for an explicit scheme. However, its truncation error is of the form

$$\begin{aligned}\varepsilon_T = &-\left(\Delta t^2/6\right)\partial^3 u/\partial t^3 + \left(\alpha\Delta x^2/12\right)\partial^4 u/\partial x^4 + \left(\alpha\Delta x^4/360\right)\partial^6 u/\partial x^6 - \left(\alpha\Delta t^2/\Delta x^2\right)\partial^2 u/\partial t^2 \\ &- \left[\alpha\Delta t^4/\left(12\Delta x^2\right)\right]\partial^4 u/\partial t^4 + \cdots.\end{aligned} \tag{3.2.22}$$

The leading terms in the truncation error are such that all would tend to zero as Δt and Δx, except for the fourth term. If the ratio ($\Delta t/\Delta x$) is not small even though Δt and Δx are individually very small (in the limit of their tending to zero), the discretization error does not reduce to zero but would tend to $-\alpha\beta^2\partial^2 u/\partial t^2$, where $\beta = (\Delta t/\Delta x)$. To put it in another way, Eq. (3.2.21) is a consistent, 2nd order accurate, unconditionally stable approximation for a different partial differential equation, namely,

$$\partial u/\partial t + \alpha\beta^2\partial^2 u/\partial t^2 = \alpha\partial^2 u/\partial x^2 \tag{3.2.23}$$

which is different from the original pde given by Eq. (3.2.17). Indeed, Eq. (3.2.23) becomes a hyperbolic equation which describes wave propagation in a dispersive medium (Coulson and Jeffrey 1977) and the solution to the linear equation is in the form of a temporally decaying wave-like solution. Thus, the DuFort-Frankel scheme is not consistent as the correct solution to it exhibits a dispersive wave-like character which is not consistent with the solution of the original parabolic equation. It should be noted however that the scheme would be consistent if one were interested only in the steady state behaviour as the extra term would then vanish. Thus, lack of consistency would amount to solving a modified differential equation and the computed solution, even if solved exactly, would therefore not be the same as the exact solution of the differential equation. This leads to a lack of convergence.

3.2.4 Analysis for Stability

Investigation of the stability of a numerical scheme is not trivial in the general case of non-linear coupled equations. Let us first formulate the stability problem by considering the specific case of the one-dimensional convection Eq. (3.2.1) and its discretization using the FTBS scheme given by Eq. (3.2.4a). Let the analytical solution of the partial differential equation (Eq. 3.2.1) be denoted by A; the exact solution of the discretized equation (Eq. 3.2.4a) be denoted by D; and the numerical solution of the discretized equation obtained with finite machine accuracy be denoted by N. Note that D can be different from A because it contains the truncation error. Then we can write

$$\text{discretization error} = \mathrm{A} - \mathrm{D} \tag{3.2.24}$$

$$\text{round-off/compounded error} = \mathrm{D} - \mathrm{N} \tag{3.2.25}$$

Since stability deals with the accuracy with which the computed solution, N, approaches the exact solution, D, of the discretized equation, we can define error ε as

$$\varepsilon = \mathrm{D} - \mathrm{N} \quad \text{or} \quad \mathrm{N} = \mathrm{D} + \varepsilon. \tag{3.2.26}$$

Now since the computed numerical solution must satisfy the discretized equation, we have, to machine accuracy,

$$(\mathrm{N}_i^{n+1} - \mathrm{N}_i^n)/\Delta t + a(\mathrm{N}_i^n - \mathrm{N}_{i-1}^n)/\Delta x = 0 \tag{3.2.27}$$

Substituting $\mathrm{N}_i^n = \mathrm{D}_i^n + \varepsilon_i^n$, etc. in Eq. (3.2.27), we obtain

$$[(D_i^{n+1} + \varepsilon_i^{n+1}) - (D_i^n - \varepsilon_i^n)]/\Delta t + a[(D_i^n + \varepsilon_i^n) - (D_{i-1}^n - \varepsilon_{i-1}^n)]/\Delta x = 0$$

Rearranging the terms, we get

$$(D_i^{n+1} - D_i^n)/\Delta t + a(D_i^n - D_{i-1}^n)/\Delta x + (\varepsilon_i^{n+1} - \varepsilon_i^n)/\Delta t + a(\varepsilon_i^n - \varepsilon_{i-1}^n)/\Delta x = 0 \quad (3.2.28)$$

Since D is the exact solution of the discretized equation, we have

$$(D_i^{n+1} - D_i^n)/\Delta t + a(D_i^n - D_{i-1}^n)/\Delta x = 0$$

Note that this will not be the case if D is the exact solution of the PDE because the discretized equation is only an approximation of the PDE. Equation (3.2.28) then reduces to

$$(\varepsilon_i^{n+1} - \varepsilon_i^n)/\Delta t + a(\varepsilon_i^n - \varepsilon_{i-1}^n)/\Delta x = 0 \quad (3.2.29)$$

Equation (3.2.29) shows how the error at i evolves in time as a result of contribution from the error at neighbouring points. If the error equation is shown to be unstable, i.e., if a perturbation of the input values at nth time step is shown to grow larger at (n + 1)th time step, then the computed solution of the discretized equation will also be unstable and hence the scheme is not useful. The true solution of the PDE, however, may have a different evolution because it is different from the discretized equation by the truncation error. Another point to note is the error at a particular spatial point evolves as result of errors made at the neighbouring locations (see Eq. 3.2.29). Thus, compounding of errors is possible. Whatever be the source of error, one can say that for stability, the error evolution should be such that

$$|\varepsilon_i^{n+1}/\varepsilon_i^n| \leq 1 \quad (3.2.30)$$

Various methods have been developed for the analysis of stability, and nearly all of them are limited to linear problems. For a linear problem with constant coefficients, the problem of stability is now well understood when the influence of boundaries can be neglected or removed. This is the case either for an infinite domain or for periodic boundary conditions on a finite domain. For such problems, a formal stability analysis procedure, known as the Fourier or von Neumann method, has been developed during the Second World War and was published later in 1950 by Charney et al. (1950). Despite its limitations, it is widely used for stability analysis and is described below.

3.2.5 von Neumann Stability Analysis

The spatial distribution of error of at time t, i.e., $\varepsilon(x, t)$, can be decomposed into a Fourier series of the form

$$\varepsilon(x, t) = \sum_m b_m(t)e^{jk_m x} \quad (3.2.31)$$

where $j = \sqrt{-1}$ and k_m is the wave number, i.e., 2π/wavelength. One may recall that

$$e^{j\theta} = \cos\theta + j\sin\theta \quad (3.2.32)$$

If the physical domain extending from x = 0 to x = L has periodic boundaries, then the above series has M + 1 wave components where $M = L/\Delta x$. The wave numbers of the wave components are given by

$$k_m = m\pi/L, \quad m = 0, 1, 2, \ldots, M. \quad (3.2.33)$$

By expanding the error in the form of Eq. (3.2.31), we are expressing the spatial distribution of error at a given time as a sum of (M + 1) sinusoidal functions, the wave numbers (or wavelength) of which are given by Eq. (3.2.33). If the amplitude, $b_m(t)$, for each wave at time t, is found, then the error is completely determined. Examples of spatial distribution of a function constructed from summation of wave components are shown in Fig. 3.11. Here, Fig. 3.11a shows two sinusoidal wave components of wavelengths of 0.1 and 0.2 m over a domain length of 1 m. Figure 3.11b shows, over the same domain length, two more sinusoidal components having wavelengths of 0.3 and 0.4 m. Figure 3.11c shows two functions f_1 and f_2 constructed as linear combinations of these wave components:

$$f_1 = 0.1\sin(\pi x/0.1) + 0.2\sin(\pi x/0.2) + 0.3\sin(\pi x/0.3) + 0.5\sin(\pi x/0.4)$$
$$f_2 = 0.25\sin(\pi x/0.1) + 0.2\sin(\pi x/0.2) + 0.15\sin(\pi x/0.3) + 0.1\sin(\pi x/0.4)$$

One can see from Fig. 3.11c that although traces of the original sine waves can be seen in f_1 and f_2, the actual variation of the two functions is quite different although both have been composed of the same wave components. Figure 3.11d shows the variation over the same domain of function f_3 composed of an arbitrary linear combination of ten sinusoidal wave components:

$$\begin{aligned} f_3 = {} & 0.1 + 0.1\sin(\pi x/0.1) + 0.2\sin(\pi x/0.2) + 0.3\sin(\pi x/0.3) \\ & + 0.4\sin(\pi x/0.4) + 0.5\sin(\pi x/0.5) + 0.6\sin(\pi x/0.6) \\ & + 0.7\sin(\pi x/0.7) + 0.8\sin(\pi x/0.8) + 0.9\sin(\pi x/0.9) + 1.0\sin(\pi x/1) \end{aligned}$$

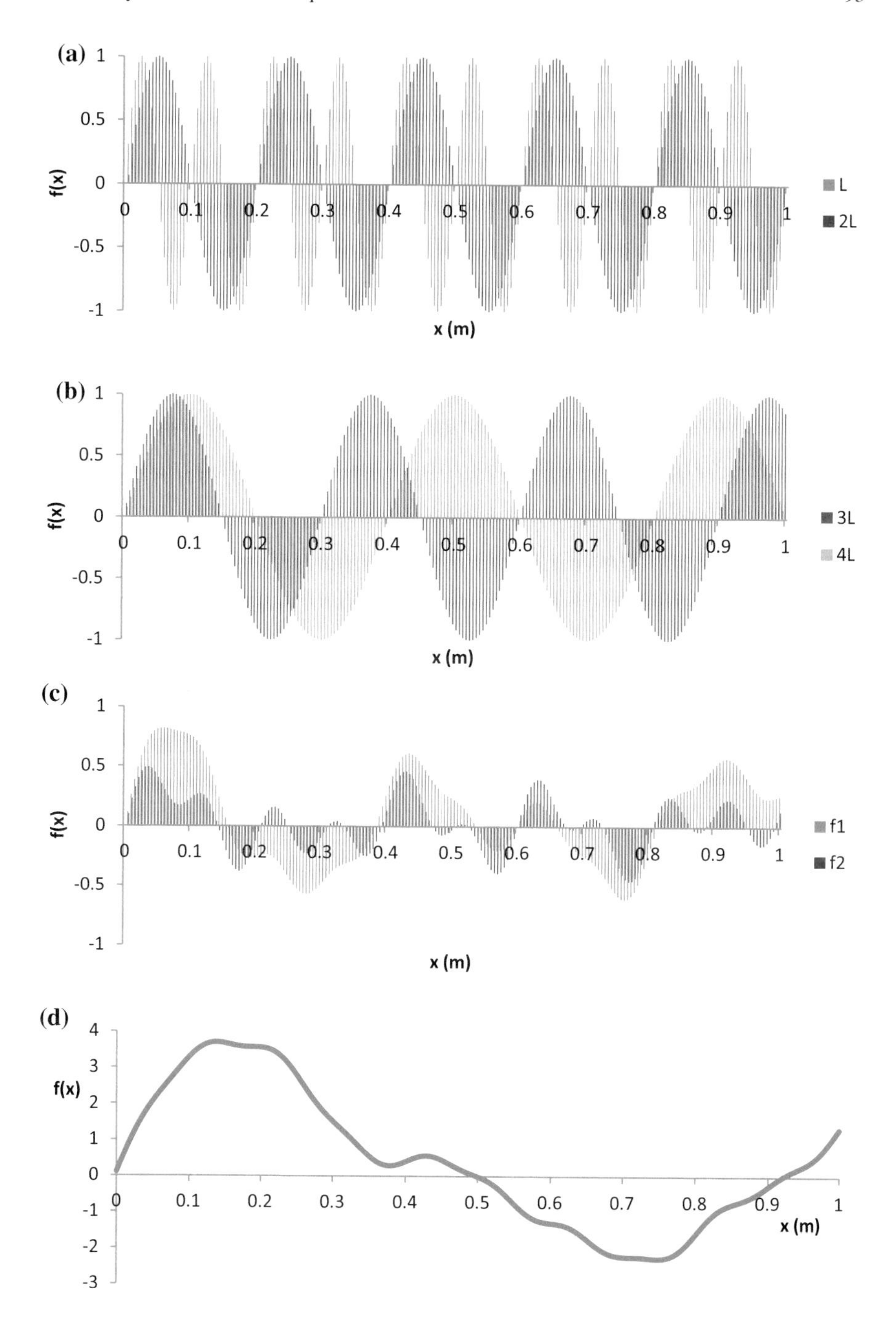

Fig. 3.11 Sinusoidal functions of wavelength **a** L and 2L, **b** 3L and 4L, **c** two superpositions f_1 and f_2 (bottom); $f_1 = 0.1\ \sin(\pi x/L) + 0.2\ \sin(\pi x/2L) + 0.3\ \sin(\pi x/3L) + 0.5\ \sin(\pi x/4L)$; $f_2 = 0.25\ \sin(\pi x/L) + 0.2\ \sin(\pi x/2L) + 0.15\ \sin(\pi x/3L) + 0.1\ \sin(\pi x/4L)$ and **d** spatial function obtained by superposition of ten wave components: $f(x) = 0.1 + 0.1\ \sin(\pi x/\lambda) + 0.2\ \sin(\pi x/(2\lambda)) + 0.3\ \sin(\pi x/(3\lambda)) + 0.4\ \sin(\pi x/(4\lambda)) + 0.5\ \sin(\pi x/(5\lambda)) + 0.6\ \sin(\pi x/(6\lambda)) + 0.7\ \sin(\pi x/(7\lambda)) + 0.8\ \sin(\pi x/(8\lambda)) + 0.9\ \sin(\pi x/(9\lambda)) + 1.0\ \sin(\pi x/(10\lambda))$

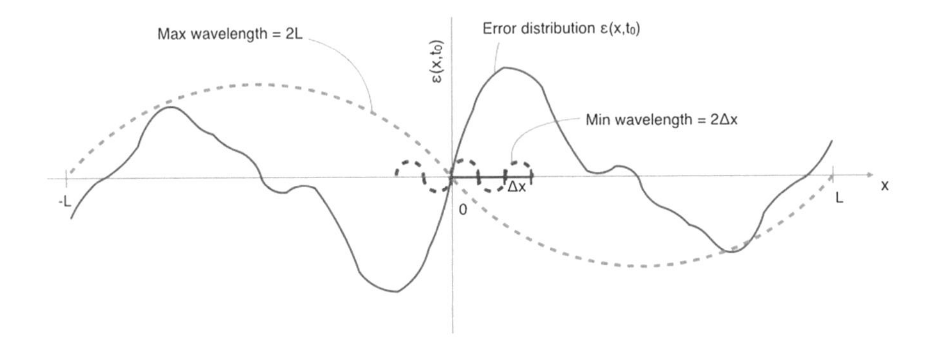

Fig. 3.12 Decomposition of the spatial distribution ε (x, t_0), $0 \leq x \leq L$ at time t_0 into its Fourier components over the domain $-L \leq x \leq L$. For a discrete spatial distribution with a spacing of Δx, the wavelengths that appear in the Fourier components range from a minimum of $2\Delta x$ to a maximum of 2L as shown in the figure

One can see that all the three functional variations are different; depending on the amplitude of the individual components, different spatial variations are obtained.

Any discrete error distribution on a periodic boundary where the error is evaluated at uniformly distributed points can be expressed as a weighted sum of the (M + 1) wave components. The wave components having the largest wavelength and the smallest wavelengths, of 2L and $2\Delta x$, are shown in Fig. 3.12. The error at a given space location and time step has contributions coming from all these (M + 1) wave components.

If the error evolution is governed by a non-linear equation, one may expect these different wave components to interact. Since the equation governing the evolution of the error (Eq. 3.2.29) is linear, such interaction does not occur and the solution may be obtained by superposition. Hence it is sufficient to study the behaviour of each wave component separately. Consider the contribution of the mth wave component:

$$\varepsilon_m(x,t) = b_m(t)e^{jk_m x} \tag{3.2.34}$$

Anticipating the need for stability, we seek a solution of the form

$$\varepsilon_m(x,t) = e^{h_m t}e^{jk_m x} \quad \text{or} \quad \varepsilon_i^n = e^{h_m n\Delta t}e^{jk_m i\Delta x} \tag{3.2.35}$$

where k_m is real and but, $e^{h_m t}$, the amplitude of wave component m at time t, may be complex. Then the stability condition (3.2.10) can be expressed as

$$\left|\varepsilon_i^{n+1}/\varepsilon_i^n\right| = \left|e^{h_m \Delta it}\right| = |G| \leq 1 \text{ for all m} \tag{3.2.36}$$

Here G is called the amplification factor and is the multiplicative factor by which the amplitude of the wave component m increases (or decreases) in time Δt. The amplification factor can be obtained as follows. Substituting Eq. (3.2.35) in Eq. (3.2.29) and dividing throughout by $e^{h_m t} e^{jk_m x}$, we obtain

$$\left(e^{h_m \Delta t} - 1\right) + a\Delta t/\Delta x\left(1 - e^{-jk_m \Delta x}\right) = 0 \tag{3.2.37}$$

Noting $e_{m,i}^{n+1}/e_{m,i}^{n} = G = e^{h_m \Delta t}$, the above equation can be written as

$$G = 1-\sigma+\sigma e^{-j\phi} = 1 - \sigma + \sigma \cos\phi - j\sigma \sin\phi \tag{3.2.38}$$

where $\sigma = a\Delta t/\Delta x$ and $\phi = k_m \Delta x$. We may note here that for a given wave component k_m, the value of phase angle is fixed. Since the error function consists of contribution from a number of wave components, the value of ϕ changes from 0 to π in discrete steps, the number of steps increasing as the grid is refined for the same domain length, L. As a result, the corresponding amplification factor also changes. The variation of G with ϕ in the complex plane is shown in Fig. 3.13. Here, the top half corresponds to $0 \leq \phi \leq \pi$; due to periodicity (see Fig. 3.12), the bottom half corresponds to $0 \geq -\phi \geq -\pi$.

Equation (3.2.38) represents a circle of radius σ centred on the real axis at $(1 - \sigma)$. The magnitude of the amplification factor, $|G|$, can be found as

$$|G|^2 = GG^* = [(1 - \sigma + \sigma \cos\phi) - (j\sigma \sin\phi)]\,[(1 - \sigma + \sigma \cos\phi) + (j\sigma \sin\phi)] \tag{3.2.39}$$

It can be seen that the condition that $|G| \leq 1$ is met if

$$0 < \sigma \leq 1 \tag{3.2.40}$$

which provides the stability condition for the scheme. For higher values of σ, there is at least one value of ϕ for which $|G| > 1$. The corresponding wave component(s) will grow exponentially with time and will soon render the solution meaningless. The parameter σ is called the Courant number and the condition given by Eq. (3.2.40) is called the CFL condition after Courant, Friedrichs and Lewy (1928).

Now let us consider the FTCS scheme given by Eq. (3.2.4b) for which the error evolution equation can be derived as

$$(\varepsilon_i^{n+1} - \varepsilon_i^n)/\Delta t + a(\varepsilon_{i+1}^n - \varepsilon_{i-1}^n)/(2\Delta x) = 0 \tag{3.2.41}$$

Substituting Eq. (3.2.35) into Eq. (3.2.41) and dividing by $e^{h_m t} e^{jk_m x}$, we obtain

$$\left(e^{h_m \Delta t} - 1\right) + \frac{a\Delta t}{(2\Delta x)}\left(e^{jk_m \Delta x} - e^{-jk_m \Delta x}\right) = 0 \tag{3.2.42}$$

Denoting $e_{m,i}^{n+1}/e_{m,i}^{n} = G = e^{h_m \Delta t}$, and noting that

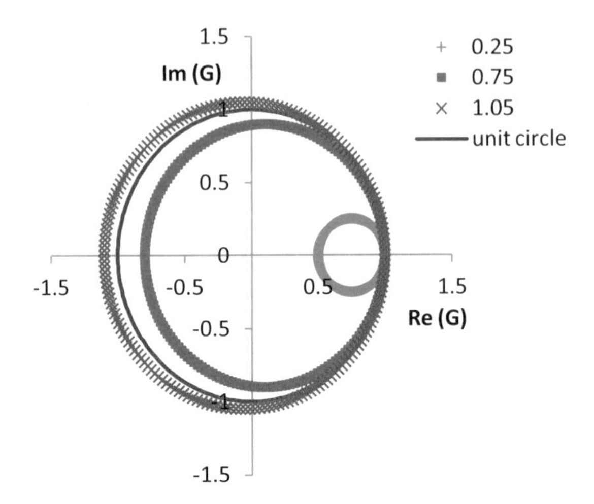

Fig. 3.13 Variation of the amplification with Courant number for the FTBS scheme for the linear wave equation

$$\left(e^{jk_m\Delta x} - e^{-jk_m\Delta x}\right) = 2j\,\sin(k_m\Delta x)$$

Equation (3.2.42) can be written as

$$G = 1 - j\sigma\sin\phi \qquad \text{where } \phi = k_m\Delta x \tag{3.2.42}$$

Evaluating the modulus of G, we note that

$$|G|^2 = 1 + \sigma^2\sin^2\phi > 0 \tag{3.2.43}$$

This implies that the FTCS scheme for the one-dimensional wave equation, Eq. (3.2.1), is unstable for all values of σ. It can be similarly shown that for the FTFS scheme for Eq. (3.2.1) has the following amplification factor G:

$$G = 1 + \sigma - \sigma e^{j\phi} = (1 + \sigma - \sigma\cos\phi) - j(\sigma\sin\phi) \tag{3.2.44}$$

Equation (3.2.44) represents a circle of radius σ centred on the real axis at $(1 + \sigma)$. Therefore, the condition that $|G| \leq 1$ is never met and the FTFS scheme is unconditionally unstable. This is borne out by the computed results shown in Fig. 3.8d.

It is interesting to note that if the convection speed, a, is negative, the roles of the FTBS and the FTFS schemes are reversed with the latter becoming unconditionally unstable and the latter becoming conditionally stable ($|\sigma| \leq 1$). This is the basis for upwind schemes, the origin of which is attributed to Courant et al. (1952), see Hirsch (1990). The FTCS scheme for the one-dimensional wave equation however remains unconditionally unstable.

Finally, let us consider an implicit scheme for the convection Eq. (3.2.1). We choose the unstable FTCS explicit scheme (3.2.4b) and make it implicit in time by writing it as

$$\text{FTCS-implicit:} \quad u_i^{n+1} = u_i^n - \sigma(u_{i+1}^{n+1} - u_{i-1}^{n+1})/2 \tag{3.2.45}$$

Proceeding with the von Neumann stability analysis as above, we find that the amplification factor is given by

$$G - 1 + \sigma/2\, G(e^{j\phi} - e^{-j\phi}) = 0 \quad \text{or} \quad G = 1/(1 + j\sigma \sin\phi) \tag{3.2.46}$$

The modulus of the amplification factor is given by

$$|G|^2 = G.G^* = 1/(1 + \sigma^2 \sin^2\phi) \tag{3.2.47}$$

which is always <0. Thus this scheme is unconditionally stable for all values of σ.

Thus, rendering the scheme implicit from explicit has made the method stable. In general, implicit methods are more stable than explicit methods and therefore larger time steps can be employed for a given Δx and convection speed. However, large time steps also introduce large discretization errors. (One may note here that stability does not imply accuracy; it only implies that the errors made in different terms of the equation do not compound quickly when continued over a number of steps. Accuracy is still determined by the influence of the truncation errors.) When the time evolution of the problem is important or when solving highly non-linear coupled equations, Δt should be restricted to about 5–10 times the maximum Δt allowed in a corresponding explicit scheme. Another point to note with the implicit scheme is that a solution of $Au^{n+1} = b$ is necessary in which both the boundary conditions, i.e., u at i = 1 and u at i = N_i, the number of grid points in the x-direction, are brought into play. This is contrary to the nature of the hyperbolic equation in which either the left boundary or the right boundary should be left open (depending on the direction of wave propagation). The implicit solution will therefore be correct only as long as the wave front is far from the open boundary.

We will end this section by considering two cases of special interest. We will first demonstrate the unconditional stability of the DuFort and Frankel (1953) scheme, Eq. (3.2.21), for the Stokes problem (Eq. 3.2.17), which has been shown to be inconsistent. The error evolution equation in this case can be written as

$$(\varepsilon_i^{n+1} - \varepsilon_i^{n-1})/(2\Delta t) = \alpha(\varepsilon_{i+1}^n - \varepsilon_i^{n+1} - \varepsilon_i^{n-1} - \varepsilon_{i-1}^n)/\Delta x^2 \tag{3.2.48}$$

Substituting Eq. (3.2.35) in Eq. (3.2.48) and dividing by $e^{h_m t} e^{jk_m x}$, we obtain

$$\left(e^{h_m \Delta t} - e^{-h_m \Delta t}\right) - \left(2\alpha \Delta t/\Delta x^2\right)\left(e^{jk_m \Delta x} + e^{-jk_m \Delta x} - e^{h_m \Delta t} - e^{-h_m \Delta t}\right) = 0 \quad (3.2.49)$$

Denoting $\varepsilon_{m,i}^{n+1}/\varepsilon_{m,i}^{n} = G = e^{h_m \Delta t}$, $\beta = \alpha\Delta t/\Delta x^2$ and $\phi = k_m \Delta x$, Eq. (3.2.49) can be written as

$$(G - 1/G) - 4\,\beta\,\cos\phi + 2\beta(G + 1/G) = 0$$

or

$$(1 + 2\beta)\,G^2 - 4\beta\cos\phi\,G - (1 - 2\beta) = 0 \quad (3.2.50)$$

Equation (3.2.50) has two roots which are given by

$$G = \left[2\beta\cos\phi + \{(1 - 4\beta^2\sin^2\phi)\}^{0.5}\right]/(1 + 2\beta) \quad (3.2.51)$$

It is left as an exercise to the reader to show that the modulus of G is ≤ 1, thus proving that the DuFort-Frankel scheme for the linear convection equation is unconditionally stable.

Finally, practical fluid flow problems are rarely one-dimensional; let us see how we can extend the analysis to a two-dimensional case. Consider the two-dimensional unsteady heat conduction equation

$$\partial u/\partial t = \alpha\left(\partial^2 u/\partial x^2 + \partial^2 u/\partial y^2\right) \quad (3.2.52)$$

and consider the explicit FTCS discretization scheme for it:

$$\left(u_{i,j}^{n+1} - u_{i,j}^{n}\right)/\Delta t = \left(\alpha/\Delta x^2\right)\left(u_{i+1,j}^{n} - 2u_{i,j}^{n} + u_{i-1,j}^{n}\right) + \left(\alpha/\Delta y^2\right)\left(u_{i,j+1}^{n} - 2u_{i,j}^{n} + u_{i,j-1}^{n}\right)$$

The corresponding error evolution equation can be readily written as

$$\left(\varepsilon_{i,j}^{n+1} - \varepsilon_{i,j}^{n}\right)/\Delta t = \left(\alpha/\Delta x^2\right)\left(\varepsilon_{i+1,j}^{n} - 2\varepsilon_{i,j}^{n} + \varepsilon_{i-1,j}^{n}\right) + \left(\alpha/\Delta y^2\right)\left(\varepsilon_{i,j+1}^{n} - 2\varepsilon_{i,j}^{n} + \varepsilon_{i,j-1}^{n}\right) \quad (3.2.53)$$

As a generalization of the formulation for one-dimensional case, the error distribution $\varepsilon(\mathbf{x}, t)$ at any time and spatial location is expressed as

$$\varepsilon(\mathbf{x}, t) = \sum_m b_m(t) e^{j\mathbf{k}_m \cdot \mathbf{x}} = \sum_m b_m(t) e^{jk_{m,x} x} e^{jk_{m,y} y} e^{jk_{m,z} z} \quad (3.2.54)$$

We seek a two-dimensional solution of the form for the mth wave component:

$$\varepsilon_m(x, y, t) = e^{h_m t} e^{jk_{m,x}x} e^{jk_{m,y}y} \quad \text{or} \quad \varepsilon^n_{p,q} = e^{h_m n\Delta t} e^{jk_{mx} p\Delta x} e^{jk_{my} q\Delta y} \tag{3.2.55}$$

where x = pΔx and y = qΔx. Substituting Eq. (3.2.55) into the error evolution Eq. (3.2.53), dividing throughout by $e^{h_m n\Delta t} e^{jk_{mx} p\Delta x} e^{jk_{my} q\Delta y}$, and denoting $\beta_x = \alpha\ \Delta t/\Delta x^2$, $\beta_y = \alpha\ \Delta t/\Delta y^2$, $\phi_x = k_{mx}\Delta x$ and $\phi_y = k_{my}\Delta y$, we obtain the amplification factor as

$$G = 1 + 2\,\beta_x(\cos\phi_x - 1) + 2\,\beta_y(\cos\phi_y - 1)$$

Substituting cos $\phi = 1 - 2\sin^2(\phi/2)$ etc. and simplifying, we have

$$G = 1 - 4\,\beta_x \sin^2(\phi_x/2) - 4\,\beta_y \sin^2(\phi_y/2) \tag{3.2.56}$$

Thus, for stability we should have

$$\left|1 - 4\,\beta_x \sin^2 \phi_x/2 - 4\,\beta_y \sin^2 \phi_y/2\right| \le 1 \tag{3.2.57}$$

There are two possibilities. If $(1 - 4\beta_x\sin^2(\phi_x/2) - 4\beta_y\sin^2(\phi_y/2)) > 0$, then $G < 1$ as β_x and β_y are positive and Eq. (3.2.57) is always satisfied. If $(1 - 4\ \beta_x \sin^2(\phi_x/2) - 4\ \beta_y\sin^2(\phi_y/2)) < 0$, then Eq. (3.2.57) requires that

$$4\,\beta_x\sin^2(\phi_x/2) + 4\,\beta_y\sin^2(\phi_y/2) - 1 \le 1 \quad \text{or}$$
$$4\,\beta_x\sin^2(\phi_x/2) + 4\,\beta_y\sin^2(\phi_y/2) \le 2$$

which is satisfied only if $(\beta_x + \beta_y) \le ½$. Thus, the condition for stability of the FTCS scheme for the two-dimensional heat conduction can be written as

$$(\alpha\Delta t/\Delta x^2 + \alpha\Delta t/\Delta y^2) \le ½ \tag{3.2.58}$$

It can be readily shown that the stability condition for the one-dimensional equation using the same scheme is $\alpha\Delta t/\Delta x^2 \le ½$. Thus, the effective time step in the two-dimensional case is reduced and will be even more reduced in three-dimensional cases. In general, the time step required to satisfy the stability condition such as Eq. (3.2.58) may change from point to point within the calculation domain. In such a case, the maximum time step that one can take corresponds to the largest allowable at all the grid points.

3.3 Application to the Generic Scalar Transport Equation

So far, we have looked at equations involving the convective and the diffusive terms separately. In the generic scalar transport equation, both the terms appear together; this can influence the stability of a scheme. We illustrate this by considering the one-dimensional case without source term. The corresponding advection-diffusion equation can be written as

$$\partial u/\partial t + u\partial u/\partial x = (\Gamma/\rho)\partial^2 u/\partial x^2 \tag{3.3.1}$$

Equation (3.3.1) is known as viscous Burgers' equation and has been introduced by Burgers (1948) as a simple equation capable of exhibiting the non-linear convective and the linear diffusive properties associated with fluid flow. In the general case, u is a function of x and t, as are Γ and ρ. We therefore linearize the equation about the point (x_0, t_0) and write it as

$$\partial u/\partial t + u_0\partial u/\partial x = \gamma_0\partial^2 u/\partial x^2 \tag{3.3.2}$$

where u_0 and γ_0 are the velocity and diffusivity (Γ/ρ) of u at (x_0, t_0) and are treated as constants in the stability analysis. Any source term needs to be linearized and included in the analysis in a similar way. Using the explicit FTCS scheme for discretizing Eq. (3.3.2), we can write

$$\left(u_i^{n+1} - u_i^n\right)/\Delta t + u_0/(2\Delta x)\left(u_{i+1}^n - u_{i-1}^n\right) = \left(\gamma_0/\Delta x^2\right)\left(u_{i+1}^n - 2u_i^n + u_{i-1}^n\right) \tag{3.3.3}$$

Formal application of the von Neumann stability analysis to the above scheme gives the amplification factor as

$$G = 1 + 2\beta(\cos\phi - 1) - i\sigma\sin\phi \tag{3.3.4}$$

where $\beta = \gamma_0\Delta t/\Delta x^2$, $\sigma = u_0\ \Delta t/\Delta x$ and $\phi = k_m\Delta x$. Equation (3.3.4) describes an ellipse centred on the positive real axis at $(1 - 2\beta)$ with semi-major and semi-minor axes given by 2β and σ, respectively (Fig. 3.14). Also, for $\phi = 0$, $G = 1$ and $\partial G/\partial\phi = 0$. Thus, the ellipse is tangent to the unit circle at the point where the unit circle intersects the positive real axis. Thus, for stability, i.e., for $|G| \leq 1$, the ellipse should be entirely within the unit circle. This requires that

$$\sigma \leq 1 \quad \text{and} \quad \beta \leq 1/2 \tag{3.3.5}$$

This appears to hold good for the case of $\beta = 0.45$, $\sigma = 0.75$ shown in Fig. 3.14. However, for the case of $\beta = 0.25$, $\sigma = 0.95$, the erstwhile semi-minor axis is longer than the semi-major axis, and satisfaction of (3.3.5) may still yield $|G| > 1$ for some values of ϕ. Hence for complete stability, we require that

$$\beta \leq 1/2 \quad \text{and} \quad \sigma^2 \leq 2\beta \quad \text{or} \quad \sigma^2 \leq 2\beta \leq 1 \tag{3.3.6}$$

It is interesting to note that the introduction of diffusion term has made the FTCS scheme conditionally stable; without the diffusion term, it would have been unconditionally unstable. Even though stability is guaranteed when conditions (3.3.6) are satisfied, the solution may still show bounded spatial oscillations for some

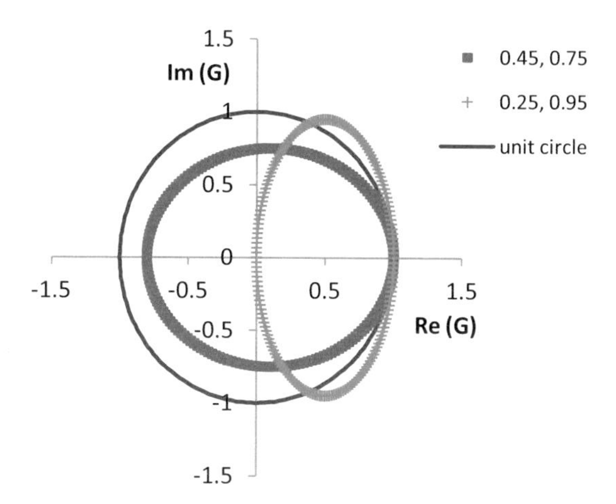

Fig. 3.14 Amplification factor for the linearized Burger's equation using the FT-CS-CS scheme for two sets of values of (β, σ) = (0.45, 0.75) and (0.25, 0.95)

combinations of β and σ. Defining a mesh Peclet number, $Pe_{\Delta x}$, as $Pe_{\Delta x} = u_0\,\Delta x/\nu_0$, bounded oscillations are produced when $2 < Pe_{\Delta x} < 2/\sigma$. Rearranging the terms, the FTCS scheme can be rewritten as

$$u_i^{n+1} = (\beta/2)\,(2 - Pe_{\Delta x})\,u_{i+1}^n + (1 - 2\beta)\,u_i^n + (\beta/2)(2 + Pe_{\Delta x})\,u_{i-1}^n \qquad (3.3.7)$$

The evolution of the solution for three cases, viz., $Pe_1 = 1.95$, $Pe_2 = 2.1$ and $Pe_3 = 2.2$, is shown in Fig. 3.15 with β = 0.45 in all cases, and σ = 0.8775, 0.945 and 0.99, respectively. The initial and boundary conditions (noting that Eq. 3.3.2 is parabolic in nature) are given by

$$\begin{array}{lll} u(x,0) = 0 & \text{except for } 0.4 \le x \le 0.5 & \text{where } u(x,0) = 1.0 \\ u(0,t) = 0 & \text{for } t > 0 & \\ u(2,t) = 0 & \text{for } t > 0 & \end{array} \qquad (3.3.8)$$

The variation in the Peclet number in the three cases is obtained by changing the value of only u_0; Δx and Δt have been kept the same in all the three cases. It can be seen that no oscillations are produced in the first case where the mesh Peclet number is less than two (Fig. 3.15a) while a spatially-oscillatory solution is obtained for the second case (Fig. 3.15b) with Pe greater than two but less than 2/σ. The third case (Fig. 3.15c) with Peclet number greater than 2/σ is unstable, as expected.

The oscillations produced by the FTCS scheme are unphysical and can lead to problems if the variable is physically bounded (e.g., mass fraction which is bounded

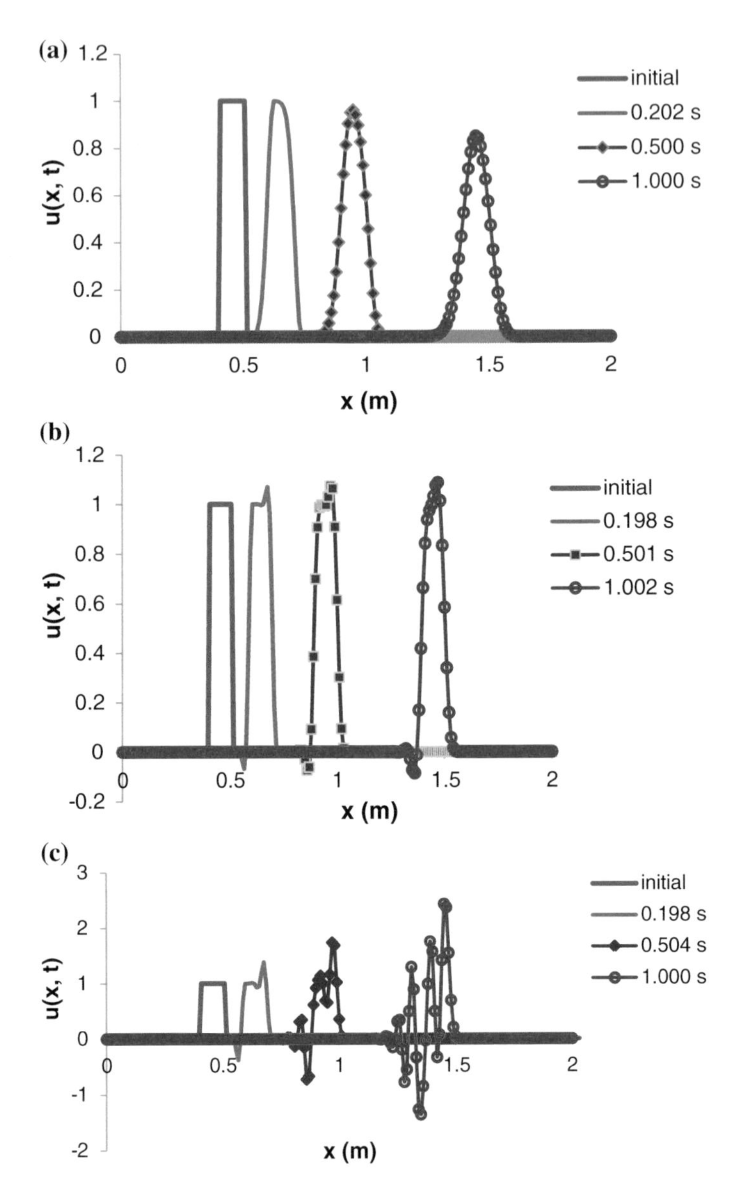

Fig. 3.15 Response of the FTCS scheme with (β, σ, Pe) of **a** (0.45, 0.8775, 1.95), **b** (0.45, 0.945, 2.1) and **c** (0.45, 0.990, 2.2)

between 0 and 1). They can be eliminated by sufficiently refining the grid so that the mesh Peclet number, which is directly proportional to the grid spacing, is less than two. This is illustrated in Fig. 3.16a where the grid spacing corresponding to the second case is halved so that the Peclet number, at the same u_0, is now 1.05. The time step is also halved to keep the same value of σ as in Fig. 3.15b. We see smoothened profiles without any over- or undershoots although the diffusion seems to be much faster than in Fig. 3.15b.

Another way of eliminating the oscillations is to use a first order "upwind" differencing scheme for the convective term, i.e., writing $\partial u/\partial x = (u_i^n - u_{i-1}^n)/\Delta x$. The resulting discretization scheme, which can be referred to as FTBSCS scheme, can be written in the form

$$u_i^{n+1} = \beta u_{i+1}^n + (1 - 2\beta - \sigma)\, u_i^n + \beta(1 + Pe_{\Delta x})\, u_{i-1}^n \tag{3.3.9}$$

The above scheme is stable for

$$(\sigma + 2\beta) < 1 \tag{3.3.10}$$

This condition can be more restrictive than the corresponding condition (Eq. 3.3.6) for the FTCS scheme in some cases. For example, for the case considered in Fig. 3.15a, the FTBSCS scheme is unstable. The solution obtained using $\beta = 0.24$ and $\sigma = 0.468$ is shown in Fig. 3.16b at nearly the same times as in Fig. 3.15a. One can see a smooth variation (some oscillations can be obtained near the peak for some time for $\beta = 0.25$ in this case). However, for the same problem parameters, namely, $u = 1$ m/s and $\alpha = 0.005168$ m^2/s, the solution obtained using FTBSCS scheme in Fig. 3.16b appears to be much more diffusive than the corresponding one in Fig. 3.15a obtained using the FTCS scheme. Thus, the upwind scheme for the advection term appears to be disadvantageous compared to the FTCS scheme. However, in cases where the diffusivity (α) is very small or velocity is very high, the condition of $\sigma^2 < 2\beta < 1$ for the FTCS scheme introduces too severe restrictions on Δt that can be used for a stable solution. A less restrictive time step condition may be obtained with the FTBSCS scheme. In order to counter the first order accuracy of the backward differencing scheme, higher order upwind schemes such as the QUICK scheme of Leonard (1979) are often used to discretize the convective terms. However, the boundedness property of the solution is often lost with higher order schemes. This is discussed further in the following section.

3.4 Dissipation and Dispersion Errors

Stability is not the only criterion that determines the satisfactory performance of a numerical scheme. Stable numerical schemes suffer from dissipative and dispersive errors which can degrade the computed solution. Dissipative error is manifested as the smearing, or loss of sharpness, of a large gradient in the computed solution (see the FTBS solution in Fig. 3.17) as compared to the correct solution. Dispersive error is manifested in the form of oscillations produced upstream and/or downstream of a sharp discontinuity as seen in the FTCS solution in Fig. 3.17. While dissipative error results in loss of accuracy of solution in regions of large gradients,

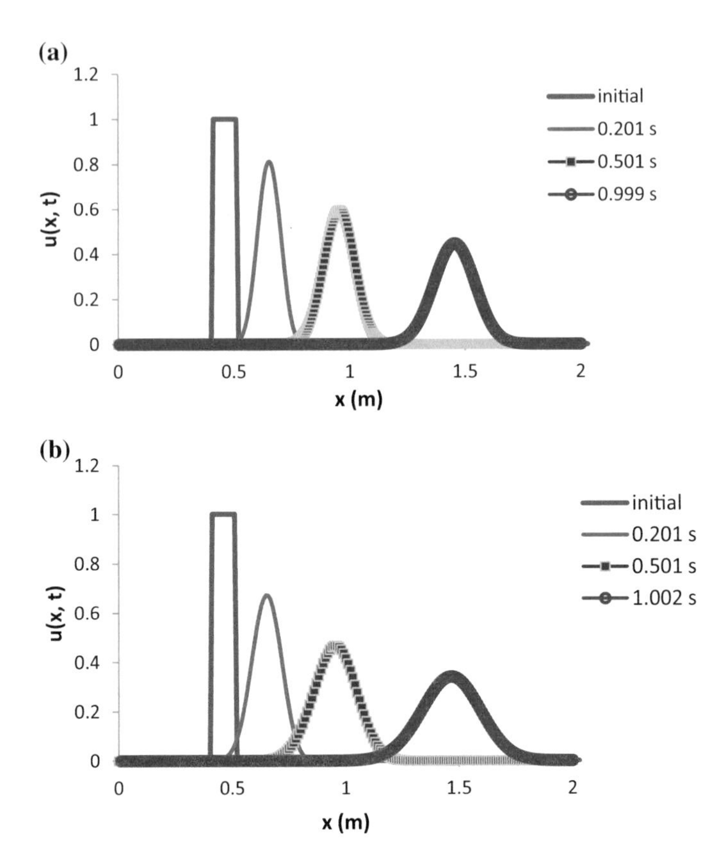

Fig. 3.16 Two ways of eliminating oscillations: **a** Reducing the mesh Peclet number to 1.05; solution obtained with $\beta = 0.45$ and $\sigma = 0.4725$. **b** Using upwinding difference for the advection term; solution obtained with $\beta = 0.24$, $\sigma = 0.468$ and $Pe = 1.95$. In both cases, smooth solution is obtained but with severe diffusion

dispersive errors can give rise to convergence problems. Basic concepts concerning these errors are discussed below.

Consider the first order wave Eq. (3.2.1) subject to the initial and boundary conditions given by Eq. (3.2.2) representing a wave propagating in the x-direction with a speed $a > 0$. Consider the explicit FTBS approximation for this equation:

$$\left(u_i^{n+1} - u_i^n\right)/\Delta t + a\left(u_i^n - u_{i-1}^n\right)/\Delta x = 0 \tag{3.4.1}$$

We can investigate the behaviour of the error involved in the above approximation by expanding u_i^{n+1} and u_{i-1}^n in terms of Taylor series about (i, n):

$$\begin{aligned} u_i^{n+1} &= u_i^n + \Delta t\, u_t + \Delta t^2/2!\, u_{tt} + \Delta t^3/3!\, u_{ttt} + \cdots. \\ u_{i-1}^n &= u_i^n - \Delta x\, u_x + \Delta x^2/2!\, u_{xx} - \Delta t^3/3!\, u_{xx} + \cdots. \end{aligned}$$

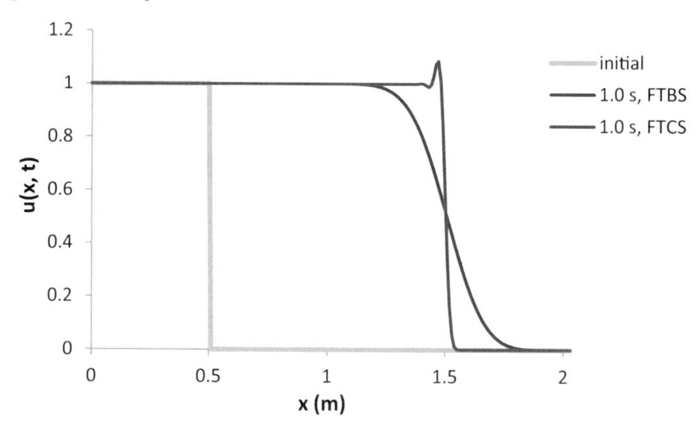

Fig. 3.17 Illustration of diffusive and dispersive errors that occur in numerical solutions: FTBS (diffusive) and FTCS (dispersive) solution to the 1-d Burger's equation for a step input

where the subscripts denote differentiation, i.e., $u_t = \partial u/\partial t$, etc. Substituting these into Eq. (3.4.1) and simplifying, we get

$$u_t + au_x = -\Delta t/2!\, u_{tt} + a\Delta x/2!\, u_{xx} - \Delta t^2/3!\, u_{ttt} - a\Delta x^3/3!\, u_{xxx} + \cdots. \quad (3.4.2)$$

As expected, the truncation error shows that the scheme is first order accurate in both space and time. Further insight into its behaviour can be obtained by replacing the time derivatives by spatial derivatives. Differentiating Eq. (3.4.2) with respect to t, we get

$$u_{tt} + au_{xt} = -\Delta t/2!\, u_{ttt} + a\Delta x/2!\, u_{xxt} - \Delta t^2/3!\, u_{tttt} - a\Delta x^3/3!\, u_{xxxt} + \cdots. \quad (3.4.3)$$

Differentiating (3.4.2) with respect to x, we get

$$u_{tx} + au_{xx} = -\Delta t/2\, u_{ttx} + a\Delta x/2!\, u_{xxx} - \Delta t^2/3!\, u_{tttx} - a\Delta x^3/3!\, u_{xxxx} + \cdots. \quad (3.4.4)$$

Multiplying (3.4.4) by a and subtracting it from (3.4.3) and rearranging, we get

$$u_{tt} = a^2 u_{xx} + \Delta t(-u_{ttt}/2 + a/2\, u_{ttx} + O(\Delta t)) + \Delta x\left(a/2\, u_{xxt} - a^2/2\, u_{xxx} + O(\Delta x)\right) \quad (3.4.5)$$

Similarly, one can obtain

$$\begin{aligned} u_{ttt} &= -a^3 u_{xxx} + O(\Delta t, \Delta x) \\ u_{ttx} &= a^2 u_{xxx} + O(\Delta t, \Delta x) \\ u_{xxt} &= -a\, u_{xxx} + O(\Delta t, \Delta x) \end{aligned}$$

Substituting these into Eq. (3.4.2), rearranging and denoting $\sigma = a\Delta t/\Delta x$, we obtain

$$u_t + au_x = (a\Delta x/2)(1-\sigma)u_{xx} - (a\Delta x)^2/3!)(2\sigma^2 - 3\sigma + 1)u_{xxx} + O(\Delta t^3, \Delta t^2\Delta x, \Delta t\Delta x^2, \Delta x^3) \tag{3.4.6}$$

Equation (3.4.6) is known as the modified equation and is the actual equation that is solved when using the FTBS approximation for the one-dimensional wave equation. Considering only the leading term in the truncation error, Eq. (3.4.6) can be written as

$$u_t + au_x = \gamma u_{xx} \tag{3.4.7}$$

where $\gamma = a\Delta x(1 - \sigma)/2$. Equation (3.4.7) represents an unsteady, linear convection-diffusion equation; thus, the leading term in the truncation error of the FTBS scheme introduces a diffusive error into the solution. Two important points can be deduced further from Eq. (3.4.7):

- Since diffusivity has to be greater than zero for a stable system, we can recover the CFL stability condition, viz., $\sigma \leq 1$, by requiring $\gamma > 0$ in Eq. (3.4.7). Indeed, this modified equation method has been used to investigate stability of finite difference approximations of some pdes (Warming and Hyett 1974).
- It can be shown that Eq. (3.4.7) admits solution of the type $u(x,t) = e^{-bt}e^{ikx}$. For example, for the initial condition that $u(x, 0) = \sin(kx)$ and for periodic boundary conditions, the exact solution of (3.4.7) is given by

$$u(x,t) = \exp(-k^2\gamma t)\sin(k(x - at)) \tag{3.4.8}$$

which represents a wave travelling in the positive x-direction with a speed of *a* but with an amplitude attenuating as time progresses. Thus, the nature of the diffusive term is to reduce the amplitude of the wave. Since a first order approximation will have a leading term containing a second derivative, first order terms suffer from severe diffusive errors and are not considered to be sufficiently accurate for time-accurate solutions.

In view of the problem of excessive numerical diffusion associated with first order schemes, let us consider the alternative of a second order approximation to the linear wave equation. The simplest of these is the leap frog method which can be written as

$$\left(u_i^{n+1} - u_i^{n-1}\right)/(2\Delta t) + a\left(u_{i+1}^n - u_{i-1}^n\right)/(2\Delta x) = 0 \tag{3.4.9}$$

The modified equation for this approximation can be derived to be

$$u_t + au_x = (a\Delta x^2/6)\left(\sigma^2 - 1\right) u_{xxx} - a(\Delta x^4/120)\left(9\sigma^4 - 10\sigma^2 + 1\right) u_{xxxxx} + \cdots. \tag{3.4.10}$$

Considering only the leading term on the right hand side and neglecting the other terms, Eq. (3.4.10) can be rewritten as

$$u_t + au_x + \mu u_{xxx} = 0 \tag{3.4.11}$$

where $\mu = a(\Delta x^2/3!\ (1 - \sigma^2)$. Equation (3.4.11) is the equation governing the propagation of wave systems exhibiting small dispersion effects. Its non-linear equivalent is the well-known Korteweg de Vries (KdV) equation

$$\partial v/\partial t + v\partial v/\partial x + \mu\, \partial^3 v/\partial x^3 = 0 \tag{3.4.12}$$

which is used to study non-linear wave systems with a small amount of dispersion. When referring to long waves with small dispersion, the wave speed given by the linear Eq. (3.4.11) can be approximated as (Lighthill 1978)

$$a_d = a\left(1 - \tfrac{1}{6} k^2 h^2\right) \tag{3.4.13}$$

where a_d is the speed of the dispersive wave, k is the wave number, h is the depth of water and $a = (gh)^{0.5}$, g being the acceleration due to gravity, is the speed in a non-dispersive system. The effect of dispersion is to make the wave speed a function of wave number (or wavelength which is equal to $2\pi/k$). Thus, the effect of a third derivative as the leading term in the truncation error of an approximation (such as the leap frog scheme) is to introduce a dispersive term to the original wave equation. Two points can be noted in this context:

- If the initial condition, namely, f(x) in Eq. (3.2.2) contains contributions from waves of several wave numbers (see Figs. 3.11 and 3.12), then a computed solution having a third derivative as the leading order term will make them travel at different speeds so that their superposition leads to a different shape. A typical solution will have overshoots or undershoots as illustrated in the FTCS solution shown in Fig. 3.17.
- The term μ in Eq. (3.4.11) representing a dispersive wave system is positive. This is possible only if $\sigma < 1$ in Eq. (3.4.10). Thus, we again recover the CFL stability condition from the modified equation for the leap frog method.

A numerical approximation to a partial differential equation may contain both dissipative and dispersive errors. Often the attenuation of the wave amplitude is also a function of the wave length and the computed solution may exhibit a complicated behaviour which depends on the wave numbers present in (contributing to) the wave form f(x, 0) and also on the attenuation and dispersion characteristics of the scheme. A formal investigation of these errors is based on the analysis of the amplification factor is explained below.

Consider the amplification factor for the FTBS scheme. This is given by

$$G = (1 - \sigma + \sigma\cos\phi) - j(\sigma\sin\phi) \quad (3.2.38)$$

In general, G is a complex variable and has a modulus, |G|, and phase angle, φ, associated with it. Thus, G can be written as

$$G = |G|\,e^{j\varphi} \quad \text{where} \quad \varphi = \tan^{-1}\{\mathrm{Im}(G)/\mathrm{Re}(G)\} \quad (3.4.14)$$

For example, for the FTBS scheme, |G| and φ are given by

$$|G| = \left[(1 - \sigma + \sigma\cos\phi)^2 + (-\sigma\sin\phi)^2\right]^{0.5} \quad (3.4.15a)$$

$$\varphi = \tan^{-1}[(-\sigma\sin\phi)/(1 - \sigma + \sigma\cos\phi)] \quad (3.4.15b)$$

The magnitude of amplification is such that $|G|^n$ represents the ratio of the wave amplitude at the end of n time steps to the original amplitude and $(1 - |G|^n)$ is the factor by which the wave amplitude is attenuated after n time steps. Similarly, $n\varphi$ is the amount of shift in the phase angle compared to the original phase. In general, these would be different from the exact amplitude and phase of the wave. The difference between these two gives rise to dissipative and dispersive errors.

In the present case, the exact solution can be found by substituting into Eq. (3.2.1) a solution of the form

$$u = e^{bt} e^{jk_m\Delta x} \quad (3.4.16)$$

This gives us $(b + jk_m\Delta x) = 0$ or $b = -jk_m\Delta x$. Thus, the exact solution is given by

$$u = e^{jk_m(x-ct)} \quad (3.4.17)$$

The exact amplification factor is therefore given by

$$\tilde{G} = u(t + \Delta t)/u(t) = \exp(-jk_m a\Delta t) = \exp(j\tilde{\varphi}) \quad (3.4.18a)$$

and the exact modulus and phase angle of the amplification factor are given by

$$|\tilde{G}| = 1 \quad \text{and} \quad \tilde{\varphi} = -k_m a\Delta t = -\sigma\phi \quad (3.4.18b, c)$$

The error in amplitude is called the dissipation or diffusion error, Φ_{dif}. Its value, relative to the amplitude of the exact solution, can be defined in terms of the ratio of the computed amplitude to the exact amplitude, viz., $|G|/|\tilde{G}|$ as

$$\Phi_{\mathrm{dif}} = (1 - |\mathrm{G}|/|\widetilde{G}|) \tag{3.4.19a}$$

The error in phase, defined here as the difference between the exact phase and the computed phase, viz., $(\widetilde{\varphi} - \varphi)$ (it is also defined sometimes as the ratio of the two, i.e., as $\varphi/\widetilde{\varphi}$), is called the dispersion error, Φ_{dis}. Thus,

$$\Phi_{\mathrm{dis}} = (\widetilde{\varphi} - \varphi) \tag{3.4.19b}$$

If the computed phase is less than the exact phase, then the computed solution lags behind and therefore the computed wave speed will be less than that of the true solution. If the computed phase is greater than the exact phase, then the computed wave travels faster than the true one. In either case, an error in wave speed occurs. For the FTBS scheme, the dissipation error and the dispersion error can be written respectively as

$$\Phi_{\mathrm{dif}} = 1 - \left[(1 - \sigma + \sigma\cos\phi)^2 + (-\sigma\sin\phi)^2\right]^{0.5} \tag{3.4.20a}$$

and

$$\Phi_{\mathrm{dis}} = -\sigma\phi - \tan^{-1}[(-\sigma\sin\phi)/(1 - \sigma + \sigma\cos\phi)] \tag{3.4.20b}$$

These are plotted in Fig. 3.18 as a function of the dimensionless wave number ϕ ($=k\Delta x$) for different values of σ. We see in Fig. 3.18a that the FTBS scheme damps out high frequency waves (corresponding to high values of ϕ) relative to low frequency components, thus leading to a smooth solution. For a given wave number, Φ_{dif} exhibits a non-monotonic variation with σ with more severe damping at $\sigma = 0.5$ than at 0.25 or 0.9. The phase error in Fig. 3.18b is small for low wave number components but is high for high wave numbers. Since these are anyway damped, the effective dispersion error is small and a non-oscillatory, smooth solution, such as the ones seen in Fig. 3.8a, is produced.

Consider now the leap frog scheme for the first order wave Eq. (3.2.1) given by Eq. (3.4.9) above. The amplification factor for this scheme can be written as

$$\mathrm{G} = +\left(1 - \sigma^2\sin^2\phi\right)^{0.5} - \mathrm{j}\,\sigma\sin\phi \tag{3.4.21}$$

Hence the modulus of the amplification factor and the phase angle are given, for $\sigma \leq 1$, by

$$|\mathrm{G}| = 1 \quad \text{and} \quad \varphi = \tan^{-1}\{-\sigma\sin\phi/[+(1 - \sigma^2\sin^2\phi)^{0.5})]\} \tag{3.4.22}$$

It is interesting to note that for $\sigma \leq 1$, $|\mathrm{G}| = 1$ for all ϕ and for all σ. This means that waves of any wave number (including contributions from rounding off errors) are not neither damped nor blown up and the scheme is said to be neutrally stable. Since the modulus of the amplification factor of the true wave, $|\widetilde{G}|$, is also

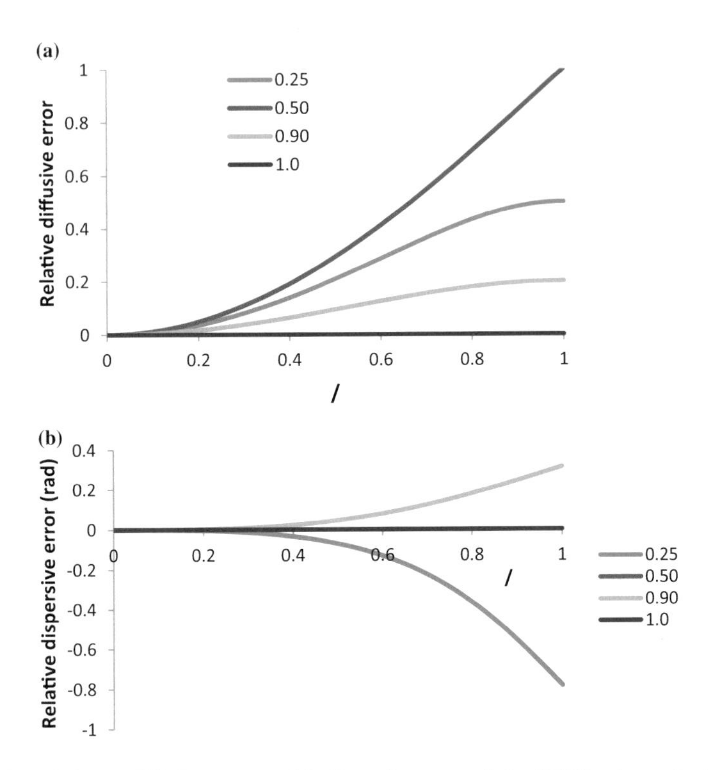

Fig. 3.18 Relative **a** diffusive error and **b** dispersive error of the FTBS scheme for the linear convection problem

equal to unity, there is no dissipation error associated with the leap frog scheme. Thus, $\Phi_{dif} = 0$. The relative phase error of the scheme is given by

$$\Phi_{dis} = -\sigma\phi - \tan^{-1}\{-\sigma \sin\phi/[+\left(1 - \sigma^2 \sin^2\phi\right)^{0.5})]\} \tag{3.4.23}$$

The dispersive error for the leap frog scheme is plotted in Fig. 3.19a as a function of σ and ϕ. One can see that there is severe dispersion of the high wave number components. Since these are not damped (unlike in the case of the FTBS scheme), one may expect these to be present in the solution. This is shown in Fig. 3.19b which shows the solution obtained for the 1-d linear wave equation for $\sigma = 0.5$. The presence and persistence of high wave number oscillations is clearly evident.

Finally let us consider the implicit FTCS approximation for the first order wave equation. This can be written as

$$\left(u_j^{n+1} - u_j^n\right)/\Delta t + a/(2\Delta x)\left(u_{j+1}^{n+1} - u_{j-1}^{n+1}\right) = 0 \tag{3.4.24}$$

The modified equation for the scheme (3.4.24) can be derived to be

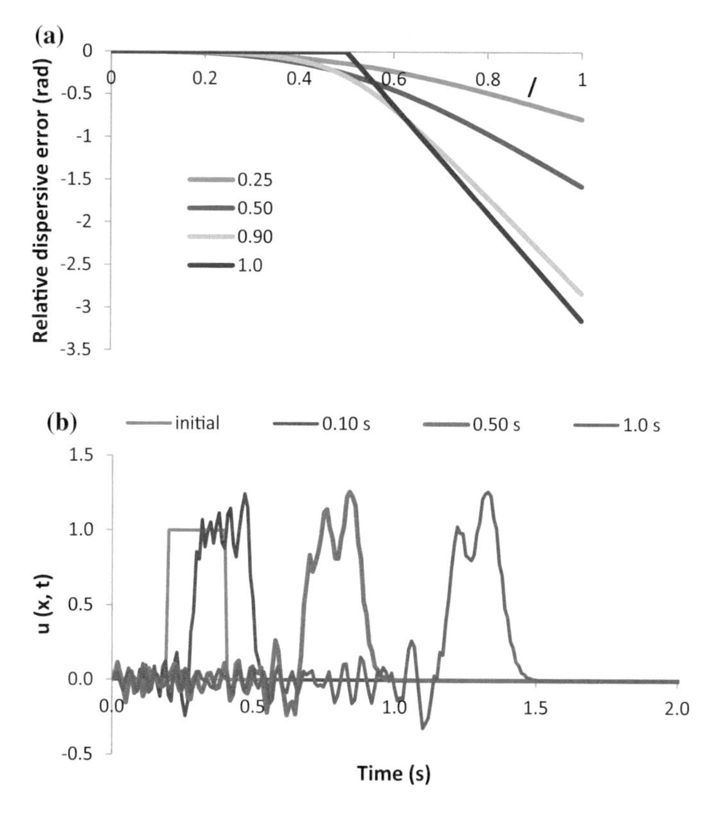

Fig. 3.19 **a** Dispersive behaviour of the leap frog scheme as a function of σ. **b** Solution obtained using the leap frog scheme for the linear wave equation for σ = 0.5 showing the persistence of high wave number oscillations

$$u_t - au_x = \left(a^2\Delta t/2\right) u_{xx} - \left(a\,\Delta x^2/6 + a^3\Delta t^2/3\right) u_{xxx} + \cdots \tag{3.4.25}$$

Examining the truncation error, we can see that the *implicit* FTCS scheme is first order accurate in time and second order accurate in space. Also, the leading diffusion-like (second derivate) term will always be positive for all values of a and Δt. The scheme is therefore stable for all time steps. The amplification factor for this scheme is given by

$$G = (1 - j\sigma\sin\phi)/\left(1 + \sigma^2\sin^2\phi\right) \tag{3.4.26}$$

Thus, the modulus and phase angle of the amplification factor for the implicit FTCS scheme are given respectively by

$$|G| = \left(1/\left(1+\sigma^2 \sin^2\phi\right)\right)^{0.5} \quad \text{and} \quad \varphi = \tan^{-1}(-\sigma \sin\phi) \qquad (3.4.27)$$

The diffusive error and the dispersive error associated with the scheme can be written as

$$\Phi_{dif} = 1 - \left(1/\left(1+\sigma^2 \sin^2\phi\right)\right)^{0.5} \quad \text{and} \quad \Phi_{dis} = -\sigma\phi - \tan^{-1}(-\sigma \sin\phi) \qquad (3.4.28)$$

These are plotted in Fig. 3.20 as a function of σ and ϕ. One can see that although the scheme is always stable, diffusive error increases as σ, or in other words, as the size of the time step, increases for intermediate wave numbers. This loss of accuracy, rather than loss of stability, would limit the maximum allowable time step when time-accurate calculations are required. Similarly, there is a severe lagging phase error for large wave numbers which increases with increasing σ. Since these are undamped, low wavelength oscillations may be expected. Thus, implicit schemes, though stable, suffer from the same type of diffusive and/or dispersive errors as explicit schemes.

Let us now consider the other model case of the linear one-dimensional parabolic unsteady heat conduction equation given by

$$\partial u/\partial t = \alpha \partial^2 u/\partial x^2 \qquad (3.4.29)$$

Well-posedness of this problem requires specification of initial conditions, i.e., u(x, 0) as well as the boundary conditions, i.e., u(0, t) and u(1, t) where x = 0 to 1 is the span of the domain under consideration. For the initial condition

$$u(x, 0) = f(x) \quad 0 \le x \le 1 \qquad (3.4.30a)$$

and the Dirichlet boundary conditions

$$u(0, t) = 0 \quad \text{and} \quad u(L, t) = 0 \qquad (3.4.30b)$$

The exact solution to the above problem can be written as

$$u(x, t) = \sum_{n=1}^{\infty} A_n \exp\left(-\alpha k^2 t\right) \sin(kx) \qquad (3.4.31)$$

where

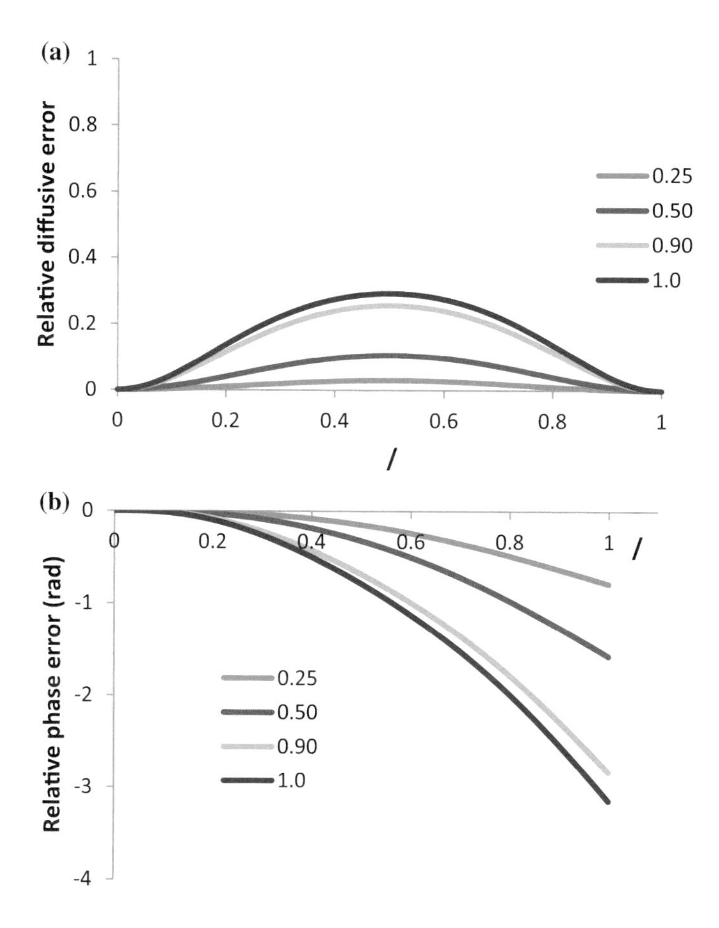

Fig. 3.20 **a** Diffusive and **b** dispersive error with the implicit FTCS scheme for the linear wave equation

$$A_n = 2\int_0^1 \{f(x)\,\sin kx\,dx\} \tag{3.4.32}$$

where $k = n\pi$. Consider a particular wave of wave number k_m. The exact solution can be written as

$$u_m(x,t) = C\exp(-\alpha k_m^2 t)\exp(jk_m x) \tag{3.4.33}$$

From this, the exact amplification factor can be obtained as

$$|\widetilde{G}| = u(t+\Delta t)/u(t) = \exp(-\alpha k_m^2 \Delta t) \tag{3.4.34}$$

Thus, for the exact solution

$$|\widetilde{G}| = \exp(-\beta\phi^2) \text{ and } \widetilde{\varphi} = 0 \quad \text{where} \quad \beta = \alpha\Delta t/\Delta x^2 \text{ and } \phi = k_m \Delta x. \tag{3.4.35}$$

Now let us investigate the dissipation and dispersion errors associated with some basic discretization schemes for this parabolic equation. Considering the explicit FTCS scheme once again, we can write the discretized equation as

$$(u_i^{n+1} - u_i^n)/\Delta t = \alpha(u_{i+1}^n - 2u_i^n + u_{i-1}^n)/\Delta x^2 \tag{3.4.36}$$

The modified equation corresponding to the above discretization can be written as

$$u_t - \alpha u_{xx} = \left(-\tfrac{1}{2}\alpha^2 \Delta t + \tfrac{1}{12}\alpha \Delta x^2\right) u_{xxxx} + \left(\tfrac{1}{3}\alpha^3 \Delta t^2 - \tfrac{1}{12}\alpha^2 \Delta t \Delta x^2 + \tfrac{1}{360}\alpha \Delta x^4\right) u_{xxxxxx} + \cdots. \tag{3.4.37}$$

Since the modified equation consists of only even derivatives of u, there is no dispersion error. For the scheme to be stable, the coefficients associated with the diffusion-like terms on the right hand side of Eq. (3.4.37) have to be positive. This means that $\alpha\Delta t/\Delta x^2 = \beta < ½$ which is the stability condition. The amplification factor corresponding to this discretization scheme is

$$G = 1 + 2\beta(\cos\phi - 1) \tag{3.4.38}$$

which is a real number as β is real. Thus, the modulus of the amplification factor is given by (3.4.38) and the phase angle is zero. Thus, the dissipation error and dispersion error associated with the explicit FTCS scheme are given as

$$\Phi_{dif} = 1 - [1 + 2\beta(\cos\phi - 1)]/[\exp(-\beta\phi^2)] \quad \text{and} \quad \Phi_{dis} = 0 \tag{3.4.39}$$

Thus, there is no dispersion error associated with this scheme and an oscillatory solution of the type shown in Fig. 3.19b cannot be obtained. One can also see from the mathematical expression (Eq. 3.4.39) that severe diffusive error will be present for high wave numbers. This diffusive error would add to the natural diffusion already present in the form of the diffusivity α in Eq. (3.4.29).

The implicit FTCS approximation for Eq. (3.4.29) can be written as

$$(u_i^{n+1} - u_i^n)/\Delta t + (\alpha/\Delta x^2)(u_{i+1}^{n+1} - 2u_i^{n+1} + u_{i-1}^{n+1}) = 0 \tag{3.4.40}$$

It has the following modified equation:

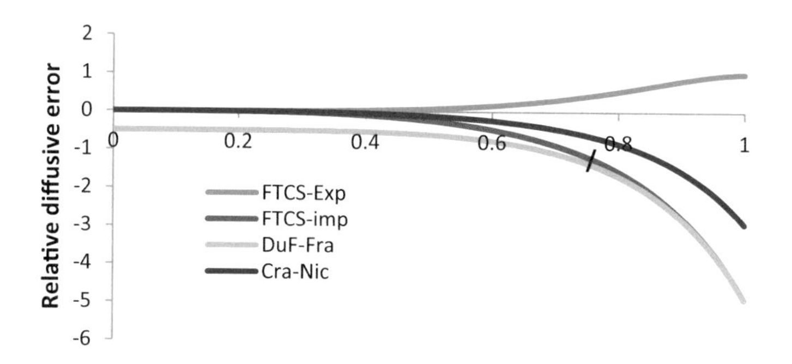

Fig. 3.21 Diffusive error with various schemes for the 1-d transient heat conduction equation for $\beta = 0.25$

$$u_t - \alpha u_{xx} = \left(\tfrac{1}{2}\alpha^2\Delta t + \tfrac{1}{12}\alpha\Delta x^2\right)u_{xxxx} + \left(\tfrac{1}{3}\alpha^3\Delta t^2 + \tfrac{1}{12}\alpha^2\Delta t\Delta x^2 + \tfrac{1}{360}\alpha\Delta x^4\right)u_{xxxxxx} + \cdots. \tag{3.4.41}$$

Once again, the modified equation has no odd derivative terms on the right hand side indicating that the scheme is purely diffusive. The coefficients of the terms on the right hand side are always positive indicating the unconditional stability of the scheme. The amplification factor given by

$$G = 1/[1 + 2\beta(1 - \cos\phi)] \tag{3.4.42}$$

is real and has a modulus which is always less than or equal to unity. The diffusive and dispersive errors associated with this scheme are given by

$$\Phi_{dif} = 1 - 1/\{[1 + 2\beta(1 - \cos\phi)]\exp(-\beta\phi^2)\} \quad \text{and} \quad \Phi_{dis} = 0 \tag{3.4.43}$$

This indicates a smaller diffusive error than that for the explicit FTCS scheme.

Finally, let us consider two schemes of second order accuracy in time. The modified equation for the Crank-Nicolson (1947) scheme can be shown to be

$$u_t - \alpha u_{xx} = \tfrac{1}{12}\alpha\Delta x^2 u_{xxxx} + \left(\tfrac{1}{12}\alpha^3\Delta t^2 + \tfrac{1}{360}\alpha\Delta x^4\right)u_{xxxxxx} + \cdots. \tag{3.4.44}$$

indicating a second order accuracy in time and space and unconditional stability. The amplification factor for this scheme is given by

$$G = [1 - \beta(1 - \cos\phi)]/[1 + \beta(1 - \cos\phi)] \tag{3.4.45}$$

which is always real thus indicating a purely diffusive behaviour. The dissipative and dispersive errors for the Crank-Nicolson method can therefore be written as

$$\Phi_{dif} = 1 - [1 - \beta(1 - \cos\phi)]/\{[1 + \beta(1 - \cos\phi)]\exp(-\beta\phi^2)\} \quad \text{and} \quad \Phi_{dis} = 0 \tag{3.4.46}$$

The diffusive error of this is compared in Fig. 3.21 with that of the earlier schemes which show that the error is considerably less than with this scheme.

Another second-order time accurate scheme for the unsteady heat conduction is the DuFort-Frankel (1954) scheme for which the amplification factor can be written as

$$G = \left[2\beta\cos\phi \pm \left(1 - 4\beta^2\sin^2\phi\right)^{0.5}\right]/(1 + 2\beta) \tag{3.4.47}$$

Here, for $\beta \leq ½$, G is real and the scheme is purely diffusive. For $\beta > ½$, G is complex, thus indicating that dispersion creeps into the solution for $\beta > ½$. The DuFort-Frankel scheme has the rather unusual property of having dispersive error for the heat conduction equation which is not in keeping with the purely diffusive nature of the original partial differential equation. Most schemes for the unsteady heat conduction are non-dispersive. High frequency components are highly damped. The negative values in Fig. 3.21 arise from partly from the definition of the diffusive error (Eq. 3.4.19a) and partly from the division by the exact amplitude (Eq. 3.4.35) which becomes small as ϕ increases.

3.5 Control of Oscillations

We have seen in the last section that a first order upwind approximation to the linear convection (wave) equation leads to a conditionally stable solution but gives rise to large diffusive error. The use of second order-accurate schemes leads to dispersive errors resulting in oscillations of the solution at large gradients (for example, in the region of shocks in highly compressible flows). Such oscillations are undesirable and may result, for example, in negative quantities for such variables as density, concentration and turbulent kinetic energy. Using even more (higher order) accurate schemes, such as the third-order accurate QUICK (quadratic upwind interpolation for convection kinematics) scheme of Leonard (1979) does not solve the problem and sustained oscillations may ensue. One approach to removing these oscillations is the use of artificial dissipation or artificial viscosity (Lax and Wendroff 1960) where a term proportional to the second or fourth derivative is added. Consider the Lax-Wendroff discretization of the first order wave equation:

$$u_i^{n+1} = u_i^n - {}^1/_2\,\sigma\left(u_{i+1}^n - u_{i-1}^n\right) + {}^1/_2\,\sigma^2\left(u_{i+1}^n - 2u_i^n + u_{i-1}^n\right) \tag{3.5.1}$$

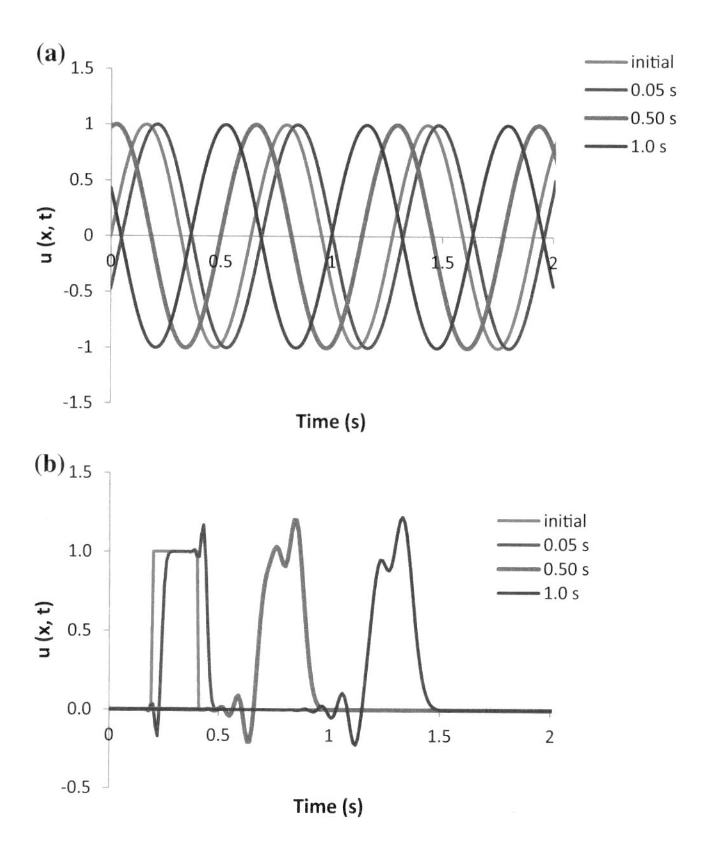

Fig. 3.22 Lax-Wendroff solution for $\sigma = 0.5$ for **a** a sinusoidal initial condition where smooth gradients do not produce new oscillations, and for **b** a rectangular pulse initial condition which produces new, unwarranted oscillations

which is second order accurate in both time and space. A typical solution (see Fig. 3.22) produces oscillations in regions of large gradients. However, if a second order dissipation term, D, of form

$$D = \mu_e \left(u_{i+1}^n - 2u_i^n + u_{i-1}^n \right) \tag{3.5.2}$$

where μ_e is the effective damping factor, is added to the right hand side of Eq. (3.5.1), then the resulting solution does not exhibit oscillations for certain values of the damping factor. It should be noted that since the damping term itself is evaluated to a second-order accuracy, the accuracy of the overall scheme is not affected. However, if μ_e is sufficiently high, then the stability condition ($\sigma \leq 1$) of the Lax-Wendroff scheme may be violated and the damping term gives rise to instability. If the damping is too low, then the oscillations may not completely die down. Also, the damping term introduces additional dissipation error resulting in the decrease of the amplitude with time.

To get around the problem of excessive dissipation introduced by damping term, Boris and Book (1973) introduced the flux corrected transport (FCT) method where some of the excessive damping is removed. The single-step Lax-Wendroff method is converted into a predictor-corrector type in which damping term is added in the predictor step and some of it is removed in the corrector step. The method is explained below for the one-dimensional form of the continuity equation, namely,

$$\partial\rho/\partial t + \partial(\rho u)/\partial x = 0 \tag{3.5.3}$$

Here, ρ is the unknown and u varies from grid point to grid point. From known nth time step values of ρ_i and u_i, ρ is advanced first by a half-step:

$$\rho^{n+\frac{1}{2}}_{i+1/2} = \frac{1}{2}\left(\rho^n_i + \rho^n_{i+1}\right) - \frac{\delta t}{(2\delta x)}\left(\rho^n_{i+1}u^n_{i+1} - \rho^n_i u^n_i\right) \tag{3.5.4}$$

These are then used to advance provisionally to the full step values:

$$\tilde{\rho}^{n+1}_i = \rho^n_i - \frac{\delta t}{\delta x}\left(\rho^{n+1/2}_{i+1/2}u^{n+1/2}_{i+1/2} - \rho^{n+1/2}_{i-1/2}u^{n+1/2}_{i-1/2}\right) \tag{3.5.5}$$

The fluxes, namely, ρu at the mid-points between grid nodes, are divided into diffusive fluxes (f^0) and anti-diffusive fluxes (f^1), and these are evaluated respectively from the previous step and the provisional full-step values:

$$f^0_{i+1/2} = \mu^0\left(\rho^n_{i+1} - \rho^n_i\right); \quad f^1_{i+1/2} = \mu^1\left(\tilde{\rho}^{n+1}_{i+1} - \tilde{\rho}^{n+1}_i\right) \tag{3.5.6a, b}$$

where μ^0 and μ^1 are parameters which need to be fixed. The diffusive flux is added to the provisional value to get the diffused, transported value denoted by an overbar, namely, $\bar{\rho}$:

$$\bar{\rho}^{n+1}_i = \tilde{\rho}^{n+1}_i + f^0_{i+1/2} - f^0_{i-1/2} \tag{3.5.7}$$

The antidiffusive fluxes are then limited based on the difference between the diffused, transported values:

$$f^C_{i+1/2} = S\max\{0, \min[S.\Delta_{i-1/2}, \left|f^1_{i+1/2}\right|, S.\Delta_{i+3/2}\} \tag{3.5.8}$$

where

$$S = \operatorname{sign}\left\{f^1_{i+1/2}\right\} \quad \text{and} \quad \Delta_{i+1/2} = \bar{\rho}^{n+1}_{i+1} - \bar{\rho}^{n+1}_i \tag{3.5.9}$$

The final solution is obtained by anti-diffusing the diffused, transported solution:

$$\rho_i^{n+1} = \overline{\rho}_i^1 - f_{i+1/2}^C + f_{i-1/2}^C \tag{3.5.10}$$

Book et al. (1975) and Boris and Book (1976) suggested a number of possibilities for the diffusivities appearing in the evaluation of the diffusive and anit-diffusive fluxes. Among those recommended are the following values for the solution of the continuity equation using the flux corrected transported method:

$$\mu^0 = 1/6\left(1 + 1/2\sigma^2\right) \quad \text{and} \quad \mu^1 = 1/6\left(1 - \sigma^2\right) \tag{3.5.11}$$

where σ is the Courant number. However, the optimum values of μ^0 and μ^1 may vary from case to case and often the damping is either too little or too excessive.

Ideally, damping terms should be used only where necessary, i.e., at locations where large gradients occur, and only as much as is necessary, i.e., just enough to damp out oscillations. Such localized treatment of dispersion effects has been made possible using the monotonicity and total variation diminishing (TVD) properties of numerical schemes, which have been analyzed systematically by Harten (1983, 1984). This led to the development of oscillation-free computation of large gradients for the linear and non-linear convection equation with second order accuracy and this is considered as one of the major achievements of the 1980s in the field of computational fluid dynamics (Hirsch 1988) of relevance to compressible flows. Here, we examine the concepts underlying these developments; a fuller treatment of the topic is given in Hirsch (1988) to which the interested reader is referred.

Consider the scalar conservation equation

$$\partial\phi/\partial t + \partial f_\phi/\partial x = 0 \tag{3.5.12}$$

where ϕ is the scalar and f_ϕ is the flux of ϕ, which consists of convective and diffusive fluxes for the generic scalar convection-diffusion equation, as discussed in Chap. 2. A numerical scheme is said to be monotonic if it does not produce oscillatory solutions. A monotonic scheme is monotonicity preserving, i.e., as time progresses, no new local extrema (minima or maxima) are created and the value of the local minimum is non-decreasing and that of a local maximum is non-increasing. This avoids the possibility of unphysical, oscillatory solution arising. Going back to the example considered in Sect. 3.2.1, an initially square pulse will remain more or less square without any wiggles around sharp corners. Satisfaction of monotonicity condition also ensures the entropy condition in inviscid but highly compressible flows is satisfied which ensures that unphysical expansion shocks are disallowed (see Sect. 4.3). Thus monotonicity is a highly desirable property of a numerical scheme.

It can be shown that a scheme of the form

$$\phi_i^{n+1} = \sum_k \left(b_k \phi_{i+k}^n\right) \tag{3.5.13a}$$

is monotonic if the following conditions are satisfied on the coefficients:

$$b_k \geq 0 \text{ for all } k \quad \text{and} \quad \sum_k b_k = 1 \tag{3.5.13b}$$

The first-order accurate upwind scheme for the linear wave equation (3.5.12) satisfying the stability condition ($\sigma \leq 1$) is monotone as it can be written as

$$\phi_i^{n+1} = \sigma\phi_{i-1}^n + (1-\sigma)\phi_i^n \tag{3.5.14}$$

The coefficients are always positive as $\sigma \leq 1$ for stability. A second order upwind scheme for the advection term results in an overall discretized equation which can be written in the form of Eq. (3.5.13) as follows:

$$\phi_i^{n+1} = \tfrac{1}{2}\sigma(1-\sigma)\phi_{i-2}^n + \sigma(2-\sigma)\phi_{i-1}^n + (1-\sigma/2)(\sigma-1)\phi_i^n \tag{3.5.15}$$

The stability condition for this scheme is $0 \leq \sigma \leq 2$. One can see that in the above expression for ϕ_i^{n+1} the first term is negative for $0 < \sigma < 1$ and the third term is negative for $1 < \sigma < 2$. Thus the condition for monotonicity, which requires all b_k to be positive, is violated for the second order upwind approximation. It can be shown similarly that higher order approximations for the advection term violate monotonicity condition. This puts a severe restriction on possible choices of discretization schemes if unphysical oscillatory solutions are to be avoided. A lesser, but one which is equally effective in avoiding oscillatory solutions, property is that of total variation diminishing (TVD).

If $\tilde{\phi}$ is a physically admissible solution of (3.5.12), then the total variation (TV) of the scalar, defined as

$$TV = \int |\partial\tilde{\phi}/\partial x| dx \tag{3.5.16}$$

is bounded and does not increase with time (Lax 1973). For a numerical solution, ϕ, where ϕ is evaluated at discrete points, the total variation can be defined as

$$TV(\phi) = \sum_i (|\phi_{i+1} - \phi_i|) \tag{3.5.17}$$

A numerical solution is said to be total variation stable or of bounded total variation if the total variation given by Eq. (3.5.17) is bounded in t and Δx, i.e., it should not increase with time or with decreasing Δx. A numerical solution is said to be total variation diminishing (TVD) if in the course of time (or iteration) marching

$$TV(\phi^{n+1}) \leq TV(\phi^n) \tag{3.5.18}$$

where n is the time step or iteration number index. In a TVD scheme, no new extrema can be produced as this would lead to an increase in TV as per Eq. (3.5.17). In this sense, TVD schemes are monotonicity preserving, i.e., if the scheme is monotonic at time step n, then it will be monotonic at (n + 1)th time step. Therefore, an oscillation-free solution for Eq. (3.5.12) can be obtained if a TVD scheme can be found for it. However, TVD schemes of higher than first order accuracy cannot be found if the coefficients b_k in Eq. (3.5.13) are linear. Consider the second-order upwind approximation for the convection term, $u\partial\phi/\partial x$, given by

$$(u\partial\phi/\partial x)_i = u/(2\Delta x)\,(3\phi_i - 4\phi_{i-1} + \phi_{i-2}) + O(\Delta x^2) \tag{3.5.19}$$

and a third order upwind approximation given by

$$(u\partial\phi/\partial x)_i = u/(6\Delta x)\,(11\phi_i - 18\phi_{i-1} + 9\phi_{i-2} - 2\phi_{i-3}) + O(\Delta x^3) \tag{3.5.20}$$

In order to develop TVD schemes of higher order accuracy, let us write a discretization of the spatial derivative in terms of spatial increments of ϕ, $\delta\phi_{i+1/2}$, defined as $\delta\phi_{i+1/2} = (\phi_{i+1} - \phi_i)$, in the following form:

$$(u\partial\phi/\partial x)_i = (u/\Delta x)\sum_{k=1}^{j}\left\{A_{i+k-1/2}\,\delta\phi_{i+k-1/2} + B_{i-k+1/2}\,\delta\phi_{i-k+1/2}\right\} \tag{3.5.21}$$

Here, A and B represent the contributions from waves with negative and positive wave speeds, respectively. A scheme of the form (3.5.21) would be TVD if and only if the coefficients A and B satisfy the condition that

$$A_{i+k-1/2} \le 0 \quad \text{and} \quad B_{i-k+1/2} \ge 0 \tag{3.5.22}$$

The second order scheme given by Eq. (3.5.19) above can be written in the above form as

$$(u\partial\phi/\partial x)_i = (u/\Delta x)\{3/2\,(\phi_i - \phi_{i-1}) - 1/2(\phi_{i-1} - \phi_{i-2})\}$$

or

$$(u\partial\phi/\partial x)_i = (u/\Delta x)\left\{3/2\,\delta\phi_{i-1+1/2} - 1/2\,\delta\phi_{i-2+1/2}\right\} \tag{3.5.23}$$

For this scheme, we have $A_1 = A_2 = 0$, $B_1 = 3u/2 > 0$ but $B_2 = -u/2 < 0$ as $u > 0$. Thus, the second-order upwind scheme does not satisfy the TVD condition. The QUICK (quadratic upwind interpolation for convection kinematics) of Leonard (1979) can be written as

$$(u\partial\phi/\partial x)_i = u/(8\Delta x)\,(3\phi_{i+1} + 3\phi_i - 7\phi_{i-1} + \phi_{i-2}) + O(\Delta x^3) \quad \text{for } u > 0 \tag{3.5.24}$$

which can be written in the form of Eq. (3.5.21) as

$$(u\partial\phi/\partial x)_i = (u/\Delta x)\{3/8(\phi_{i+1} - \phi_i) + 6/8(\phi_i - \phi_{i-1}) - 1/8(\phi_{i-1} - \phi_{i-2})\}$$

or

$$(u\partial\phi/\partial x)_i = (u/\Delta x)\left\{3/8\,\delta\phi_{i+1/2} + 6/8\,\delta\phi_{i-1+1/2} - 1/8\,\delta\phi_{i-2+1/2}\right\} \tag{3.5.25}$$

where the coefficients again do not satisfy the TVD condition given by Eq. (3.5.22). Similarly, it can be shown that any linear high order upwind scheme, for example, such as the one given by Eq. (3.5.20), is not TVD. However, a second- or higher-order accurate scheme can be made TVD by making it non-linear, i.e., making the coefficients A and B functions of ϕ_i as explained below.

Consider the second order upwind scheme given by Eq. (3.5.19). It can be rewritten as

$$(u\partial\phi/\partial x)_i = (u/2\Delta x)\,\{3 - (\phi_{i-1} - \phi_{i-2})/(\phi_i - \phi_{i-1})\}(\phi_i - \phi_{i-1})\} \tag{3.5.26}$$

which is of the form of Eq. (3.5.21) with

$$B_1 = B_{i-1+1/2} = (u/2)\{3 - (\phi_{i-1} - \phi_{i-2})/(\phi_i - \phi_{i-1})\} \tag{3.5.27}$$

The above scheme is not TVD if $B_1 < 0$ or if

$$(\phi_{i-1} - \phi_{i-2})/(\phi_i - \phi_{i-1}) > 3 \tag{3.5.28}$$

which may happen in regions of large gradients of ϕ. However, if the value of the coefficient B_1 is artificially restricted to being positive in such regions, then the TVD property can be maintained. This restriction of the value of B_1 makes the scheme locally first order accurate, but the solution would otherwise be oscillation-free. This restriction can be formally done by introducing limiting functions, Ψ, as explained below for the second order upwind scheme of Eq. (3.5.19). This equation can be rewritten as the sum of the first order upwind scheme and the extra terms associated with a second order scheme as:

$$(u\partial\phi/\partial x)_i = (u/\Delta x)\,\{(\phi_i - \phi_{i-1}) + 1/2\,(\phi_i - \phi_{i-1}) - 1/2\,(\phi_{i-1} - \phi_{i-2})\} \tag{3.5.29}$$

This equation is of the form of Eq. (3.5.21) with $B_1 = ½$ and $B_2 = -½$. In order to make the scheme (3.5.29) TVD, these coefficients are modulated by limiting functions $\Psi(r)$ as:

$$(u\partial\phi/\partial x)_i = (u/\Delta x)\{(\phi_i - \phi_{i-1}) + ½\Psi\left(r_{i-½}\right)(\phi_i - \phi_{i-1}) \\ -½\Psi\left(r_{i-3/2}\right)(\phi_{i-1} - \phi_{i-2})\} \tag{3.5.30}$$

where $r_{i-½}$ is defined as

$$r_{i-½} = (\phi_{i+1} - \phi_i)/(\phi_i - \phi_{i-1}) \tag{3.5.31}$$

Rewriting Eq. (3.5.30) as

$$(u\partial\phi/\partial x)_i = (u/\Delta x)\left\{1 + ½\Psi\left(r_{i-½}\right) - ½\Psi\left(r_{i-3/2}\right)/\left(r_{i-3/2}\right)\right\}(\phi_i - \phi_{i-1})\} \tag{3.5.32}$$

gives the TVD condition that

$$1 + ½\Psi\left(r_{i-½}\right) - ½\Psi\left(r_{i-3/2}\right)/\left(r_{i-3/2}\right) \geq 0 \tag{3.5.33}$$

While a number of functions can satisfy the above condition, the generality of $\Psi(r)$ is restricted by requiring that

$$\Psi(r) \geq 0 \text{ for } r \geq 0 \quad \text{and} \quad \Psi(r) = 0 \text{ for } r < 0. \tag{3.5.34}$$

Note that $r < 0$ indicates the occurrence of a local extremum and that setting $\Psi = 0$ makes the scheme only first order accurate. Under these conditions, the TVD condition given by Eq. (3.5.33) can be satisfied if $0 \leq \Psi(r) \leq 2$. A number of forms for $\Psi(r)$ are used, but three popular ones (Sweby 1984; Roe 1985) are given below and are shown schematically in Fig. 3.23a:

$$\text{van Leer:} \quad \Psi(r) = (r + |r|)/(1 + r) \tag{3.5.35a}$$

$$\text{Minmod:} \quad \Psi(r) = \text{Max}\{0, \min(r, 1)\} \tag{3.5.35b}$$

$$\text{Superbee:} \quad \Psi(r) = \text{Max}\{0, \min(2r, 1), \min(r, 2)\} \tag{3.5.35c}$$

The effect of the van Leer limiter on the linear convection of rectangular pulse is shown in Fig. 3.23b. One can see that compared to the Lax-Wendroff scheme shown in Fig. 3.22b, the rectangular shape is preserved well and no new maxima are produced as the pulse is convected.

The convection of a triangular initial function is compared in Fig. 3.24 with the three limiter functions. One can see that while no new maxima are created, there are differences with the three functions. The Minmod limiter gives a smooth variation but with a broader peak while the other two give a discontinuous slope in a few places. In all cases, it can be seen that no oscillations are produced, thus showing

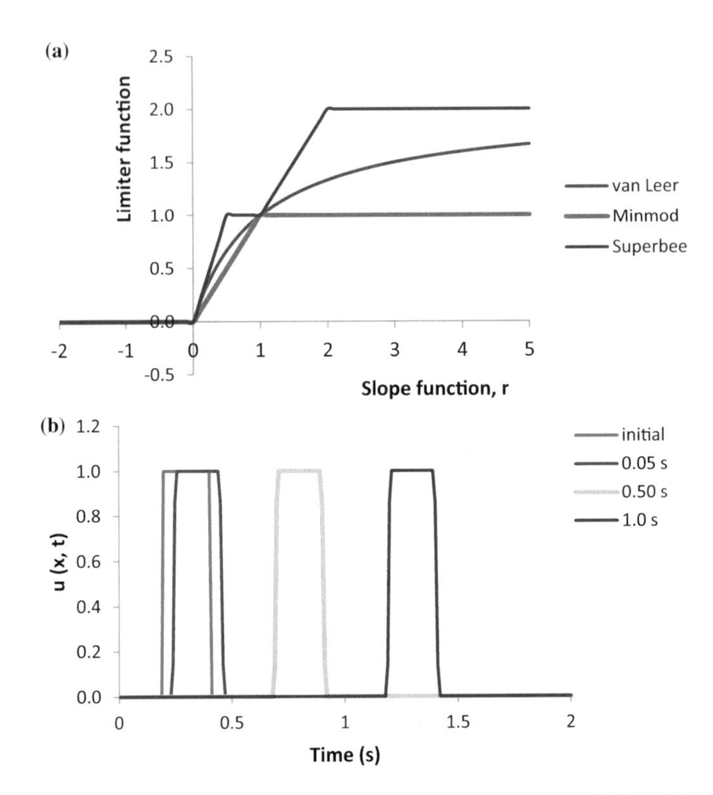

Fig. 3.23 **a** Graphical representation of three limiter functions, and **b** prediction of the computed solution for the linear advection problem

that all limiters are successful in this respect. It is left as an exercise to the reader to show that, with finer grid spacing, the step-wise variation in obtained in Fig. 3.24a, c would disappear.

While the TVD schemes have thus proved to be successful for the first order wave equation, the TVD property is valid only for scalar hyperbolic conservation equations and a general formulation is not available for non-homogeneous hyperbolic equations. Also, the formal extension of TVD schemes to multidimensions and systems of non-linear equations is yet to be established. Yet another area of development is the concept of essentially non-oscillating (ENO) schemes which attempt to preserve second-order accuracy even at extrema. For a thorough discussion of the TVD schemes, the reader is referred to the excellent reference of Hirsch (1988). Useful reviews and up-to-date results on the ENO schemes can be found in Harten et al. (1987) and Ivan and Groth (2014).

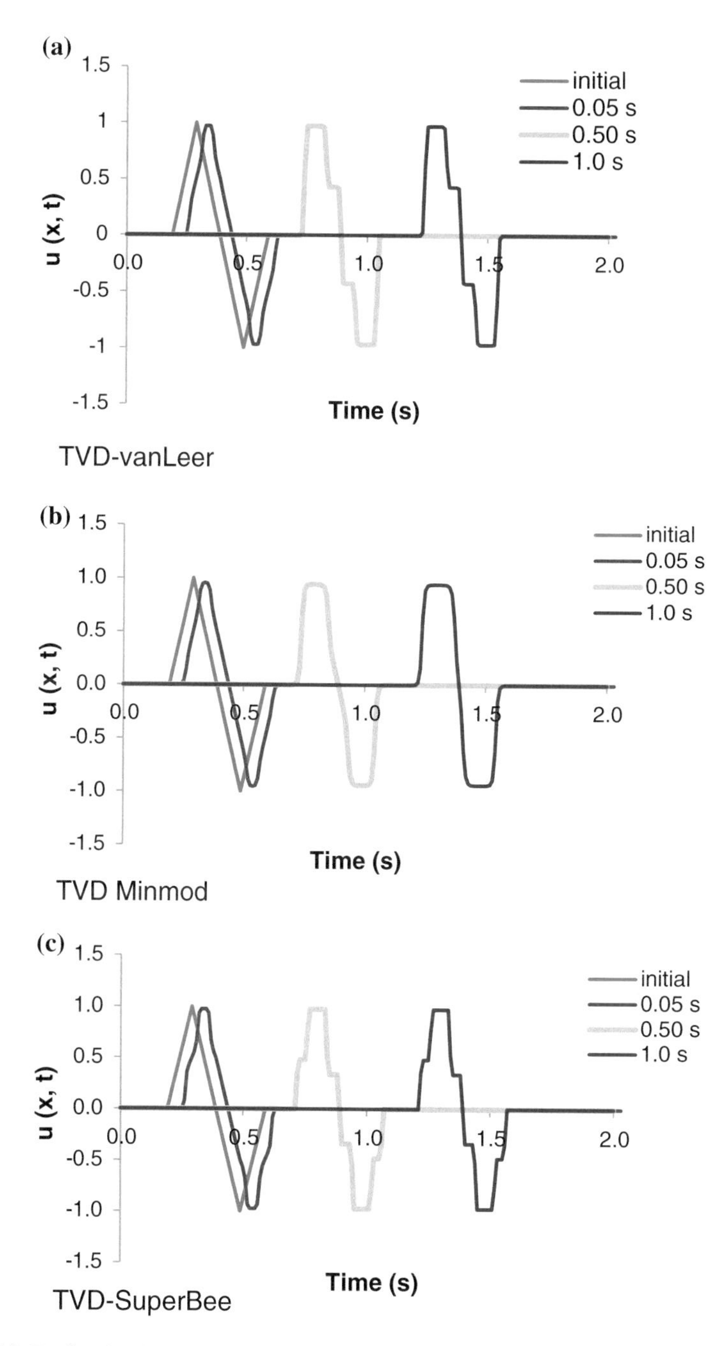

Fig. 3.24 Predicted solution for the linear convection equation for a triangular initial condition using **a** the van Leer, **b** the Minmod, and **c** the SuperBee limiter functions

3.6 Summary

In this chapter, we have seen a formal approach to deriving finite difference approximations to derivatives. When these are systematically applied to all the derivatives that occur in a differential equation, the latter can be converted into a finite difference approximation at a particular point. These difference equations can have a number of attributes such as order of accuracy of approximation, truncation error, consistency, stability, boundedness etc. These issues have been examined for typical model equations that mimic partly the nature of the equations that govern fluid flow. The point is made that getting a good discretization scheme is not a straightforward and trivial exercise and that care has to be taken to arrive at one that gives satisfactory performance. The issue of numerical errors, both diffusive and dispersive, that crop up as a collateral to finite difference approximation has also been studied at depth.

Two important concepts, namely, consistency and stability of a finite difference scheme for the solution of a partial differential equation, need to be mentioned again here for emphasis. It has been shown through the example of DuFort-Frankel scheme that making ad hoc approximations, even when these are consistent with the order of accuracy of the discretization, can lead to consistency problems. The powerful technique of von Neumann stability analysis has been presented to verify the stability of a given discretization scheme. However, this analysis is restricted to linear, well-posed initial value problems with periodic boundary conditions with uniform discretization interval, i.e., constant Δx, Δt etc. For non-linear problems, the method can be extended to a case where local linearization is made. For the more general case of non-periodic conditions and a non-uniform mesh, one can use other techniques such matrix stability analysis. The method can also be used to study the stability of system of equations. A thorough discussion of these is given in Hirsch (1988). In a number of practical cases, non-linearity and coupling preclude an accurate analysis of stability and empirical expressions are used. One such example is discussed in the next chapter.

We have also seen that even with consistent and stable schemes, we may have diffusion and dispersion errors. For flows with large gradients, as in the region of shocks in highly compressible flows, diffusive errors cause smearing of the shock while dispersion errors can give rise to wiggles and unphysical solution. We have seen that all high order schemes have the problem of wiggles and one way of suppressing these is to use non-linear flux limiter functions in regions where wiggles are likely to arise. This enables proper resolution of shocks in calculation methods incorporating shock capturing schemes.

We are now familiar with a number of issues that arise when we try to solve a single scalar transport equation. In the next chapter, we see how this knowledge can be used to solve the Navier-Stokes equations where we have to solve a set of coupled, non-linear partial differential equations.

Problems

1. Show through Taylor series expansion that the truncation error in the following finite difference approximations is as given for each case:

$$\text{(a)}\quad d^2u/\Delta x^2|_i = (-u_{i+2} + 16u_{i+1} - 30u_i + 16u_{i-1} - u_{i-2})/\left(12\,\Delta x^2\right) + \Delta x^4/90\, d^6u/\Delta x^6 + \cdots \tag{Q3.1}$$

$$\text{(b)}\quad d^3u/\Delta x^3|_i = (-3u_{i+4} + 14u_{i+3} - 24u_{i+2} + 18u_{i+1} - 5u_i)/\left(2\Delta x^3\right) + 21\Delta x^2/12\, d^5u/\Delta x^5 + \cdots \tag{Q3.2}$$

$$\text{(c)}\quad du/\Delta x|_i = (u_{i-2} - 7u_{i-1} + 3u_i + 3u_{i+2})/(8\Delta x) - \left(\Delta x^2/24\right)d^3u/\Delta x^3 + \cdots \tag{Q3.3}$$

2. Construct a finite difference approximation for the derivatives as per the following requirements:
   (a) forward, 2nd order accurate expression for the second derivative
   (b) central, 4th order accurate approximation for the first derivative
   (c) backward, 2nd order accurate expression for the third derivative
3. Consider a non-uniform grid in which the grid spacing varies in geometric progression, i.e., $\Delta x_i/\Delta x_{i-1} = \beta$, where, typically, $0.7 < \beta < 1.4$. For this case, derive second order accurate central difference formulae for (a) the first and (b) the second derivative.
4. Repeat Problem #3 by deriving second order accurate backward and forward difference approximations for the first and the second derivatives.
5. Consider the case of fully developed flow through a rectangular duct considered in Chap. 1. You take advantage of the symmetry of the problem and wish to do the computations only for one-quarter of the duct. Write the difference approximations for the grid points on all the four boundaries preserving second order accuracy of discretization.
6. Write a fourth order-accurate discretization for the Laplace equation.
7. Repeat Problem #6 with a fourth order-accurate treatment of the boundary conditions.
8. The FTCS approximation for the one-dimensional wave equation was rendered conditionally stable by Lax (1954) by substituting $u_i^n = (u_{i+1}^n + u_{i-1}^n)/2$. Investigate the accuracy, consistency and stability of the resulting Lax scheme.
9. The Lax-Wendroff (1960) scheme for the 1-d wave equation can be derived by expanding u_i^{n+1} in terms time derivatives at u_i^n:

$$u_i^{n+1} = u_i^n + \Delta t\, \partial u/\partial t|_i^n + \Delta t^2/2 \partial^2 u/\partial t^2|_i^n + \cdots, \tag{Q3.4}$$

substituting

$$\partial u/\partial t|_i^n = -a\, \partial u/\partial x|_i^n \quad \text{and} \quad \partial^2 u/\partial t^2|_i^n = a^2 \partial^2 u/\partial x^2 \tag{Q3.5}$$

and finally using central difference approximations for the spatial derivatives. Show that the resulting scheme is

$$u_i^{n+1} = u_i^n - \sigma/2\left(u_{i+1}^n - u_{i-1}^n\right) + \sigma^2/2\left(u_{i+1}^n - 2u_i^n + u_{i-1}^n\right) \tag{Q3.6}$$

Investigate its accuracy, consistency and stability.

10. Consider an initially hot rod of rectangular cross-section which is getting cooled by both axial conduction at the two ends and by convective heat loss along the length. Assume that for this thin, long rod, $L/D \gg 1$, so that the problem can be treated as one-dimensional in space. Assume a constant convective heat transfer coefficient between the hot rod and the ambient air. Formulate an unsteady one-dimensional heat conduction problem. Discretize the equation using FTCS-explicit, FTCS-implicit and Crank-Nicolson schemes and determine stability conditions.

11. Consider the one-dimensional non-linear wave equation

$$\partial\varphi/\partial t + \partial(u\varphi)/\partial x = \partial\varphi/\partial t + \partial f^c/\partial x = 0 \tag{Q3.7}$$

where f^c is the convective flux. Write the advection term at grid point i as

$$\partial(u\varphi)/\partial x|_i = \left(f_{i+1/2}^c - f_{i-1/2}^c\right)/\Delta x \tag{Q3.8}$$

where, assuming $u_{i+1/2} > 0$

$$f_{i+1/2}^c = u_{i+1/2}\left[\left(\varphi_i + \varphi_{i+1}\right)/2 + \{(1-\kappa)/4\}\left(-\varphi_{i-1} + 2\varphi_i - \varphi_{i+1}\right)\right] \quad \text{for} \quad u_{i+1/2} > 0 \tag{Q3.9}$$

$$f_{i+1/2}^c = u_{i+1/2}\left[\left(\varphi_i + \varphi_{i+1}\right)/2 + \{(1-\kappa)/4\}\left(-\varphi_i + 2\varphi_{i+1} - \varphi_{i+2}\right)\right] \quad \text{for} \quad u_{i+1/2} < 0 \tag{Q3.10}$$

This κ-scheme introduced by van Leer (1977) is such that $\kappa = 1$ corresponds to the central scheme, $\kappa = -1$ corresponds to the second order upwind scheme of Warming and Beam (1976), $\kappa = 0$ corresponds to the zero-average phase error scheme of Fromm (1968), κ = 1/2 gives the QUICK scheme of Leonard (1979) and κ = 1/3 gives the third order upwind scheme of Anderson et al. (1986). Repeat the case study given in Fig. 3.8 with each of these schemes with first order upwind scheme for the time derivative.

12. Time accuracy of solutions can be improved by using multistep methods borrowed from the solution techniques for ordinary differential equations. A Runge-Kutta procedure for the linear wave equation can be constructed as follows. Write the wave equation as

$$\partial u/\partial t = -a\partial u/\partial x = f(u). \tag{Q3.11}$$

Use the fourth-order Runge-Kutta (RK4) method to evaluate u_i^{n+1} from u_i^n as follows:

$$\begin{aligned}
\text{Step I} : u^{(1)} &= u^n + \Delta t/2\, f^n \\
\text{Step II} : u^{(2)} &= u^n + \Delta t/2\, f^{(1)} \\
\text{Step III} : u^{(3)} &= u^n + \Delta t\, f^{(2)} \\
\text{Step IV} : u^{n+1} &= u^n + \Delta t/6 \left\{ f^n + 2f^{(1)} + 2f^{(2)} + f^{(3)} \right\}
\end{aligned} \tag{Q3.12}$$

Write a computer program to evaluate the case discussed in Fig. 3.8 and explore the accuracy of the solution using the κ-schemes given in Problem #11.

13. Explore stability conditions of the RK4 method by trying different time steps in Problem #12.
14. Derive the TVD flux limiter conditions for different κ-values and repeat Problem #11.
15. Use as initial condition wave trains of the type shown in Fig. 3.11 over a part of the domain to study the dispersion and dissipation characteristics of the FTBS, FTCS-implicit, Lax, Lax-Wendroff and RK4 schemes for the wave equation. Note that if there is dispersion error, then the computed wave speed would depend on the wavelength.
16. Repeat the 1-d unsteady heat conduction case (Fig. 3.9) using the RK4 method.
17. Use the wave trains of Problem 15 as initial conditions and study the dissipation and dispersion characteristics for the FTCS-explicit, the DuFort-Frankel scheme and the Crank-Nicolson scheme for the 1-d unsteady equation.
18. Consider the 1-d unsteady heat conduction equation with the boundary condition that T(t, 0) = T(t, L) = 0 and the initial condition that

$$T(0, x) = C \sin(\pi x/L) \tag{Q3.13}$$

The exact solution for this problem is

$$T(t, x) = C \exp\left(-\alpha\pi^2 t/L^2\right) \sin(\pi x/L) \tag{Q3.14}$$

Compare the computed solution obtained using FTCS-explicit, FTCS-implicit and the Crank-Nicolson schemes with the exact solution. Explore the stability limits and accuracy of the solutions using different values of Δt and Δx.

19. Solve the 1-d unsteady convection-diffusion given by Eq. (3.4.7) using RK4 method and compare with the exact solution given by Eq. (3.4.8) for the corresponding initial and boundary conditions.
20. Several exact solutions of the one-dimensional Burgers equation are known (Benton and Platzman 1972). For the nonlinear partial differential equation

$$\partial u/\partial t + u\,\partial u/\partial x = \nu \partial^2 u/\partial x^2 \tag{Q3.15}$$

one exact solution is

$$u(t, x) = 1/\left[1 + \exp\{(2x - t)/(4\nu)\}\right] \tag{Q3.16}$$

Let us consider the solution in the interval [a, b]. The initial condition is then given by u(0, x); the left boundary condition by u(t, a) and the right boundary condition by u(t, b). For any such function u(t, x) satisfying Burgers' equation and a given interval [a, b], a corresponding initial-boundary value problem for which u(t, x) is the solution can be obtained by taking f(x) = u(0, x), g(t) = u(t, a), and h(t) = u(t, b).

Set up a problem, i.e., choose ν, a and b, to investigate, through computer simulation, to see how the dissipation and dispersion characteristics of the FTCSCS and the FTBSCS schemes affect the computed solution.

Chapter 4
Solution of Navier Stokes Equations

In the previous chapter, we have dealt with the discretization of a generic scalar transport equation. The solution of fluid flow requires simultaneous solution of the momentum and the continuity equations. For the general case, these constitute four second-order partial differential equations which are coupled in the sense that none of the equations can be solved independently. For example, if we wish to solve the conservation of momentum in the x-direction for u(i, j, k), we need to know the values of v and w at neighbouring grid points because these appear in the advection terms. This coupling poses some problems, especially in dealing with incompressible or nearly incompressible flows, which are of primary interest to the process engineer. A different type of problem, namely, a discontinuous solution, appears at the other end of the spectrum of highly compressible, hypersonic flows. These difficulties and some remedial measures are discussed in this chapter leading in each case to algorithms which enable the solution of the Navier Stokes equations for a time-dependent, three-dimensional problem.

We begin by first looking at the extension of the methods developed in the last chapter for the stability of a discretization scheme to the solution of a system of partial differential equations. Following this, we will discuss methods wherein these are directly applied for the solution of coupled equations, then discuss the difficulties associated with extension of these methods to flows with shocks and incompressible flows, and finally discuss special methods developed to deal with them.

4.1 Extension of Stability Analysis to Coupled Non-linear Equations

4.1.1 Solution of Coupled Equations

In practical fluid flow problems, we often solve a system of equations, for example, the continuity and the momentum equations for an isothermal flow problem. In

S. Jayanti, *Computational Fluid Dynamics for Engineers and Scientists*,
https://doi.org/10.1007/978-94-024-1217-8_4

order to see how we can apply the concepts of the previous chapter to such cases, let us consider the following set of first order equations:

$$\begin{aligned} \partial u/\partial t + c\partial v/\partial x &= 0 \\ \partial v/\partial t + c\partial u/\partial x &= 0 \end{aligned} \tag{4.1.1}$$

These equations look simple but contain the essential elements of coupling between the two equations, which is the point of interest. (One can show that, taken together, they represent the second order wave equation, namely, $\partial^2 u/\partial t^2 - c^2\partial^2 u/\partial x^2 = 0$.) Let us write the system of equations in a matrix form as follows:

$$\partial \mathbf{U}/\partial t + \mathbf{A}\,\partial \mathbf{U}/\partial x = 0 \text{ where } \quad \mathbf{U} = [u \quad v]^T; A = \begin{pmatrix} 0 & c \\ c & 0 \end{pmatrix} \tag{4.1.2}$$

For a given discretization scheme, Eq. (4.1.2) can be written as

$$\mathbf{U}^{n+1} = \mathbf{C}\mathbf{U}^n + \mathbf{Q} \tag{4.1.3}$$

where C is the discretization operator of the scheme. For example, choosing the explicit FTCS discretization scheme, we can write Eq. (4.1.2) as

$$\mathbf{U}_i^{n+1} = \mathbf{U}_i^n - \mathbf{A}\,\Delta t/(2\Delta x)\left(\mathbf{U}_{i+1}^n - \mathbf{U}_{i-1}^n\right) \tag{4.1.4}$$

where $\mathbf{U}_i^n = [u_i^n \ v_i^n]^T$. Using the shift (or displacement) operator E defined as

$$Eu_i = u_{i+1} \quad \text{and} \quad E^{-1}u_i = u_{i-1} \tag{4.1.5}$$

Equation (4.1.4) can be written in the form of Eq. (4.1.3) with

$$\mathbf{C} = \mathbf{I} - \mathbf{A}\Delta t/(2\Delta x)\left(E - E^{-1}\right) \quad \text{and} \quad Q = 0 \tag{4.1.6}$$

In order to check for stability, we follow the earlier von Neumann analysis and seek a solution to (4.1.3) of the form

$$\mathbf{U}_i^n = \sum_m \left[\mathbf{U}_m^{n\Delta t} \exp(jk_m i\Delta x)\right] \tag{4.1.7}$$

where k_m is the wave number of the mth harmonic in the Fourier expansion. The stability of Eq. (4.1.3) is governed by its homogeneous part. Substituting Eq. (4.1.7) into it and examining the mth wave component, we obtain

$$\mathbf{U}_m^{(n+1)\Delta t} \exp(jk_m i\Delta x) = \mathbf{C}\mathbf{U}_m^{n\Delta t} \exp(jk_m i\Delta x) = \mathbf{G}(\varphi)\mathbf{U}_m^{n\Delta t} \exp(ji\varphi)$$

where $\varphi = k_m\Delta x$. Dividing by $\exp(jk_m i\Delta x)$ and dropping the subscript m for the wave component, we get

$$\mathbf{U}^{n+1} = \mathbf{G}(\varphi)\mathbf{U}^n \tag{4.1.8}$$

If U^0 represents the initial solution at time equal to zero, then Eq. (4.1.8) implies

$$\begin{aligned}
\mathbf{U}^1 &= \mathbf{G}\cdot\mathbf{U}^0 \\
\mathbf{U}^2 &= \mathbf{G}\cdot\mathbf{U}^1 = \mathbf{GGU}^0 = (\mathbf{G})^2\cdot\mathbf{U}^0 \\
&. \\
&. \\
\mathbf{U}^n &= (\mathbf{G})^n\cdot\mathbf{U}^0
\end{aligned}$$

Stability requires that $(\mathbf{G})^n$ should remain bounded for all values of ϕ, i.e.,

$$||\mathbf{G}|| < K \text{ for all } \varphi$$

where $||\mathbf{G}||$ is some norm of the matrix $\mathbf{G}$. It can be shown that if $\rho(\mathbf{G})$ is the spectral radius, the largest eigenvalue, of $\mathbf{G}$, then

$$||\mathbf{G}||^n \geq ||\mathbf{G}^n|| \geq \rho^n.$$

Thus, a *necessary* condition for stability is that the spectral radius of the amplification matrix $\mathbf{G}$ should be less than unity, or

$$\rho(\mathbf{G}) \leq 1 \quad \text{for all } \varphi \tag{4.1.9}$$

Let us apply this to the set of equations given by Eq. (4.1.1) discretized using the FTCS scheme as given by (4.1.4). The discretized equations can be written as

$$u_i^{n+1} = u_i^n - c\Delta t/(2\Delta x)\left(v_{i+1}^n - v_{i-1}^n\right) \tag{4.1.10a}$$

and

$$v_i^{n+1} = v_i^n - c\Delta t/(2\Delta x)\left(u_{i+1}^n - u_{i-1}^n\right) \tag{4.1.10b}$$

Putting Eqs. (4.1.10a, b) in the form of Eq. (4.1.3), i.e., as

$$\mathbf{U} = \left[u_i^{n+1}\; v_i^{n+1}\right]^T = \mathbf{C}\left[u_i^n\; v_i^n\right]^T + \mathbf{Q} \tag{4.1.11a}$$

we obtain

$$\mathbf{C} = \begin{pmatrix} 1 & -\frac{c\Delta t}{2\Delta X}\left(E - E^{-1}\right) \\ -\frac{c\Delta t}{2\Delta X}\left(E - E^{-1}\right) & 1 \end{pmatrix} \quad \text{and} \quad \mathbf{Q} = 0 \tag{4.1.11b}$$

Putting $u_i^n = u^n \exp(jk_m i\Delta x)$ and $v_i^n = v^n \exp(jk_m i\Delta x)$ and substituting in (4.1.10a, b), we get

$$u^{n+1}\exp(jk_m i\Delta x) = u^n \exp(jk_m i\Delta x) - c\Delta t/(2\Delta x) v^n \exp(jk_m i\Delta x)[\exp(jk_m \Delta x) - \exp(-jk_m \Delta x)] \quad (4.1.12a)$$

$$v^{n+1}\exp(jk_m i\Delta x) = v^n \exp(jk_m i\Delta x) - c\Delta t/(2\Delta x) u^n \exp(jk_m i\Delta x)[\exp(jk_m \Delta x) - \exp(-jk_m \Delta x)] \quad (4.1.12b)$$

Dividing (4.1.12a, b) by $\exp(jk_m i\Delta x)$, and putting it in matrix form, we obtain

$$\mathbf{U}^{n+1} = \mathbf{G}\,\mathbf{U}^n \quad (4.1.13a)$$

where $\mathbf{U}^{n+1} = [u^{n+1}\ v^{n+1}]^T$; $\mathbf{U}^n = [u^n\ v^n]^T$ and

$$\mathbf{C} = \begin{pmatrix} 1 & -\frac{c\Delta t}{2\Delta X}\left(e^{j\varphi} - e^{-j\varphi}\right) \\ -\frac{c\Delta t}{2\Delta X}\left(e^{j\varphi} - e^{-j\varphi}\right) & 1 \end{pmatrix} \quad \text{and} \quad \mathbf{Q} = 0 \quad (4.1.13b)$$

Noting that $(e^{j\varphi} - e^{-j\varphi}) = 2j\sin\varphi$ etc. and setting $\sigma = c\Delta t/\Delta x$, the amplification matrix can be written as

$$G = \begin{pmatrix} 1 & -j\sigma \sin\varphi \\ -j\sigma \sin\varphi & 1 \end{pmatrix} \quad (4.1.14)$$

The stability of the numerical scheme is governed by the eigenvalues of the amplification matrix **G**. These can be determined by setting det $|\mathbf{G} - \lambda\mathbf{I}| = 0$ where λ is the eigenvalue and **I** is the unit matrix. Evaluating these, we find that

$$\begin{vmatrix} 1-\lambda & -j\sigma\sin\varphi \\ -j\sigma\sin\varphi & 1-\lambda \end{vmatrix} = 0 \quad \text{or } (1-\lambda)^2 + \sigma^2\sin^2\varphi = 0.$$

Hence the two eigenvalues are

$$\lambda = 1 \pm j\sigma\sin\varphi \quad (4.1.15)$$

The spectral radius then is

$$\rho(\mathbf{G}) = 1 + \sigma^2\sin^2\varphi \geq 1 \quad (4.1.16)$$

which is to be expected since a central scheme is being used to discretize the spatial derivative. It can be readily verified that for an FTBS scheme, the amplification matrix would be

$$G = \begin{pmatrix} 1 & -\sigma(1-\cos\varphi + j\sin\varphi) \\ -\sigma(1-\cos\varphi + j\sin\varphi) & 1 \end{pmatrix} \tag{4.1.17}$$

the two eigenvalues of which are

$$\lambda = [1 \pm \sigma(1-\cos\varphi)] + j\,\sigma\sin\varphi \tag{4.1.18}$$

and spectral radius is

$$\rho(\mathbf{G}) = [1 \pm \sigma(1-\cos\varphi)]^2 + \sigma^2\sin^2\varphi \tag{4.1.19}$$

Since the spectral radius is <1 only if $\sigma < 1$, the scheme is conditionally stable. A number other equations and their stability have been investigated in Hirsch (1988). The matrix method of stability analysis can also be used for assessing the effect of non-periodic boundary conditions and non-uniform grid spacing. Here the spatially and temporally discretized equation with boundary conditions and non-uniform grid relevant to the problem is posed in the form of Eq. (4.1.3), and the stability is investigated by finding the eigenvalues of the corresponding **C** matrix. It may be noted that while this is possible in theory, determination of the eigenvalues of a large matrix is not a trivial exercise, and the computational effort for finding these may preclude a stability analysis of the fully-discretized flow problem. Non-linearity of equations also limits the scope of stability analysis. In such cases, results from von Neumann analysis can be taken as guidelines for assessing stability of the discretization schemes in the neighbourhood of u_i^n.

4.1.2 Solution of Non-linear Equations

The Navier-Stokes equations are quasi-linear partial differential equations in the sense that the highest derivative terms, namely, the second derivative terms representing the diffusion of momentum, are linear when the diffusivities are constant. The principal non-linearity for laminar flow arises from the advection term. Consider the x-momentum equation for a constant-property flow in Cartesian coordinates:

$$\begin{aligned} &\partial u/\partial t + \partial(u^2)/\partial x + \partial(uv)/\partial y + \partial(uw)/\partial z = \\ &\quad -(1/\rho)\partial p/\partial x + (\mu/\rho)\left(\partial^2 u/\partial x^2 + \partial^2 u/\partial y^2 + \partial^2 w/\partial z^2\right) \end{aligned} \tag{4.1.20}$$

Here, the second term on the left hand side is non-linear while the third and the fourth terms, which require knowledge of v and w when we are solving for u, present coupling problems. For non-constant properties, the evaluation of the fluid properties brings in coupling with other equations, such as the energy equation if the viscosity (or density) is a strong function of temperature or if the composition of the fluid is changing because of some chemical reaction, etc. As we shall see in

Chap. 6, this coupling may become non-linear for the diffusion terms on the right hand side in case of turbulent flows, where the effective viscosity depends on velocity gradients in a highly non-linear way. The continuity equation for a species may also become a highly non-linear function of the energy equation due to the strong coupling that exists between the temperature and the chemical reaction rate. There are three standard ways of dealing with these issues in CFD:

- In the *deferred correction* approach, the unknown terms are shifted to the right hand side and are evaluated based on previous iteration value. When this is done, the corresponding term does not enter the computation molecule and has no role to play in the stability of the discretization scheme. Usually source terms are treated in this way, especially when they otherwise lead to loss of diagonal dominance of the system of equations.
- In the *Picard substitution* approach, one ignores non-linearity and coupling by using estimated values for the unknown variables. Given that we often employ a sequential solution approach for the solution of NSE, we can rewrite Eq. (4.1.20) as

$$\begin{aligned}&\partial u/\partial t+\partial(u^*u)/\partial x+\partial(uv^*)/\partial y+\partial(uw^*)/\partial z=\\&\quad-(1/\rho)\partial p/\partial x+(1/\rho)[\partial/\partial x\ (\mu^*\partial u/\partial x)\\&\quad+\partial/\partial y\ (\mu^*\partial u/\partial y)+\partial/\partial z(\mu^*\partial u/\partial z)]\end{aligned} \tag{4.1.21}$$

where the starred quantities are estimated quantities which take the value of the previous iteration or are initialized using some simple estimates to begin the iterative calculation.

- In *Newton linearization*, the non-linear terms are formally linearized as follows. Consider a non-linear algebraic function f(u). It can be expanded in terms of its value in the neighbourhood of u* as

$$f(u)=f(u^*)+(\partial f/\partial u)|_{u*}\Delta u \text{ where } \quad \Delta u=u-u^* \tag{4.1.22}$$

For the specific case of

$$f(u)=u^2 \tag{4.1.23}$$

we have

$$u^2=u^{*2}+2u^*\Delta u \tag{4.1.24a}$$

or

$$u^2=2u^*u-u^{*2} \tag{4.1.24b}$$

The approximation of (4.1.24a) can be used where the variable to be solved for is Δu (as in the case of the Beam-Warming scheme, see Sect. 4.2.2). Equation (4.1.24b) can be used when the variable to be solved for is u itself. The linearization can be extended to coupled terms. For example, for the case where f is a function of both u and v, we have

$$f(u, v) = f(u^*, v^*) + (\partial f/\partial u)\big|_{u^*,v^*} \Delta u + (\partial f/\partial v)\big|_{u^*,v^*} \Delta v \tag{4.1.25}$$

For the specific case of

$$f = uv \tag{4.1.26a}$$

we have

$$uv = u^* v^* + v^* \Delta u + u^* \Delta v \tag{4.1.26b}$$

or

$$uv = v^* u + u^* v - u^* v^*$$

Newton's linearization would be effective in coupled solution approaches (see Sect. 4.4). For sequential solvers (to be discussed later), it is simpler to use Picard's iterative update approach, especially in turbulent flows. The deferred correction approach is often used for dealing with cross-derivative or curvature terms appearing in non-orthogonal coordinate systems (see Chap. 6) or for source terms of a certain type. In these cases, the inclusion of certain terms in the governing equation may lead to loss of diagonal dominance of the coefficient matrix of the set of discretized equations. Diagonal dominance is a desirable feature for many linear equation solvers (this will be discussed in detail in Chap. 5); by employing deferred correction, the diagonal dominance feature can be preserved.

4.2 Solution of Coupled Equations for Compressible Flows

At the outset, we note that compressible flow is a flow property than a fluid property. That is, there can be incompressible flows of a compressible fluid such as air. The distinguishing feature is that, in compressible flows, the flow velocity changes are large enough to cause a corresponding pressure variation (for example, through the Bernoulli equation) which in turn leads to significant density changes. If the characteristic velocity of the flow is fairly small, then only small changes in pressure may be created in the flow field and the induced changes in the density of the fluid may be negligibly small. As a rough indication, the characteristic flow velocity has to be to high enough for the Mach number, which is the ratio of the flow velocity to the speed of sound in the fluid medium, to be 0.3 or higher. In ambient air conditions, flows involving air velocities of 100 m/s or higher can be considered as compressible; for water, the flow velocity has be of the order of 300 m/s for compressibility effects to be felt significantly. Therefore liquid flows are usually considered as incompressible flows.

Compressible flows enjoy an advantage over incompressible flows in terms of the coupling of the governing equations. In flows with varying density, the continuity and the momentum equations taken together contain five variables, namely, the three velocity components, the pressure and the density. (A proper formulation of compressible flow also includes the energy equation which brings in an additional variable such as the enthalpy or the temperature.) These are determined by four partial differential equations expressing the conservation of mass and momentum and one equation of state, often of algebraic form (for example, the ideal gas law) which links the density, pressure and temperature of the fluid. This provides a natural linkage between the continuity equation (which does not contain pressure) and the momentum balance equations in which pressure is the main driving force. This enables us to develop methods for the coupled system of equations which are natural extensions of the methods discussed in Chap. 3. In this section, we will study a couple of classical methods for the solution of the compressible N-S equations.

4.2.1 Explicit MacCormack Method

This method was originally developed for solving the non-linear wave equation (MacCormack 1969), and is a two-step method consisting of a predictor step and a corrector step. For the first order one-dimensional wave equation, namely, $\partial u/\partial t + c\partial u/\partial x = 0$, it can be written as

$$\text{Predictor step}: \overline{u_i^{n+1}} = u_i^n - c\Delta t/\Delta x\left(u_{i+1}^n - u_i^n\right) \tag{4.2.1a}$$

$$\text{Corrector step}: u_i^{n+1} = \frac{1}{2}\left(\overline{u_i^{n+1}} + u_i^n\right) - c\Delta t/\Delta x\left(\overline{u_i^{n+1}} - \overline{u_{i-1}^{n+1}}\right) \tag{4.2.1b}$$

The predictor step gives an estimate of u_i^{n+1}; since this is only an estimate, it is denoted as $\overline{u_i^{n+1}}$. Once this is computed for the entire field, i.e., for all i, then a corrected value of u_i^{n+1} is computed using Eq. (4.2.1b). It may also be noted that in the predictor step, the spatial derivative term is approximated using a forward difference formula while in the corrector step, this is done using a backward differencing formula. On the whole, the scheme is second order accurate in both time and space and has the stability condition that the Courant number (defined as $c\Delta t/\Delta x$) ≤ 1. The evaluation of the velocity at the intermediate time step makes it suitable for non-linear problems. Consider the one-dimensional inviscid Burgers equation which is the non-linear form of the first order wave equation:

$$\partial u/\partial t + u\partial u/\partial x = 0 \tag{4.2.2}$$

This can be written in terms of momentum flux f (which contains the non-linear term) as

$$\partial u/\partial t + \partial f/\partial x = 0 \quad \text{where } f = u^2/2. \tag{4.2.3}$$

The application of second-order accurate schemes such as the Lax-Wendroff scheme (see Sects. 3.4 and 3.5) to the above equation would have required the evaluation of the Jacobian $A = \partial f/\partial u$. While this is not a problem when dealing with a single equation, it can be trickier when we have to extend the method to coupled equations such as the Euler equations or the Navier-Stokes equations. In order to get round this difficulty, Richtmyer and Morton (1967) introduced a two-step procedure in which the first step evaluates u at (i + ½, n + ½) and this is used in the second step to evaluate the fluxes at i + ½ and i − ½ which are then used to evaluate u at (i, n + 1). The overall scheme is second order accurate in both space and time and can be summarized as below:

$$u_{i+1/2}^{n+1/2} = \frac{1}{2}\left(u_i^n + u_{i+1}^n\right) - \left(\frac{\Delta t}{2\Delta x}\right)\left(f_{i+1}^n - f_i^n\right) \tag{4.2.4a}$$

$$u_i^{n+1} = u_i^n - \left(\frac{\Delta t}{\Delta x}\right)\left(f_{i+1/2}^{n+1/2} - f_{i-1/2}^{n+1/2}\right) \tag{4.2.4b}$$

The overall scheme is of second order accuracy in both time and space for u_i^{n+1}. The MacCormack scheme belongs to this Richtmyer class of schemes; its explicit version for Eq. (4.2.3) can be written as

$$\text{Predictor}: \overline{u_i^{n+1}} = u_i^n - \left(\frac{\Delta t}{\Delta x}\right)\left(f_{i+1}^n - f_i^n\right) \tag{4.2.5a}$$

$$\text{Corrector step}: u_i^{n+1} = \frac{1}{2}\left(\overline{u_i^{n+1}} + u_i^n\right) - \frac{\Delta t}{2\Delta x}\left(\overline{f_i^{n+1}} - \overline{f_{i-1}^{n+1}}\right) \tag{4.2.5b}$$

Here, the fluxes in the corrector step are evaluated using the velocities obtained in the predictor step. The scheme can also be written as follows to bring out the predictor and corrector nature of the steps:

$$\overline{u_i^{n+1}} = u_i^n - \left(\frac{\Delta t}{\Delta x}\right)\left(f_{i+1}^n - f_i^n\right) \tag{4.2.6a}$$

$$\overline{\overline{u_i^{n+1}}} = u_i^n - \left(\frac{\Delta t}{\Delta x}\right)\left(f_i^{\bar{n}} - f_{i-1}^{\bar{n}}\right) \tag{4.2.6b}$$

$$u_i^{n+1} = \frac{1}{2}\left(\overline{u_i^{n+1}} + \overline{\overline{u_i^{n+1}}}\right) \quad (4.2.6c)$$

Here, Eq. (4.2.6c) can be considered as the updating of the values of u_i obtained in the predictor and the corrector steps. Now let us consider the extension of the MacCormack scheme to the solution of the Navier-Stokes equations for compressible flows. One may recall that compressible flows often require the solution of the energy conservation equation along with the mass and momentum conservation equations. Thus, one needs to solve a set of four coupled partial differential equations in two dimensions and a set of five coupled partial differential equations in three dimensions. In addition, an equation of state, linking the density, pressure and temperature, also needs to be used along with the partial differential equations. Let us see how we can use the explicit MacCormack method for the solution of these coupled equations.

Considering an unsteady, 3-D flow through a domain described in Cartesian coordinates, we can write the set of partial differential equations as follows

$$\partial U/\partial t + \partial E/\partial x + \partial F/\partial y + \partial G/\partial z = 0 \quad (4.2.7)$$

where U, E, F and G are vectors given by

$$U = \begin{bmatrix} \rho \\ \rho u \\ \rho v \\ \rho w \\ \rho e_t \end{bmatrix}; \quad E = \begin{bmatrix} \rho u \\ \rho u^2 + p - \tau_{xx} \\ \rho uv - \tau_{yx} \\ \rho uw - \tau_{zx} \\ (\rho e_t + p)u - u\tau_{xx} - v\tau_{yx} - w\tau_{zx} + q_x \end{bmatrix} \quad (4.2.8)$$

$$F = \begin{bmatrix} \rho v \\ \rho u v - \tau_{xy} \\ \rho v^2 + p - \tau_{yy} \\ \rho vw - \tau_{zy} \\ (\rho e_t + p)v - u\tau_{xy} - v\tau_{yy} - w\tau_{zy} + q_y \end{bmatrix};$$

$$G = \begin{bmatrix} \rho w \\ \rho E - \tau_{xz} \\ \rho vw - \tau_{yz} \\ \rho w^2 + p - \tau_{zz} \\ (\rho e_t + p)w - u\tau_{xz} - v\tau_{yz} - w\tau_{zz} + q_z \end{bmatrix}$$

The elements of the U vector, namely, ρ, ρu, ρv, ρw and ρe_t, are the five conserved variables of the equations of conservation of mass, x-, y- and z-momentum and energy, respectively. Here, e_t is the total energy per unit mass and is defined as $e_t = e + (u^2 + v^2 + w^2)/2$. It may be recalled that here τ_{ij} is the shear stress component which is given, assuming the fluid to be Newtonian, as before, by $\tau_{ij} = \mu(\partial u_i/\partial x_j + \partial u_j/\partial x_i) + \lambda \partial u_k/\partial x_k$. The term q_i is the conductive heat flux in the ith direction and is given by Fourier's law of heat conduction as $q_i = -k\partial T/\partial x_i$

where k is the thermal conductivity of the fluid. Equation (4.2.7) can be viewed as a three-dimensional version of Eq. (4.2.3) although the former has vectors (U, E, F, G) as its variables. The MacCormack method for the solution of Eq. (4.2.7) can thus be written in a predictor-corrector format as

$$\begin{aligned}\text{Predictor step}: \overline{U^{n+1}_{i,j,k}} &= U^{n}_{i,j,k} - \left(\frac{\Delta t}{\Delta x}\right)\left(E^{n}_{i+1,y,z} - E^{n}_{i,j,k}\right) - \left(\frac{\Delta t}{\Delta y}\right)\left(F^{n}_{i,y+1,z} - F^{n}_{i,j,k}\right)\\ &\quad - \left(\frac{\Delta t}{\Delta z}\right)\left(G^{n}_{i,y,z+1} - G^{n}_{i,j,k}\right)\end{aligned} \tag{4.2.9}$$

$$\begin{aligned}\text{Corrector step}: U^{n+1}_{i,j,k} &= \frac{1}{2}\left[U^{n}_{i,j,k} + \overline{U^{n+1}_{i,j,k}}\right] - \left(\frac{\Delta t}{2\Delta x}\right)\left(\overline{E^{n+1}_{i,y,z}} - \overline{E^{n+1}_{i-1,y,z}}\right)\\ &\quad - \left(\frac{\Delta t}{2\Delta y}\right)\left(\overline{F^{n+1}_{i,y,z}} - \overline{F^{n+1}_{i,y-1,z}}\right) - \left(\frac{\Delta t}{2\Delta z}\right)\left(\overline{G^{n+1}_{i,y,z}} - \overline{G^{n+1}_{i,y,z-1}}\right)\end{aligned} \tag{4.2.10}$$

It may be noted that the "fluxes", E, F and G are evaluated using forward differencing in the predictor step and using backward differencing in the corrector step or vice versa. While in the simple case considered in Eq. (4.2.3) only advective flux is present, diffusive (viscous) fluxes, resulting from second derivative terms (such as $\partial^2 u/\partial x^2$), will also be present in the Navier-Stokes equations. These will have to be evaluated carefully in order to preserve the second order accuracy of the discretization. Thus, the normal derivative terms of the viscous stresses will have to be evaluated using the combination of forward and backward differences while the cross derivative terms will have to be evaluated using central differencing approximation in both the predictor and the corrector steps. In other words, the x-derivative terms in E, the y-derivative terms in F and the z-derivative terms in G are evaluated using alternately forward and backward differences in the predictor and the corrector steps. The other terms, for example, the y- and z-derivative terms in E, the x- and the z-derivative terms in F etc., are evaluated using central difference approximations in both the predictor and the corrector steps. As a specific example, let us consider terms of F appearing in the x-momentum equation in (4.2.7). These can be written as

$$(F)_x = \rho uv - \mu\partial u/\partial y - \mu\partial v/\partial x. \tag{4.2.11}$$

Since the F-term appears in Eq. (4.2.7) as $\partial u/\partial y$, the second term of Eq. (4.2.11) is a normal derivative (the overall term becomes $-\mu\partial^2 u/\partial^2 y$ when μ is constant) when it appears in the F-term and the last term is the cross-derivative (it is equivalent to $-\mu\partial^2 v/\partial x\partial y$). Therefore the second term should be approximated using forward and backward difference approximations while the last term should be evaluated using central differences in both the predictor and the corrector steps.

The first term, which does not contain any derivative, is evaluated as the local value. Thus, the term $(F)_x$ of Eq. (4.2.11) corresponding to the x-momentum equation is thus approximated as follows:

$$\text{Predictor step}: (F_x)^n_{i,j,k} = (\rho uv)^n_{i,j,k} - \mu\frac{u^n_{i,j,k} - u^n_{i,j-1,k}}{\Delta y} - \mu\frac{v^n_{i+1,j,k} - v^n_{i-1,j,k}}{2\Delta x} \tag{4.2.12}$$

$$\text{Corrector step}: (F_x)^{\overline{n+1}}_{i,j,k} = (\rho uv)^{\overline{n+1}}_{i,j,k} - \mu\frac{u^{\overline{n+1}}_{i,j+1,k} - u^{\overline{n+1}}_{i,j,k}}{\Delta y} - \mu\frac{v^{\overline{n+1}}_{i+1,j,k} - v^{\overline{n+1}}_{i-1,j,k}}{2\Delta x} \tag{4.2.13}$$

It may be noted that the approximation for $\partial u/\partial y$ in the predictor step corresponds to a backward difference approximation at the point (i, j, k) while that in the corrector step corresponds to a forward difference approximation at the point (i, j − 1, k).

As a further example, consider the F-term appearing in the y-momentum equation of Eq. (4.2.7) which is given by

$$(F)_y = \rho v^2 + p - 2\mu\partial v/\partial y \tag{4.2.14}$$

The first two terms on the right hand side are evaluated as point values at the appropriate grid node while the third term is evaluated using forward and backward approximations in the predictor and the corrector steps as it is a normal derivative. Thus, it is approximated as

$$\text{Predictor step}: \left(F_y\right)^n_{i,j,k} = \left(\rho v^2 + p\right)^n_{i,j,k} - 2\mu\frac{v^n_{i,j,k} - v^n_{i,j-1,k}}{\Delta y} \tag{4.2.15}$$

$$\text{Corrector step}: \left(F_y\right)^{\overline{n+1}}_{i,j,k} = \left(\rho v^2 + p\right)^{\overline{n+1}}_{i,j,k} - 2\mu\frac{v^{\overline{n+1}}_{i,j+1,k} - v^{\overline{n+1}}_{i,j,k}}{\Delta y} \tag{4.2.16}$$

The overall approximation given by Eq. (4.2.7) is second order accurate in both time and space. Since the method is explicit, stability limit on the time step can be expected. Because of the complexity of the method arising out of non-linear terms and coupling among the equations, it is not possible to make an exact calculation of the stability condition. Tannehill et al. (1975) derived the following empirical formula for the allowable time step value:

$$\Delta t \leq \sigma\Delta t_{CFL}/(1 + 2/Re_\Delta) \tag{4.2.17a}$$

where Δt_{CFL} is the inviscid CFL condition given by

$$\Delta t_{CFL} = \frac{1}{\frac{|u|}{\Delta x} + \frac{|v|}{\Delta y} + \frac{|w|}{\Delta z} + \frac{a}{\sqrt{\frac{1}{\Delta x^2} + \frac{1}{\Delta y^2} + \frac{1}{\Delta z^2}}}} \tag{4.2.17b}$$

and Re_Δ in the mesh Reynolds number given by

$$Re_\Delta = \min\{\rho|u|\Delta x/\mu, \rho|v|\Delta y/\mu, \rho|w|\Delta z/\mu\} \tag{4.2.17c}$$

with a being the speed of sound given by $a = \sqrt{\gamma p/\rho}$ for a perfect gas.

The solution of Eq. (4.2.7) subject to the above time-step limit enables the evaluation of ρ, ρu, ρv, ρw and ρe_t at each grid point at each time. Using the density, the local velocities can be evaluated. These are then used to find the internal energy from the computed ρe_t and known velocities and density. Since two thermodynamic quantities, namely, the density and the internal energy are now known, the pressure and the temperature can be obtained from a constitutive equation of state for the fluid. The method thus enables the numerical solution of the unsteady, coupled, three-dimensional, compressible N-S equations. In its original form, it has been successfully used to compute viscous flows at low and moderate Reynolds numbers. Improvements, in the form of adding fourth-order smoothing functions to damp out oscillations in regions large gradients, a hybrid of explicit-implicit scheme for high Reynolds numbers, have been suggested to extend its range of applicability.

4.2.2 *Implicit Beam-Warming Schemes*

One of the difficulties of the MacCormack explicit method is the time-step limit given by Eqs. (4.2.17a, b). The maximum allowable time step decreases as the mesh size decreases. High Reynolds number flows exhibit boundary layer type of behaviour near solid walls, and have a very thin viscous region. In order to resolve velocity variations within this thin layer accurately, it is necessary to use a fine mesh in the direction normal to the wall. This reduces the mesh Reynolds number (which is the minimum of the Reynolds number in the three directions) and makes the time step unacceptably small. In order to get round this difficulty, a number of implicit methods, which do not suffer from time step constraints as far as stability is concerned, have been developed. We will look at the class of methods known as Beam-Warming schemes as an example of this approach.

One of the difficulties with an implicit method is the non-linearity of the Navier-Stokes equations. These non-linear terms arise from the advection terms which become important at high Reynolds numbers. In explicit methods, these can be treated using Picard substitution method where a term such u^2 can be written approximately as u^* u where u^* is the value from the previous time step. Since explicit methods have a limit on the time step, the time step is small and the loss of accuracy may not be significant. On the other hand, implicit methods are typically used with large time steps and significant inaccuracies may ensue. The implicit Beam-Warming class of schemes (Briley and McDonald 1975; Beam and Warming 1976, 1978, 1982) thus has a scheme for the linearization of the equations together with second-order accurate discretization of the spatial derivatives and multi-step integration of the time derivative. Consider the one-dimensional non-linear wave equation given by Eq. (4.2.3):

$$\partial u/\partial t + \partial f/\partial x = 0. \tag{4.2.3}$$

A two-time level, second-order accurate and unconditionally stable scheme for time integration of Eq. (4.2.3) can be derived as follows. We start with the Taylor series expansion of u(x, t + Δt) around the point (x, t):

$$u_i^{n+1} = u_i^n + \Delta t \partial u/\partial t|_{i,n} + \Delta t^2/2! \partial^2 u/\partial t^2|_{i,n} + O(\Delta t^3) \tag{4.2.18a}$$

Similarly, u(x, t) can be expanded about point (x, t + Δt):

$$u_i^n = u_i^{n+1} - \Delta t \partial u/\partial t|_{i,n+1} + \Delta t^2/2!\, \partial^2 u/\partial t^2|_{i,n+1} + O(\Delta t^3) \tag{4.2.19a}$$

Subtracting Eq. (4.2.19a) from Eq. (4.2.18a) and rearranging, one can get

$$\begin{aligned} u_i^{n+1} = {} & u_i^n + \tfrac{1}{2}\Delta t\left(\partial u/\partial t|_{i,n} + \partial u/\partial t|_{i,n+1}\right) \\ & + \tfrac{1}{2}\Delta t^2/2!\left(\partial^2 u/\partial t^2|_{i,n} - \partial^2 u/\partial t^2|_{i,n+1}\right) + O\left(\Delta t^3\right) \end{aligned}$$

or

$$u_i^{n+1} = u_i^n + \tfrac{1}{2}\Delta t\left(\partial u/\partial t|_{i,n} + \partial u/\partial t|_{i,n+1}\right) + O\left(\Delta t^3\right) \tag{4.2.20}$$

as $\partial^2 u/\partial t^2|_{i,n+1} = \partial^2 u/\partial t^2|_{i,n} + \frac{\partial}{\partial t}\left[(\partial^2 u/\partial t^2|_{i,n}\right]\Delta t + O(\Delta t^2)$. Also, since $\partial u/\partial t = -\partial f/\partial x$ from Eq. (4.2.3), we can write Eq. (4.2.20) as

$$\left(u_i^{n+1} - u_i^n\right)/\Delta t = -\tfrac{1}{2}\left(\partial f/\partial x|_{i,n} + \partial f/\partial x|_{i,n+1}\right) + O\left(\Delta t^2\right) \tag{4.2.21}$$

The flux term f is usually non-linear and coupled to other equations in a multi-dimensional case. The gradients of flux are evaluated as follows:

$$f^{n+1} = f^n + (\partial f/\partial t)\Delta t + O\left(\Delta t^2\right) = f^n + (\partial f/\partial u)(\partial u/\partial t)\Delta t + O\left(\Delta t^2\right)$$

Writing ∂f/∂u = A, we have

$$f^{n+1} = f^n + [A\left(u^{n+1} - u^n\right)/\Delta t]\Delta t + O(\Delta t^2) = f^n + A\left(u^{n+1} - u^n\right) + O(\Delta t^2)$$

$$\text{Thus, } (\partial f/\partial x)^{n+1} = (\partial f/\partial x)^n + \partial\left[A\left(u^{n+1} - u^n\right)\right]/\partial x \tag{4.2.22}$$

Substituting this into Eq. (4.2.21), we get

$$\left(u_i^{n+1} - u_i^n\right)/\Delta t = -\tfrac{1}{2}\left(\partial f/\partial x\big|_{i,n} + \partial f/\partial x\big|_{i,n}\right) + \partial\left[A\left(u_i^{n+1} - u_i^n\right)\right]/\partial x + O(\Delta t^2)$$

Evaluating the space derivatives using central differences and evaluating A, which is a function of u, using u^n, we can get a final expression for u_i^{n+1} as

$$\begin{aligned} u_i^{n+1} = u_i^n &- \Delta t(f_{i+1}^n - f_{i-1}^n)/(2\Delta x) - \tfrac{1}{2}\Delta t\left[\left(A_{i+1}^n u_{i+1}^{n+1} - A_{i-1}^n u_{i-1}^{n+1}\right)/(2\Delta x)\right. \\ &\left. -(A_{i+1}^n u_{i+1}^n - A_{i-1}^n u_{i-1}^n)/(2\Delta x)\right] + O\left(\Delta x^2, \Delta t^2\right) \end{aligned} \quad (4.2.23)$$

Equation (4.2.23) can be put in a tridiagonal matrix form for implicit solution of the variables for all the grid points at a time step before marching on to the next time step:

$$\begin{aligned} &-\{\Delta t A_{i-1}^n/(4\Delta x)\}u_{i-1}^{n+1} + u_i^{n+1} + \{\Delta t A_{i+1}^n/(4\Delta x)\}u_{i+1}^{n+1} = -\{\Delta t/(2\Delta x)\}(f_{i+1}^n - f_{i-1}^n) \\ &-\{\Delta t A_{i-1}^n/(4\Delta x)\}u_{i-1}^n + u_i^n + \{\Delta t A_{i+1}^n/(4\Delta x)\}u_{i+1}^n \end{aligned} \quad (4.2.24)$$

Equation (4.2.24) can be written in a simpler "delta" form as follows:

$$\begin{aligned} &-\{\Delta t A_{i-1}^n/(4\Delta x)\}\Delta u_{i-1} + \Delta u_i + \{\Delta t A_{i+1}^n/(4\Delta x)\}\Delta u_{i+1} \\ &= -\{\Delta t/(2\Delta x)\}\left(f_{i+1}^n - f_{i-1}^n\right) \end{aligned} \quad (4.2.25)$$

where

$$\Delta u_i = u_i^{n+1} - u_i^n \quad (4.2.26)$$

Now consider its extension to a two-dimensional case of the form of Eq. (4.2.7):

$$\partial U/\partial t + \partial E/\partial x + \partial F/\partial y = 0 \quad (4.2.27)$$

where U, E and F are vectors. For the inviscid form of these equations, the Beam-Warming discretization scheme of second order accuracy can be written as

$$\begin{aligned} &[[I] + \Delta t/2\{\partial([A]^n)/\partial x + \partial([B]^n)/\partial y\}]U^{n+1} \\ &= [[I] + \Delta t/2\{\partial([A]^n)/\partial x + \partial([B]^n)/\partial y\}]U^n - \Delta t(\partial E/\partial x + \partial F/\partial y)^n \end{aligned} \quad (4.2.28)$$

Here, [A] and [B] are Jacobians given by

$$[A] = \partial E/\partial U \quad \text{and} \quad [B] = \partial F/\partial U \quad (4.2.29)$$

It may be noted that [A] and [B] are lagged in the sense that these are evaluated at the nth time step. Using central difference approximations for the space derivatives, and linearizing where necessary, Eq. (4.2.28) can be converted into a set of coupled equations. Often these are not solved directly but are converted into sets of tridiagonal equations by approximate factorization (see Sect. 5.5) which can be solved using efficient algorithms. Accordingly, the delta form of Eq. (4.2.28) is written approximately as

$$\{[I] + \Delta t/2\, \partial/\partial x([A]^n)/\partial x\}\{[I] + \Delta t/2\, \partial/\partial y([B]^n)/\partial y\}\Delta U^n = -\Delta t(\partial E/\partial x + \partial F/\partial y)^n \tag{4.2.30}$$

where $\Delta U^n = U^{n+1} - U^n$. Equation (4.2.30) is solved iteratively in two alternating steps (see Sect. 5.5.2) as

$$\{[I] + \Delta t/2\; \partial([A]^n)/\partial x\}\Delta U^* = -\Delta t(\partial E/\partial x + \partial F/\partial y)^n \tag{4.2.31a}$$

$$\{[I] + \Delta t/2\, \partial([B]^n)/\partial y\}\Delta U^n = \Delta U^* \tag{4.2.31b}$$

If viscous effects are to be included, then the full Navier-Stokes equations have to be solved. In this case, E and F vectors will include derivatives of the velocities. When $\partial E/\partial x$ and $\partial F/\partial y$ are evaluated, some of these will be normal derivatives and some will be cross-derivatives, see Eqs. (4.2.11) and (4.2.14). In order to come up with a factorizable time-integration scheme, these need to be evaluated differently as explained below.

The governing equations for 2-d compressible flow are written in vector form as follows:

$$\partial U/\partial t + \partial E/\partial x + \partial F/\partial y = \partial(E_{v1} + E_{v2})/\partial x + \partial(F_{v1} + F_{v2})/\partial y \tag{4.2.32}$$

Here, E and F are functions of U alone while the viscous terms E_{v1} and F_{v1} are functions of U and $\partial U/\partial x$ and E_{v2} and F_{v2} are functions of U and $\partial U/\partial y$. The backward, 2nd order accurate, implicit time marching scheme of Beam-Warming can be written as

$$\Delta U^n = 2/3\Delta t\, \partial(\Delta U^n)/\partial t + 2/3\Delta t\, \partial U^n/\partial t + 1/3\Delta U^{n-1} + O(\Delta t^3) \tag{4.2.33}$$

Substituting $\partial U/\partial t$ from Eq. (4.2.32), we get

$$\begin{aligned}\Delta U^n = {} & 2/3\Delta t\{\partial(-\Delta E^n + \Delta E_{v1}{}^n + \Delta E_{v2}{}^n)/\partial x + \partial(-\Delta F^n + \Delta F_{v1}{}^n + \Delta F_{v2}{}^n)/\partial y\} \\ & + 2/3\Delta t\{\partial(-E^n + E_{v1}{}^n + E_{v2}{}^n)/\partial x + \partial(-F^n + F_{v1}{}^n + F_{v2}{}^n)/\partial y\} + 1/3\Delta U^{n-1} + O(\Delta t^3)\end{aligned} \tag{4.2.34}$$

We take account of non-linearity and coupling through lagging in the following way:

The two convective terms, E and F, are evaluated as

$$\Delta E^n = [A]^n \Delta U^n; \qquad \Delta F^n = [B]^n \Delta U^n \tag{4.2.35}$$

The normal derivatives of the viscous terms, E_{v1} and F_{v2}, are treated as follows:

$$\begin{aligned}\Delta E_{v1}{}^n &= (\partial E_{v1}/\partial U)^n \Delta U^n + (\partial E_{v1}/\partial U_x)^n \Delta U_x{}^n \\ &= [P]^n \Delta U^n + [R]^n \Delta U_x{}^n \\ &= ([P]-[R_x])^n \Delta U^n + \partial([R]^n \Delta U^n)/\partial x\end{aligned} \tag{4.2.36}$$

where $[P] = \partial E_{v1}/\partial U; [R] = \partial E_{v1}/\partial U_x; [R_x] = \partial[R]/\partial x$
Similarly,

$$\begin{aligned}\Delta F_{v2}{}^n &= (\partial F_{v2}/\partial U)^n \Delta U^n + \left(\partial F_{v2}/\partial U_y\right)^n \Delta U_x{}^n \\ &= [Q]^n \Delta U^n + [S]^n \Delta U_y{}^n \\ &= \left([Q]-\left[S_y\right]\right)^n \Delta U^n + \partial([S]^n \Delta U^n)/\partial y\end{aligned} \tag{4.2.37}$$

where $[Q] = \partial F_{v2}/\partial U; [S] = \partial F_{v2}/\partial U_y; \left[S_y\right] = \partial[S]/\partial y$
The cross-derivatives of the viscous terms, E_{v2} and F_{v1}, are treated as follows

$$\Delta E^n_{v2} = \Delta E^{n-1}_{v2} \quad \Delta F^n_{v1} = \Delta F^{n-1}_{v1} \tag{4.2.38}$$

Substituting Eqs. (4.2.35) to (4.2.38) into Eq. (4.2.34) and factorizing it, we finally the Beam-Warming algorithm for compressible Navier-Stokes equations as

$$\begin{aligned}&\left\{[I] + \frac{2}{3}\Delta t\left[\frac{\partial}{\partial x}([A]-[P]+[R_x])^n - \frac{\partial^2}{\partial x^2}(R)^n\right]\right\}\left\{[I] + \frac{2}{3}\Delta t\left[\frac{\partial}{\partial y}\left([B]-[Q]+\left[S_y\right]\right)^n\right.\right. \\ &\left.\left. -\frac{\partial^2}{\partial y^2}(S)^n\right]\right\}\Delta U^n = \frac{2}{3}\Delta t\left[\frac{\partial}{\partial x}(-E + E_{v1} + E_{v2})^n + \frac{\partial}{\partial y}(-F + F_{v1} + F_{v2})^n\right] \\ &+ \frac{2}{3}\Delta t\left\{\frac{\partial}{\partial x}(\Delta E_{v2})^{n-1} + \frac{\partial}{\partial y}(\Delta F_{v1})^{n-1}\right\} + \frac{1}{3}\Delta t \Delta U^{n-1}\end{aligned} \tag{4.2.39}$$

This is factorized and converted into a three-step time integration:

$$\begin{aligned}&\left\{[I] + \frac{2}{3}\Delta t\left[\frac{\partial}{\partial x}([A]-[P]+[R_x])^n - \frac{\partial^2}{\partial x^2}(R)^n\right]\right\}\Delta U^{*n} \\ &= \frac{2}{3}\Delta t\left[\frac{\partial}{\partial x}(-E + E_{v1} + E_{v2})^n + \frac{\partial}{\partial y}(-F + F_{v1} + F_{v2})^n\right] \\ &+ \frac{2}{3}\Delta t\left\{\frac{\partial}{\partial x}(\Delta E_{v2})^{n-1} + \frac{\partial}{\partial y}(\Delta F_{v1})^{n-1}\right\} + \frac{1}{3}\Delta t \Delta U^{n-1}\end{aligned} \tag{4.2.40a}$$

$$\left\{[I] + \frac{2}{3}\Delta t\left[\frac{\partial}{\partial y}\left([B]-[Q]+\left[S_y\right]\right)^n - \frac{\partial^2}{\partial y^2}[S]^n\right]\right\}\Delta U^n = \Delta U^{*n} \tag{4.2.40b}$$

$$U^{n+1} = U^n + \Delta U^n \tag{4.2.40c}$$

The spatial derivatives are evaluated using central differences to give an overall scheme that is implicit, stable and second order accurate in both time and space. Oscillatory solution may be expected near large gradients; the common practice is to add artificial damping in the form of a fourth-order explicit dissipation term of the form

$$-D\{\Delta x^4 \partial^4 U^n/\partial x^4 + \Delta y^4 \partial^4 U^n/\partial y^4\}$$

to the right hand side of Eq. (4.2.40a).

Although the Beam-Warming method can be unconditionally stable when applied to simple equations, the approximation factorization of Eq. (4.2.39) and linearization of the terms can lead to convergence problems in cases of strong coupling. The method has been widely used in computer codes for the solution of compressible flows.

4.3 Computation of Supersonic Flows

Supersonic flow, in which the characteristic flow speed is higher than the speed of sound in that medium at the prevailing conditions, occurs in high-speed aerospace applications as well as in turbomachinery applications. In the case of the former, both external flows, e.g., flow over the wing of an aircraft, and internal flows, e.g., the intake nozzle of a scramjet engine, occur. In turbomachinery applications, the flow is, in many cases, internal and has strong influences from rotation, turbulence and wall interaction. While turbomachinery applications are often limited to a Mach number (Ma) less than 2, external flows can be hypersonic ($Ma > 5$) in some cases. A prominent feature of such flows is the occurrence of shocks caused by under-expansion in a nozzle, or by changes in surface slope or due to the presence of downstream boundary conditions forcing the flow to become subsonic. These shocks can take the form of normal shocks, oblique shocks, compression and expansion waves (see Fig. 4.1), and can have significant impact on flow and flow-dependent phenomena.

Shocks pose special problems for prediction through CFD simulations. Physically, a shock is a sharp discontinuity across which flow variables such as pressure and velocity can change significantly. The distance over which these vary, or in effect the thickness of the shock layer, is of the order of a fraction of millimeter, and is beyond the resolution of usual CFD computations. Perforce, therefore, a shock in a numerical computation is a discontinuous solution across a line in 2-d flows and a surface in 3-d flows within the computational domain. Determination of the location and strength of the discontinuity (or "shock capturing") as part of the solution has been the subject of much debate and intense research for a number of decades, especially in the second half of the last century. In this section, we take a brief look at the development of principal ideas and current approaches to the computation of high speed flows.

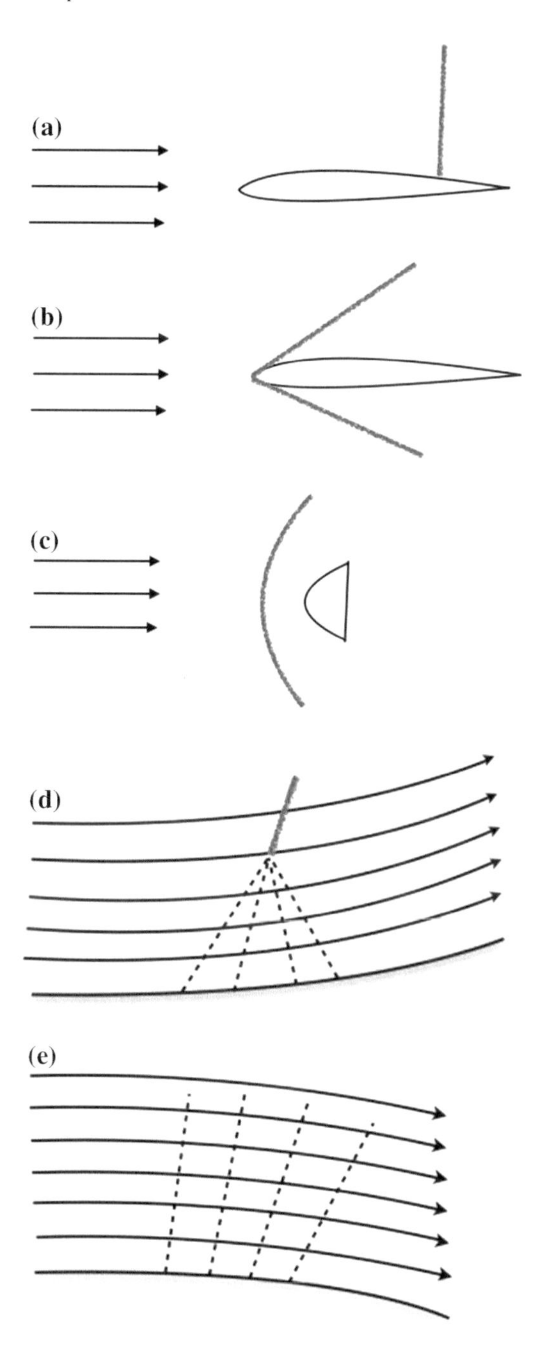

Fig. 4.1 Compressible flow with **a** normal shock, **b** oblique shock, **c** bow shock, **d** compression wave, and **e** expansion wave

4.3.1 Structure of a Shock

The simplest case of a problem where shocks occur is that of the one-dimensional, hyperbolic, non-linear wave equation (also known as the inviscid Burgers equation):

$$\frac{\partial u}{\partial t} + u\frac{\partial u}{\partial x} = 0 \qquad (4.3.1a)$$

Without losing generality, we consider an infinite domain (thus eliminating the influence of boundary conditions), and consider an initial distribution of *u* given by

$$u(x, 0) = g(x) \qquad -\infty < x < \infty, t = 0 \qquad (4.3.1b)$$

Equations (4.3.1) is already familiar to us in its linear form and its solution in many different ways has been discussed in Chap. 3. It can also be considered as the inviscid limit of the momentum conservation equation in the x-direction for nearly constant density fluids such as liquids. In contrast to the linear wave equation where all points on the wave front propagate at a constant speed, the points on a non-linear wave front can travel at different speeds. A distinctive feature of its solution is the occurrence of a discontinuity or a shock when $g(x_0)$, i.e., the initial distribution of u, has a negative slope. The amplitude of variation across the discontinuity, which is equal to $(u_1 - u_2)$, and the speed of propagation of the discontinuity are linked by the Rankine-Hugoniot conditions (Rankine 1870; Hugoniot 1889), which can be written in the general case as

$$[\mathbf{F}] \cdot \mathbf{I}_n - c[\mathbf{U}] \cdot \mathbf{I}_n = 0 \qquad (4.3.2)$$

where $\mathbf{F}$ is the inviscid flux, c is the velocity of propagation of the discontinuity, $\mathbf{I}_n$ is the direction vector normal to the discontinuity, and the term in square brackets is the difference in the value of the variable across the discontinuity, e.g., $[u] = u_1 - u_2$. Four idealized cases are shown in Fig. 4.2. For the constant negative slope case given in Fig. 4.2a, shock formation occurs after a time t_s at a location x_s given by

$$t_s = X/(u_1 - u_2) \quad \text{and} \quad x_s = u_1 t_s = u_2 t_s + X \qquad (4.3.3)$$

For this case, the solution for $t > t_s$ is given by

$$\begin{aligned} u(x, t) &= u_1 \quad \text{for} \quad x < t(u_1 + u_2)/2 \quad \text{and} \\ &= u_2 \quad \text{for} \quad x > t(u_1 + u_2)/2 \end{aligned} \qquad (4.3.4)$$

The discontinuity at $x = (u_1 + u_2)/t$ will be travelling at a velocity of $(u_1 + u_2)/2$.

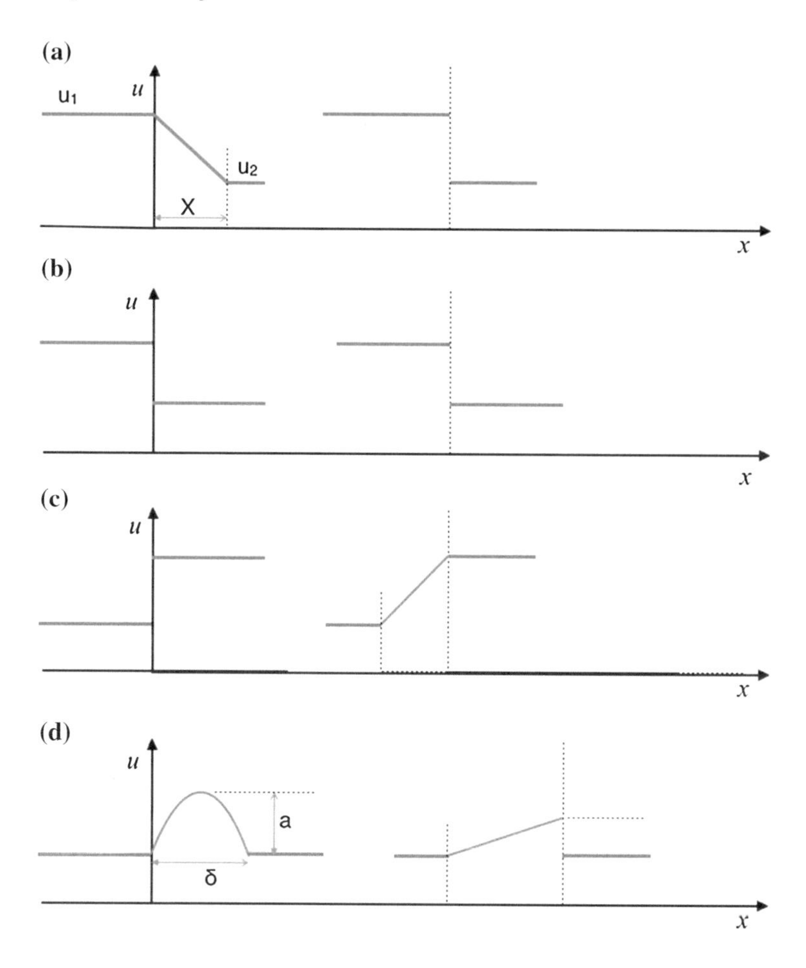

Fig. 4.2 Four idealized cases of g(x, 0) leading to shock formation in **a**, **b** and **d** but not in **c**

Consider the case shown in Fig. 4.2b which has a discontinuity (shock) at $x = 0$ such that

$$u(x,0) = u_1 \quad \text{for } x < 0, \quad u(x,0) = u_2 \quad \text{for } x > 0 \quad \text{and} \quad u_1 > u_2 \qquad (4.3.5)$$

In this case, the same shock solution will propagate along the x-axis at a speed of $(u_1 + u_2)/2$. If $u_1 = -u_2$, then the shock will be stationary. For the case where $u_1 < u_2$ shown in Fig. 4.2c, a *continuous*, shock-free solution, called an expansion fan, is possible:

$$\begin{aligned} u(x,t) &= u_1 \quad \text{for} \quad x < u_1 t, \\ &= x/t \quad \text{for} \quad u_1 t < x < u_2 t \quad \text{and} \\ &= u_2 \quad \text{for} \quad x > u_2 t \end{aligned} \qquad (4.3.6)$$

For the case where the initial condition is like a solitary wave described by a sinusoidal hump with base u_0, width δ and amplitude a (see Fig. 4.2d), the asymptotic solution would consist of a combination of a linear variation of u with x corresponding to the expansion part (the left half of the hump with positive slope) followed by a sudden discontinuity (corresponding to the right half of the hump with a negative slope) which moves at a speed given by

$$c = u_0 t + (\beta t)^{1/2}, \quad \beta = 4a\delta/\pi \tag{4.3.7}$$

where β is equal to twice the area under the hump. The amplitude of the discontinuity is given by $\sqrt{(\beta/t)}$ and therefore diminishes with time.

Discontinuous solutions in fluid mechanics can be of two types, namely, a contact discontinuity and a shock surface. A contact discontinuity is one across which the density and tangential velocity can exhibit a jump but the pressure and normal velocity are continuous. An example of a contact discontinuity is the interface between a liquid and its vapour or that between two immiscible liquids. In the case of a shock surface, all the variables, including pressure and normal velocity, are discontinuous across the shock. A discontinuous initial condition need not always lead to a sustained discontinuity. An example of this is the case shown in Fig. 4.2c, where an initial discontinuity with $u_1 < u_2$ leads to a continuous solution (the expansion fan). A well-known problem in which all the three arise simultaneously is the "shock tube" experiment in which an infinitely long straight tube is divided into two compartments separated by a diaphragm (see Fig. 4.3a). At time = 0, the two compartments are filled with the same fluid but are maintained at different pressures and thus at two different densities. One may note that here the fluid is at rest in both the compartments. At time = 0, the diaphragm is ruptured and the evolution of pressure, density, velocity in the tube are of interest.

This problem, when idealized as a one-dimensional flow of an inviscid fluid (thus neglecting effects related to viscosity, external heat transfer and boundary conditions), is known as the Riemann problem. Its solution consists of three parts: an expansion wave with a smoothly varying pressure confined to the high pressure region, a contact discontinuity moving to the right at velocity v_c which separates the high/low pressure regions, and a shock wave which propagates into the undisturbed region on the right at a velocity c_s. Thus, for $t > 0$, the tube length can be divided into five distinct sections (see Fig. 4.3b): the region marked A which is the undisturbed region to the left, the region B further to the right which is affected by the expansion fan, the region C to right of the expansion fan and to the left of the contact discontinuity, the region D between the contact discontinuity and the shock, and finally the region marked E which is the undisturbed region to the right side of the shock. For one-dimensional, inviscid, ideal flow, important features of the flow can be worked out in each region (Sod 1978). Figure 4.4 shows the typical spatial variation of pressure, velocity and density for this case (Sod 1978). One can see that the pressure and velocity remain continuous across the contact discontinuity at the

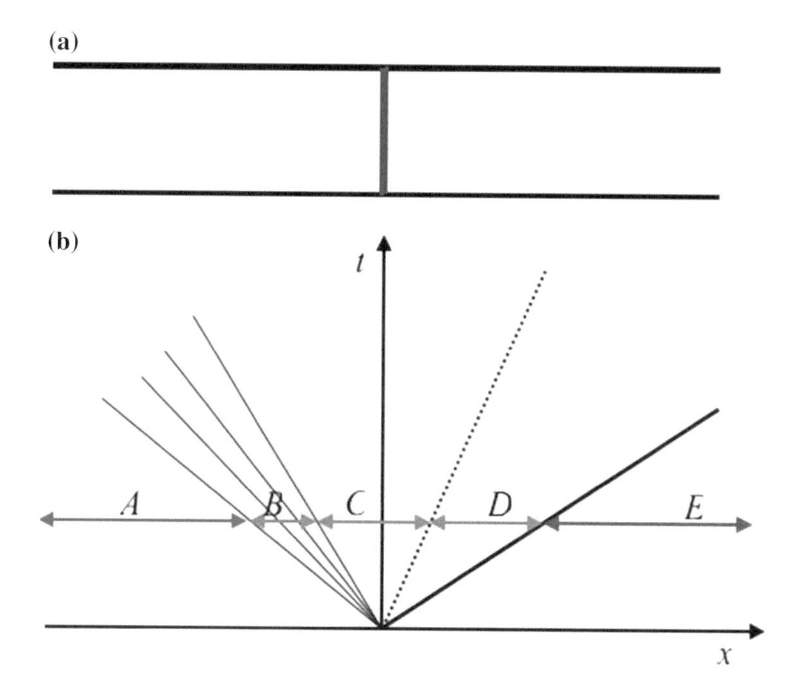

Fig. 4.3 Shock tube experiment: **a** at t = 0, the diaphragm separates the left and the right regions, $p_L > p_R$. **b** Five distinct zones of subsequent spatial variation of variables: A is the undisturbed region on the left, B is region influenced by the expansion wave, C is the region between the expansion wave and the contact discontinuity, D is the region between the contact discontinuity and the shock, and E is the undisturbed region to the right

interface between regions C and D while the density shows an abrupt transition. Across the shock, which is at the interface between regions D and E, all the quantities show an abrupt discontinuity. In the expansion region (region B), all the variables show a continuous variation.

All the jumps have to obey the Rankine-Hugoniot condition given by Eq. (4.3.4). For given initial pressures p_L and p_R to the left and right of the diaphragm, the pressure jump across the shock can be computed from the following implicit relation for P, the ratio of pressure upstream to downstream (i.e., $P = p_D/p_R$):

$$\sqrt{\frac{2}{\gamma(\gamma-1)}}\frac{P-1}{\sqrt{1+\alpha P}}=\frac{2}{\gamma-1}\frac{c_L}{c_R}\left\{\left(\frac{p_L}{p_R}\right)^{\frac{\gamma-1}{2\gamma}}-P^{\frac{\gamma-1}{2\gamma}}\right\} \quad (4.3.8)$$

Here, c_L and c_R are the speeds of sound in the medium at pressures p_L and p_R, respectively, γ is the ratio of specific heats, and $\alpha = (\gamma + 1)/(\gamma - 1)$.

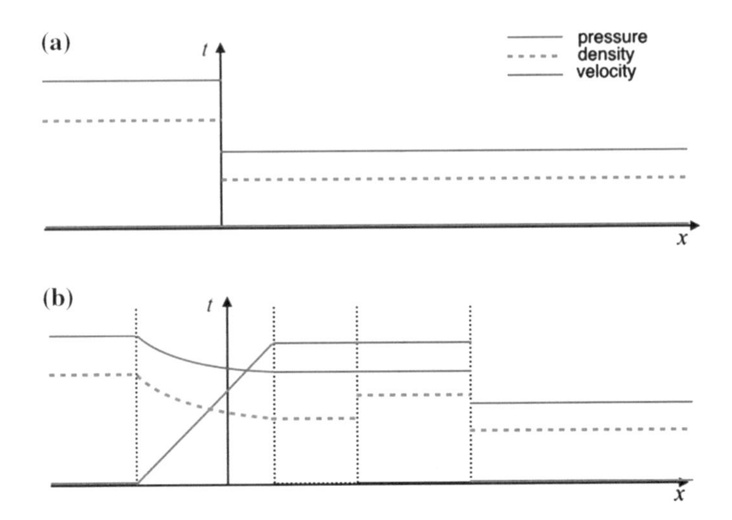

Fig. 4.4 Spatial variation of variables in a shock tube experiment. Top: Initial conditions with a discontinuous, step-down variation of pressure and density but constant velocity. Bottom: Variation of pressure, density and velocity for t > 0. Note that pressure and velocity are continuous across the contact discontinuity (between C and D) but density is discontinuous. All the three are discontinuous at the shock interface (between D and E)

The speed at which the contact discontinuity moves, v_c, is given by

$$v_c = u_D = u_C = \frac{c_R(P-1)}{\sqrt{1+\alpha P}}\sqrt{\frac{2}{\gamma(\gamma-1)}} \tag{4.3.9}$$

The speed of the shock, c_S, in the undisturbed region to the right, i.e., region E, is given by

$$c_S = \frac{(P-1)c_R^2}{\gamma v_c} \tag{4.3.10}$$

The position of the two discontinuities and the pressure, density and velocity variations in region B can also be determined theoretically. The solution for the specific case of $p_L = 1$, $p_R = 0.1$, $\rho_L = 1$, $\rho_R = 0.125$ and $u_L = u_R = 0$ at $t = 0.15$ is shown in Fig. 4.5. Because of the change in density, the Mach number shows a discontinuity across the contact discontinuity, although the velocity is continuous. The entropy decreases sharply across the shock. This case has served as a useful benchmark exercise for testing the accuracy of many numerical schemes. Several of these are discussed in advanced CFD books such as those of Hirsch (1990), Tannehill et al. (1997), Wesseling (2004). As we shall see later, the Riemann

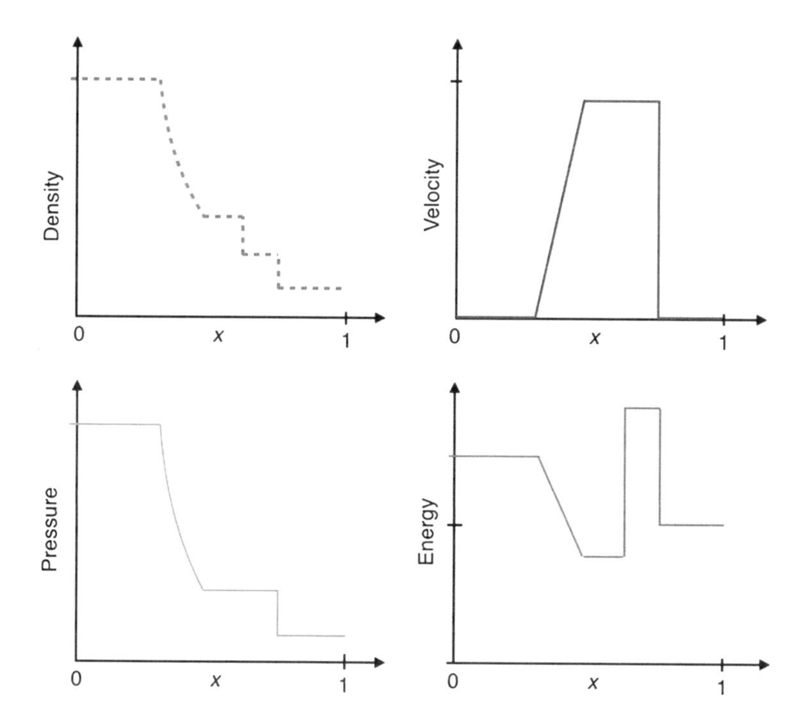

Fig. 4.5 Solution of the one-dimensional Euler equations for the shock tube problem (Sod 1978) at time = 0.15 for a gas with $\gamma = 1.4$ with initial conditions of $p_L = 1$, $p_R = 0.1$, $\rho_L = 1$, $\rho_R = 0.125$ and $u_L = u_R = 0$ with the diaphragm at $x = 0.5$

problem also formed an integral part of an early method for the computation of solutions with shocks.

One can now introduce the concept of a weak solution to the hyperbolic Eqs. (4.3.1a, b). This equation can be considered as the limiting case of one-dimensional flow of a real fluid with vanishing viscosity. Such a fluid flow will exhibit a shock with u varying continuously across the shock width. Also, only a compression shock would occur (i.e., for the case of $u_1 > u_2$), and an expansion shock (case of $u_1 < u_2$) would not occur. A strong solution to Eqs. (4.3.1a, b) would be such a physically correct solution. A weak solution is one which is discontinuous across the shock but is otherwise continuous; it would not exhibit an expansion shock and would obey the jump conditions given by the Rankine-Hugoniot equations. A physically inadmissible solution is one which allows an expansion shock to form or one which does not obey the jump conditions across the discontinuity. In a numerical solution, one cannot resolve the flow variables across the thin shock layer; one can therefore only hope to obtain a weak solution.

4.3.2 Computation of Shocks with Central Schemes

The computation of discontinuous solutions poses problems for finite difference methods which are based on Taylor series approximations for derivatives, and which are therefore strictly applicable only for continuously differentiable functions. Another aspect of the solution of Eqs. (4.3.1a, b) pointed out by Lax (1954) is that a conservation formulation of the equation, in the form of Eq. (4.2.7) is required for proper solution. Thus, Eqs. (4.3.1a, b) is written in terms of fluxes as

$$\frac{\partial u}{\partial t} + \frac{\partial f}{\partial x} = 0 \tag{4.3.11}$$

and can be conservatively discretized around point (i, n) as

$$\int_{x-\Delta x/2}^{x+\Delta x/2} \frac{\partial u}{\partial t} dx = - \int_{x-\Delta x/2}^{x+\Delta x/2} \frac{\partial f}{\partial x} dx = f^{*n}_{i+1/2} - f^{*n}_{i-1/2} \tag{4.3.12a}$$

where f* are numerical fluxes at the mid-points, and are to be evaluated consistently, i.e.,

$$f^*(u_i, u_{i+1}, \ldots) = f(u) \text{ when all } u_i = u \tag{4.3.12b}$$

Using a single time-step discretization scheme, Eq. (4.3.12a) can be written as

$$u_i^{n+1} = u_i^n - \frac{\Delta t}{\Delta x}\left[f^{*n}_{i+1/2} - f^{*n}_{i-1/2}\right] \tag{4.3.12c}$$

Equations (4.3.12a, b) can thus be considered as a finite volume discretization of Eq. (4.3.11) centred around spatial location i, and can be considered as a weak solution that allows for a discontinuity to be present within the spatial domain $[x_{i-½\,\Delta x}, x_{i+½\,\Delta x}]$.

Despite the limitation on the applicability of Taylor series approximations in the presence of discontinuities, it is possible to obtain the correct speed of propagation of the discontinuity using a central scheme for the discretization of the spatial derivative. We write $\partial f/\partial x$ as

$$\partial f/\partial x = \partial f/\partial u\, \partial u/\partial x = A\, \partial u/\partial x \tag{4.3.13a}$$

where

$$A = \partial f/\partial u = u \tag{4.3.13b}$$

The Lax scheme (Lax 1954), which is also referred to as Lax-Friedrichs scheme, for Eqs. (4.3.12a, b, c) takes the form

$$u_i^{n+1} = \frac{\left(u_{i+1}^n + u_{i-1}^n\right)}{2} - \frac{\Delta t}{\Delta x}\frac{\left(f_{i+1}^n - f_{i-1}^n\right)}{2} \tag{4.3.14}$$

This can be put in the form of Eq. (4.3.12c) by defining the numerical flux as

$$f_{i+\frac{1}{2}}^{*n} = \frac{f_i^n + f_{i+1}^n}{2} - \frac{1}{2}\frac{\Delta x}{\Delta t}\left(u_{i+1}^n - u_i^n\right) \tag{4.3.15a}$$

or

$$f_{i+\frac{1}{2}}^{*} = f_{i+\frac{1}{2}} - \frac{1}{2}\frac{\Delta x}{\Delta t}\left(u_{i+1} - u_i\right) \tag{4.3.15b}$$

We note that the above definition of numerical flux satisfies the condition of consistent evaluation of flux given by Eq. (4.3.12b). However, being only first order accurate in space and time, it is strongly dissipative and a highly smeared solution is soon obtained as the shock moves.

The Lax-Wendroff scheme (Lax and Wendroff 1964) is second order accurate in space and takes the form

$$u_i^{n+1} = u_i^n - \frac{\Delta t}{\Delta x}\frac{f_{i+1}^n - f_{i-1}^n}{2} + \frac{1}{2}\left(\frac{\Delta t}{\Delta x}\right)^2\left\{A_{i+\frac{1}{2}}^n\left(f_{i+1}^n - f_i^n\right) - A_{i-\frac{1}{2}}^n\left(f_i^n - f_{i-1}^n\right)\right\} \tag{4.3.16}$$

Here, $A_{i+1/2}^n$ may be evaluated either as $A_{i+1/2}^n = A\left\{\left(u_i^n + u_{i+1}^n\right)/2\right\}$ or as $A_{i+1/2}^n = \left(A_i^n + A_{i+1}^n\right)/2$. Rewriting it in the form of Eq. (4.3.12c), one can show that the numerical fluxes, f*, are defined in this case as

$$f_{i+1/2}^{*} = (f_i + f_{i+1})/2 - 1/2(\Delta t/\Delta x)A_{i+1/2}(f_{i+1} - f_i) \tag{4.3.17}$$

The Lax-Wendroff scheme is a central scheme with second order accuracy in space and is highly dispersive producing oscillatory solution near discontinuities. These unwanted but unavoidable high frequency spatial oscillations can be controlled to some extent by the addition of artificial dissipation in the sense of von Neumann and Richtmayer (1950). This additional dissipation term should be such that it would act on the scale of the mesh so as to suppress local oscillations, and be of an order equal to or higher than the truncation error in regions where the solution is smooth. The formulation of the artificial dissipation term should also satisfy the consistency condition (see Eq. 4.3.12b). For the Lax-Wendroff discretization, a suitable form of the numerical flux with the artificial dissipation term would be as follows:

$$f^*_{i+1/2} = (f_i + f_{i+1})/2 - 1/2(\Delta t/\Delta x)A_{i+1/2}(f_{i+1} - f_i) - D(u_{i+1} - u_i) \quad (4.3.18)$$

Here, D at i + ½ depends on u_i and u_{i+1} only, and it should be a positive function of (u_{i+1}-u_i) that goes to zero at least linearly with ($u_{i+1} - u_i$). The entire last term constitutes the artificial dissipation term; substituting it into Eq. (4.3.12c), one would obtain the following expression for u_i^{n+1}:

$$u_i^{n+1} = u_i^n - \frac{1}{2}\frac{\Delta t}{\Delta x}\left(f_{i+1}^n - f_{i-1}^n\right) + \frac{1}{2}\left(\frac{\Delta t}{\Delta x}\right)^2\left\{A_{i+\frac{1}{2}}^n\left(f_{i+1}^n - f_i^n\right) - A_{i-\frac{1}{2}}^n\left(f_i^n - f_{i-1}^n\right)\right\}$$
$$+ \frac{\Delta t}{\Delta x}\left\{D_{i+\frac{1}{2}}^n\left(u_{i+1}^n - u_i^n\right) - D_{i-\frac{1}{2}}^n\left(u_i^n - u_{i-1}^n\right)\right\} \quad (4.3.19)$$

When u in Eqs. (4.3.1a, b) has the dimensions of velocity, the product ($D\Delta x$) can be interpreted as kinematic viscosity, and D has to be positive in order to have a stabilizing effect. It can be modelled, for example, as

$$D = \alpha\Delta x\left|\frac{\partial u}{\partial x}\right| \quad (4.3.20a)$$

In this case, the term $D_{i+1/2}(u_{i+1} - u_i)$ in Eq. (4.3.19) can be written as

$$D_{i+\frac{1}{2}}^n\left(u_{i+1}^n - u_i^n\right) = \alpha\left|\left(u_{i+1}^n - u_i^n\right)\right|\left(u_{i+1}^n - u_i^n\right) \quad (4.3.20b)$$

Here, α is a positive coefficient of a value of around unity and can be tuned to introduce sufficient amount of artificial dissipation.

4.3.3 Godunov-Type Upwinding Schemes for Computation of Shocks

While central scheme-based approaches such the Lax-Wendroff methods and the MacCormack method can be used, the solutions suffer from unwanted dispersion in regions of slope discontinuity, and can be smoothened only to some extent using artificial viscosity. Importantly, they also suffer from not being able to distinguish between the physically admissible compression shocks and the inadmissible expansion shocks. The first method which implicitly accounted for this and allowed only physically admissible solutions was proposed by Godunov (1959). He assumed the conserved variable to be piecewise constant, i.e., constant over each mesh but varying from mesh to the neighbouring mesh point. The time evolution of u was determined by integrating the piecewise constant function over the cell and over the time step. This would lead to an expression for the cell- and time-step averaged value of u_i^{n+1} in terms of the value at the previous time step and time-averaged fluxes at the two cell interfaces. Consider the conservation form of the one-dimensional non-linear wave equation given by Eq. (4.3.11). At a grid point (x_i, t_n), integrating over the domain $(x_i - \Delta x/2, x_i + \Delta x/2)$, we get

$$\frac{d}{dt}\left\{\int_{x_i-\Delta x/2}^{x_i+\Delta x/2} u(x,t)dx\right\} = f(x_i - \Delta x/2, t) - f(x_i + \Delta x/2, t) \tag{4.3.21}$$

Integrating Eq. (4.3.21) over the time interval (t^n, $t^{n+\Delta t}$), we get

$$\left\{\int_{x_i-\Delta x/2}^{x_i+\Delta x/2} u(x,t^n+\Delta t)dx\right\} - \left\{\int_{x_i-\Delta x/2}^{x_i+\Delta x/2} u(x,t^n-\Delta t)dx\right\} = \left\{\int_{t^n-\Delta t/2}^{t^n+\Delta t/2} f(x_i-\Delta x/2,t)dt\right\} - \left\{\int_{t^n-\Delta t/2}^{t^n+\Delta t/2} f(x_i+\Delta x/2,t)dt\right\} \tag{4.3.22a}$$

Denoting time-average by an overbar (–) and space average by $\langle\cdots\rangle$, and using the index notation, we can rewrite Eq. (4.3.22a) as

$$\langle u_i\rangle^{n+1} - \langle u_i\rangle^n = \frac{\Delta t}{\Delta x}\left(\bar{f}^n_{i-1/2} - \bar{f}^n_{i+1/2}\right) \tag{4.3.22b}$$

In the Godunov (1959) scheme, u_i, u_{i+1} are assumed to be piecewise constant over the cell, and the time-averaged fluxes at the cell interfaces are obtained by solving the local Riemann problem involving u_i^n and u_{i-1}^n at the left boundary (i.e., $i - ½$) and u_{i+1}^n and u_i^n at the right boundary at ($i + ½$). The method can be readily understood by considering the linear wave equation:

$$\frac{\partial u}{\partial t} + a\frac{\partial u}{\partial x} = 0 \tag{4.3.23}$$

As per this equation, the velocity, a, is constant, and an initial condition moves to the right (for $a > 0$) at speed a. Consider the linear spatial distribution at points $i - 1$, i and $i + 1$ shown in Fig. 4.6a at time step n on a uniform mesh of spacing Δx. This is made piecewise constant as shown in Fig. 4.6a giving it a step-ladder type of variation. Over a time duration of Δt, the step ladder moves to the right by a distance $a\Delta t$ (such that $a\Delta t < \Delta x$), as shown by the dashed line in Fig. 4.6b. The new value of u_i at (n + 1)th time step is now computed by averaging the exact u^{n+1} over each cell giving a new piecewise constant shape shown in Fig. 4.6c. The linear shape can be recovered at (n + 1)th time step by replacing it with a piecewise linear variation. With reference to Godunov's method, the average flux entering through the left face of cell i over time step n is less than that leaving from the right face. As a result, the value of u_i decreases, as shown in Fig. 4.6c.

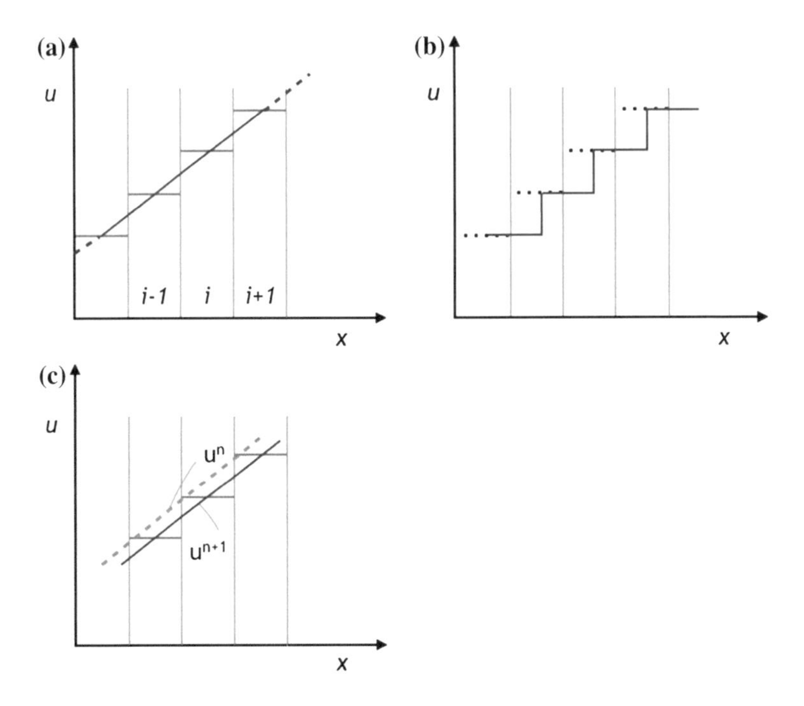

Fig. 4.6 Application of the method of Godunov (1959) to a linear advection problem with a constant wave velocity of $a > 0$. **a** The solution at nth time step; the linear variation of u^n is made piecewise-constant. **b** The piecewise-constant values in each cell are moved to the right by an amount $a\Delta t$. **c** The averaged, piecewise constant cell values at (n + 1)th time step. For this linear case with constant a, the average cell value changes by the same amount over a time step for the three cells. The flux entering from the left is less than that leaving from the right leading to a reduction in the cell value

For the non-linear wave equation, each step at the cell interfaces can be seen as a local Riemann problem, and the variation of u with time as the discontinuity passes through the cell is given by its solution. It can then be integrated over the distance travelled to obtain the time-averaged flux through that cell interface. Considering cell *i*, the right side interface is a step function from u_i to u_{i+1}. The Riemann problem for this interface can be described in terms of a local distance coordinate ξ (such that $\xi = 0$ at right boundary):

$$u = u_i \text{ for } \xi < 0 \quad \text{and} \quad u = u_{i+1} \quad \text{for } \xi > 0 \tag{4.3.24}$$

The exact solution to the Riemann problem depends on the relative values u_i and u_{i+1} and can involve a discontinuity moving at a speed of $(u_i + u_{i+1})/2$ if $u_i > u_{i+1}$ or an expansion fan if $u_i < u_{i+1}$. Thus, the local Riemann solution, u_R, which is a function of ξ/t, u_i and u_{i+1}, is given as follows.

$$\text{For } u_i > u_{i+1}, \tag{4.3.25a}$$

$$u_R = u_i \text{ for } \quad \xi/t < C_{i+1/2} \tag{4.3.25b}$$

$$u_R = u_{i+1} \text{ for } \quad \xi/t > C_{i+1/2} \tag{4.3.25c}$$

$$\text{where } C_{i+1/2} = \text{shock speed} = (u_i + u_{i+1})/2 \tag{4.3.25d}$$

$$\text{For } u_i < u_{i+1}, \tag{4.3.26a}$$

$$u_R = u_i \text{ for } \quad \xi/t < u_i \tag{4.3.26b}$$

$$u_R = \xi/t \text{ for } \quad u_i < \xi/t < u_{i+1} \tag{4.3.26c}$$

$$u_R = u_{i+1} \text{ for } \quad \xi/t > u_{i+1} \tag{4.3.26d}$$

Since the interface is located at $\xi = 0$, we need to evaluate $f_{i+½} = u^2/2|_{\xi = 0}$. Therefore, the flux that goes into Eq. (4.3.22b) at interface i + ½ is given by

$$\bar{f}_{i+½} = ½\, u_{i+1}{}^2 \quad \text{if both } u_i \text{ and } u_{i+1} \text{ are negative} \tag{4.3.27a}$$

$$\bar{f}_{i+½} = ½\, u_i{}^2 \quad \text{if both } u_i \text{ and } u_{i+1} \text{ are positive} \tag{4.3.27b}$$

$$\bar{f}_{i+½} = 0 \quad \text{if } u_i < 0 < u_{i+1} \text{ from eqn. (4.3.26c)} \tag{4.3.27c}$$

$$\bar{f}_{i+½} = ½\, u_i{}^2 \quad \text{if } u_i > 0 > u_{i+1} \text{ and } C_{i+1/2} > 0 \tag{4.3.27d}$$

$$\bar{f}_{i+½} = ½\, u_{i+1}{}^2 \text{if } u_i > 0 > u_{i+1} \text{ and } C_{i+1/2} < 0 \tag{4.3.27e}$$

It may be noted that the last two relations stem from Eqs. (4.3.25b, c), respectively. The flux at the left interface, $\bar{f}_{i-1/2}$ can be similarly evaluated from the corresponding local Riemann problem involving u_{i-1} and u_i. One may also note that the numerical flux, $f^*_{i+½}$, in Eq. (4.3.12a) is equal to the exact flux in the case of Godunov scheme.

4.3.4 Approximate Riemann Solvers for Computation of Shocks

As an explicit method, the Godunov scheme has a stability limit. In order to satisfy the condition that the waves from the interface must not interfere, the time step should be such that $|u_{max}\, \Delta t/\Delta x| < ½$. Although the Godunov method works well and produces no oscillatory solution, it has two major disadvantages. Firstly, it is an upwind scheme with a two-point stencil and is therefore only first order accurate in space. Secondly, for Euler equations, the computation of the exact solution for the Riemann problem requires iterative solution. For example, for one-dimensional Euler equations, the pressure ratio, P, is given by an implicit non-linear relation given by Eq. (4.3.8) above. The first difficulty has been addressed by many researchers by using higher order upwinding schemes together with approaches such as the TVD and ENO schemes (see Chap. 3.5) to minimize spurious oscillations that are produced around slope discontinuities. In order to address the problem of having to solve non-linear equations for the solution of the local Riemann problem, Roe (1981), among others, proposed the use of approximate Riemann solvers in place of the exact Riemann solvers. Writing Eq. (4.3.11) in the form of Eqs. (4.3.13a, b), Roe proposed its approximate evaluation as

$$\frac{\partial u}{\partial t} + \bar{A}\frac{\partial f}{\partial x} = 0 \tag{4.3.28}$$

Here $\bar{A}$ is a constant matrix that depends on the local conditions and satisfies the consistency condition:

$$\bar{A}(\mathrm{u_i}, \mathrm{u_{i+1}}) = \bar{A}(\mathrm{u}, \mathrm{u}) = \partial \mathrm{f}/\partial \mathrm{u} = \mathrm{u} \quad \text{if } \mathrm{u_i} = \mathrm{u_{i+1}} = \mathrm{u} \tag{4.3.29a}$$

as well as the jump condition:

$$\mathrm{f_{i+1}} - \mathrm{f_i} = \bar{A}(\mathrm{u_{i+1}} - \mathrm{u_i}) \tag{4.3.29b}$$

Since in this case $\partial \mathrm{u}/\partial \mathrm{t} = \mathrm{A} = \mathrm{u}$, Eq. (4.3.28) can be written as

$$\partial \mathrm{u}/\partial \mathrm{t} + \bar{\mathrm{u}}\, \partial \mathrm{u}/\partial \mathrm{x} = 0 \tag{4.3.30}$$

As per Eq. (4.3.29b), $\bar{u}$ is evaluated for cell interface (i + ½) between i and i + 1 as

$$\bar{\mathrm{u}}_{\mathrm{i}+1/2} = {}^1\!/\!_2(\mathrm{u}_{\mathrm{i}+1}^2 - \mathrm{u}_{\mathrm{i}}^2)/(\mathrm{u_{i+1}} - \mathrm{u_i}) \text{ or}$$

$$\bar{\mathrm{u}}_{i+1/2} = (\mathrm{u_i} + \mathrm{u_{i+1}})/2 \quad \text{if } \mathrm{u_i} \neq \mathrm{u_{i+1}} \tag{4.3.31a}$$

$$\bar{\mathrm{u}}_{\mathrm{i}+1/2} = \mathrm{u_i} \quad \text{if } \mathrm{u_i} = \mathrm{u_{i+1}} \tag{4.3.31b}$$

In order to account for the possibility that $\bar{\mathrm{u}}_{i+1/2}$ may be travelling in the positive or negative direction, the numerical flux, $\mathrm{f}^*_{\mathrm{i}+1/2}$ of the discretized equation (Eq. 4.3.12a) is evaluated as

$$\mathrm{f}^*_{\mathrm{i}+1/2} = (\mathrm{f_i} + \mathrm{f_{i+1}})/2 - {}^1\!/\!_2\left|(\bar{u}_{\mathrm{i}+1/2})\right|(\mathrm{u_{i+1}} - \mathrm{u_i}) \tag{4.3.32}$$

where $\mathrm{f_i} = \mathrm{f(u_i)} = \mathrm{u_i^2}/2$.

The stability limit for Roe's scheme is that the Courant number should be less than unity. The scheme is first-order accurate upwind scheme and is therefore highly diffusive. There is another well-known problem with Roe's scheme, namely, the sonic glitch. If there is a sonic transition from Ma < 1 to Ma > 1 between two successive grid points, Roe's scheme will produce a discontinuous, step-like variation through the sonic point. This type of "expansion shock" is unphysical and cannot occur. In this sense, Roe's scheme is said to violate the entropy condition (Lax 1973; see Hirsch 1990; Wesseling 2004 for a thorough discussion of this condition). Harten and Hyman (1983) proposed the following cure for this sonic glitch (which can be quite strong in strong expansion fans) in the form of a correction to the $\bar{u}_{i+1/2}$ evaluated using Eq. (4.3.32) above:

$$\bar{u}_{i+1/2} = \bar{u}_{i+1/2} \text{ if } \quad \bar{u}_{i+1/2} \geq \varepsilon \tag{4.3.33a}$$

$$= \varepsilon \text{ if } \quad \bar{u}_{i+1/2} < \varepsilon \tag{4.3.33b}$$

where

$$\varepsilon = \max[0, (u_{i+1} - u_i)/2] \tag{4.3.33c}$$

It may be observed that the above modification does not come into force in case of a compression shock (for which $\varepsilon = 0$) but acts only for an expansion shock. This intervention is said to be equivalent to adding dissipation to the expansion shock thereby smearing it intentionally!

The scheme of Enquist and Osher (1980, 1981) has an in-built fix for the detection of the sonic transition and thereby avoids the expansion shock and the ensuing entropy violation problem. In this method, the flux at any location is expressed as the sum of contribution from left and right moving waves:

$$f = f^+ + f^- \tag{4.3.34}$$

These flux components, which are functions of u, are evaluated independently by integrating over u as follows:

$$f^+ = \int_0^u \mu(u) \frac{\partial f}{\partial u} du \tag{4.3.35a}$$

$$f^- = \int_0^u \{1 - \mu(u)\} \frac{\partial f}{\partial u} du \tag{4.3.35b}$$

$$\text{with } \mu(u) = 1 \text{ if } \partial f/\partial u \geq 0 \quad \text{and} \quad = 0 \text{ if } \partial f/\partial u < 0 \tag{4.3.36}$$

With this formulation, if the sonic transition falls between cells i and i + 1 with $u_i < u_{i+1}$, then Eq. (4.3.35b) will be used up to the sonic point and Eq. (4.3.35a) from the sonic point to u_{i+1}. The numerical flux at i + ½ is thus evaluated as

$$f^*_{i+1/2} = \frac{(f_i + f_{i+1})}{2} - \int_{u_i}^{u_{i+1}} (|u|du) \tag{4.3.37}$$

This results in the following values for the numerical flux:

$$f^*_{i+1/2} = 1/2\, u^2_{i+1} \quad \text{if } u_i < 0 \quad \text{and} \quad u_{i+1} < 0 \tag{4.3.38a}$$

$$= 1/2\, u_i^2 \quad \text{if } u_i > 0, u_{i+1} > 0 \tag{4.3.38b}$$

$$= 0 \quad \text{if } u_i < 0 < u_{i+1} \tag{4.3.38c}$$

$$= 1/2\left(u_i^2 + u^2_{i+1}\right) \text{if } u_i > 0 > u_{i+1} \tag{4.3.38d}$$

The Enquist-Osher scheme avoids entropy violation and prevents the occurrence of an expansion shock. The explicit treatment of the sonic transitions results in a bounded, small deviation in the slope at the sonic point (Chakravarthy and Osher 1985).

4.3.5 Approximate Riemann Solvers for the Shock Tube Problem

The advantage of the use of approximate Riemann solver schemes can be appreciated when one considers the shock tube problem. For this one-dimensional case, neglecting viscous effects, one can write the governing Euler equations in conservation form as follows:

$$\frac{\partial \rho}{\partial t} + \frac{\partial}{\partial x}(\rho u) = 0 \tag{4.3.39a}$$

$$\frac{\partial}{\partial t}(\rho u) + \frac{\partial}{\partial x}\left(\rho u^2 + p\right) = 0 \tag{4.3.39b}$$

$$\frac{\partial}{\partial t}(\rho \epsilon) + \frac{\partial}{\partial x}(\rho u H) = 0 \tag{4.3.39c}$$

These can be written compactly in vector form as

$$\frac{\partial U}{\partial t} + \frac{\partial F}{\partial x} = 0 \tag{4.3.40a}$$

$$\text{with} \quad U = \begin{Bmatrix} \rho \\ m \\ \epsilon \end{Bmatrix} \text{ and } F = \begin{Bmatrix} m \\ \frac{m^2}{\varrho} + p \\ mH \end{Bmatrix} \tag{4.3.40b}$$

where

$$m = \rho u; \quad \epsilon = \rho e + 1/2\left(m^2/\rho\right); \quad H = h + 1/2\, u^2; \quad h = e + p/\rho \tag{4.3.40c}$$

In case of a shock, the shock speed, s, is given by the Rankine-Hugoniot jump conditions across the shock, namely,

$$s[U] = [F(U)] \tag{4.3.41a}$$

For the three conserved variables, these take the following form:

$$s(\rho_R - \rho_L) = m_R - m_L \tag{4.3.41b}$$

$$s(m_R - m_L) = m_R^2/\rho_R + p_R - m_L^2/\rho_L - p_L \tag{4.3.41c}$$

$$s(\rho_R \epsilon_R - \rho_L \epsilon_L) = m_R H_R - m_L H_L \tag{4.3.41d}$$

Assuming a perfect gas, one can further write that $p = \rho RT = (\gamma - 1)\rho e$, $e = e(T)$, $c_v = de/dT$, $h = h(T)$, $c_p = dh/dT$, $R = c_p - c_v$, $\gamma = c_p/c_v$ and $p = (\gamma - 1)\rho e$.

Let us now consider the solution of Euler equations using the methods discussed above. In the Godunov scheme, Eq. (4.3.40a) can be numerically solved as

$$U_i^{n+1} = U_i^n - \frac{\Delta t}{\Delta x}\left(\bar{F}_{i+1/2}^n - \bar{F}_{i-1/2}^n\right) \tag{4.3.42a}$$

$$= U_i^n - \frac{\Delta t}{\Delta x}\{\bar{F}_R(0, U_i, U_{i+1}) - \bar{F}_L(0, U_{i-1}, U_i)\} \tag{4.3.42b}$$

Here, $\bar{F}_R$, which is a function of U, is the flux evaluated based on the U from the solution of the local Riemann problem. For the case with $\rho = \rho_L$, $u = u_L$, and $p = p_L$ for $x < x_0$ and $t = 0$, and $\rho = \rho_R$, $u = u_R$, $p = p_R$ for $x > x_0$ and $t = 0$, and with an initial discontinuity at x_0, the pressure ratio, P, across the shock is given by

$$\sqrt{\frac{2}{\gamma(\gamma-1)}}\frac{P-1}{\sqrt{1+\alpha P}} = \frac{2}{\gamma-1}\, {}^{c_L}\!/_{c_R}\left\{\left(\frac{p_L}{p_R}\right)^{\frac{\gamma-1}{2\gamma}} - P^{\frac{\gamma-1}{2\gamma}}\right\} + \frac{u_L - U_R}{c_R} \tag{4.3.43}$$

Here c is the speed of sound and $\alpha = (\gamma + 1)/(\gamma - 1)$. The corresponding density and velocity are given by

$$\frac{\rho_2}{\rho_R} = \frac{1 + \alpha P}{\alpha + P} \tag{4.3.44}$$

$$\frac{u_2 - u_R}{c_R} = \frac{P - 1}{\sqrt{1 + \alpha P}} \sqrt{\frac{2}{\gamma(\gamma - 1)}} \tag{4.3.45}$$

The velocity of the contact discontinuity, V, is given by

$$V - u_L = \frac{2}{\gamma - 1} c_L \left\{ 1 - \left(\frac{p_3}{p_L} \right)^{\frac{\gamma-1}{2\gamma}} \right\} \tag{4.3.46}$$

The velocity and the pressure in the expansion fan are given by

$$u_5 = \frac{2}{\gamma - 1} \left(\frac{x}{t} + c_L \right) + u_L \tag{4.3.47a}$$

$$p_5 = p_L \left(\frac{u_5}{c_L} \right)^{\frac{2\gamma}{\gamma-1}} \tag{4.3.47b}$$

The expressions for u_5 and p_5 are applicable over the following range:

$$-\left(\frac{\gamma - 1}{2} u_L + c_L \right) < x/t < \left(\frac{\gamma + 1}{2} V - c_L - \frac{\gamma - 1}{2} u_L \right) \tag{4.3.47c}$$

The three eigenvalues of the Euler equations (Eq. 4.3.40a) are u, u + c and u − c where c is the speed of sound given by $\sqrt{\gamma p/\rho}$. The wave associated with u is a contact discontinuity while those associated with (u − c) and (u + c) can be shock or expansion fan. The transition between u_i and u_{i+1} has four regions where the variables are constant (see Figs. 4.3 and 4.4). These regions separate the three wave families. If the constant pressure in the region between u and (u − c) is greater than p_i, then (u − c) is a shock; otherwise it is an expansion fan. Similarly, if the pressure in the region between u and (u + c) is greater than p_{i+1}, then (u + c) is a shock and it is an expansion otherwise. Using the above relations, it is possible to find the appropriate value of each component of U, and these can be used to determine the local $\overline{f_R}$ and $\overline{f_L}$ at each interface. Substitution of these into Eq. (4.3.42b) will yield U_i^{n+1}.

While the Godunov method yields accurate shock transitions, the local Riemann solution of Eq. (4.3.43) for the pressure ratio across the shock requires iterative solution and these equations need to be solved at each interface. In order to reduce the computational effort required for the solution, Roe (1981) sought to solve a quasi-linear form of the Euler equations. His method can be best understood by studying the solution of a *linear* hyperbolic system. Consider the linear system of hyperbolic equations with constant coefficients given by

$$\frac{\partial U}{\partial t} + A \frac{\partial U}{\partial x} = 0 \tag{4.3.48}$$

The eigenvalues (λ_p) and the corresponding eigenvectors ($\mathbf{e}_p$) of the coefficient matrix A (for a three-equation hyperbolic system, there will be three real and distinct eigenvalues) are given by

$$A\,\mathbf{e}_p = \lambda_p\,\mathbf{e}_p \tag{4.3.49}$$

If the eigenvectors are normalized, i.e., if $|\mathbf{e}_p| = 1$, then they constitute an orthonormal basis so that U can written as

$$U = \sum_p w_p \mathbf{e}_p \tag{4.3.50}$$

Substituting the above into Eq. (4.3.48), one obtains

$$\sum_p \left\{ \left(\frac{\partial w_p}{\partial t}\right) \mathbf{e}_p \right\} + \sum_p \left\{ \left(\frac{\partial w_p}{\partial t}\right) A\mathbf{e}_p = 0 \right\} \tag{4.3.51}$$

Taking note of Eq. (4.3.49) and since the eigenvectors are orthonormal, Eq. (4.3.51) can be seen to consist of three independent hyperbolic equations with constant coefficients:

$$\frac{\partial w_p}{\partial t} + \lambda_p \frac{\partial w_p}{\partial x} = 0 \tag{4.3.52}$$

with the solution

$$w_p(x,t) = w_{p0}\left(x - \lambda_p t\right) \tag{4.3.53}$$

For a local Riemann problem with $w = w_L$ for $x < 0$ and $w = w_R$ for $x > 0$, one can write

$$\begin{aligned} w_p \lambda_p &= w_{pL} \text{ if } \quad \lambda_p > 0 \quad \text{and} \\ &= w_{pR} \text{ if } \quad \lambda_p < 0 \end{aligned} \tag{4.3.54a}$$

Or

$$\begin{aligned} w_p\,\lambda_p &= 1/2\left\{ \left(\lambda_p + |\lambda_p|\right) w_{pL} + \left(\lambda_p + |\lambda_p|\right) w_{pL} \right\} \\ &= 1/2\left\{ \lambda_p\left(w_{pL} + w_{pR}\right) - |\lambda_p|\left(w_{pR} - w_{pL}\right) \right\} \end{aligned} \tag{4.3.55}$$

The overall solution can be obtained as the sum of the three solutions.

For the non-linear problem involving Euler equations for the flow of a perfect gas, using the homogeneous property of the flux vector, namely,

$$F(\lambda U) = \lambda F(U) \text{ for any constant value of } \lambda \tag{4.3.56}$$

Equation (4.3.40a) is rewritten in the form of Eq. (4.3.48) with

$$A = \partial F/\partial U \tag{4.3.57a}$$

For a perfect gas for the case of Euler equations, A is given (see Hirsch 1990) as

$$A = \begin{bmatrix} \frac{\partial F}{\partial \rho} \\ \frac{\partial F}{\partial m} \\ \frac{\partial F}{\partial \epsilon} \end{bmatrix} = \begin{bmatrix} 0 & 1 & 0 \\ (\gamma-3)\frac{u^2}{2} & (3-\gamma)u & \gamma-1 \\ (\gamma-1)u^3-\gamma uE & \gamma E-3(\gamma-1)\frac{u^2}{2} & \gamma u \end{bmatrix} \tag{4.3.57b}$$

It may be noted that for a perfect gas, the Euler equations can also be written in terms of primitive variables ρ, u and p as (see for example Hirsch 1990)

$$\frac{\partial \mathbf{V}}{\partial t} + \tilde{A}\frac{\partial \mathbf{V}}{\partial x} = 0 \tag{4.3.58a}$$

$$\text{with} \quad V = \begin{bmatrix} \rho \\ u \\ p \end{bmatrix} \quad \text{and} \quad \tilde{A} = \begin{Bmatrix} u & \rho & 0 \\ 0 & u & \frac{1}{\rho} \\ 0 & \rho c^2 & u \end{Bmatrix} \tag{4.3.58b}$$

Both A and $\tilde{A}$ have three real eigenvalues given by u, u − c and u + c, and they therefore represent hyperbolic systems. Roe (1981) sought to write a quasi-linear approximation for the flux F at i + ½ such that

$$F_{i+1} - F_i = \{(\partial F/\partial U)\Delta U\}|_{i+1/2} = \tilde{A}(U_i, U_{i+1})\Delta U = \tilde{A}_{i+1/2}(U_{i+1} - U_i) \tag{4.3.59}$$

such that for any (U_i, U_{i+1}), $\tilde{A}$ (U_i, U_{i+1}) satisfies the consistency condition given by Eq. (4.3.29a) and further that

- $F_{i+1} - F_i = \tilde{A}(U_i, U_{i+1}) - \tilde{A}(U_{i+1}, U_i)$ (4.3.60a)
- $\tilde{A}$ has real eigenvalues and linearly independent eigenvectors (4.3.60b)

The above two conditions are such that if λ_p are the eigenvalues, then they satisfy the Rankine-Hugoniot condition, namely,

$$F_{i+1} - F_i = \lambda_p(U_{i+1} - U_i) \tag{4.3.61}$$

with λ_p being the speed of the discontinuity. Roe (1981) found that $\tilde{A}$ would be identical to the Jacobian A given by Eq. (4.3.57b) with E being replaced by H provided these variables, namely, ρ, u and H, are replaced by average values weighted by the square root of densities. Thus, for $\tilde{A}_{i+1/2}$ which depends on (U_i, U_{i+1}), we redefine the variables as

$$\tilde{\rho}_{i+1/2} = \sqrt{\rho_{i+1}\rho_i} \tag{4.3.62a}$$

$$\tilde{u}_{i+1/2} = \frac{\left\{\left(u\sqrt{\rho}\right)_{i+1} + \left(u\sqrt{\rho}\right)_i\right\}}{\left(\sqrt{\rho_{i+1}} + \sqrt{\rho_i}\right)} \tag{4.3.62b}$$

$$\tilde{H}_{i+1/2} = \frac{\left\{\left(H\sqrt{\rho}\right)_{i+1} + \left(H\sqrt{\rho}\right)_i\right\}}{\left(\sqrt{\rho_{i+1}} + \sqrt{\rho_i}\right)} \tag{4.3.62c}$$

The linearized Jacobian matrix $\tilde{A}$ will have its three eigenvalues given by

$$\tilde{\lambda}_1 = \tilde{u}; \quad \tilde{\lambda}_2 = \tilde{u} + \tilde{c}; \quad \tilde{\lambda}_3 = \tilde{u} - \tilde{c} \tag{4.3.63a}$$

where $\tilde{c}$, the speed of sound, is given by

$$\tilde{c}^2 = (\gamma - 1)\left(\tilde{H} - \frac{\tilde{u}^2}{2}\right) \tag{4.3.63b}$$

The corresponding eigenvectors, $\tilde{\boldsymbol{e}}_i$, are given by

$$\tilde{e}_1 = \begin{Bmatrix} 1 \\ \tilde{u} \\ \frac{\tilde{u}^2}{2} \end{Bmatrix}; \quad \tilde{e}_2 = \frac{\tilde{\varrho}}{2\tilde{c}}\begin{Bmatrix} 1 \\ \tilde{u} + \tilde{c} \\ \tilde{H} + \tilde{u}\,\tilde{c} \end{Bmatrix}; \quad \tilde{e}_3 = \frac{\tilde{\varrho}}{2\tilde{c}}\begin{Bmatrix} 1 \\ \tilde{u} - \tilde{c} \\ \tilde{H} - \tilde{u}\,\tilde{c} \end{Bmatrix} \tag{4.3.64}$$

In order to determine the numerical flux at the interface i + ½ as per Roe's scheme, one evaluates the three wave amplitudes δw_1, δw_2 and δw_3, which are defined as follows:

$$\delta w_1 = \delta p - \delta \rho / \tilde{c}^2 \tag{4.3.65a}$$

$$\delta w_2 = \delta u + \delta p \,/\, (\tilde{p}\tilde{c}) \tag{4.3.65b}$$

$$\delta w_3 = \delta u - \delta p / (\rho\tilde{c}) \tag{4.3.65c}$$

where $\delta u_{i+1/2} = u_{i+1} - u_i$; $\delta \rho_{i+1/2} = \rho_{i+1} - \rho_i$; $\delta p_{i+1/2} = p_{i+1} - p_i$

The numerical flux is then evaluated as the sum of the three wave components:

$$\bar{F}^{*OE}_{i+1/2} = \frac{1}{2}(F_i + F_{i+1}) - \frac{1}{2}\sum_p |\lambda_p| \delta w_p \bar{e}_p \tag{4.3.66}$$

One can see that in this procedure, it is not necessary to solve non-linear algebraic equations as part of the local Riemann problem. As mentioned earlier, in order to eliminate the expansion shock, an entropy fix needs to be used. The one suggested by Harten and Hymen (1983), see Eqs. (4.3.33a, b, c), takes the following form:

$$|\tilde{\lambda}_{mod}| = |\tilde{\lambda}_{i+1/2}| \text{ if } \tilde{\lambda}_{i+1/2} \geq \varepsilon \tag{4.3.67a}$$

$$|\tilde{\lambda}_{mod}| = \varepsilon \text{ if } \tilde{\lambda}_{i+1/2} < \varepsilon \tag{4.3.67b}$$

where

$$\varepsilon = \max\left\{0, \left(\tilde{\lambda}_{i+1/2} - \lambda_i\right), \left(\lambda_{i+1} - \tilde{\lambda}_{i+1/2}\right)\right\} \tag{4.3.67c}$$

The scheme of Enquist and Osher (1980, 1981) for a scalar quantity given by Eqs. (4.3.47a, b, c) was extended to the hyperbolic system of Euler equations by Osher and Chakravarthy (1983) by writing it as

$$F^*_{i+1/2} = \frac{F_i + F_{i+1}}{2} - \frac{1}{2}\int_{U_i}^{U_{i+1}} \{A(U)dU\} \tag{4.3.68}$$

The path of integration of the last term is parametrized in such a way that

$$U = U(\sigma) \quad \text{where } 0 \leq \sigma \leq 1 \quad \text{with} \quad U(0) = U_i \quad \text{and} \quad U(1) = U_{i+1} \tag{4.3.69}$$

With this parametrization, the numerical flux can be written as

$$F^*_{i+1/2} = \frac{F_i + F_{i+1}}{2} - \frac{1}{2}\int_0^1 \left\{A(U)\frac{dU}{d\sigma}d\sigma\right\} \tag{4.3.70}$$

For ease of computation of the numerical flux, the evaluation of the integral should be easy. To this end, Osher and Chakravarthy (1983) decomposed $|A|$ into

$$|A(U)| = R|\Lambda|R^{-1} \tag{4.3.71}$$

where $|\Lambda|$ is a diagonal matrix with the absolute values of the three eigenvalues of A as its diagonal elements, and R is a 3×3 matrix containing the three eigenvectors (R_1, R_2 and R_3) as its three columns. They divided the path of the integral into three parts, namely, Λ_1 ($0 \le \sigma < 1/3$), Λ_2 ($1/3 \le \sigma < 2/3$) and Λ_3 ($2/3 \le \sigma \le 1$), and chose the following functional relationship between U and σ:

$$dU/d\sigma = R_3 \text{ on } \Lambda_1; \quad dU/d\sigma = R_2 \text{ on } \Lambda_2 \quad \text{and} \quad dU/d\sigma = R_1 \text{ on } \Lambda_3 \qquad (4.3.72)$$

The choice of these paths is such that, on each path, two Riemann invariants are constants (details of the derivation are given in Wesseling 2004). Thus,

$$du - dp/(\rho c) = 0 \quad \text{and} \quad ds = 0 \text{ on } \Lambda_1 \qquad (4.3.73a)$$

$$du + dp/(\rho c) = 0 \quad \text{and} \quad du - dp/(\rho c) = 0 \text{ on } \Lambda_2 \qquad (4.3.73b)$$

$$p\rho^{-\gamma} = \text{constant} \quad \text{and} \quad u - 2c/(\gamma - 1) = \text{constant on } \Lambda_3 \qquad (4.3.73c)$$

Using these invariants, the values of $\{\rho, u, p\}$ can be determined at the end points of Λ_1, Λ_2 and Λ_3. It is also possible to determine the value of σ at which sonic point occurs on Λ_1 and Λ_3, if at all it occurs. Let these be denoted, if they exist, by σ_1 and σ_3. The values of ρ, u and p can also be evaluated at σ_1 and σ_3 using the Riemann invariants. There cannot be a sonic point on Λ_2 for Euler equations. With these values ρ, u and p become available at all end points, the values of components of the flux vector, F, can also be determined at the corresponding end points. Now, the integration in Eq. (4.3.70) on a given path can be carried out in the following way. Taking the integration from $\sigma = 0$ to $\sigma = \sigma_1$ on Λ_1, one can write

$$\begin{aligned}
\int_0^{\sigma_1} |A| dU &= \int_0^{\sigma_1} |A| \frac{dU}{d\sigma} d\sigma = \int_0^{\sigma_1} |A| R_3 d\sigma = \int_0^{\sigma_1} |A| R_3 d\sigma = \text{sign}\{\lambda_3(0)\} \int_0^{\sigma_1} \lambda_3 R_3 d\sigma \\
&= \text{sign}\{\lambda_3(0)\} \int_0^{\sigma_1} A(U) dU = \text{sign}\{\lambda_3(0)\} \int_0^{\sigma_1} \frac{\partial F}{\partial U} dU \\
&= \text{sign}\{\lambda_3(0)\}[F\{U(\sigma_1)\} - F\{U(0)\}] = \text{sign}\{\lambda_3(0)\}(F_{\sigma_1} - F_0)
\end{aligned} \qquad (4.3.74)$$

Here, $F\{U(\sigma_1)\}$ is written as F_{σ_1}. Evaluating the remaining the parts of the integral similarly, the numerical flux in the Osher scheme for the Euler equations can be written as

$$F^*_{i+1/2} = \frac{1}{2}[1+\text{sign}\{\lambda_3(0)\}]F_0 + \frac{1}{2}\left[\text{sign}\left\{\lambda_3\left(\frac{1}{3}\right)\right\} - \text{sign}\{\lambda_3(0)\}\right]F_{\sigma_1}$$
$$+\frac{1}{2}\left[\text{sign}\{u_{1/3}\} - \text{sign}\left\{\lambda_3\left(\frac{1}{3}\right)\right\}\right]F_{1/3} + \frac{1}{2}\left[\text{sign}\left\{\lambda_1\left(\frac{2}{3}\right)\right\} - \text{sign}\{u_{1/3}\}\right]F_{2/3}$$
$$+\frac{1}{2}\left[\text{sign}\{\lambda_1(1)\} - \text{sign}\left\{\lambda_1\left(\frac{2}{3}\right)\right\}\right]F_{\sigma_3} + \frac{1}{2}[1-\text{sign}\{\lambda_1(1)\}]F_1 \quad (4.3.75)$$

4.3.6 Flux Vector Splitting Schemes

The Osher-Chakrvarthy scheme gives good predictions of shocks while inherently satisfying the Rankine-Hugoniot jump conditions and the entropy condition for prevention of expansion shock. All the three schemes, namely, the Godunov scheme, the entropy-fixed Roe scheme and the Osher-Chakravarthy scheme, are known as flux difference schemes. These upwinding schemes are accurate in shock resolution, but require considerable computational effort. In response to this, a different class of upwinding schemes, known as flux vector splitting schemes, has also been developed for the solution of Euler equations. The best of known of these is the Steger and Warming (1981) approach and its derivatives. This scheme, which is applicable for a system of equations with a homogeneous flux vector [see Eq. (4.3.48)], is briefly explained below.

Following Eq. (4.3.71), we write the flux vector, F, as

$$F = AU = R\{\Lambda\}R^{-1}U = \left[R\{\Lambda^+\}R^{-1} + R\{\Lambda^-\}R^{-1}\right]U = [\{A^+\} + \{A^-\}]U$$
$$= F^+ + F^- \quad (4.3.76)$$

where the diagonal matrix of eigenvalues is divided into one containing only the positive elements ($\{\Lambda^+\}$) and one containing only the negative elements ($\{\Lambda^-\}$), and F^+ and F^- denote the flux associated with the waves propagating in the positive and the negative x-directions, respectively. For the one-dimensional Euler equations with eigenvalues of u, u + c and u − c, where u can be positive or negative, the three diagonal elements of the positive diagonal matrix can be written as

$$\Lambda_1^+ = (u+|u|)/2, \Lambda_2^+ = [(u+c)+|u+c|]/2, \quad \text{and} \quad \Lambda_3^+ = [(u-c)+|u-c|]/2 \quad (4.3.77)$$

$$\Lambda_1^- = (u-|u|)/2, \Lambda_2^- = [(u+c)-|u+c|]/2, \Lambda_3^- = [(u-c)-|u-c|]/2 \quad (4.3.78)$$

For supersonic flow, for which all the three eigenvalues are positive for positive u, the above splitting gives

$$\{\Lambda^{-}\} = 0 \quad \text{and} \quad \{\Lambda^{+}\} = \{\Lambda\} \tag{4.3.79}$$

This leads to

$$\mathrm{F}^{+} = \mathrm{F} \quad \text{and} \quad \mathrm{F}^{-} = 0 \quad \text{for} \quad \mathrm{u} > \mathrm{c} \tag{4.3.80}$$

For subsonic flow, for the case of positive u, the first two eigenvalues are positive and the third one is negative. For this case,

$$\{\Lambda^{+}\} = \begin{Bmatrix} u & 0 & 0 \\ 0 & u+c & 0 \\ 0 & 0 & 0 \end{Bmatrix} \quad \{\Lambda^{-}\} = \begin{Bmatrix} 0 & 0 & 0 \\ 0 & 0 & 0 \\ 0 & 0 & u-c \end{Bmatrix} \tag{4.3.81}$$

The corresponding split flux vectors are as follows:

$$\mathrm{F}^{+} = \frac{\rho}{2\gamma}\begin{Bmatrix} 2\gamma \mathrm{u} + \mathrm{c} - \mathrm{u} \\ 2(\gamma-1)\mathrm{u}^2 + (\mathrm{u}+\mathrm{c})^2 \\ (\gamma-1)\mathrm{u}^3 + \frac{(\mathrm{u}+\mathrm{c})^3}{2} + \frac{(3-\gamma)(\mathrm{u}+\mathrm{c})\mathrm{c}^2}{2(\gamma-1)} \end{Bmatrix} \quad \text{if } 0 \le \mathrm{u} \le \mathrm{c} \tag{4.3.82a}$$

$$\mathrm{F}^{-} = \frac{\rho}{2\gamma}\begin{Bmatrix} \mathrm{u} - \mathrm{c} \\ (\mathrm{u}-\mathrm{c})^2 \\ \frac{(\mathrm{u}-\mathrm{c})^3}{2} + \frac{(3-\gamma)(\mathrm{u}-\mathrm{c})\mathrm{c}^2}{2(\gamma-1)} \end{Bmatrix} \quad \text{if } 0 \le \mathrm{u} \le \mathrm{c} \tag{4.3.82b}$$

$$\mathrm{F}^{+} = \mathrm{F}, \quad \mathrm{F}^{-} = 0 \quad \text{if } \mathrm{u} > \mathrm{c} \tag{4.3.82c}$$

The formulations given in Eqs. (4.3.77) and (4.3.78) for $\{\Lambda^{+}\}$ and $\{\Lambda^{-}\}$ will always give positive eigenvalues in $\{\Lambda^{+}\}$ and negative values in $\{\Lambda^{-}\}$ irrespective of the sign of u. This splitting of the eigenvalues is not unique; other possibilities have been suggested by Steger & Warming themselves. They therefore derived a generalized flux vector $\mathcal{F}$ for one-dimensional Euler equations for a perfect gas in terms of generalized eigenvalues $\widehat{\lambda}_k$:

$$\mathcal{F} = \frac{\rho}{2\gamma}\begin{Bmatrix} 2(\gamma-1)\widehat{\lambda}_1 + \widehat{\lambda}_2 + \widehat{\lambda}_3 \\ 2(\gamma-1)\widehat{\lambda}_1 u + \widehat{\lambda}_2(\mathrm{u}+\mathrm{c}) + \widehat{\lambda}_3(\mathrm{u}-\mathrm{c}) \\ (\gamma-1)\widehat{\lambda}_1\mathrm{u}^2 + \frac{\widehat{\lambda}_2}{2}(\mathrm{u}+\mathrm{c})^2 + \frac{\widehat{\lambda}_3}{2}(\mathrm{u}-\mathrm{c})^2 + \frac{(3-\gamma)\left(\widehat{\lambda}_2 + \widehat{\lambda}_3\right)\mathrm{c}^2}{2(\gamma-1)} \end{Bmatrix} \tag{4.3.83}$$

The generalized flux vector $\mathcal{F}$ of Eq. (4.3.83) is such that it can be used to compute F^{+} and F^{-} by substituting the corresponding values of $\{\Lambda^{+}\}$ and $\{\Lambda^{-}\}$

from Eqs. (4.3.77) and (4.3.78), respectively, into Eq. (4.3.83). Steger and Warming also gave the corresponding generalized flux vector matrices for two- and three-dimensional Euler equations, and the interested reader is referred to the original paper.

Once the split flux vectors are determined, the governing system of equations (Eq. 4.3.41a) can be written as

$$\frac{\partial U}{\partial t} + \frac{\partial F^{+}}{\partial x} + \frac{\partial F^{-}}{\partial x} = 0 \tag{4.3.84}$$

Using a single-step forward differencing scheme for the time derivative, and first-order upwind schemes for the two flux derivatives, one can write the discretized equation as

$$U_i^{n+1} = U_i^n - \Delta t\left[\left\{(F^+)_i^n - (F^+)_{i-1}^n\right\}/\Delta x + \left\{(F^-)_{i+1}^n - (F^-)_i^n\right\}/\Delta x\right] \tag{4.3.85}$$

It may be noted that backward differencing has been used to evaluate $\partial F^+/\partial x$ and forward differencing for $\partial F^-/\partial x$. Multi-step time integration schemes and higher order upwind differencing schemes can be readily incorporated to discretize and solve Eq. (4.3.84).

The Steger and Warming scheme (Steger 1978; Steger and Warming 1981; van Leer 1982) does not incorporate the Rankine-Hugoniot jump conditions and cannot distinguish between a compression shock and an expansion shock. It also has problems at sonic transitions and stagnation points due to slope discontinuities in the split mass flux. However, it is easier to program and compute in comparison with the flux difference splitting methods. Combinations of these two approaches, namely, the flux difference splitting and the flux vector splitting approaches, such as the advection upstream splitting method (Liou and Steffen 1993; Liou 1996; Liou 2006; Kim et al. 2001), have also been developed. These methods have spawned a number of variants and improvements over the years, and the interested reader is referred to the current literature for a discussion and assessment of these. Shock/boundary layer, shock/wall and shock/turbulence interactions are further aspects that are important in practical flows, especially in turbomachinery applications. For a physical perspective on the shock wave/boundary layer interaction, the reader is referred to Délery and Dussauge (2009).

4.4 Solution Methods for Incompressible Flows

As mentioned earlier, incompressibility is a flow property rather than a fluid property. Thus, the flow of a compressible fluid such as air can be incompressible if the air velocity is less than about 100 m/s in ambient conditions. In other words, the case of a car travelling at 100 m/s or 360 kmph, can therefore be treated as incompressible flow. The solution of the Navier-Stokes equations for incompressible flows is complicated by the lack of an independent equation for the pressure, the gradient of which appears in each of the three momentum equations. Further, the continuity equation does not have a dominant variable in incompressible flows unlike in compressible flows, where it can be used to determine the density from which pressure is obtained using an equation of state. In incompressible flows, the density is constant and it therefore drops out of the continuity equation altogether. The continuity equation is thus reduced to imposing a divergence-free condition on the velocity field:

$$\nabla \cdot \mathbf{u} = 0 \tag{4.4.1}$$

In addition, in flows with constant thermophysical properties (the fluid density, viscosity, specific heat and thermal conductivity), the momentum equations get decoupled from the energy balance equation. The continuity and the momentum balance equations need to be solved together; however, the energy balance equation, usually expressed in terms of temperature as the main variable, can be solved after the velocity field is computed.

The absence of the density-enabled linkage between the continuity and the momentum equations means that methods developed for compressible flows, such as those discussed in Sect. 4.2, cannot be applied to incompressible flows. For example, consider the stability limits of the MacCormack method given by Eqs. (4.2.17a, b). For nearly incompressible flows, the speed of sound is very large, and the allowable time step becomes very small. The latter tends to zero for incompressible flows. Implicit formulations such as the Beam-Warming scheme discussed in Sect. 4.2.2 will suffer time step limitations arising from approximations made in linearization and coupling. Special methods are therefore required for incompressible flows to evaluate pressure and to provide a linkage between the continuity and the momentum equations. Several methods are available. Four methods, which encompass the range of approaches developed, are discussed below.

4.4.1 Artificial Compressibility Approach

As the name implies, in this method, which was proposed originally by Chorin (1967), an artificial linkage between the continuity and the momentum equations is created by introducing an additional term in the continuity equation as follows:

$$\partial(\rho^*)/\partial \mathrm{t} + \nabla \cdot \mathrm{u} = 0 \tag{4.4.2}$$

The fictitious density, ρ^*, is linked to the pressure through a fictitious compressibility factor, β, defined as $\rho^* = p/\beta^2$. The pressure linkage can be readily seen by writing the continuity and the momentum equations as

$$(1/\beta^2)\partial p/\partial t + \nabla \cdot \mathbf{u} = 0 \tag{4.4.3a}$$

$$\partial \mathbf{u}/\partial t + \nabla \cdot (\mathbf{uu}) = -(1/\rho)\,\nabla p + (\mu/\rho)\nabla^2 \mathbf{u} \tag{4.4.3b}$$

where ρ and μ are the true properties of the fluid. Equation (4.4.2) can now be solved for ρ* using the methods described earlier. Using the known compressibility factor (β), the pressure can be evaluated. This can then be used in the momentum equations to find the velocities. The velocities are of course initially incorrect as they satisfy the modified continuity equation. The solution is marched forward in time until a steady state is reached at which point the additional term in the continuity equation (the first term in Eq. 4.4.3a), which is a time derivative, becomes zero and the velocity and the pressure satisfy the exact N-S equations. The value of the parameter β has thus no influence on the steady state solution but does influence the evolution to steady state. As it determines the effective speed of propagation of the pressure wave, it may also have an effect on the stability of a given scheme. For truly incompressible flows, the wave speed is infinite, but with the artificial compressibility introduced into the equations, the speed of propagation, *c*, becomes $c = \sqrt{(u^2 + \beta^2)}$. The value of β should be chosen to be high so that the propagation of the pseudo-compressibility waves does not affect the convergence of the solution. Based on the ratio of the time scales for the propagation of pressure and viscous diffusion effects, Chang and Kwak (1984) suggested $\beta^2/u_{ref}^2 = 5$ to 10 for duct flows. The calculation of external flows is less sensitive to β at high Reynolds numbers as the boundary layer thickness is expected to be small for such flows. While the method has been successfully applied for the computation of incompressible flows (Choi and Merkle 1985; Rizzi and Eriksson 1985), it is clear that it does not generate a time-accurate solution and that it can therefore be applied only for steady state flow fields.

4.4.2 Streamfunction-Vorticity Approach

Another way to deal with the difficulty that pressure appears in the momentum equations but not in the continuity equation for incompressible flows has been to eliminate pressure from the momentum equation. For two dimensional flows, this is possible by the introduction of a new variable, called streamfunction (lines of constant streamfunction values are the streamlines), which enables the three coupled equations involving u, v and p to be rewritten as two coupled equations involving streamfunction and vorticity and one decoupled Poisson equation for pressure. We can illustrate this as follows for the case of a two-dimensional flow in Cartesian coordinates for which the governing equations, assuming constant properties, become

$$\partial u/\partial x + \partial v/\partial y = 0 \tag{4.4.4}$$

$$\partial u/\partial t + \partial(u^2)/\partial x + \partial(uv)/\partial y = -1/\rho\, \partial p/\partial x + \mu/\rho\left(\partial^2 u/\partial x^2 + \partial^2 u/\partial y^2\right) \tag{4.4.5}$$

$$\partial v/\partial t + \partial(uv)/\partial x + \partial(v^2)/\partial y = -1/\rho\, \partial p/\partial y + \mu/\rho\left(\partial^2 v/\partial x^2 + \partial^2 v/\partial y^2\right) \tag{4.4.6}$$

We introduce the new variable streamfunction, ψ, defined such that the continuity equation is satisfied. For 2-d Cartesian coordinates, ψ is such that

$$u = \partial\psi/\partial y \quad \text{and} \quad v = -\partial\psi/\partial x \tag{4.4.7}$$

One may check by substitution that the continuity equation, $\nabla \cdot \mathbf{u} = 0$, is satisfied. The other variable, vorticity, defined as $\boldsymbol{\omega} = \nabla \times \mathbf{u}$, is already introduced in Chap. 2 as one which defines the rate of rotation of a fluid element. In the Cartesian x-y plane, the rotational rate component in the z-direction can be written as

$$\omega_z = \partial v/\partial x - \partial u/\partial y \tag{4.4.8}$$

Pressure can now be eliminated from the momentum equations by taking the curl of the momentum equation, i.e., by evaluating $\partial/\partial y$(Eq. 4.4.5) − $\partial/\partial x$(Eq. 4.4.6). After algebraic manipulations, one can derive the following vorticity transport equation:

$$\partial\omega_z/\partial t + u\partial\omega_z/\partial x + v\partial\omega_z/\partial y = (\mu/\rho)\left(\partial^2\omega_z/\partial x^2 + \partial^2\omega_z/\partial y^2\right) \tag{4.4.9a}$$

or

$$\partial\omega_z/\partial t + (\partial\psi/\partial y)\partial\omega_z/\partial x - (\partial\psi/\partial x)\partial\omega_z/\partial y = (\mu/\rho)\left(\partial^2\omega_z/\partial x^2 + \partial^2\omega_z/\partial y\right) \tag{4.4.9b}$$

The definition of ω_z (Eq. 4.4.8) provides one more equation involving the two variables, namely,

$$\partial^2\psi/\partial x^2 + \partial^2\psi/\partial y^2 = -\omega_z \tag{4.4.10}$$

Equations (4.4.9a, b) and (4.4.10) can be solved together to get $\psi(x, y, t)$ from which the velocity field can be obtained using Eq. (4.4.7). Thus, the velocity field is obtained without solving explicitly for pressure. If pressure is to be evaluated, then the following Poisson equation can be derived for pressure by taking divergence of the momentum equation, i.e., by evaluating $(\partial/\partial x)(4.4.5) + (\partial/\partial y)(4.4.6)$, and simplifying:

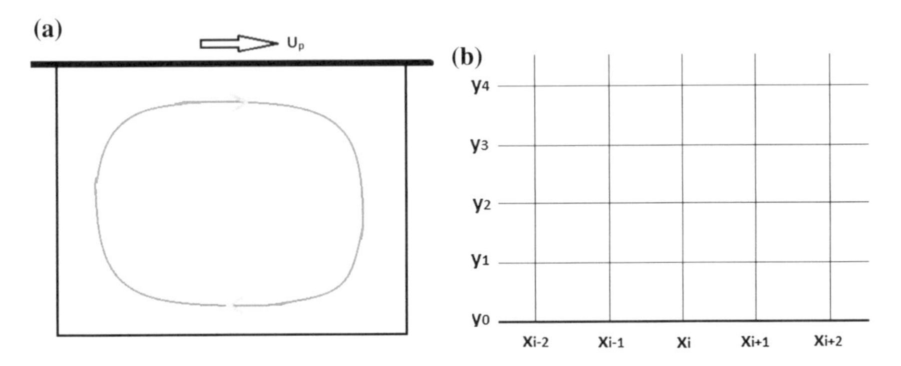

Fig. 4.7 **a** Schematic diagram of the lid-driven cavity problem. The infinitely long top plate moves horizontally at a constant velocity of U_p. The circulation pattern induced inside the cavity varies significantly with U_p and the viscosity of the fluid. **b** A simple grid near the bottom wall

$$\partial^2 p/\partial x^2 + \partial^2 p/\partial y^2 = 2\rho[(\partial u/\partial x)(\partial v/\partial y) - (\partial u/\partial y)(\partial v/\partial x)] \qquad (4.4.11)$$

Unlike the artificial compressibility method, this approach can be used to generate time-accurate solutions, but only for 2-dimensional flows. The solution is not trivial as Eqs. (4.4.9a, b) is a second order partial differential equation with highly non-linear coefficients. The specification of the boundary conditions too is not straightforward as it requires boundary conditions for ψ and ω which are in terms of gradients of the velocity components. Consider the lid-driven cavity problem of square cross-section shown in Fig. 4.7 where the top surface plates moves horizontally at a constant velocity U_p while the other bounding surfaces are the walls of a cavity and thus have zero velocity.

In a "primitive variables" approach where we solve directly for u and v, the specification of the velocity boundary conditions is straightforward. In the streamfunction-vorticity approach, one needs to specify the values of the streamfunction and vorticity on these walls. These can be derived as follows. Since walls are impermeable, a constant/ reference value, usually zero, is specified for the streamfunction on all the walls. With reference to Fig. 4.1b, the zero velocity boundary condition at the bottom wall ($y = y_0$) can be incorporated as follows by expressing $\psi(x_i, y_1)$ in terms of $\psi(x_i, y_0)$:

$$\psi(x_i, y_1) = \psi(x_i, y_0) + \Delta y\, \partial\psi/\partial y|_{y0} + \Delta y^2/2!\, \partial^2\psi/\partial y^2|_{y0} + O(\Delta y^3) \qquad (4.4.12)$$

We have $\psi(x_i, y_0) = \psi_{ref} = 0$ and $\partial\psi/\partial y|_{y0} = u_{wall} = 0$. Also, since $v = 0$ along the wall, $(\partial v/\partial x)_{wall} = 0$. Thus,

$$\omega_z(x_i, y_0) = [\partial v/\partial x]|_{(xi,y0)} - [\partial u/\partial y]|_{(xi,y0)} = -[\partial u/\partial y]|_{(xi,y0)} = -\partial^2\psi/\partial y^2|_{y0} \qquad (4.4.13)$$

Substituting Eq. (4.4.13) into Eq. (4.4.12), we get a boundary condition for Eq. (4.4.10), namely,

$$\psi(x_i, y_1) = -\Delta y^2/2!\,\omega_z(x_i, y_0) \quad (4.4.14)$$

Similarly, the boundary condition for Eqs. (4.4.9a, b) is given by Eq. (4.4.13), namely,

$$\omega_z(x_i, y_0) = -\partial^2\psi/\partial y^2(x_i, y_0) \quad (4.4.15)$$

Using a second order accurate one-sided difference approximation for the second derivative $\partial^2\psi/\partial y^2|_{y0}$, one can arrive at the following boundary condition (assuming uniform spacing in the y-direction) for Eq. (4.4.9a, 9b) at the bottom wall:

$$\omega_z(x_i, y_0) = (1/\Delta y^2)[-3\psi(x_i, y_0) + 4\psi(x_i, y_1) + \psi(x_i, y_2)] \quad (4.4.16)$$

The boundary condition for pressure can be derived from writing the x-momentum conservation equation (Eq. 4.4.5) at the wall:

$$\partial p/\partial x|_{wall} = \mu\partial^2 u/\partial y^2|_{wall} = -\mu\partial\omega_z/\partial y|_{wall} \quad (4.4.17)$$

The treatment of the left and the right walls in Fig. 4.7 can be done in an analogous way. For the top wall, we make use of the condition that $\partial\psi/\partial y|_{yp} = U_p$ and that $\psi\ (x_i,\ y_p) = 0$ where y_p refers to the height of the plate. An iterative procedure is required for the solution of Eqs. (4.4.9b) and (4.4.10). Starting with an assumed velocity field, we solve Eq. (4.4.9a) subject to vorticity boundary conditions such as that given by Eq. (4.4.15). This is then used to solve the Poisson equation for the streamfunction given by Eq. (4.4.10) subject to streamfucntion boundary conditions such as those given by Eq. (4.4.14). The updated streamfunction is then used to solve the vorticity transport equation once again, and so on until convergence. A flow chart for the iterative solution is given in Fig. 4.8.

The definition of the streamfunction varies with the type of flow. For example, the steady flow around a stationary sphere (Fig. 4.9) can be reduced to a two-dimensional flow if one takes advantage of the axisymmetry of the problem. In this case, the problem is described in axisymmetric spherical coordinates with r and θ being the coordinate directions in which the flow variables (velocity components and pressure) are expected to change. In this case, the streamfunction is expressed as

$$u_r = \frac{1}{r^2 \sin\theta}\frac{\partial\psi}{\partial\theta} \quad \text{and} \quad u_\theta = -\frac{1}{r\sin\theta}\frac{\partial\psi}{\partial r} \quad (4.4.18)$$

By substitution, one can verify that this definition satisfies the continuity equation in spherical coordinates, namely,

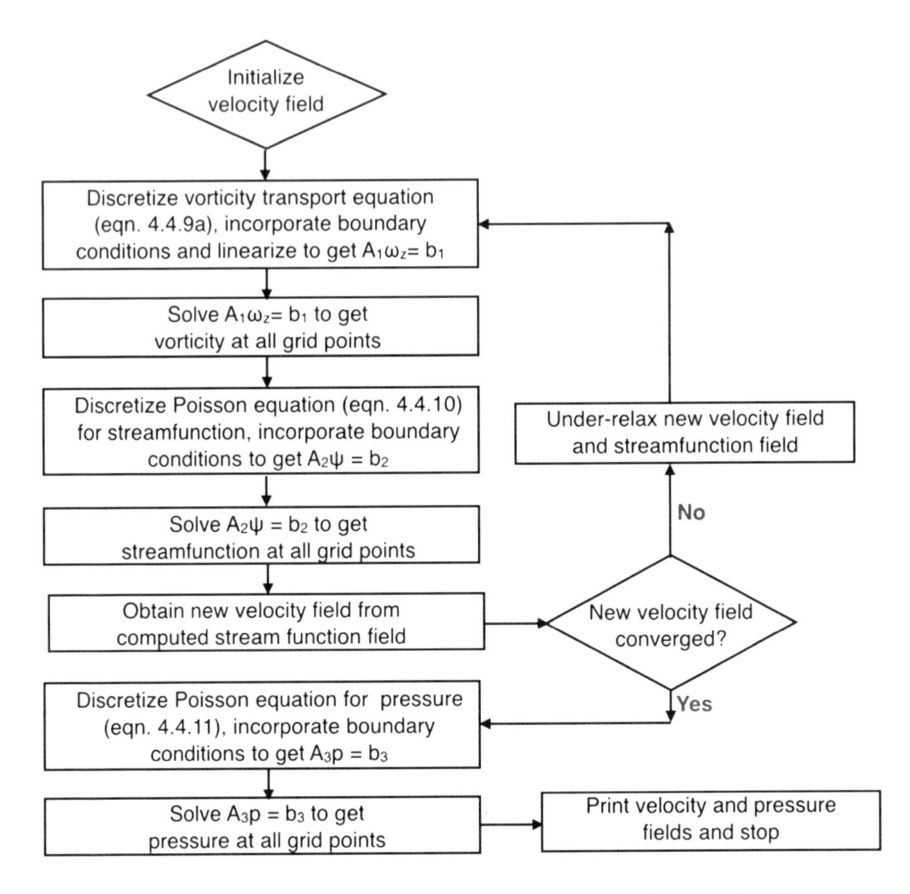

Fig. 4.8 Flow chart for the calculation of the flow field using the streamfunction-vorticity approach

$$\frac{1}{r^2}\frac{\partial}{\partial r}\left(r^2 u_r\right) + \frac{1}{r\,\sin\theta}\frac{\partial}{\partial\theta}\left(u_\theta \sin\theta\right) = 0 \tag{4.4.19}$$

The vectorial forms of the governing equations (Eqs. 4.4.9a, b and 4.4.10) should be used in spherical coordinates. The definition of streamfunction is similarly different for two-dimensional flows in other coordinates.

4.4.3 *Pressure Equation Approach*

As can be seen from the vorticity-streamfunction approach, it is possible to compute the velocity field without explicitly computing pressure. It is sufficient that the velocity field computed from the momentum equation also satisfies the continuity

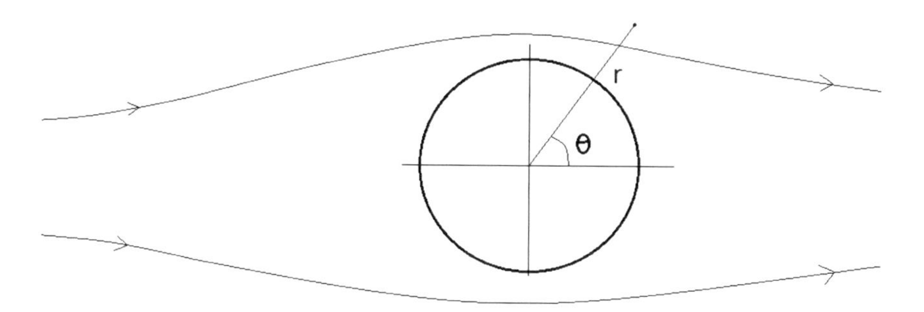

Fig. 4.9 The case of flow over a stationary sphere

equation or vice versa. Thus, one can see a specific role for pressure in incompressible flow, namely, that of enforcing continuity. The various pressure-based computations of incompressible Navier Stokes equations use this principle to generate a linkage between the continuity equation and pressure. Unlike the streamfunction-vorticity method, this linkage enables solution directly in terms of "primitive" variables, i.e., velocity components themselves. Several pressure-based methods still in use today have their roots in ideas developed in the 1960s (Harlow and Welch 1965, 1966; Chorin 1968; Temam 1969). Methods based on the evaluation of pressure through a Poisson equation are discussed here while those based on a Poisson equation for pressure correction are discussed in Sect. 4.4.4.

For a 2-d flow, we have derived in Sect. 4.4.2 a Poisson equation for pressure by taking divergence of the momentum equation. For three-dimensional, unsteady incompressible flow with constant properties, this equation can be written as

$$\frac{\partial}{\partial x_i}\left(\frac{\partial p}{\partial x_i}\right) = -\rho \frac{\partial}{\partial x_i}\left[\frac{\partial(u_i u_j)}{\partial x_j}\right] \tag{4.4.20}$$

This means that if the velocity field is known, then the pressure can be evaluated. For the solution of {u, v, w, p} to be the true solution, the velocity field used to evaluate the pressure field should satisfy the continuity equation while the pressure field so derived should, when substituted back into the momentum equation, produce a velocity field that satisfies the continuity equation. Using these principles, a sequential time-marching method of successive evaluations of the pressure and the velocity fields can be constructed as follows. Let us assume that the velocity field is known, for example, as part of the specification of the initial condition. We need to evaluate the corresponding pressure field which when substituted into the momentum will give a continuity-satisfying velocity field. For this, consider the discretized form of the momentum equation

$$\delta(u_i)/\delta t = -\delta(u_i u_j)/\delta x_j + \nu\delta\left[(\delta u_i/\delta x_j)\right]/\delta x_j - (1/\rho)\,\delta p/\delta x_i \tag{4.4.21}$$

which can be written in an explicit scheme where the right hand side is evaluated at time step n as

$$u_i^{n+1} - u_i^n = \Delta t[\delta\{-u_i u_j + \nu(\delta u_i/\delta x_j)\}/\delta x_j]_i^n - (\Delta t/\rho)(\delta p/\delta x_i)_i^n \tag{4.4.22}$$

Taking divergence of (4.4.22), we get

$$\begin{aligned}\partial(u_i^{n+1})/\partial x_i - \partial(u_i^n)/\partial x_i = {} & \Delta t\,\partial\{[\delta\{-u_i u_j + \nu(\delta u_i/\delta x_j)\}/\delta x_j]_i^n\}/\partial x_i \\ & - (\Delta t/\rho)\partial\left[(\delta p/\delta x_i)_i^n\right]/\partial x_i\end{aligned} \tag{4.4.23}$$

From the given (or computed) velocity field u_i^n, we wish to evaluate p such that the u_i^{n+1} satisfies the continuity equation. Therefore, we set $\partial(u_i^{n+1})/\partial x_i = 0$ in Eq. (4.4.23). Since the velocity field at u_i^n satisfies continuity, the second term on the LHS of Eq. (4.4.23) is also zero. The resulting equation can now be written as an equation for pressure:

$$(1/\rho)\partial\left[(\delta p/\delta x_i)_i^n\right]/\partial x_i = \partial\{[\delta\{-u_i u_j + \nu(\delta u_i/\delta x_j)\}/\delta x_j]_i^n\}/\partial x_i \tag{4.4.24}$$

Equation (4.4.24) permits the evaluation of p_i^n from continuity-satisfying u_i^n. This pressure field can then be used in Eq. (4.4.22) to evaluate the velocity field at the next time step, i.e., u_i^{n+1}. This can then be used to obtain p_i^{n+1}, and then u_i^{n+2}, p_i^{n+2}, and so on. We thus start with a continuity-satisfying velocity field u^0, solve Eq. (4.4.24) to get p^0, then use Eq. (4.4.19) to get u^1, then use Eq. (4.4.24) to p^1, then use Eq. (4.2.22) to get u^2, and progress until we reach the end of the simulation time. In this way, a time-accurate, three-dimensional solution can be obtained for the N-S equations.

An implicit pressure-equation can be constructed by writing the implicit version of Eq. (4.4.21) as

$$u_i^{n+1} - u_i^n = \Delta t\,[\delta\{-u_i u_j + \nu(\delta u_i/\delta x_j)\}/\delta x_j]_i^{n+1} - (\Delta t/\rho)(\delta\, p/\delta x_i)_i^{n+1} \tag{4.4.25}$$

Taking divergence of (4.4.25), we get

$$\begin{aligned}\partial(u_i^{n+1})/\partial x_i - \partial(u_i^n)/\partial x_i = {} & \Delta t\,\partial\left\{[\delta\{-u_i u_j + \nu(\delta u_i/\delta x_j)\}/\delta x_j]_i^{n+1}\right\}/\partial x_i \\ & - (\Delta t/\rho)\partial\left[(\delta p/\delta x_i)_i^{n+1}\right]/\partial x_i\end{aligned} \tag{4.4.26}$$

Setting the left hand side to zero in order to impose the condition of continuity at both nth and (n + 1)th time steps, the equation for pressure now becomes

$$(1/\rho)\partial\left[(\delta p/\delta x_i)_i^{n+1}\right]/\partial x_i = \partial\left\{[\delta\{-u_i u_j + \nu(\delta u_i/\delta x_j)\}/\delta x_j]_i^{n+1}\right\}/\partial x_i \tag{4.4.27}$$

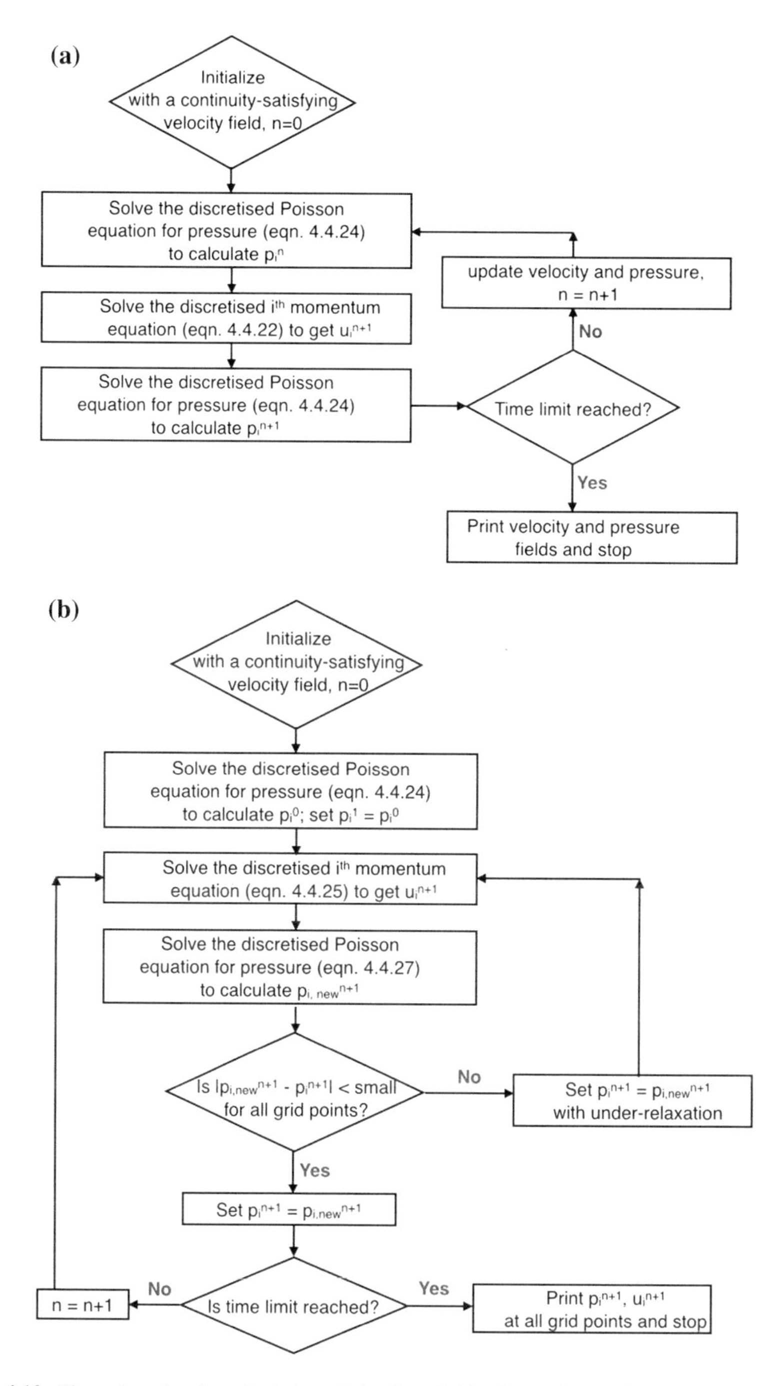

Fig. 4.10 Flow chart for the calculation of the flow field using **a** the explicit and **b** the implicit pressure equation approach

Unlike in the explicit method, Eq. (4.4.27) cannot be used to compute the pressure field as the velocity field at u_i^{n+1} is not known. Hence, Eqs. (4.4.25) and (4.4.27) need to be solved iteratively to get u_i^{n+1} and p_i^{n+1}, which will then obviously satisfy the discretized form of the momentum equation and the discretized continuity equation. The explicit and the implicit pressure equation approaches are contrasted in Fig. 4.10 through a flow chart of the calculation procedure in each case.

It can be seen that it is thus possible to generate the velocity and pressure fields while working in this primitive variable form. Both the explicit and the implicit versions can be used to compute time-accurate solutions in a three-dimensional flow situation and thus they do not suffer from the limitations of limited applicability of the methods outlined in Sects. 4.4.1 and 4.4.2. It may be noted that the pressure equation is a Poisson equation and thus requires boundary conditions. At inlets and outlets, one uses a Neumann boundary condition for pressure, namely, that the normal pressure gradient is zero. At the walls, using the no-slip boundary condition reduces the normal pressure gradient to $\partial p/\partial n = -(\nabla \cdot \tau)_n$. At high Reynolds numbers, where a boundary layer-like situation prevails near the walls, the boundary layer condition of zero normal pressure gradient can be applied at the walls too.

Harlow and Welch (1965) introduced the idea of staggered grid system (see Fig. 4.11a) in which the locations at which the pressure and the velocity components are evaluated are displaced relative to each other. Thus, if (i, j, k) are the locations at which the pressure is evaluated, then u is evaluated at (i + ½, j, k), v at (i, j + ½, k) and w at (i, j, k + ½). Using the staggered grid, it is possible to choose the boundaries of the computational mesh to coincide with planes at which the normal velocity components are evaluated (see Fig. 4.11b). This obviates the need for the determination of pressure at the boundaries. Use of a staggered grid gives rise to strong linkage between the local velocity field and the local pressure gradients, and eliminates the possibility of occurrence of chequerboard oscillations in the solution. This will be discussed further in Sect. 4.4.6 below.

So far, we have not said anything about the evaluation of the velocity field, i.e., the simultaneous solution of the three momentum conservation equations in the case of incompressible flows which are once again non-linear and coupled. This is discussed in detail in the following section in connection with the pressure correction approach.

4.4.4 Pressure Correction Approach

One of the earliest methods of the pressure correction approach is that of Chorin (1968), Temam (1969) who suggested a different way of constructing the pressure equation. In this fractional step method, a provisional velocity $u{*}_i^n$ is computed from u_i^n by advancing the momentum equation without the pressure gradient. The provisional u* may not satisfy the continuity equation. Therefore, a velocity correction is now sought so that the new velocity u_i^{n+1} includes the influence of the

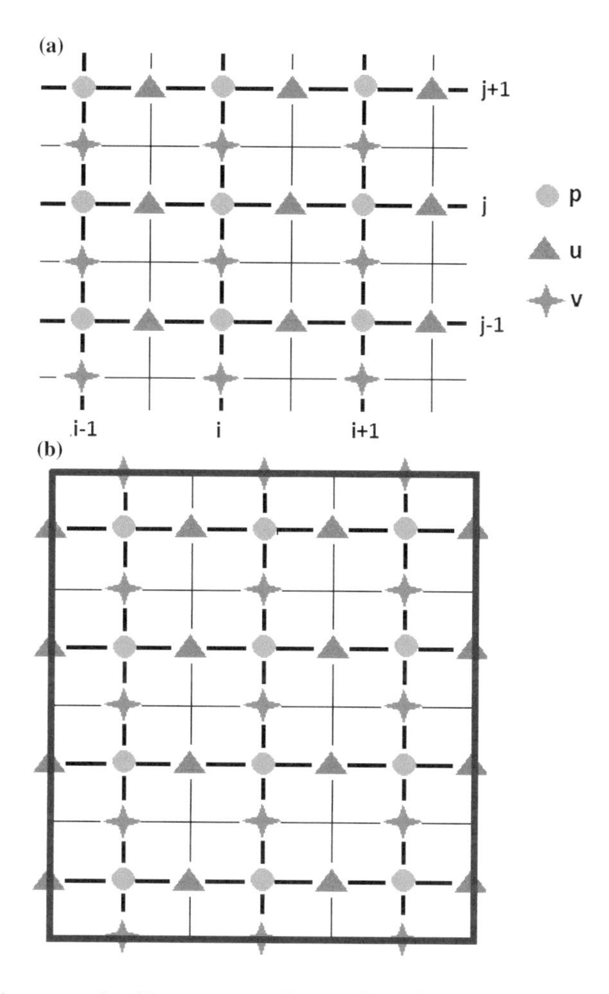

Fig. 4.11 **a** A staggered grid arrangement in two-dimensions; locations of evaluation of p (circles), u (triangles) and v (stars). **b** Staggered grid arrangement for a rectangular domain showing location u and v components of velocity on the left/right and top/bottom boundaries, respectively

pressure gradient and also satisfies the continuity. The influence of the pressure gradient can be included in u_i^{n+1} by evaluating it as

$$\left(u_i^{n+1} - u^*_i\right)/\Delta t = -(1/\rho)(\nabla p)_i \tag{4.4.28}$$

The satisfaction of the continuity equation can now be enforced by taking divergence of Eq. (4.4.28) and setting $\nabla.\ (u_i^{n+1}) = 0$. This now gives a Poisson equation for pressure as

$$\nabla \cdot \left[(\Delta t/\rho)\, \nabla p^{n+1}\right]_i = (\nabla \cdot u^*)_i^{\,n} \tag{4.4.29}$$

Chorin (1968) proposed this method on a non-staggered grid but the method has since been used on both non-staggered and staggered grids. The method can be seen to belong to the class of approaches known as operator splitting or fractional step methods.

Another popular variant of the pressure correction approach is the semi-implicit method for pressure linked equations (SIMPLE) developed in the early 1970s by Spalding and co-workers (Patankar and Spalding 1972). This approach was developed for steady flows but can be extended to unsteady flows by following a slightly modified procedure at each time step. In the SIMPLE approach, a provisional velocity field u* is obtained from the momentum conservation equations using a guessed pressure field p* using implicit evaluation of the velocities. The provisional velocity field does not satisfy the continuity equation in the general case (as it has not been used in evaluating it). Therefore, a pressure correction, p', is sought such that the corrected pressure field (p = p* + p'), when substituted in the momentum equation, will give an improved velocity field (u = u* + u') which will satisfy the continuity equation. These steps can be summarized as follows for a steady flow. The provisional velocity field u* is calculated from the discretized and linearized momentum equation solved implicitly:

$$[\nabla \cdot (\hat{u}u^*)]_{i,j,k} = -(1/\rho)\,[\nabla p^*]_{i,j,k} + \nu[\nabla^2 u^*]_{i,j,k} \tag{4.4.30}$$

where the u* is the velocity component which is solved for implicitly, while û, the advection velocity component and p*, the guessed pressure field are known (guessed or estimated or lagged from previous iteration). On a staggered grid, using upwind differencing for the advection term and central differencing for the pressure gradient and the viscous diffusion terms, Eq. (4.4.30) takes the following seven-point molecule for the x-momentum balance at grid point (i + ½, j, k):

$$\begin{aligned} &a^u_{i+1/2,j,k}u^*_{i+1/2,j,k} + a^u_{i-1/2,j,k}u^*_{i-1/2,j,k} + a^u_{i+3/2,j,k}u^*_{i+3/2,j,k} \\ &\quad + a^u_{i+1/2,j-1,k}u^*_{i+1/2,j-1,k} + a^u_{i+1/2,j+1,k}u^*_{i+1/2,j+1,k} \\ &\quad + a^u_{i+1/2,j,k-1}u^*_{i+1/2,j,k-1} + a^u_{i+1/2,j,k+1}u^*_{i+1/2,j,k+1} = b^u_{i,j,k}p^*_{i,j,k} - b^u_{i+1,j,k}p^*_{i+1,j,k} \end{aligned} \tag{4.4.31}$$

where the asterisk (*) on the variable indicates provisional values of the variables, and the superscript "u" on the coefficients a and b indicates the corresponding coefficients for the x-momentum balance equation. These coefficients include interpolated values of u and other variables as well as geometric information. Consider the two-dimensional case in Fig. 4.12 which shows the staggered grid used for the discretization of the x-component of Eq. (4.4.30) which can be written as

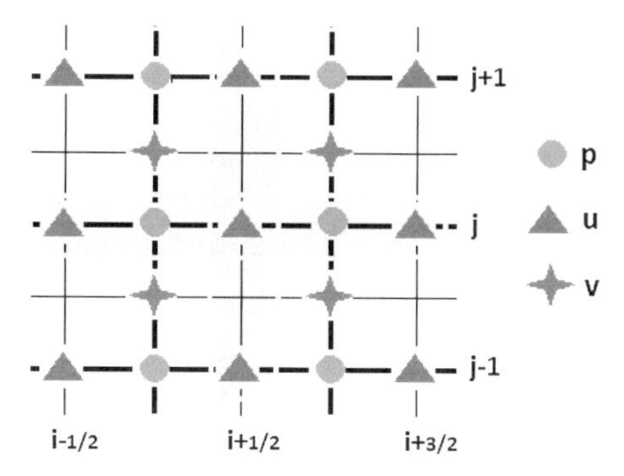

Fig. 4.12 Staggered grid for the discretization of Eq. (4.4.30)

$$\partial(\hat{u}u^*)/\partial x + \partial(\hat{v}u^*)/\partial y = -(1/\rho)\partial p^*/\partial x + \nu(\partial^2 u^*/\partial x^2 + \partial^2 u^*/\partial y^2) \quad (4.4.32)$$

When discretized around grid point (i + ½, j) which is the point where u* is evaluated in a staggered grid, we get

$$\left[(\hat{u}u^*)_{i+1,j} - (\hat{u}u^*)_{i,j}\right]/\Delta x + [(\hat{v}u^*)_{i+\frac{1}{2},j+\frac{1}{2}} - (\hat{v}u^*)_{i+\frac{1}{2},j-\frac{1}{2}}]/\Delta y = -(1/\rho)(p^*_{i+1,j} - p^*_{i,j})]/\Delta x$$
$$+ \nu(u^*_{i+\frac{3}{2},j} - 2u^*_{i+\frac{1}{2},j} + u^*_{i-\frac{1}{2},j})/\Delta x^2 + \nu(u^*_{i+\frac{1}{2},j+1} - 2u^*_{i+\frac{1}{2},j} + u^*_{i+\frac{1}{2},j-1})/\Delta y^2 \quad (4.4.33)$$

As can be seen from Fig. 4.12, the u velocity component is not evaluated at i and the v component is not evaluated at i + ½. Thus, $\hat{u}$ and u* at (i, j) and (i + 1, j) and $\hat{v}$ and u* at (i + ½, j-½) and (i + ½, j + ½) need to be estimated to evaluate the advection terms. Assuming u and v to be positive at the respective locations, and using the upwind scheme for the advection terms, these can be evaluated as follows:

$$u^*(i,j) = u^*(i - \tfrac{1}{2}, j); \quad u^*(i+1, j) = u^*(i + \tfrac{1}{2}, j) \text{ for } \hat{u} > 0 \quad (4.4.34a)$$

$$u^*(i+\tfrac{1}{2}, j+\tfrac{1}{2}) = u^*(i+\tfrac{1}{2}, j); u^*(i+\tfrac{1}{2}, j-\tfrac{1}{2}) = u^*(i+\tfrac{1}{2}, j-1) \text{ for } \hat{v} > 0 \quad (4.4.34b)$$

The values of $\hat{u}$ and $\hat{v}$ can be evaluated using central differences and Picard substitution of available latest values of variables to avoid non-linear terms in the discretized equations. Thus,

$$\hat{u}(i+1,j) = \tfrac{1}{2}[\hat{u}(i+\tfrac{1}{2},j) + \hat{u}(i+\tfrac{3}{2},j)];$$
$$\hat{u}(i,j) = \tfrac{1}{2}[\hat{u}(i-\tfrac{1}{2},j) + \hat{u}(i+\tfrac{1}{2},j)] \quad (4.4.35a)$$

$$\widehat{v}(i+\tfrac{1}{2}, j+\tfrac{1}{2}) = \tfrac{1}{2}[\widehat{v}(i, j+\tfrac{1}{2},) + \widehat{v}(i+1, j+\tfrac{1}{2},)];$$
$$\widehat{v}(i+\tfrac{1}{2}, j-\tfrac{1}{2}) = \tfrac{1}{2}[\widehat{v}(i, j-\tfrac{1}{2},) + \widehat{v}(i+1, j-\tfrac{1}{2},)] \quad (4.4.35b)$$

Substitution of these into Eq. (4.4.33) and rearranging will give an algebraic equation of the form

$$a^u_{i+1/2,j}u^*_{i+1/2,j} + a^u_{i-1/2,j}u^*_{i-1/2,j} + a^u_{i+3/2,j}u^*_{i+3/2,j} + a^u_{i+1/2,j-1}u^*_{i+1/2,j-1} + a^u_{i+1/2,j+1}u^*_{i+1/2,j+1} = b^u_{i,j}p^*_{i,j} - b^u_{i+1,j}p^*_{i+1,j} \tag{4.4.36}$$

with

$$\begin{aligned}
a^u_{i+1/2,j} &= 1/2\,[\hat{u}(i+1/2,j)+\hat{u}(i+3/2,j)]/\Delta x + 1/2\,[\hat{v}[(i,j+1/2,)\\
&\quad + \hat{v}(i+1,j+1/2,)]/\Delta y + 2\nu/\Delta x^2 + 2\nu/\Delta y^2\\
a^u_{i-1/2,j} &= -1/2[\hat{u}(i-1/2,j)+\hat{u}(i+1/2,j)]/\Delta x - \nu/\Delta x^2\\
a^u_{i+3/2,j} &= -\nu/\Delta x^2\\
a^u_{i+1/2,j-1} &= -1/2\,[\hat{v}(i,j-1/2,)+\hat{v}(i+1,j-1/2,)]/\Delta y - \nu/\Delta y^2\\
a^u_{i+1/2,j+1} &= -\nu/\Delta y^2\\
b^u_{i,j} &= 1/(\rho\Delta x)\\
b^u_{i+1,j} &= 1/(\rho\Delta x)
\end{aligned} \tag{4.4.37}$$

By inspection, one can write the coefficients for the three-dimensional case (see Eq. 4.4.31), using upwind differencing for the advection terms and assuming $\{u, v, w\}_{i,j,k} > 0$, as

$$\begin{aligned}
a^u_{i+1/2,j,k} &= 1/2[\hat{u}(i+1/2,j,k)+\hat{u}(i+3/2,j,k)]/\Delta x + 1/2[\hat{v}(i,j+1/2,k)\\
&\quad + \hat{v}(i+1,j+1/2,k)]/\Delta y + 1/2[\hat{w}(i,j,k+1/2)+\hat{w}(i+1,j,k+1/2,)]/\Delta z\\
&\quad + 2\nu/\Delta x^2 + 2\nu/\Delta y^2 + 2\nu/\Delta z^2\\
a^u_{i-1/2,j,k} &= -1/2[\hat{u}(i-1/2,j,k)+\hat{u}(i+1/2,j,k)]/\Delta x - \nu/\Delta x^2\\
a^u_{i+3/2,j,k} &= -\nu/\Delta x^2\\
a^u_{i+1/2,j-1,k} &= -1/2[\hat{v}(i,j-1/2,k)+\hat{v}(i+1,j-1/2,k)]/\Delta y - \nu/\Delta y^2\\
a^u_{i+1/2,j+1,k} &= -\nu/\Delta y^2\\
a^u_{i+1/2,j,k-1} &= 1/2[\widehat{w}(i,j,k-1/2)+\widehat{w}(i+1,j,k-1/2,)]/\Delta z\\
a^u_{i+1/2,j,k+1} &= -\nu/\Delta z^2\\
b^u_{i,j,k} &= 1/(\rho\Delta x)\\
b^u_{i+1,j,k} &= 1/(\rho\Delta x)
\end{aligned} \tag{4.4.38}$$

Writing this equation for all grid points (and incorporating boundary conditions as appropriate) at which u^* needs to be evaluated gives a set of linear algebraic equations which can be solved (see Chap. 5) to obtain u^* at all the grid points. Similarly, the discretized and linearized y-momentum at grid point (i, j + ½, k) and z-momentum equation at (i, j, k + ½) are:

$$\begin{aligned} & a^v_{i,j+1/2,k} v^*_{i,j+1/2,k} + a^v_{i,j-1/2,k} v^*_{i,j-1/2,k} + a^v_{i,j+3/2,k} v^*_{i,j+3/2,k} + a^v_{i-1,j+1/2,k} v^*_{i-1,j+1/2,k} \\ & \quad + a^v_{i+1,j+1/2,k} v^*_{i+1,j+1/2,k} + a^v_{i,j+1/2,k-1} v^*_{i,j+1/2,k-1} + a^v_{i,j+1/2,k+1} v^*_{i,j+1/2,k+1} \\ & \quad = b^v_{i,j,k} p^*_{i,j,k} - b^v_{i,j+1,k} p^*_{i,j+1,k} \end{aligned} \tag{4.4.39}$$

$$\begin{aligned} & a^w_{i,j,k+1/2} w^*_{i,j,k+1/2} + a^w_{i,j,k-1/2} w^*_{i,j,k-1/2} + a^w_{i,j,k+3/2} w^*_{i,j,k+3/2} \\ & \quad + a^w_{i-1,j,k+1/2} w^*_{i-1,j,k+1/2} + a^w_{i+1,j,k+1/2} w^*_{i+1,j,k+1/2} \\ & \quad + a^w_{i,j-1,k+1/2} w^*_{i,j-1,k+1/2} + a^w_{i,j+1,k+1/2} w^*_{i,j+1,k+1/2} = b^w_{i,j,k} p^*_{i,j,k} - b^w_{i,j,k+1} p^*_{i,j,k+1} \end{aligned} \tag{4.4.40}$$

Solution of Eqs. (4.4.31), (4.4.39) and (4.4.40) will give values of u^*, v^* and w^* throughout the domain. Since u^*, v^* and w^* have been obtained with an assumed p^*, the continuity equation is not satisfied. Only a correct pressure field would produce a correct velocity field that would also satisfy the continuity equation. A correction to the velocity and pressure fields can be obtained in the following way. The exact velocity and pressure fields, u, v, w and p, satisfy the same discretized linearized equations to within errors arising out of discretization and interpolation; thus,

$$\begin{aligned} & a^u_{i+1/2,j,k} u_{i+1/2,j,k} + a^u_{i-1/2,j,k} u_{i-1/2,j,k} + a^u_{i+3/2,j,k} u_{i+3/2,j,k} + a^u_{i+1/2,j-1,k} u_{i+1/2,j-1,k} \\ & \quad + a^u_{i+1/2,j+1,k} u_{i+1/2,j+1,k} + a^u_{i+1/2,j,k-1} u_{i+1/2,j,k-1} \\ & \quad + a^u_{i+1/2,j,k+1} u_{i+1/2,j,k+1} = b^u_{i,j,k} p_{i,j,k} - b^u_{i+1,j,k} p_{i+1,j,k} \end{aligned} \tag{4.4.41}$$

Substracting Eq. (4.4.31) from Eq. (4.4.41), and defining velocity and pressure corrections as

$$u' = u - u^*;\ v' = v - v^*;\ w' = w - w^* \quad \text{and} \quad p' = p - p^* \tag{4.4.42}$$

we get

$$
\begin{aligned}
& a^{u}_{i+1/2,j,k}u'_{i+1/2,j,k} + a^{u}_{i-1/2,j,k}u'_{i-1/2,j,k} + a^{u}_{i+3/2,j,k}u'_{i+3/2,j,k} \\
& \quad + a^{u}_{i+1/2,j-1,k}u'_{i+1/2,j-1,k} + a^{u}_{i+1/2,j+1,k}u'_{i+1/2,j+1,k} + a^{u}_{i+1/2,j,k-1}u'_{i+1/2,j,k-1} \\
& \quad + a^{u}_{i+1/2,j,k+1}u'_{i+1/2,j,k+1} = b^{u}_{i,j,k}p'_{i,j,k} - b^{u}_{i+1,j,k}p^{u}_{i+1,j,k}
\end{aligned} \tag{4.4.43}
$$

Similar equations can be written for v' and w' in terms of pressure corrections; however, we are no closer to a solution as the pressure correction is not known. A measure of the velocity correction, $u'_{i+½,j,k}$, can be obtained by neglecting the contribution from all terms on the LHS of Eq. (4.4.43) except for the first one. With this assumption, one can rewrite Eq. (4.4.43) as

$$
u'_{i+1/2,j,k} \approx \left(b^{u}_{i,j,k}p'_{i,j,k} - b^{u}_{i+1,j,k}p'_{i+1,j,k}\right)/a^{u}_{i+1/2,j,k} \tag{4.4.44}
$$

Similarly, one can derive the following approximate expressions for v' and w':

$$
v'_{i,j+1/2,k} \approx \left(b^{v}_{i,j,k}p'_{i,j,k} - b^{v}_{i,j+1,k}p'_{i,j+1,k}\right)/a^{v}_{i,j+1/2,k} \tag{4.4.45}
$$

and

$$
w'_{i,j,k+1/2} \approx \left(b^{w}_{i,j,k}p'_{i,j,k} - b^{w}_{i,j,k+1}p'_{i,j,k+1}\right)/a^{w}_{i,j,k+1/2} \tag{4.4.46}
$$

The continuity equation is discretized at grid point (i, j, k) and can be written as

$$
\begin{aligned}
& \left(u_{i+1/2,j,k} - u_{i-1/2,j,k}\right)/\Delta x + \left(v_{i,j+1/2,k} - v_{i,j-1/2,k}\right)/\Delta y \\
& \quad + \left(w_{i,j,k+1/2} - w_{i,j,k-1/2}\right)/\Delta z = 0
\end{aligned} \tag{4.4.47}
$$

Using Eq. (4.4.42) and the estimated velocity corrections given by Eqs. (4.4.44) to (4.4.46), one can rewrite Eq. (4.4.42) as

$$
\begin{aligned}
& \left[\left(b^{u}_{i,j,k}p'_{i,j,k} - b^{u}_{i+1,j,k}p'_{i+1,j,k}\right)/a^{u}_{i+1/2,j,k} - \left(b^{u}_{i-1,j,k}p'_{i-1,j,k} - b^{u}_{i,j,k}p'_{i,j,k}\right)/a^{u}_{i-1/2,j,k}\right]/\Delta x \\
& \quad + \left[\left(b^{v}_{i,j,k}p'_{i,j,k} - b^{v}_{i,j+1,k}p'_{i,j+1,k}\right)/a^{v}_{i,j+1/2,k} - \left(b^{v}_{i-1,j,k}p'_{i-1,j,k} - b^{v}_{i,j,k}p'_{i,j,k}\right)/a^{v}_{i,j-1/2,k}\right]/\Delta y \\
& \quad + \left[\left(b^{w}_{i,j,k}p'_{i,j,k} - b^{w}_{i,j,k+1}p'_{i,j,k+1}\right)/a^{w}_{i,j,k+1/2} - \left(b^{w}_{i,j,k-1}p'_{i,j,k-1} - b^{w}_{i,j,k}p'_{i,j,k}\right)/a^{w}_{i,j,k-1/2}\right]/\Delta z \\
& \quad = \left(u^{*}_{i-1/2,j,k} - u^{*}_{i+1/2,j,k}\right)/\Delta x + \left(v^{*}_{i,j-1/2,k} - v^{*}_{i,j+1/2,k}\right)/\Delta y + \left(w^{*}_{i,j,k-1/2} - w^{*}_{i,j,k+1/2}\right)/\Delta z
\end{aligned} \tag{4.4.48}
$$

The above equation can be rewritten in the form of a pressure correction equation

$$a^{p}_{i,j,k}p'_{i,j,k} + a^{p}_{i-1,j,k}p'_{i-1,j,k} + a^{p}_{i+1,j,k}p'_{i+1,j,k} + a^{p}_{i,j-1,k}p'_{i,j-1,k} + a^{p}_{i,j+1,k}p'_{i,j+1,k} + a^{p}_{i,j,k-1}p'_{i,j,k-1} + a^{p}_{i,j,k+1}p'_{i,j,k+1} = b^{p}_{i,j,k} \tag{4.4.49}$$

where the RHS is the same as that of Eq. (4.4.48) and will tend to zero as the method converges. Simultaneous solution of Eq. (4.4.49) written for all the grid points gives the pressure correction at all (i, j, k) from which velocity corrections are evaluated using Eqs. (4.4.44) to (4.4.46). Thus, the SIMPLE method requires iterative and sequential solution of u^*, v^*, w^* and p' using Eqs. (4.4.31), (4.4.39), (4.4.40) and (4.4.49), respectively. The velocity corrections, u', v' and w', are then evaluated using Eqs. (4.4.44) to (4.4.46). The velocity and pressure fields are then updated using an under-relaxation factor (α):

$$\begin{aligned} p^{*m+1} &= p^{*m} + \alpha_p p'^{m} \\ u^{*m+1} &= u^{*m} + \alpha_u u'^{m} \\ v^{*m+1} &= v^{*m} + \alpha_v v'^{m} \\ w^{*m+1} &= w^{*m} + \alpha_w w'^{m} \end{aligned} \tag{4.4.50}$$

Typical values of these under-relaxation factors are $\alpha_u = \alpha_v = \alpha_w = 0.7$ and $\alpha_p = (1 - \alpha_u)$. The updated variables are then used to continue with iteration until convergence is achieved which is indicated by the RHS of Eq. (4.4.49) becoming very small. Use of a staggered grid allows the solution domain to be chosen in such a way that the boundaries coincide with locations where the velocities are evaluated. The pressure correction equation is solved with Neumann boundary condition at the walls.

The above SIMPLE method has proved to be popular for a range of internal flows. It allows a stationary (steady) problem to be solved in a steady formulation. The implicitness of the velocity and pressure fields allows a measure of stability which can be improved to some extent by a suitable choice of the under-relaxation factors. Much of its popularity can also be linked to its incorporation in commercially available CFD software programs which enable ready computation of practical flows involving turbulence, heat transfer, mass transfer and chemical reactions. A discussion of the sensitivity of the convergence to the under-relaxation factors is given in Ferziger and Peric (1999).

A number of variants of SIMPLE scheme have been proposed to improve the convergence rate which is affected by neglecting the contribution of neighbouring velocity corrections while deriving Eq. (4.4.44) from Eq. (4.4.43). In the SIMPLEC or SIMPLE-Consistent method, van Doormal and Raithby (1984) suggested approximating all the neighbouring u' in Eq. (4.4.43) by $u'_{i+\frac{1}{2},j,k}$ so that it reduces to

$$\begin{aligned}(a^u_{i+1/2,j,k} + a^u_{i-1/2,j,k} + a^u_{i+3/2,j,k} + a^u_{i+1/2,j-1,k} + a^u_{i+1/2,j+1,k} + a^u_{i+1/2,j,k-1} \\ + a^u_{i+1/2,j,k+1})u'_{i+1/2,j,k} = b^u_{i,j,k}p'_{i,j,k} - b^u_{i+1,j,k}p'_{i+1,j,k}\end{aligned} \tag{4.4.51}$$

resulting in a modified velocity correction formula of

$$\begin{aligned}u'_{i+1/2,j,k} = \left(b^u_{i,j,k}p'_{i,j,k} - b^u_{i+1,j,k}p'_{i+1,j,k}\right)/(a^u_{i+1/2,j,k} + a^u_{i-1/2,j,k} + a^u_{i+3/2,j,k} \\ + a^u_{i+1/2,j-1,k} + a^u_{i+1/2,j+1,k} + a^u_{i+1/2,j,k-1} + a^u_{i+1/2,j,k+1})\end{aligned} \tag{4.4.52}$$

In the SIMPLER (or SIMPLE-Revised) method, Patankar (1980) suggested using the corrected velocity field obtained from the pressure correction equation (Eq. 4.4.49) to solve a pressure equation obtained by taking the divergence of the discretized momentum equation. Thus, in this method, two evaluations of pressure are needed for one update of the velocity field. The PISO (pressure-implicit with splitting of operators) method proposed by Issa (1986), Issa et al. (1986) calls for two evaluations of the pressure correction. After obtaining the pressure and velocity corrections as in the SIMPLE method (by solving Eq. (4.4.49) for the pressure correction and Eqs. (4.4.44) to (4.4.46) for the velocity corrections, a second evaluation of these is made, this time by including the contribution of neighbouring velocity corrections in Eq. (4.4.43). Because of a more accurate treatment of the influence of the velocity corrections from the neighbouring grid points, there is no need to under-relax (see Eq. 4.4.50) the pressure correction in the SIMPLEC or SIMPLER or PISO methods. A concise description of these approaches can be found in the review paper of Acharya et al. (2007). The relative merits of the methods has been tested for some benchmark cases. Jang et al. (1986) reported direct comparison for the 2-dimensional cases of sudden expansion in a plane channel, turbulent flow in an axisymmetric channel, swirling reacting flow in a furnace and free convection in a rectangular enclosure. In general, the PISO algorithm showed the most robust convergence; however, where the scalar variable was weakly linked to the momentum equations, as in the case of the free convection, SIMPLEC and SIMPLER showed better convergence behaviour.

A note on the calculation of time-dependent flows using SIMPLE and its variants is necessary. It may be seen from Eq. (4.4.31) that the calculation of the velocity field for a given pressure field is implicit. An implicit treatment of the time derivatives in the momentum equations can be readily incorporated into the SIMPLE scheme. Consider the implicitly discretized momentum equation:

$$(\delta u/\delta t)_{i,j,k}{}^{n+1} + [\nabla.(\hat{u}u^*)]_{i,j,k}{}^{n+1} = -(1/\rho)[\nabla p^*]_{i,j,k}{}^{n+1} + \nu\left[\nabla^2 u^*\right]_{i,j,k}{}^{n+1} \tag{4.4.53}$$

where n indicates the time step. Using forward differencing for the time derivative, one can rewrite the above equation as

$$\begin{aligned}(1/\Delta t)\left(u_{i,j,k}{}^{n+1} - u_i{}^n\right) + [\nabla \cdot (\hat{u}u^*)]_{i,j,k}{}^{n+1} \\ = -(1/\rho)[\nabla p^*]_{i,j,k}{}^{n+1} + \nu\left[\nabla^2 u^*\right]_{i,j,k}{}^{n+1}\end{aligned} \quad (4.4.54)$$

The problem is now reduced to finding u^{n+1} and p^{n+1};
Following the formulation of Eq. (4.4.30), Eq. (4.4.54) can be rewritten as

$$(1/\Delta t)u^*{}_{i,j,k}{}^{n+1} + [\nabla \cdot (\hat{u}u^*)]_{i,j,k}{}^{n+1} = -(1/\rho)[\nabla p^*]_{i,j,k}{}^{n+1} + \nu\left[\nabla^2 u^*\right]_{i,j,k}{}^{n+1} + (1/\Delta t)\,u_i{}^n \quad (4.4.55)$$

Comparing Eq. (4.4.55) with the equation for the steady state case (Eq. 4.4.30) and its discretized form (Eq. 4.4.31), we find that in the discretized form of Eq. (4.4.55) is similar to Eq. (4.4.31) with an additional coefficient of (1/Δt) appearing in the term $a^u_{i+½,j,k}$ and the term ($1/\Delta t\ u_i^n$) appearing in $b^u_{i,j,k}$ in Eq. (4.4.38). Similar corrections will need to be made in the discretized y- and z-momentum balance equations, namely, Eqs. (4.4.39) and (4.4.40), respectively. The continuity equation remains unaffected directly as it has no time derivative. With these changes, the SIMPLE method can be applied to evaluate the velocity and pressure fields at that time step. The calculations are repeated for the next time step. The flow charts for the steady state and unsteady state calculations are compared in Figs. 4.13 and 4.14, respectively.

4.4.5 Extension of SIMPLE to Flows of All Speeds

Although the SIMPLE scheme was originally developed for incompressible flows, the method can be extended to compressible flows by introducing a density correction in addition to pressure correction (van Doormal et al. 1987; Karki and Patankar 1989). Consider the steady state form of continuity equation for compressible flow:

$$\partial(\rho u)/\partial x + \partial(\rho v)/\partial y + \partial(\rho w)/\partial z = 0 \quad (4.4.56)$$

Writing finite difference approximations about grid point (i, j, k), we have

$$\begin{aligned}\left[(\rho u)_{i+1/2,j,k} - (\rho u)_{i-1/2,j,k}\right]/\Delta x + \left[(\rho v)_{i,j+1/2,k} - (\rho v)_{i,j-1/2,k}\right]/\Delta y \\ + \left[(\rho w)_{i,j,k+1/2} - (\rho w)_{i,j,k-1/2}\right]/\Delta z = 0\end{aligned} \quad (4.4.57)$$

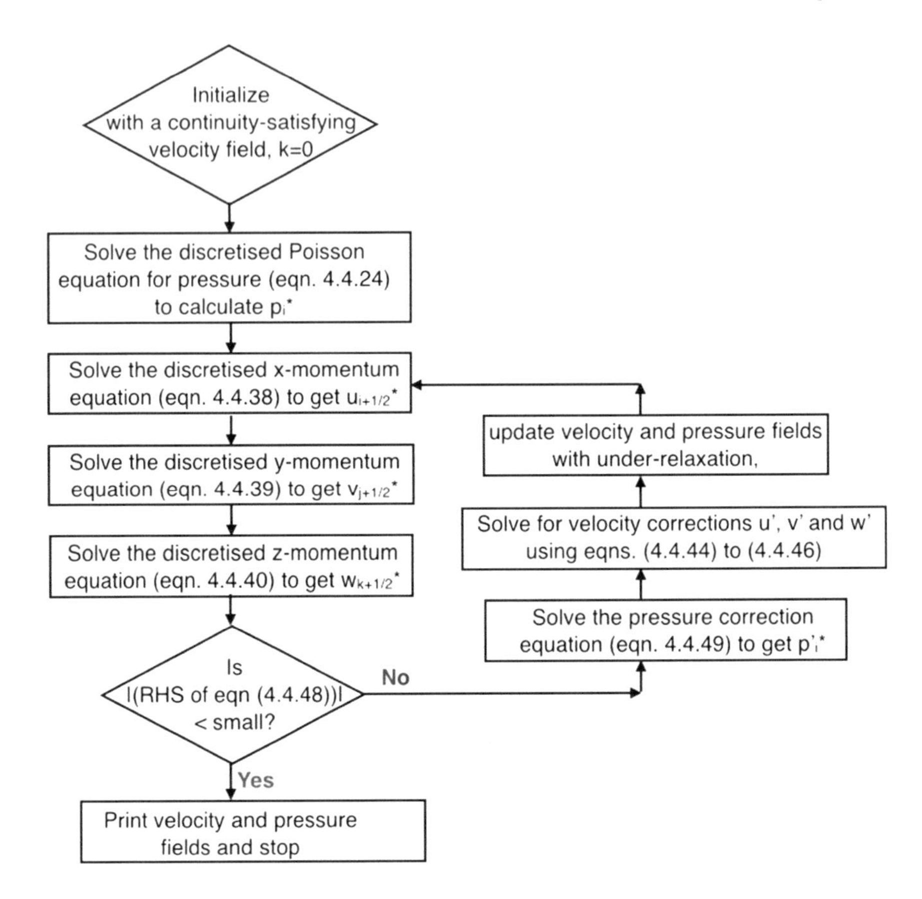

Fig. 4.13 Flow chart for the calculation of the flow field using the SIMPLE scheme for a steady flow case

Resolving $u = u^* + u'$, $\rho = \rho^* + \rho'$, etc., and neglecting terms involving products of velocity-velocity or velocity-pressure corrections as these are expected to be small, the above equation can be written as

$$\left[(\rho^* u' + u^* \rho')_{i+1/2,j,k} - (\rho^* u' + u^* \rho')_{i-1/2,j,k}\right]/\Delta x + \left[(\rho^* v' + v^* \rho')_{i,j+1/2,k} - (\rho^* v' + v^* \rho')_{i,j-1/2,k}\right]/\Delta y$$
$$+ \left[(\rho^* w' + w^* \rho')_{i,j,k+1/2} - (\rho * w' + w^* \rho')_{i,j,k-1/2}\right]/\Delta z = -\left\{\left[(\rho^* u^*)_{i+1/2,j,k} - (\rho^* u^*)_{i-1/2,j,k}\right]/\Delta x\right.$$
$$\left. + \left[(\rho^* v^*)_{i,j+1/2,k} - (\rho^* v^*)_{i,j-1/2,k}\right]/\Delta y + \left[(\rho^* w^*)_{i,j,k+1/2} - (\rho^* w^*)_{i,j,k-1/2}\right]/\Delta z\right\} \tag{4.4.58}$$

Consider the first two terms on the LHS. Here $(\rho^* u')_{i+½,j,k}$ can be treated as in the conventional SIMPLE method (see Eq. 4.4.44) as

$$(\rho^* u')_{i+1/2,j,k} \approx \rho^*_{i+1/2,j,k}\left(b^u_{i,j,k} p'_{i,j,k} - b^u_{i+1,j,k} p'_{i+1,j,k}\right)/a^u_{i+1/2,j,k} \tag{4.4.59}$$

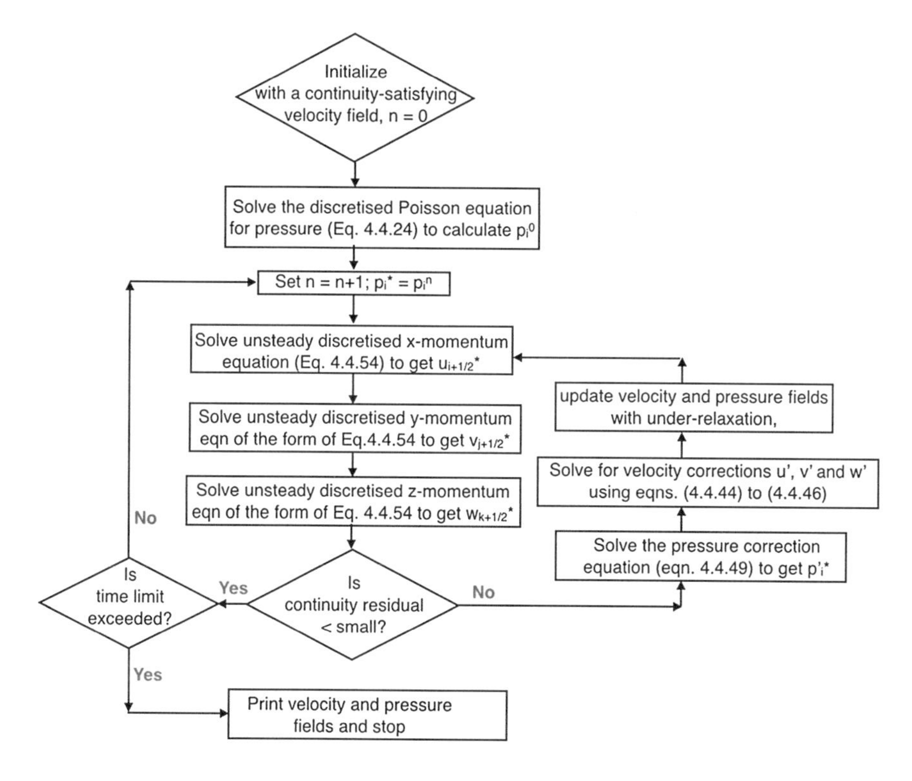

Fig. 4.14 Flow chart for the calculation of the flow field using the SIMPLE scheme for an unsteady flow case

The second term, $(u^*\rho')_{i+1/2,j,k}$, requires the evaluation of density correction in terms of pressure correction; this can be done using the equation of state for the gas, namely,

$$\rho' \approx p'(\partial\rho/\partial p)_T = C_p p' = (1/RT)p' \text{ for a perfect gas} \tag{4.4.60}$$

where the derivative of density with respect to pressure is at the local temperature at the grid, C_p is the specific heat of the fluid, and R is the universal gas constant. For gases that do not obey the perfect gas assumption and for other fluids, the equation of state must be used to evaluate specific heat. With this modelling, the second term on the LHS in Eq. (4.4.58) can be written as

$$(u^*\rho')_{i+1/2,j,k} = \left(u^*_{i+1/2,j,k}/\left(RT_{i+1/2,j,k}\right)\right] p'_{i+1/2,j,k} \tag{4.4.61}$$

On a staggered grid, p′ and T are not evaluated at grid point (i + ½, j, k) and they need to be interpolated from neighbouring nodes. The same procedure must be used to evaluate other terms on the LHS of Eq. (4.4.58) in order to derive the pressure correction equation for compressible flows. It may be noted that the contributions to the pressure corrections from velocity and density corrections are fundamentally different; in the former, the contribution is proportional to the pressure gradient whereas in the latter, it is proportional to the absolute value itself. The contribution from the latter will dominate at high Mach numbers and the pressure correction effectively becomes a means of finding the correct density, which is the role the continuity equation has in compressible flows (see Sect. 4.2.1). van Doormal et al. (1987), Karki and Patankar (1989), Demirdžic et al. (1993), Lilek (1995) have presented examples of application of the extended SIMPLE method for the computation of subsonic and supersonic compressible flows.

4.4.6 Computation of Pressure on a Collocated Grid

Any discussion of pressure-based methods for the solution of NS equations for incompressible flows would be incomplete without a mention of the possibility of the occurrence of chequerboard type oscillations of the pressure and velocity fields. Consider the discrete form of the momentum balance equation given by Eq. (4.4.30):

$$[\nabla \cdot (\hat{u}u^*)]_{i,j,k} = -1/\rho\,[\nabla p^*]_{i,j,k} + \nu[\nabla^2 u^*]_{i,j,k} \tag{4.4.30}$$

On a staggered grid system, the x-component of the balance equation takes the form given by Eq. (4.4.31) wherein the u velocity component is evaluated mid-way between two pressure evaluation points. This enables a correct treatment of pressure as a surface force as the pressure difference between the left and right faces of a control volume centred around the grid point (i + ½, j, k) contributes to the change in the linear momentum in the x-direction. Similarly, the discretization of the continuity equation around grid point (i, j, k) using central differences enables the evaluation of the divergence term without having to interpolate velocities. These advantages would be lost in a collocated grid in which all the flow variables are evaluated at grid nodes (i, j, k). When the x-momentum is discretized around grid point (i,j,k), the pressure gradient term would have to be expressed in terms of p(i + 1, j, k) and p (i-1, j, k), e.g.

$$\partial p/\partial x|_{i,j,k} = \left(p_{i+1,j,k} - p_{i-1,j,k}\right)/(2\Delta x) \tag{4.4.62}$$

This leads to the decoupling of the velocity and the pressure at point (i, j, k) in the sense that the value of pressure at the point does not influence the velocity. For the discretization of the continuity equation in the collocated grid case, Eq. (4.4.47) will have to be written as

$$\left(u_{i+1,j,k} - u_{i-1,j,k}\right)/(2\Delta x) + \left(v_{i,j+1,k} - v_{i,j-1,k}\right)/(2\Delta y) + \left(w_{i,j,k+1} - w_{i,j,k-1}\right)/(2\Delta z) = 0 \tag{4.4.63}$$

If this is used to evaluate p' (i, j, k), then the velocity field at (i, j, k) gets decoupled from the evaluation of pressure at the point. If the formulation of Eq. (4.4.47) is retained for discretization of the continuity equation, then the velocity components at the mid-way points, i.e., u (i + ½, j, k), v (i, j + ½, k) and w (i, j, k + ½) will have to be interpolated. If this is done through linear interpolation based on grid spacing, then the discretized equation will reduce to Eq. (4.4.63) on a uniform grid. Thus, the calculation of at p (i, j, k) gets decoupled from the calculation of its immediate neighbours and two parallel pressure fields, leading to a possible chequerboard type oscillations in both pressure and velocities, as shown schematically shown in Fig. 4.15. This makes the use of collocated grid unattractive for incompressible flow calculations when the primitive variables solution approach is desired.

While the use of a staggered grid imposes no significant penalty on simple grids, their use becomes cumbersome and untenable for non-orthogonal body-fitted grids (see Chap. 6.1) which are needed for tackling cases of flow through irregular but practically relevant flow geometries. Using a staggered grid in such cases would mean working with four different grids. Also, terms involving cross-derivatives, such as $\partial^2\varphi/\partial x\partial y$, would appear in governing equations when non-orthogonal grids are used (this is discussed in Chap. 6), and treatment of these on a staggered grid would be even more cumbersome. Several cures, such as filtering out the oscillations and adding a fourth derivative of pressure to the Poisson equation for pressure to smooth out the oscillations, have been explored to deal with the chequerboard oscillations in the non-staggered grid approach. One of the most successful approaches is that of Rhie and Chow (1983) who proposed an interpolation scheme for determining the velocities at mid-way points that brings in an effective coupling between the local velocity and the local pressure gradient. Consider the discretization of the continuity equation at grid point (i, j, k) using velocities at mid-points:

$$\left(u_{i+1/2,j,k} - u_{i-1/2,j,k}\right)/\Delta x + \left(v_{i,j+1/2,k} - v_{i,j-1/2,k}\right)/\Delta y + \left(w_{i,j,k+1/2} - w_{i,j,k-1/2}\right)/\Delta z = 0 \tag{4.4.47}$$

The velocity field computed from the momentum equations satisfies the following discretized equation for the x-momentum balance:

$$a^u_{i,j,k}u^*_{i,j,k} + a^u_{i-1,j,k}u^*_{i-1,j,k} + a^u_{i+1,j,k}u^*_{i+1,j,k} + a^u_{i,j-1,k}u^*_{i,j-1,k} + a^u_{i,j+1,k}u^*_{i,j+1,k} + a^u_{i,j,k-1}u^*_{i,j,k-1} + a^u_{i,j,k+1}u^*_{i,j,k+1} = b^u_{i-1,j,k}p^*_{i-1,j,k} - b^u_{i+1,j,k}p^*_{i+1,j,k} \tag{4.4.64}$$

Interpolation of $u_{i+½,j,k}$ from $u_{i,j,k}$ obtained using Eq. (4.4.64) would effectively mean a coarse-grid discretization of the continuity equation (Eq. 4.4.63) which could lead to chequerboard oscillations. In order to get a good estimate for $u_{i+1/2,j,k}$ in the collocated grid approach, Rhie and Chow (1983) sought to evaluate

Eq. (4.4.31) at (i + ½, j, k) using coefficients interpolated from the collocated grid, i.e.,

$$\bar{a}^{u}_{i+1/2,j,k}u^{*}_{i+1/2,j,k}+\bar{a}^{u}_{i-1/2,j,k}u^{*}_{i-1/2,j,k}+\bar{a}^{u}_{i+3/2,j,k}u^{*}_{i+3/2,j,k}+\bar{a}^{u}_{i+1/2,j-1,k}u^{*}_{i+1/2,j-1,k}$$
$$+\bar{a}^{u}_{i+1/2,j+1,k}u^{*}_{i+1/2,j+1,k}+\bar{a}^{u}_{i+1/2,j,k-1}u^{*}_{i+1/2,j,k-1}=\bar{b}^{u}_{i,j,k}p^{*}_{i,j,k}-\bar{b}^{u}_{i+1,j,k}p^{*}_{i+1,j,k} \tag{4.4.65}$$

where the overbar over the coefficient indicates a value interpolated from the neighbouring values. Writing the x-momentum balance equation (4.4.64) on the non-staggered grid at (i + 1, j, k), and using these to interpolate each term to the mid-point (since Eq. (4.4.64) is linear), one can obtain the interpolated x-momentum equation in terms of the interpolated u-velocity components, $\bar{u}_{i+1/2,j,k}$:

$$\bar{a}^{u}_{i+1/2,j,k}\bar{u}_{i+1/2,j,k}+\bar{a}^{u}_{i-1/2,j,k}\bar{u}_{i-1/2,j,k}+\bar{a}^{u}_{i+3/2,j,k}\bar{u}_{i+3/2,j,k}+\bar{a}^{u}_{i+1/2,j-1,k}\bar{u}_{i+1/2,j-1,k}$$
$$+\bar{a}^{u}_{i+1/2,j+1,k}\bar{u}_{i+1/2,j+1,k}+\bar{a}^{u}_{i+1/2,j,k-1}\bar{u}_{i+1/2,j,k-1}+\bar{a}^{u}_{i+1/2,j,k+1}\bar{u}_{i+1/2,j,k+1}$$
$$=\bar{b}^{u}_{i,j,k}\bar{p}_{i,j,k}-\bar{b}^{u}_{i+1,j,k}\bar{p}_{i+1,j,k} \tag{4.4.66}$$

Subtracting (4.4.66) from (4.4.65) and assuming that the off-diagonal terms on the LHS cancel out, we have

$$\bar{a}^{u}_{i+1/2,j,k}\left(u^{*}_{i+1/2,j,k}-\bar{u}_{i+1/2,j,k}\right)\approx\bar{b}^{u}_{i+1/2,j,k}\left[\left(p^{*}_{i,j,k}-p^{*}_{i+1,j,k}\right)-\left(\bar{p}_{i,j,k}-\bar{p}_{i+1,j,k}\right)\right] \tag{4.4.67}$$

or

$$u^{*}_{i+1/2,j,k}\approx\bar{u}_{i+1/2,j,k}+\bar{b}^{u}_{i+1/2,j,k}/\bar{a}^{u}_{i+1/2,j,k}\left[\left(p^{*}_{i,j,k}-p^{*}_{i+1,j,k}\right)-\left(\bar{p}_{i,j,k}-\bar{p}_{i+1,j,k}\right)\right] \tag{4.4.68}$$

It can be shown that the term within the square brackets on the RHS of Eq. (4.4.68) is equivalent to a fourth-order dissipation term which has the role of suppressing the high frequency oscillations associated with chequerboard oscillations. In a smoothly varying pressure field, the effect of this term would be negligible and the velocity given by Eq. (4.4.68) would be equal to the interpolated velocity. In a rapidly varying velocity field such as the one shown in Fig. 4.15, the difference between the two estimates would act to introduce a velocity correction which is proportional to the pressure difference between immediate grid neighbours. This would not have been possible if the interpolated pressure gradient alone is used. The Rhie-Chow interpolation formula has been used widely in many internal flow computations involving collocated grids.

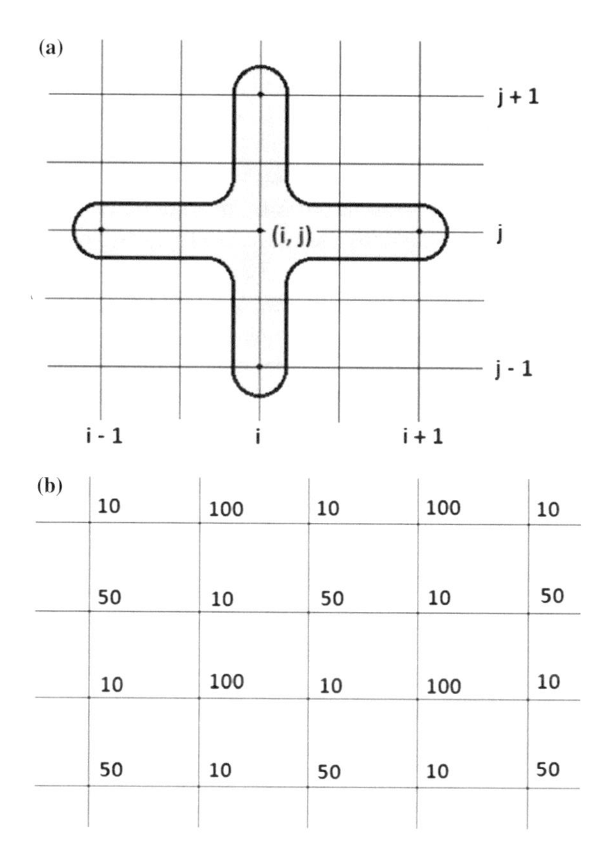

Fig. 4.15 **a** A collocated or non-staggered grid in which all variables are evaluated at grid points (i, j, k). **b** Chequerboard oscillations in pressure. The discretized momentum equation will read this pressure field as having $\partial p/\partial x = 0$ and $\partial p/\partial y = 0$ everywhere

4.5 Coupled and Sequential or Segregated Solvers

The methods described in Sects. 4.2–4.4 differ from each other in one fundamental way. For compressible flows, a strong coupling exists among the continuity, the momentum and the energy equations. For supersonic flows, the equations are hyperbolic in nature. Because of these factors, the conservation equations are often solved in a coupled manner, i.e., all the variable values are evaluated simultaneously at a grid node before the solution is found at the next grid node. In the case of incompressible flow, this natural coupling does not exist. For flows with constant thermophysical properties, the energy conservation equation can be decoupled from the mass and the momentum conservation equations. The pressure can also be decoupled completely from the velocity field computation in the streamfunction-vorticity approach. Also, the governing equations are elliptic and need to be solved implicitly. Due to complex geometry, the flows are three-dimensional and often turbulent and require solution of an extended system of equations (see Chap. 6). For these reasons, the momentum conservation equation is expressed in terms of the scalar components of the velocity vector, and evaluation of the scalar variables is done sequentially using the Picard substitution method to take account of unknown variables. For example, in the SIMPLE method of solution of the governing equations, the three momentum equations are individually solved over the entire computational domain one after the other. Thus, the value of the u-velocity component gets evaluated first for all the grid points using guessed/approximated values of other variables that appear in the x-momentum conservation equation. This is followed by independent evaluation of the v-velocity component at each grid point and that of w. These velocity components are then used to solve the pressure correction equation. This whole process is repeated with updated coefficients. Methods using this kind of approach are called segregated or sequential solvers. For time-dependent flows, the sequential and iterative solution of the flow variables is carried out at every time step.

A coupled solution takes better account of the interaction of the various terms (which often represent momentum, energy and other fluxes in a control volume or cell centred around the grid node), and therefore the iterative evaluation converges faster. A pressure-based method for the coupled solution of incompressible Navier-Stokes equations was developed by Caretto et al. (1972a), i.e., around the same time as the SIMPLE method, under the acronym of SIVA (for simultaneous variable arrangement). But it was the sequential solver SIMPLE that found favour with the research community due to its lower memory requirement (by a factor of two to three) and coding simplicity. There has been renewed interest in coupled methods due to two factors. Firstly, the cost of computer memory has come down drastically rendering the higher memory requirements less of a disadvantage. Secondly, increasing the number of grid points in geometrically complex domains is known to reduce the convergence rate of segregated solvers due to the strong coupling that exists among the flow variables. Darwish et al. (2007) give a brief review of the different coupled solution approaches for pressure-based Navier-Stokes equations. In order to gain an understanding of the coupled solution method, we can study the unstructured grid 2-d, pressure equation-based formulation of Darwish et al. (2009). Using the cell-centred finite volume method, they discretize the steady-state momentum equation around a triangular cell P as

$$\sum\nolimits_{f=nb(P)}\{(\rho\mathbf{vv}-\mu\nabla\mathbf{v})_f\cdot\mathbf{S}_f+p_f\mathbf{S}_f\}=\mathbf{b}_P\Omega_P \tag{4.5.1}$$

Here, the subscript f stands for a value of the variable at the face, the subscript nb refers to the common faces of cell P with neighbouring cells, $\mathbf{S}_f$ is the outward normal vector of the face and Ω_P is the volume of the cell P. It may be noted that the pressure gradient that appears in the momentum balance equation is evaluated in strong conservation form as a surface force. The quantity in curly brackets is summed over all the neighbouring (enclosing) faces of cell P. In 2-d with triangulated flow domain, there will be three cell faces at which the quantity in curly brackets needs to be evaluated. The vector form of Eq. (4.5.1) is then converted into algebraic equations for the two velocity components in the usual way. These can be written as

$$a_P^{uu} u_P + a_P^{uv} v_P + a_P^{up} p_P + \sum_{F=nb(P)} a_F^{uu} u_F + \sum_{F=nb(P)} a_F^{uv} v_F + \sum_{F=nb(P)} a_F^{up} p_F = b_P^u \quad (4.5.2)$$

$$a_P^{vu} u_P + a_P^{vv} v_P + a_P^{vp} p_P + \sum_{F=nb(P)} a_F^{vu} u_F + \sum_{F=nb(P)} a_F^{vv} v_F + \sum_{F=nb(P)} a_F^{vp} p_F = b_P^v \quad (4.5.3)$$

Here, a_P^{uu} stands for the coefficient for the u-velocity component at cell centre in the discretized x-momentum balance equation. Similarly, a_F^{uv} stands for the coefficient for the v-velocity component at cell face in the discretized x-momentum balance equation while a_P^{vu} and a_F^{vv} stands for the coefficient for the respective quantities in the discretized y-momentum balance equations. These values depend on the method of discretization employed for various terms, such as simple upwinding for the advection terms and central differencing for the diffusion terms, etc.

An equation for pressure is derived from the discretized continuity equation which can be written for the cell as

$$\sum_{f=nb(P)} \rho_f \mathbf{v}_f \cdot \mathbf{S}_f = 0 \quad (4.5.4)$$

Pressure at cell P is brought into the continuity equation by using the Rhie-Chow interpolation scheme (1983) to determine the face velocities:

$$\sum_{f=nb(P)} \rho_f \{ \bar{\boldsymbol{v}}_f - \bar{\boldsymbol{D}}_f (\nabla p_f - \overline{\nabla p_f}) \} \cdot \mathbf{S}_f = 0 \quad (4.5.5)$$

Here, the overbar indicates a geometrically interpolated quantity. This results in an algebraic equation for pressure in cell P in terms of pressures and velocities in the neighbouring cells:

$$a_P^{pp} p_P + a_P^{pu} u_P + a_P^{pv} v_P + \sum_{f=nb(P)} a_f^{pp} p_f + \sum_{f=nb(P)} a_f^{pu} u_f + \sum_{f=nb(P)} a_f^{pv} v_f = b_P^p \quad (4.5.6)$$

Combining Eqs. (4.5.2), (4.5.3) and (4.5.6), the discretized conservation equations for cell P can be written as

$$\begin{bmatrix} a_P^{uu} & a_P^{uv} & a_P^{up} \\ a_P^{vu} & a_P^{vv} & a_P^{vp} \\ a_P^{pu} & a_P^{pv} & a_P^{pp} \end{bmatrix} \begin{bmatrix} u_P \\ v_P \\ p_P \end{bmatrix} + \sum_{f=nb(P)} \begin{bmatrix} a_f^{uu} & a_f^{uv} & a_f^{up} \\ a_f^{vu} & a_f^{vv} & a_f^{vp} \\ a_f^{pu} & a_f^{pv} & a_f^{pp} \end{bmatrix} \begin{bmatrix} u_f \\ v_f \\ p_f \end{bmatrix} = \begin{bmatrix} b_P^u \\ b_P^v \\ b_P^p \end{bmatrix} \quad (4.5.7)$$

When written for all the cells and with boundary conditions incorporated appropriately, the resulting equations can be represented in a matrix equation of the form

$$\mathbf{A}\,\mathbf{\Phi} = \mathbf{B} \tag{4.5.8}$$

where $\mathbf{\Phi}$ is the vector, $\{\varphi_1, \varphi_2, \varphi_3\}$, containing all the variables, i.e., u_P, v_P and p_P for all the cells. For a single variable case, as in the segregated solver case, Eq. (4.4.8) would take the following form for a grid with n cells:

$$\begin{bmatrix} a_{11} & a_{12} & \cdots & \cdots & \cdots & a_{1n} \\ a_{21} & a_{22} & \cdots & \cdots & \cdots & a_{1n} \\ . & . & & & & . \\ . & . & & & & . \\ . & . & & & & . \\ a_{1n} & a_{2n} & & & & a_{nn} \end{bmatrix} \begin{bmatrix} \varphi_1 \\ \varphi_2 \\ \cdots \\ \cdots \\ \cdots \\ \varphi_n \end{bmatrix} = \begin{bmatrix} b_1 \\ b_2 \\ \cdots \\ \cdots \\ \cdots \\ b_n \end{bmatrix} \tag{4.5.9}$$

For a 2-d, coupled case with n number of cells and three variables per cell, Eq. (4.5.8) takes the following form:

$$\begin{bmatrix} \begin{bmatrix} a_{11}^{11} & a_{11}^{12} & a_{11}^{13} \\ a_{11}^{21} & a_{11}^{22} & a_{11}^{23} \\ a_{11}^{31} & a_{11}^{32} & a_{11}^{33} \end{bmatrix} & \cdots & \begin{bmatrix} a_{1n}^{11} & a_{1n}^{12} & a_{1n}^{13} \\ a_{1n}^{21} & a_{1n}^{22} & a_{1n}^{23} \\ a_{1n}^{31} & a_{1n}^{32} & a_{1n}^{33} \end{bmatrix} \\ . & \cdots & . \\ . & \cdots & . \\ . & \cdots & . \\ \begin{bmatrix} a_{n1}^{11} & a_{n1}^{12} & a_{n1}^{13} \\ a_{n1}^{21} & a_{n1}^{22} & a_{n1}^{23} \\ a_{n1}^{31} & a_{n1}^{32} & a_{n1}^{33} \end{bmatrix} & \cdots & \begin{bmatrix} a_{nn}^{11} & a_{nn}^{12} & a_{nn}^{13} \\ a_{nn}^{21} & a_{nn}^{22} & a_{nn}^{23} \\ a_{nn}^{31} & a_{nn}^{32} & a_{nn}^{33} \end{bmatrix} \end{bmatrix} \begin{bmatrix} \begin{pmatrix} \varphi_1^1 \\ \varphi_1^2 \\ \varphi_1^3 \end{pmatrix} \\ . \\ . \\ . \\ \begin{pmatrix} \varphi_n^1 \\ \varphi_n^2 \\ \varphi_n^3 \end{pmatrix} \end{bmatrix} = \begin{bmatrix} \begin{pmatrix} b_1^1 \\ b_1^2 \\ b_1^3 \end{pmatrix} \\ . \\ . \\ . \\ \begin{pmatrix} b_n^1 \\ b_n^2 \\ b_n^3 \end{pmatrix} \end{bmatrix} \tag{4.5.10}$$

On a structured grid, the coefficient matrix contains a block diagonal structure which is similar to the one obtained from a segregated solver, with a 3 × 3 block tridiagonal system replacing the single coefficient of the segregated approach. Solution of these simultaneous algebraic equations is not as straightforward as in the case of the algebraic equations resulting from the segregated approach because some of the diagonal elements may be zero in some cases leading to an ill-conditioned matrix (Hanby et al. 1996). Special techniques such as coupled line Gauss-Seidel method resulting in a form of alternating direction implicit (ADI) method with a penta diagonal matrix algorithm (PDMA) as the base method (Galpin et al. 1985), coupled strongly implicit procedures (Rubin and Khosla 1982), symmetrical coupled Gauss-Seidel procedure (Vanka 1986), LU decomposition, incomplete LU decomposition (ILU), quasi minimal residual algorithm (Freund and

Nachtigal 1991), generalized minimal residual (GMRES) method (Saad and Schultz 1986) are among the methods that have been used by several researchers. The basis of some of these methods is discussed in Chap. 5. Several researchers, including Ghia et al. (1982), Vanka (1986), Ammara and Masson (2004), Darwish et al. (2007, 2009), have used multigrid techniques (see Chap. 5) together with coupled solution algorithms, and have reported impressive convergence rates for some test cases. Most of the gains in computational efficiency in such cases can be attributed to the use of multigrids. Direct comparisons of coupled and segregated solvers have also been reported (see for example Galpin et al. 1985; Hanby et al. 1996; Darwish et al. 2009; Chen and Przekwas 2010). These show that the number of outer iterations linking the continuity and the momentum equations needed to achieve a defined level of convergence is far less with the coupled solvers, although the computing time per iteration is significantly higher. The overall result is that typically the overall computational time for a segregated approach is two to three times higher than that taken by coupled solvers. For five 2-d laminar flow cases, Darwish et al. (2009) reported gains of more than an order of magnitude in some fine mesh cases. The gain in improved performance is attributed to not having to solve the pressure correction equation, which is the most time-consuming step in the segregated solution approach. The use of multigrid approach (which is discussed in Chap. 5) helps in creating an additional coupling in the segregated solvers and improves the situation somewhat.

It may be noted that the coupled solution often extends only up to the solution of the continuity and the momentum conservation equations. In turbulent flows and flows with heat transfer and chemical reactions, the four conservations equations for u, v, w and p can be solved in a coupled manner. Additional equations for turbulence, heat transfer, chemical reactions etc. need to be solved sequentially after the solution of the velocity field. If these have a strong bearing on the velocity field, then the convergence rate of the outer iterations for the determination of u, v, w and p gets compromised. The simplicity of coding and solution offered by the segregated approach makes it more convenient to use in such cases.

4.6 Summary

In this chapter, we have looked at several ways solving a set of simultaneous, coupled and nonlinear partial differential equations representing fluid flow using the techniques of CFD. To begin with, we discussed some aspects related to the stability of solution schemes for coupled equations. Ways of dealing with non-linearity have also been discussed. We have seen that the coupling of the continuity equation with the momentum equations is important. When a natural coupling exists, as in the case of compressible flow, then the solution is relatively simple. The focus will then be on capturing gradients accurately and suppressing oscillations that may arise at discontinuities and strong gradients which are normally associated with shocks in highly compressible flows. With the use of the TVD and ENO concepts, it is now possible to develop upwind discretization schemes with built-in, high-fidelity shock capturing capability. For incompressible flows, the coupling through the density between the continuity and the momentum equations is no longer present, and special methods have to be constructed to recover pressure from the continuity equation. Invariably, these flows are also strongly influenced by physical complexities such as turbulence and chemical reactions. These render the computation more difficult and time-consuming than for the case of compressible flows, especially if the latter are restricted to the solution of Euler equations and not the full Navier-Stokes equations. The strong diffusive contribution to flux transport leads to the use of central schemes except for the advection terms which are handled through upwind schemes. The state-of-the-art is that robust algorithms are available for routine calculations of three-dimensional, unsteady compressible or incompressible single-phase flows for general engineering purposes.

Questions

1. One-dimensional shallow-water equations with height, h, and velocity, u, as the variables are given by

$$\partial h/\partial t + u\partial h/\partial x + h\partial u/\partial x = 0 \qquad (Q4.1)$$

$$\partial u/\partial t + u\partial u/\partial x + g\partial h/\partial x = 0. \qquad (Q4.2)$$

   (a) Write these coupled non-linear equations in matrix form as

$$\partial U/\partial t + A\partial U/\partial x = 0 \qquad (Q4.3)$$

   where $U = \{h\ u\}^T$, $A = \begin{pmatrix} u & h \\ g & u \end{pmatrix}$

   (b) Show that the characteristic velocities, a_1, a_2, given by

$$\begin{vmatrix} u - a & h \\ g & u - a \end{vmatrix} = 0 \qquad (Q4.4)$$

   are real and distinct and the system is therefore hyperbolic.

   (c) Linearize the system around $\{u_0, h_0\}$ and show that for the FTCS-explicit discretization scheme, also known as the Euler scheme, given by

$$U_i^{n+1} = U_i^n - A\Delta t/\Delta x\left(U_{i+1}^n - U_{i-1}^n\right) \qquad (Q4.5)$$

   has the spectral radius is given by

$$\rho(G) = 1 + (\Delta t/\Delta x)^2 \{u_0 + \sqrt{gh_0}\}^2 \sin^2 \varphi \qquad \text{(Q4.6)}$$

and that the scheme is therefore unstable.

(d) Show the Lax-Friedrichs (Lax, 1954; Hirsch, 1990) scheme, in which we substitute $U_i^n = (U_{i+1}^n + U_{i-1}^n)/2$ in Eq. (Q4.6), has a spectral radius given by

$$\rho(G_{LF}) = \left\{\cos^2 \varphi + (\sigma_0 + \sigma)^2 \sin^2 \varphi\right\}^{1/2} \qquad \text{(Q4.7)}$$

and that the Lax-Friedrichs scheme is therefore conditionally stable for

$$\{u_0 + \sqrt{gh_0}\}\Delta t/\Delta x \leq 1 \qquad \text{(Q4.8)}$$

2. Consider the CTCS-explicit approximation, also known as the leapfrog scheme, to the one-dimensional wave equation:

$$\left(u_i^{n+1} - u_i^{n-1}\right)/(2\Delta t) = -a\left(u_{i+1}^n - u_{i-1}^n\right)/(2\Delta x) \qquad \text{(Q4.9)}$$

(a) Put $v^n = u^{n-1}$ so that $u_i^{n-1} = v_i^n$. Substituting this into equation (Q4.9), write the latter in the form of equation (Q4.3) with $U = (u\ v)^T$.
(b) Show that the amplification factor for this scheme is given by

$$G = -j\sigma \sin \varphi \pm \left(1 - \sigma^2 \sin^2 \varphi\right)^{1/2} \qquad \text{(Q4.10)}$$

(c) Find the stability condition, dissipation and dispersion errors.

This problem illustrates the use of the matrix method of stability analysis for multi-time step schemes.

3. Show that the leapfrog scheme is unconditionally unstable for the one-dimensional transient heat conduction equation.
4. Consider the unsteady, one-dimensional heat conduction equation over the domain [0,1]:

$$\partial u/\partial t = \alpha \partial^2 u/\partial x^2 \qquad \text{(Q4.11)}$$

subject to Dirichlet boundary conditions that

$$u(0) = a \text{ and } u(1) = b \qquad \text{(Q4.12)}$$

(a) For uniform spacing and with second order accurate central differencing of the space derivative, show that Eq. (Q4.11) can be written as

$$dU/dt = SU + Q \tag{Q4.13}$$

$$\text{where } S = \frac{\alpha}{\Delta x^2}\begin{bmatrix} -2 & 1 & & & & \\ 1 & -2 & 1 & & & \\ & 1 & -2 & 1 & & \\ & & \cdot & \cdot & \cdot & \\ & & & 1 & -2 & 1 \\ & & & & 1 & -2 \end{bmatrix}, U = \begin{bmatrix} u_1 \\ u_2 \\ u_3 \\ \cdot \\ u_{n-1} \\ u_n \end{bmatrix} \text{ and } Q = \begin{bmatrix} \frac{\alpha a}{\Delta x^2} \\ 0 \\ 0 \\ \cdot \\ 0 \\ \frac{\alpha b}{\Delta x^2} \end{bmatrix} \tag{Q4.14}$$

(b) The stability of the time marching scheme (Q4.13) is governed by the homogeneous part of the equation, see Hirsch (1988) for details. A necessary but not sufficient condition for stability is that the real part of all the eigenvalues of S must be either negative or zero. Take 10 equal divisions for the interval [a, b], determine the elements of the (tridiagonal) coefficient matrix S, and find its nine eigenvalues. Show that all the eigenvalues are real and negative and that the scheme given by eqn. (Q4.13) and (Q4.14) is likely to be stable (note that it satisfies only the necessary condition for stability).

Since the matrix S includes the boundary conditions, this method presents a way of investigating the influence of non-periodic boundary conditions on the stability. For further details, see Hirsch (1988).

5. Consider steady, compressible but subsonic flow between two parallel plates of length L and separation distance of H. You wish to solve this as the eventual steady flow condition of an unsteady flow using methods like the MacCormack method. Write down the governing equations, initial and boundary conditions for this flow. Treat the flow as laminar even though it is likely to be turbulent.
6. Take a rectangular grid of constant mesh size of Δx and Δy, discretize the equations in Problem #5 for solution using the explicit MacCormack method. Make a flow chart for the calculation procedure.
7. Repeat Problem #6 using the Beam-Warming method.
8. You wish to solve Problem #5 as a steady flow problem using the SIMPLE formulation extended to compressible flows. Discretize the governing equations and make a flow chart for the calculation procedure.
9. Consider the case of incompressible Poiseuille flow between two parallel plates of length L and separation distance of H. Formulate the problem for solution using the artificial compressibility approach. Assume uniform velocity profile at the inlet and fully developed flow at the outlet. Note that the flow is steady and 2-d as the centerline velocity increases from the uniform value to a value about 50% higher. Since $\partial u/\partial x \neq 0$, by continuity, $\partial v/\partial y \neq 0$ and $v \neq 0$.

10. Repeat Problem #9 for solution using the SIMPLE scheme.
11. Repeat Problem #9 for solution using PISO scheme of Issa (1986).
12. Formulate the problem of supersonic flow through a nozzle supersonic and subsonic exit conditions. Neglect viscous effects and treat the problem as quasi-one-dimensional with variations in the flow direction. Note that a supersonic flow at entry requires specification of three boundary conditions, for example, pressure, density and Mach number (from which u and e_t are evaluated). If the flow remains supersonic at the exit, then no boundary conditions need to be specified at the outlet. The outlet values of the flow variables need to be extrapolated from interior grid points near the boundary. If the flow is subsonic at the outlet, one boundary condition needs to be specified, for example, the exit density or velocity and the other two must extrapolated from interior nodes.
13. Evaluate the Sod (1978) solution for the shock tube problem for different upstream conditions, see Wesseling (2004).
14. Obtain a numerical solution to the shock tube problem for the different cases studied in Problem #13 using the Lax-Wendroff scheme and the Godunov scheme.
15. Repeat Problem #14 using the scheme of Roe (1981) with the correction of Harten and Hyman (1983) for the sonic glitch.
16. Repeat Problem #14 using the Enquist-Osher (1981) scheme.
17. Repeat Problem #14 using the flux vector splitting method of Steger and Warming (1981).
18. Solve the lid-driven cavity problem (see Problem #2.20h) using the streamfunction-vorticity method and compare your solution with that of Ghia et al. (1982).
19. Repeat Problem #18 using the artificial compressibility approach.
20. Repeat Problem #18 using the SIMPLE scheme.
21. Solve using the SIMPLE scheme the parabolized form of the Navier-Stokes equations (see Problem #2.20g) for flow over a flat plate.
22. Solve the incompressible flow over a backward facing step problem (see Problem #2.20i) using the SIMPLE scheme and compare your results with those given in the literature (see Gresho et al.,1993).
23. Solve the problem of compressible flow over a forward facing step discussed in Woodward and Colella (1984) and compare your results with the experimental data for shock location and strength.
24. Consider the case of natural convection in 2-d enclosure in which one side wall is located at 70 deg C and the other side wall is located at 30 deg C. The top and the bottom walls are adiabatic. Compute the buoyancy-induced circulation using the streamfunction-vorticity approach and the SIMPLE scheme.
25. You wish to simulate the onset of natural convection in a 2-d enclosure. For this case, repeat Problem #24 with an initial condition of zero velocity throughout the enclosure and a uniform temperature of 30 deg C of the air inside the enclosure. At time = 0, the right side wall temperature is increased to 70 deg C and held constant subsequently. This sets up a transient convection which

eventually leads to the steady state convection simulated in Problem #24. Resolve the time-varying velocity and temperature field inside the 2-d enclosure using the streamfunction-vorticity approach, the MAC (or pressure equation) approach of Harlow and Welch (1965) and the PISO scheme of Issa (1986).

Chapter 5
Solution of Linearized Algebraic Equations

As seen in Chaps. 1 and 3, the accuracy of a CFD solution depends on Δx; only on fine grids can we expect the finite difference approximations to be adequately accurate. We have also seen that implicit schemes have higher stability and these are invariably used in the algorithms such as the Beam-Warming method (Sect. 4.2) and the SIMPLE scheme (Sect. 4.4) for the solution of Navier-Stokes equations. We have also seen in order to account for coupling and non-linearity, one would have to solve the discretized equations several times. In this chapter, we will look at methods that are available for the solution of a system of linear algebraic equations and discuss the special methods that have been developed or adopted for CFD applications; without these, the overnight solutions that we have come to expect from CFD simulations would not have been possible.

5.1 Need for Speed

In the last two chapters, we have seen the principles underlying the discretization of the governing equations using schemes of the desired accuracy and stability behaviour. The result of the discretization of an equation is a system of coupled, often non-linear, algebraic equations. In this chapter, we will look at methods available to solve these discretized equations. Before discussing the details of the algebraic equation solvers, let us first examine the nature of the discretized equations by taking the example of the generic scalar transport equation:

$$\partial(\rho\phi)/\partial t + \nabla.(\rho u\phi) = \nabla.(\Gamma\nabla\phi) + S_\phi \tag{5.1.1}$$

where ϕ is the scalar and the other terms and variables have the usual significance. Without much loss of generality, we restrict our attention to a two-dimensional case in Cartesian coordinates with constant properties and no source terms. For this case, Eq. (5.1.1) can be written as

S. Jayanti, *Computational Fluid Dynamics for Engineers and Scientists*,
https://doi.org/10.1007/978-94-024-1217-8_5

$$\partial(\rho\phi)/\partial t + \partial(\rho u\phi)/\partial x + \partial(\rho v\phi)/\partial y = \Gamma(\partial^2\phi/\partial x^2 + \partial^2\phi/\partial y^2) \quad (5.1.2)$$

Considering a rectangular grid with uniform spacing of Δx and Δy in the x- and y-directions and using an implicit first order upwind scheme (assuming u and v to be positive, known and constant) for the advection term, central scheme for the diffusion term and a first order scheme for the time derivative, the discretized equation corresponding to Eq. (5.1.2) can be written as

$$\begin{aligned} &1/\Delta t\ (\phi_{i,j}^{n+1} - \phi_{i,j}^{n}) + u/\Delta x\ (\phi_{i,j}^{n+1} - \phi_{i-1,j}^{n+1}) + v/\Delta y\ (\phi_{i,j}^{n+1} - \phi_{i,j-1}^{n+1}) \\ &= \Gamma[(\phi_{i+1,j}^{n+1} - 2\phi_{i,j}^{n+1} + \phi_{i-1,j}^{n+1})/\Delta x^2 + (\phi_{i,j+1}^{n+1} - 2\phi_{i,j}^{n+1} + \phi_{i,j-1}^{n+1})/\Delta y^2] \end{aligned} \quad (5.1.3)$$

The above equation can be rearranged to cast it in the following form:

$$a_{i,j}\,\phi_{i,j}^{n+1} + a_{i-1,j}\,\phi_{i-1,j}^{n+1} + a_{i+1,j}\,\phi_{i+1,j}^{n+1} + a_{i,j-1}\,\phi_{i,j-1}^{n+1} + a_{i,j+1}\,\phi_{i,j+1}^{n+1} = b_{i,j} \quad (5.1.4)$$

where

$$\begin{aligned} a_{i-1,j} &= -\Gamma\Delta t/\Delta x^2 - u\,\Delta t/\Delta x \\ a_{i+1,j} &= -\Gamma\Delta t/\Delta x^2 \\ a_{i,j-1} &= -\Gamma\Delta t/\Delta y^2 - v\Delta t/\Delta y \\ a_{i,j+1} &= -\Gamma\Delta t/\Delta y^2 \\ a_{i,j} &= (1/\Delta t) + u\Delta t/\Delta x + v\,\Delta t/\Delta y + 2\Gamma\Delta t/\Delta x^2 + 2\Gamma\Delta t/\Delta y^2 \\ b_{i,j} &= \phi_{i,j}^{n}/\Delta t \end{aligned} \quad (5.1.5)$$

Incorporating boundary conditions and arranging the algebraic equations in "lexicographic" order, i.e., in a permutation involving (i, j, k), going through all the possible values of i before changing j and k and continuing this way until all the permutations are over, results in a set of algebraic equations which can be represented in matrix form as

$$[A][\phi] = [b] \quad (5.1.6)$$

where coefficient matrix A is a square matrix of size $N \times N$, where N is the number of variables and b is a column matrix with N elements.

Equations of the form (5.1.6) have to be solved at each time step in order to obtain the values of $\phi_{i,j}$. Before we address the question of how to solve these equations, a number of remarks are in order. In general, the flow may be characterized by a number of variables. For example, laminar, Newtonian, incompressible, non-reacting, isothermal two-dimensional flow requires the specification of u, v and p to completely characterize the flow field. Unsteady, three-dimensional laminar flow with heat transfer requires u, v, w, p and T as variables which will also depend on time. A number of additional variables are required for the description of turbulent reacting flows as will be seen in Chap. 6. Each such variable will have a set of equations of the form (5.1.6). Often, these equations are coupled and the coefficients $a_{i,j}$ are non-linear and/or unknown (as they may involve variables which have not yet been solved for). In the general case, these are linearized and solved iteratively. The process of linearization results in a set of equations of the form of Eq. (5.1.6) for each variable with constant (estimated) coefficients which are regularly updated in the iterative scheme. These have to be repeatedly solved in order to arrive at the desired solution. Therefore, an efficient method for the solution of the linear algebraic equations is necessary in order to keep the computational requirements to within reasonable limits. In the present chapter, we discuss various options for doing so.

It is pertinent, at this stage, to consider the general features of the set of linearized algebraic equations that we wish to solve. The number of equations in the set is equal to the number of grid points at which the variable is to be evaluated. For a typical CFD problem, the number of grid points is very large and can be of the order of 10,000–1,000,000 or even higher for complicated three-dimensional flows. Hence, the set of equations we are dealing with is very large. Hence speed of solution and memory requirements are important considerations in choosing an efficient method. Another general feature of the equations is the sparseness of the coefficient matrix A. Since the value of a variable $\phi_{i,j}$ is typically expressed in terms of its immediate neighbours, each equation will have only a few non-zero coefficients. For example, in the example considered, $\phi_{i,j}$ is expressed in terms of its four neighbours and the "computational molecule", see Fig. 5.1a, involves five nodal points and correspondingly, each algebraic equation contains at most five non-zero coefficients.

If schemes of higher order accuracy are used, or if a three-dimensional case is considered in a non-orthogonal coordinate system, more number of nodes may be involved. Schemes with a 19-node computational molecule have been proposed. This implies that each equation of the system of equations given by Eq. (5.1.6) may contain up to 19 non-zero coefficients. For a problem with 10,000 grid points, the non-zero coefficients could therefore be 190,000 out of a total number of

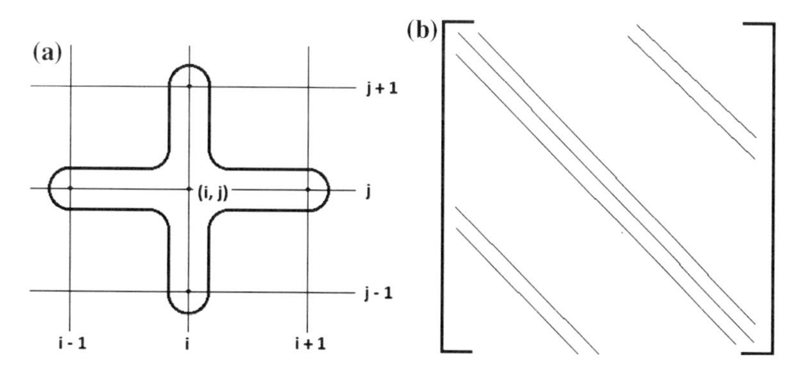

Fig. 5.1 **a** The five-point computational molecule of the discretized Eq. (5.1.4). **b** The 7-diagonal structure of matrix A for the discretization of Eq. (5.1.1) in three dimensions

$10{,}000 \times 10{,}000 = 10^8$ coefficients of matrix A, i.e., less than 0.2%. Thus, the coefficient matrix A is generally very sparse. Also, the non-zero coefficients may lie, in the case of structured grids, along certain diagonals (see Fig. 5.1b). The number of these diagonals depends on the terms present in the governing partial differential equation and the discretization and linearization schemes used to obtain the linearized algebraic equations. This diagonal structure of the coefficient matrix is not in general present in an unstructured mesh. Finally, the coefficient matrix usually exhibits the diagonal dominance feature, i.e., the magnitude of the diagonal element is greater than (at least for one equation) or equal to the sum of the magnitudes of the off-diagonal elements. While this is not a requirement for the general case, this is certainly a desirable feature, and special care is taken in the discretization and linearization process to ensure it in its weak form, i.e., that diagonal dominance is present for at least one of the set of equations while for the others the magnitude of the diagonal element is at least equal to the sum of the magnitudes of the off-diagonal elements.

These are the general features of the set of linear algebraic equations, namely, equations of the type $A\phi = b$ where the coefficient matrix is large, sparse and possibly structured. Two large classes of methods are available for the solution of these linear algebraic equations: direct and iterative methods. Direct methods use a series of arithmetic operations to reach the exact solution (except for round-off errors that are inevitable in a computed solution) in a countably finite number of operations. Iterative methods, on the other hand, are based on producing a succession of approximate solutions which asymptotically converges to the exact solution. Thus, if one uses infinite precision arithmetic, then one reaches the exact solution only after an infinite number of steps! However, in practice, we cannot avoid round-off errors, and an exact solution may any way not be needed because the equations that we are solving are themselves approximate. There are other considerations too for requiring only a sufficiently accurate solution: in many cases, the coefficients that appear in matrix A are themselves estimated values and incorporate uncertainties arising from linearization, coupling and lagging of variables and terms. Iterative improvement of these coefficients is needed. In such cases, one may wish to solve the equations approximately to begin with and seek much higher accuracy as one approaches the exact solution. Iterative methods make sense in such cases of seeking a solution with a tolerably small error. Also, iterative methods take account of the sparseness of the coefficient matrix while direct methods, except in special cases, do not do so. For all these reasons, iterative methods are invariably used to solve the linearized algebraic equations in CFD problems. However, many techniques used for accelerating the convergence rate of iterative methods are based on approximations derived from direct methods. Also, when multigrid methods or coupled solution methods are used, both iterative and direct methods may be used in the overall scheme of solution. In view of this, we discuss both direct and iterative methods. The basic methods of each class are discussed in Sects. 5.1 and 5.2, respectively. These are followed in Sect. 5.3 by more advanced iterative methods and in Sect. 5.4 by multigrid methods.

5.2 Direct Methods

5.2.1 *Cramer's Rule*

This is one of the most elementary methods and is often taught in school-level algebra courses. For a system of equations described as

$$\Sigma_j\left(a_{ij}\phi_j\right) = b_i \tag{5.2.1}$$

the solution for ϕ_i can be obtained as the ratio of determinant of matrix A_i to that of A, i.e.,

$$\phi_i = |A_i| \,/\, |A| \tag{5.2.2}$$

where the matrix A_i is obtained by replacing the ith column by the column vector b_i. Thus, for the system of three equations given by

$$\begin{aligned} 2\,\phi_1 + 3\,\phi_2 + 4\,\phi_3 &= 5 \\ 6\,\phi_1 + 7\,\phi_2 + 8\,\phi_3 &= 9 \\ 10\,\phi_1 + 13\,\phi_2 + 14\,\phi_3 &= 12 \end{aligned}$$

ϕ_1, ϕ_2 and ϕ_3 are given by

$$\phi_1 = \frac{\begin{vmatrix} 5 & 3 & 4 \\ 9 & 7 & 8 \\ 12 & 13 & 14 \end{vmatrix}}{\begin{vmatrix} 2 & 3 & 4 \\ 6 & 7 & 8 \\ 10 & 13 & 14 \end{vmatrix}} = \frac{12}{8} = 1.5$$

$$\phi_2 = \frac{\begin{vmatrix} 2 & 5 & 4 \\ 6 & 9 & 8 \\ 10 & 12 & 14 \end{vmatrix}}{\begin{vmatrix} 2 & 3 & 4 \\ 6 & 7 & 8 \\ 10 & 13 & 14 \end{vmatrix}} = \frac{-32}{8} = -4$$

$$\phi_3 = \frac{\begin{vmatrix} 2 & 3 & 5 \\ 6 & 7 & 9 \\ 10 & 13 & 12 \end{vmatrix}}{\begin{vmatrix} 2 & 3 & 4 \\ 6 & 7 & 8 \\ 10 & 13 & 14 \end{vmatrix}} = \frac{28}{8} = 3.5$$

While this method is elementary and yields a solution for any non-singular matrix A, it is extremely inefficient when the size of the matrix is large. For large n, where n is the number of unknowns (or equations), the number of arithmetic operations required to obtain the solution varies as (n + 1)!. Thus, the number of arithmetic operations for 10 equations is of the order of 4×10^7 and the solution can be obtained in a fraction of a second using a Gigaflop personal computer (a Gigaflop machine can do 10^9 floating point arithmetic operations per second (flops); fast computers can have teraflop speeds, i.e., 10^{12} flops). For the solution of 20 equations, the number of computations required is of the order of 5×10^{19} which increases to 8×10^{33} for a system of only 30 equations. Even the fastest computer on earth ($\sim$30 peta flops or 30×10^{15} flops in the year 2015) would require millions of years to get a solution in this case involving the equivalent of only 30 grid points. Obviously, the Cramer's rule is not meant for CFD solutions.

5.2.2 *Gaussian Elimination*

In contrast to the Cramer's rule, Gaussian elimination is a very useful and efficient way of solving a general system of algebraic equations, i.e., one which does not have any structural simplifications such as bandedness, symmetry and sparseness, etc. The method consists of two parts. In the first part, chosen variables are systematically and successively eliminated from the subsequent system of equations resulting in equations containing fewer and fewer variables. This process, known as forward elimination, ultimately results in a coefficient matrix in the form of an upper triangular matrix. This is solved easily in the second part using a process known as back substitution. These steps are explained below.

Consider the set of equations given by

$$\begin{aligned} a_{11}u_1 + a_{12}u_2 + \cdots &= c_1 \\ a_{21}u_1 + a_{22}u_2 + \cdots &= c_2 \\ &\cdot \\ &\cdot \\ &\cdot \\ a_{n1}u_1 + a_{n2}u_2 + \cdots &= c_n \end{aligned} \tag{5.2.3}$$

The objective of the forward elimination is to transform the coefficient matrix $\{a_{ij}\}$ into an upper triangular array by eliminating some of the unknowns from the equations by algebraic operations. This can be initiated by choosing the first equation as the "pivot" equation and using it to eliminate u_1 from each of the subsequent equations. This is done by multiplying the first equation by a_{21}/a_{11} and subtracting it from the second equation. Multiplying the pivot equation by a_{31}/a_{11} and subtracting it from the third equation eliminates u_1 from the third equation. This procedure is continued until u_1 is eliminated from all the equations except the first one. This results in a system of equations in which the first equation remains unchanged and the subsequent (n − 1) equations form a subset with modified coefficients (as compared to the original a_{ij}) in which u_1 does not appear:

$$\begin{aligned} a_{11}u_1 + a_{12}u_2 + \cdots &= c_1 \\ a'_{22}u_2 + a'_{23}u_3 + \cdots &= c'_2 \\ &\cdot \\ &\cdot \\ &\cdot \\ a'_{n2}u_2 + a'_{n3}u_3 + \cdots &= c'_n \end{aligned} \tag{5.2.4}$$

Now, the first equation of this subset is used as pivot to eliminate u_2 from all the equations below it. The third equation in the altered system is then used as the next pivot equation and the process is continued until only an upper triangular form remains:

$$\begin{aligned} a_{11}u_1 + a_{12}u_2 + \cdots\cdots\cdots &= c_1 \\ a'_{22}u_2 + a'_{23}u_3 + \cdots\cdots\cdots &= c'_2 \\ a''_{33}u_3 + a''_{34}u_4 + \cdots\cdots\cdots &= c''_3 \\ &\cdot \\ a''_{n-1,n} - 1\ u_{n-1} + a''_{n-1,n}u_n &= c''_{n-1} \\ a''_{n,n}u_n &= c''_n \end{aligned} \tag{5.2.5}$$

This completes the forward elimination process. The back substitution process consists of solving the set of equations given by (5.2.5) by successive substitution starting from the bottom-most equation. Since this equation contains only one variable, namely, u_n, it can be readily calculated as $u_n = c''_n/a''_{n,n}$. Knowing u_n, the equation immediately above it can be solved by substituting the value of u_n into it. This process is repeated until all the variables are obtained.

Of the two steps, the forward elimination step is the most time consuming and requires about $n^3/3$ arithmetic operations for large n. The back substitution process requires only about $n^2/2$ arithmetic operations. Thus, for large n, the total number of arithmetic operations required to solve a linear system of n equations by Gaussian elimination varies as n^3 which is significantly less than the (n + 1)! operations required by Cramer's rule. While Gaussian elimination is the most efficient method for full matrices without any specific structure, it does not take advantage of the sparseness of the matrix. Also, unlike in the case of iterative methods, there is no possibility of getting an approximate solution involving fewer number of arithmetic operations. Thus, the full solution, and only the full solution, is possible at the end of the back substitution process. This feature of lack of an intermediate, approximation solution is shared by most direct methods. This is a disadvantage when solving a set of non-linear algebraic equations as the coefficient matrix needs to be updated repeatedly in an overall iterative scheme. Finally, for large systems that are not sparse, Gaussian elimination, when done using finite-precision arithmetic, is susceptible to accumulation of round-off errors and a proper pivoting strategy is required to reduce it. Pivoting strategy is also required to eliminate the possibility of zero-diagonal element as this will lead to a division by zero. A number of pivoting strategies have been discussed in Marion (1982).

5.2.3 Gauss-Jordon Elimination

A variant of the Gaussian elimination method is the Gauss-Jordan elimination method. Here, variables are eliminated from rows both above and below the pivot equation. The resulting coefficient matrix is a diagonal matrix containing non-zero elements only along the diagonal. This eliminates the back substitution process. However, no computational advantage is gained as the number of arithmetic operations required for the elimination process is about three times higher than that required for the Gaussian elimination method. Gauss-Jordon elimination can be used to find the inverse of the coefficient matrix efficiently. It is also particularly attractive when solving for a number of right hand side vectors [b] in Eq. (5.1.6). Thus, Gauss-Jordon method can be used to simultaneously solve the following sets of equations:

$$[A] \cdot [x_1|\_|\ x_2|\_|\ x_3|\_|\ Y] = [b_1|\_|b_2|\_|b_3|\_|I] \tag{5.2.6}$$

which is a compact notation for the following system of equations:

$$Ax_1 = b_1 \quad Ax_2 = b_2 \quad Ax_3 = b_3 \quad \text{and} \quad AY = I$$

where I is the identity matrix and Y is obviously the inverse of A, i.e., A^{-1}. The computed A^{-1} can be used later, i.e., not at the time of solving Eq. (5.2.6) to solve for an additional right hand side vector b_4, i.e., to solve $Ax_4 = b_4$ as $x_4 = A^{-1}b_4$ at very little additional cost. Since computations are usually done with finite-precision arithmetic, the computed solution, x_4, may be affected by round-off error.

5.2.4 LU Decomposition

LU decomposition is one of several factorization techniques used to decompose the coefficient matrix A into a product of two special matrices so that the resulting equation is easier to solve. In LU decomposition, the matrix A is written as the product of a lower (L) and an upper (U) triangular matrix, i.e., [A] = [L] [U] or $a_{ij} = \Sigma_k(l_{ik}\ u_{kj})$. Such decomposition is possible for any non-singular square matrix.

Before we discuss the algorithm for this decomposition, i.e., the method of finding l_{ij} and u_{ij} for a given a_{ij}, let us examine the simplification of the solution resulting from this factorization.

With LU decomposition, the matrix equation [A] [ϕ] = [b] can be written as

$$[L][U][\phi] = [b] \tag{5.2.7}$$

which can be solved in two steps as

$$[L][y] = [b] \quad \text{and} \tag{5.2.8a}$$

$$[U][\phi] = [y] \tag{5.2.8b}$$

Solution of Eqs. (5.2.8a) and (5.2.8b) can be obtained easily by forward and backward substitution, respectively. Thus, the LU decomposition renders the solution of (5.2.7) easy. The key to the overall computational efficiency lies in the effort required to find the elements of [L] and [U]. For a given non-singular matrix A, the LU decomposition is not unique. This can be seen readily by writing the decomposition in terms of the elements as

$$\begin{bmatrix} a_{11} & a_{12} & a_{13} & a_{14} \\ a_{21} & a_{22} & a_{23} & a_{24} \\ a_{31} & a_{32} & a_{33} & a_{34} \\ a_{41} & a_{42} & a_{43} & a_{44} \end{bmatrix} = \begin{bmatrix} l_{11} & 0 & 0 & 0 \\ l_{21} & l_{22} & 0 & 0 \\ l_{31} & l_{32} & l_{33} & 0 \\ l_{41} & l_{42} & l_{43} & l_{44} \end{bmatrix} \begin{bmatrix} u_{11} & u_{12} & u_{13} & u_{14} \\ 0 & u_{22} & u_{23} & u_{24} \\ 0 & 0 & u_{33} & u_{34} \\ 0 & 0 & 0 & u_{44} \end{bmatrix} \tag{5.2.9}$$

Since $a_{ij} = \Sigma_k (l_{ik}\, u_{kj})$, Eq. (5.2.9) gives n^2 equations, where n is the number of rows (or columns) in matrix A, while the number of unknowns, i.e., the elements of L and U matrices, are $n^2 + n$. An efficient algorithm, known as Crout's decomposition, can be developed if the n diagonal elements of L, i.e., l_{ii}, are set to unity. This makes the number of remaining unknowns equal to the number of available equations which can be solved rather trivially by rearranging the equations in a certain order. The Crout's algorithm for finding the l_{ij} and u_{ij} can be summarized as follows:

- Set $l_{kk} = 1 \quad \text{for } k = 1, 2, \ldots n$
- For each $j = 1, 2, \ldots n$, evaluate
  - $u_{ij} = a_{ij} - \Sigma_{(k=1 \text{ to } i-1)} \left(l_{ik} u_{kj}\right) \quad \text{for } i = 1, 2, \ldots j$
  - $l_{ij} = \{a_{ij} - \Sigma_{(k=1 \text{ to } j-1)} \left(l_{ik} u_{kj}\right)\}/u_{jj} \quad \text{for } i = j+1, j+2, \ldots n$

(5.2.10)

Pivoting is necessary to avoid division by zero (which can be achieved by basic pivoting) as well as to reduce round-off error (which can be achieved by partial pivoting involving row-wise permutations of the matrix A). A variation of the Crout's decomposition is the Doolittle decomposition where the diagonal elements of the upper triangular matrix are all set to unity and the rest of u_{ij} (i not equal to j) and l_{ij} are found by an algorithm similar to that in Eq. (5.2.10). It can be shown that this procedure is equivalent to the forward elimination step of the Gaussian elimination method described in Sect. 5.2.2. Thus, the number of arithmetic operations to perform the LU decomposition is $n^3/3$. However, the final solution is obtained by a forward substitution step followed by a back substitution step, each of which would take about $n^2/2$ number of arithmetic operations. Thus, for large n, the Gaussian elimination and the LU decomposition are nearly equal in terms of the number of arithmetic operations required to obtain the solution. The principal advantage of the LU decomposition method lies in the fact that the LU decomposition step does not require manipulation of the right hand side vector, b, of Eq. (5.2.1). If matrix A is symmetric, then LU decomposition can be performed in half the number of computations using the Cholesky decomposition algorithm.

While LU decomposition is rarely used on its own in large CFD problems, an approximate or incomplete form of it is used to accelerate the convergence of some iterative methods, as will be discussed later.

5.2.5 Direct Methods for Banded Matrices

Banded coefficient matrices often arise while solving CFD problems using structured mesh algorithms (see Chap. 6). In addition to allowing for more efficient storage of the coefficients (wherein the non-zero values of the coefficient matrix are not usually stored), banded matrices often permit simplification of the general elimination/ decomposition techniques described above to such an extent that very efficient solution methods may be found for simple, but not uncommon, banded matrices. Here, we examine the specialized algorithms for three such banded matrices: tridiagonal, pentadiagonal and block tridiagonal systems.

Consider the set of algebraic equations represented by

$$Tx = s \tag{5.2.11}$$

where T is an $n \times n$ tridiagonal matrix with elements $T_{i,i-1} = c_i$, $T_{i,i} = a_i$ and $T_{i,i+1} = b_i$. The standard LU decomposition of T then produces L and U matrices which are bidiagonal. This process can be simplified to produce an equivalent upper bidiagonal matrix of the form

$$Ux = y \tag{5.2.12}$$

where the diagonal elements of U are all unity, i.e., $U_{i,i} = 1$ for all i. Denoting the superdiagonal elements of U, namely, $U_{i,i+1}$ by d_i, Eq. (5.2.12) implies conversion of the ith equation of the tridiagonal system of Eq. (5.2.11), namely,

$$c_i x_{i-1} + a_i x_i + b_i x_{i+1} = s_i \tag{5.2.13}$$

into an equation of the form

$$x_i + d_i x_{i+1} = y_i \tag{5.2.14}$$

where the coefficients d_i and y_i are yet to be determined. This is done as follows. Solving (5.2.14) for x_i, we have

$$x_i = y_i - d_i x_{i+1}$$

Thus,

$$x_{i-1} = y_{i-1} - d_{i-1} x_i$$

Substituting the above expression for x_{i-1} into (5.2.13) and rearranging, we have

$$x_i + b_i / (a_i - c_i d_{i-1})\ x_{i+1} = (s_i - y_{i-1} c_i) / (a_i - c_i d_{i-1}) \tag{5.2.15}$$

Comparing Eqs. (5.2.14) and (5.2.15), we get the following recurrence relations to determine d_i and y_i:

$$d_i = b_i / (a_i - c_i d_{i-1}) \quad y_i = (s_i - c_i y_{i-1}) / (a_i - c_i d_{i-1}) \tag{5.2.19}$$

Also, for the first row, i.e., i = 1, we have $d_1 = b_1/a_1$ and $y_1 = s_1/a_1$. Using Eq. (5.2.15), the tridiagonal matrix can be converted into an upper bidiagonal matrix as shown schematically in Fig. 5.2. It can be seen that the resulting matrix equation, Ux = y, can be solved readily by back-substitution.

The above procedure, involving a simplification of the Gaussian elimination procedure, to solve the tridiagonal system given by Eq. (5.2.13) is known as the Thomas algorithm or the tridiagonal matrix algorithm (TDMA). It can be summarized as follows:

Step I: Determine the coefficients, d_i and y_i, of the bidiagonal system (5.2.14) using the following formulae:

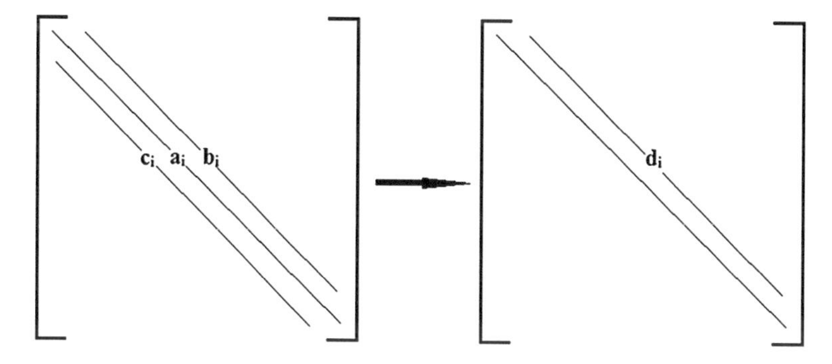

Fig. 5.2 Conversion of the tridiagonal matrix into an upper bidiagonal matrix

$$\begin{aligned} d_1 &= b_1/a_1 \text{ and } y_1 = s_1/a_1 \\ d_{i+1} &= b_{i+1}/(a_{i+1} - c_{i+1}d_i) \quad \text{for } i = 1 \text{ to } n-1 \\ \text{and} \quad y_{i+1} &= (s_{i+1} - c_{i+1}y_i)/(a_{i+1} - c_{i+1}d_i) \end{aligned} \tag{5.2.20}$$

Step II: Solve the bidiagonal system by back-substitution:

$$x_N = y_N \quad \text{and} \quad x_i = y_i - d_i x_{i+1} \quad \text{for } i = n-1,\ n-2,\ \ldots 2, 1 \tag{5.2.21}$$

The Thomas algorithm, encapsulated in Eqs. (5.2.20) and (5.2.21), is very efficient and the number of arithmetic operations required for solution of Eq. (5.2.13) varies as n (compared to $\sim n^3$ for Gaussian elimination and LU decomposition techniques). No pivoting strategy is incorporated in the Thomas algorithm, and it may fail therefore if a division by zero is encountered. It can be shown that this possibility does not arise if the matrix T is diagonally dominant, which is usually the case in many CFD applications. In such cases, the Thomas algorithm provides a very efficient method for solving the set of linear equations.

In some cases, the elements a_i, b_i and c_i of the matrix T in Eq. (5.2.13) may not be scalars but may themselves be matrices and Eq. (5.2.13) may take the following form:

$$\begin{bmatrix} [B_1] & [C_1] & & & & \\ [A_2] & [B_2] & [C_2] & & & \\ & [A_3] & [B_3] & [C_3] & & \\ & & & & & \\ & & & & [A_n] & [B_n] \end{bmatrix} \begin{bmatrix} [X_1] \\ [X_2] \\ [X_3] \\ \\ [X_n] \end{bmatrix} = \begin{bmatrix} [Y_1] \\ [Y_2] \\ [Y_3] \\ \\ [Y_n] \end{bmatrix} \tag{5.2.22}$$

where [A], [B], [C], [X] and [Y] are square matrices. Matrix of the form of the coefficient matrix in Eq. (5.2.22) are called block-tridiagonal matrix. The Thomas algorithm for scalar coefficients can be used to solve block-tridiagonal matrices also. The only modification required is that the division by the scalar coefficient should be replaced by multiplication with the inverse of the corresponding matrix. Thus, the Thomas algorithm for block-tridiagonal systems can be written as

$$d_1 = a_1^{-1} b_1 \tag{5.2.23a}$$

$$y_1 = a_1^{-1} s_1 \tag{5.2.23b}$$

$$d_{i+1} = (a_{i+1} - c_{i+1}d_i)^{-1}(b_{i+1}) \quad \Big| \quad \text{for } i = 1,\ n-1 \tag{5.2.23c}$$

$$y_{i+1} = (a_{i+1} - c_{i+1}d_i)^{-1}(s_{i+1} - c_{i+1}y_i) \Big| \tag{5.2.23d}$$

and the back substitution step is given by

$$x_n = y_n \tag{5.2.24a}$$

$$x_i = y_i - d_i x_{i+1} \tag{5.2.24b}$$

If the sub-matrices [A], [B] etc. are large, the matrix inversion steps in Eq. (5.2.23) may be performed without explicitly evaluating the inverse, for example, by solving $a_1 d_1 = b_1$ instead of as given in Eq. (5.2.23a).

The Thomas algorithm for tridiagonal matrices can be extended to the solution of pentadiagonal systems. Consider the set of equations $Px = s$, where P is a pentadiagonal matrix, i.e., it has non-zero elements only along five adjacent diagonals, for example, $P_{i,i-2}$, $P_{i,i-1}$, $P_{i,i}$, $P_{i,i+1}$ and $P_{i,i+2}$. The pentadiagonal system is first converted into a quadradiagonal system, $Qx = s'$, where Q contains non-zero elements in (i, i − 1), (i, i), (i, i + 1) and (i, i + 2), and then to an upper tridiagonal system, $Tx = s''$, where T contains non-zero elements in (i, i), (i, i + 1) and (i, i + 2) which can be solved by back-substitution. The procedure adopted for the tridiagonal system can be extended to obtain recurrence relations to determine the elements of Q, T, s′ and s″ resulting in a pentadiagonal matrix algorithm (PDMA). It should be noted that the TDMA and PDMA schemes given above are valid only when the three or five diagonals, respectively, are adjacent to each other. If there are zero diagonal elements in between, then the above schemes may not work. Specifically, the discretization of the transient heat conduction equation in two-dimensions results in a pentadiagonal system but one in which the diagonals are not adjacent to each other. The PDMA scheme cannot be used to solve this system, and thus has limited application in CFD-related problems. Methods such as cyclic reduction techniques are more efficient than TDMA but these are applicable only for specialized matrices (Pozrikidis 1998) and are not discussed here.

5.3 Basic Iterative Methods

Iterative methods adopt a completely different approach to the solution of the set of linear algebraic equations. Instead of solving the original equation

$$Ax = b \tag{5.1.6}$$

they solve

$$x = Px + q \tag{5.3.1}$$

where the matrix P and the vector q are constructed from A and b in Eq. (5.1.6). Equation (5.3.1) is solved by the method of successive approximations, also known as Picard's method. Starting with an arbitrary initial vector x^0, a sequence of vectors x^k, $k \geq 0$ is produced from the recurrence formula

$$x^{k+1} = Px^k + q \quad k \geq 0 \tag{5.3.2}$$

The iterative method is said to be convergent if

$$\lim_{k \to \infty}(x^k) = \tilde{x} \quad \text{for every initial vector } x^0 \tag{5.3.3}$$

Whether or not an iterative method converges depends on the choice of the matrix P, known as iteration matrix, in Eq. (5.3.1). Even for convergent methods, the rate of convergence is not necessarily the same. (For non-linear algebraic equations, additional considerations arise but the choice of the initial vector, x^0, is also important. However, for linear problems, this is not the case). Thus, an iterative method for the linear algebraic Eq. (5.1.6) is characterized by the construction of P and q; by the conditions for convergence of the sequence (5.3.2) and by the rate of convergence. Once these are determined, the implementation of an iterative scheme is rather simple compared to the direct methods so much so that Gauss, in 1823, was supposed to have written in reference to the iterative method (see Axelsson 1994) "I recommend this modus operandi. You will hardly eliminate anymore, at least not when you have more than two unknowns. The indirect method can be pursued while half asleep or while thinking about other things".

This advantage of simplicity is negated by the theoretical limit of infinite number of iterations needed to get the exact solution. However, there is a practical limit—the machine accuracy or round-off error—to the accuracy that can be attained when solving equations using modern computers and it is sufficient to undertake only a finite number of iterations to achieve this for any converging iterative method. Finally, when the algebraic equations are solved in a CFD context involving non-linear equations, it is not necessary to solve the equations even to machine accuracy. Some of these and other characteristics—both advantageous and disadvantageous in comparison with direct methods—are discussed below for three classical iterative methods, namely, the Jacobi method, the Gauss-Seidel method and the relaxation or more specifically, the successive overrelaxation (SOR) method. We first discuss how the iteration matrix P is constructed from the coefficient matrix A in each case, and follow this up with an analysis of their convergence behaviour.

All the three methods involve splitting of the matrix A into two matrices as

$$A = M - N \tag{5.3.4}$$

where M is an easily invertible matrix, i.e., of diagonal or triangular or block diagonal or block triangular structure. It is noted that diagonal systems, i.e., systems in which the coefficient matrix is diagonal, can be solved ("inverted", although the inverse of the matrix may not be explicitly computed in practice) readily, triangular matrices can be solved efficiently by forward or back-substitution. Substituting the above splitting into Eq. (5.1.6), we obtain

$$(M - N)x = b \quad \text{or} \quad Mx = Nx + b$$

or

$$x = M^{-1}Nx + M^{-1}b \tag{5.3.5}$$

which is in the form of Eq. (5.3.1) with $P = M^{-1}N$ and $q = M^{-1}b$. In practice, M^{-1} is not computed and the iterative scheme derived from Eq. (5.3.5) is written as

$$Mx^{k+1} = Nx^{k} + b \tag{5.3.6}$$

We can now discuss the above-mentioned three classical schemes in this framework.

5.3.1 *Jacobi Method*

In the Jacobi method, the set of equations comprising Eq. (5.1.6) are reordered, if necessary, in such a way that the diagonal elements of the coefficient matrix A are not zero and M is taken as a diagonal matrix containing all the diagonal elements of A. Thus, if the matrix A is split into three matrices, namely, D, E, and F such that

$$\begin{aligned} D_{ij} &= a_{ij}\delta_{ij} \\ E_{ij} &= -a_{ij} \quad \text{if } i<j \text{ or 0 otherwise} \\ \text{and} \quad F_{ij} &= -a_{ij} \quad \text{if } i>j \text{ and 0 otherwise} \end{aligned} \tag{5.3.7}$$

then A can be written as

$$A = D - E - F \tag{5.3.8}$$

In the Jacobi method,

$$M = D \quad N = E + F \tag{5.3.9}$$

resulting in the iteration scheme

$$\begin{aligned} &Dx = (E+F)x + b \\ \text{or} \quad &x = D^{-1}(E+F) + D^{-1}b = D^{-1}(D - A) + D^{-1}b \\ \text{or} \quad &x = \left(I - D^{-1}A\right)x + D^{-1}b \end{aligned} \tag{5.3.10}$$

The implementation of the Jacobi scheme follows the iterative formula

$$Dx^{k+1} = (E+F)x^k + b \tag{5.3.11}$$

Due to the diagonal form of D, Eq. (5.3.11) is quite simple to solve. Denoting the iteration index k by a superscript, one iteration of the Jacobi scheme for the linear system given by Eq. (5.3.11) takes the following form:

$$\begin{aligned}
a_{11}x_1^{k+1} &= b_1 - a_{12}x_2^k - a_{13}x_3^k \cdots - a_{1n-1}x_{n-1}^k - a_{1n}x_n^k \\
a_{22}x_2^{k+1} &= b_2 - a_{21}x_1^k - a_{23}x_3^k \cdots - a_{2n-1}x_{n-1}^k - a_{2n}x_n^k \\
&\cdot \\
&\cdot \\
a_{n-1,n-1}x_{n-1}^{k+1} &= b_{n-1} - a_{n-1,1}x_1^k - a_{n-1,2}x_2^k - a_{n-1,3}x_3^k \cdots - a_{n-1,n}x_n^k \\
a_{n,n}x_n^{k+1} &= b_n - a_{n1}x_1^k - a_{n2}x_2^k - a_{n3}x_3^k \cdots - a_{n-1,n}x_{n-1}^k
\end{aligned} \tag{5.3.12}$$

Its convergence behaviour will be analyzed later in the section.

5.3.2 *Gauss-Seidel Method*

In the Gauss-Seidel method, the matrices M and N of Eq. (5.3.9) are taken as

$$M = D - E \quad N = F \tag{5.3.13}$$

where D, E and F are given by Eq. (5.3.7) as above. The resulting iteration scheme is

$$(D - E)x = Fx + b \quad \text{or} \quad x = (D - E)^{-1}Fx + (D - E)^{-1}b \tag{5.3.14}$$

The implementation of the Gauss-Seidel scheme follows the iterative formula

$$(D - E)x^{k+1} = Fx^k + b \tag{5.3.15}$$

where the iteration index k is denoted as a superscript. Although Eq. (5.3.15) appears to be more complicated than the corresponding Jacobi formula given by Eq. (5.3.11), it can be rearranged to give the following equally simple formula:

$$Dx^{k+1} = Ex^{k+1} + Fx^k + b \tag{5.3.16}$$

which allows sequential evaluation of x_i in a form very similar to that of the Jacobi method. One iteration of the Gauss-Seidel scheme for the linear system given by Eq. (5.1.6) takes the following form:

$$
\begin{aligned}
a_{11}x_1^{k+1} &= b_1 - a_{12}x_2^k - a_{13}x_3^k \ldots\ldots - a_{1n-1}x_{n-1}^k - a_{1n}x_n^k \\
a_{22}x_2^{k+1} &= b_2 - a_{21}x_1^{k+1} - a_{23}x_3^k \ldots\ldots - a_{2n-1}x_{n-1}^k - a_{2n}x_n^k \\
&\cdot \\
&\cdot \\
a_{n-1,n-1}x_{n-1}^{k+1} &= b_{n-1} - a_{n-1,1}x_1^{k+1} - a_{n-1,2}x_2^{k+1} - a_{n-1,3}x_3^{k+1} \ldots\ldots - a_{n-1,n}x_n^k \\
a_{n,n}x_n^{k+1} &= b_n - a_{n1}x_1^{k+1} - a_{n2}x_2^{k+1} - a_{n3}x_3^{k+1} \ldots - a_{n-1,n}x_{n-1}^{k+1}
\end{aligned}
\tag{5.3.17}
$$

5.3.3 *Successive Over-Relaxation (SOR) Method*

The rate of convergence of either the Jacobi method or the Gauss-Seidel method can be changed by a simple technique known as relaxation in which the value of the variable at (k + 1)th iteration is taken as

$$x^{k+1} = x^k + \omega \Delta x^k \tag{5.3.18}$$

where Δx^k is the estimated improvement in x^k, i.e. $(x^{k+1} - x^k)$ where x^{k+1} is the value estimated by Jacobi or Gauss-Seidel method. If $\omega < 1$, then Eq. (5.3.18) leads to under-relaxation and if $\omega > 1$, then to over-relaxation. It can be shown (Ciarlet 1989) that convergence is possible only for $0 < \omega < 2$, and that too under certain other conditions. Under-relaxation is often used to solve non-linear algebraic equations where divergence or non-convergence of the iterative scheme is often a real possibility (see Sect. 4.4.4). In the context of solution of linear algebraic equations, over-rclaxation is used to improve the convergence rate resulting in the method known as successive over-relaxation (SOR). When applied to the Jacobi method, the splitting of the matrix A corresponds to

$$M = D/\omega \quad N = (1 - \omega)D/\omega + E + F \tag{5.3.19}$$

resulting in the iteration scheme

$$(D/\omega)\, x^{k+1} = [(1 - \omega)D/\omega + E + F]\, x^k + b$$

or

$$x^{k+1} = (D/\omega)^{-1}[(1 - \omega)D/\omega + E + F]\, x^k + (D/\omega)^{-1} b \tag{5.3.20}$$

When SOR is applied to the Gauss-Seidel method, the splitting of the matrix A corresponds to

$$M = D/\omega - E \quad N = (1 - \omega)D/\omega + F \tag{5.3.21}$$

resulting in the iteration scheme

$$\begin{aligned}(D/\omega - E)\, x^{k+1} &= [(1 - \omega)D/\omega + F]\, x^k + b \\ \text{or} \quad x^{k+1} &= (D/\omega - E)^{-1}[(1 - \omega)D/\omega + F]\, x^k + (D/\omega - E)^{-1} b\end{aligned} \tag{5.3.22}$$

It is not necessary to invert the matrix. One iteration of the Gauss-Seidel scheme with SOR for the linear system given by Eq. (5.1.6) takes the following form:

$$\begin{aligned}a_{11}x_1^{k+1} &= a_{11}x_1^k - \omega\{a_{11}x_1^k + a_{12}x_2^k + a_{13}x_3^k + \cdots + a_{1n}x_n^k - b_1\} \\ a_{22}x_2^{k+1} &= a_{22}x_2^k - \omega\{a_{21}x_1^{k+1} + a_{22}x_2^k + a_{23}x_3^k + \cdots + a_{2n}x_n^k - b_2\} \\ &\cdot \\ a_{n,n}x_n^{k+1} &= a_{n,n}x_n^k - \omega\{a_{n1}x_1^{k+1} + a_{n2}x_2^{k+1} + a_{n3}x_3^{k+1} + \cdots + a_{n,n}x_n^k - b_n\}\end{aligned} \tag{5.3.23}$$

It can be seen that the implementation of the successive over-relaxation technique requires little extra overhead in per-iteration work while significant speed-up of convergence rate (up to an order of magnitude) can be obtained with an optimum value of the relaxation parameter, ω. However, as will be shown later, the optimum value is not known a priori in many cases and may have to be estimated on a trial and error basis in the initial stages of the computation.

5.3.4 Block Iterative Methods

The above iterative methods have been discussed so far in the context of evaluation of x_i, i = 1 to n, each of which is a scalar, which corresponds to the value of a variable at a grid point. In such a case, the iterative schemes are called point methods, for example, point Jacobi method and point Gauss-Seidel method with SOR. These methods can be readily extended to the case when the coefficient matrix has a block structure such as the block tridiagonal structure given by Eq. (5.2.22). In this case, the matrix A can still be split into D, E and F each of which may consist of further sub-matrices. The Jacobi or Gauss-Seidel iterative schemes are then applicable according to this block decomposition. The corresponding iterative scheme is then called block-Jacobi or block-Gauss-Seidel scheme.

Consider the block-decomposition of the matrix A where the original n × n system of linear equations is decomposed into a 4 × 4 system of equations consisting of elements A_{pq}. If this is split into matrices D, E, F, then D consists of the sub-matrices $\{A_{11}, A_{22}, A_{33}, A_{44}\}$, E consists of $\{A_{21}, A_{31}, A_{32}, A_{41}, A_{42}, A_{43}\}$ and F of $\{A_{12}, A_{13}, A_{14}, A_{23}, A_{24}, A_{34}\}$. A block-Gauss-Seidel method with SOR for this 4 × 4 system can be written as

$$\begin{aligned}
A_{11}x_1^{k+1} &= A_{11}x_1^k - \omega\{A_{11}x_1^k + A_{12}x_2^k + A_{13}x_3^k + A_{14}x_4^k - b_1\} \\
A_{22}x_2^{k+1} &= A_{22}x_2^k - \omega\{A_{21}x_1^{k+1} + A_{22}x_2^k + A_{23}x_3^k + A_{24}x_4^k - b_2\} \\
A_{33}x_3^{k+1} &= A_{11}x_1^k - \omega\{A_{31}x_1^{k+1} + A_{32}x_2^{k+1} + A_{33}x_3^k + A_{34}x_4^k - b_3\} \\
A_{44}x_4^{k+1} &= A_{11}x_1^k - \omega\{A_{41}x_1^{k+1} + A_{42}x_2^{k+1} + A_{43}x_3^{k+1} + A_{44}x_4^{k+1} - b_4\}
\end{aligned} \tag{5.3.24}$$

where x_i and b_i themselves are vectors. Thus, each of the above four equations themselves represents a further set of linear equations to be solved either by direct or iterative methods. Block matrices arise in electrical or hydrodynamic network problems as well as in the coupled solution of Navier-Stokes equations.

5.4 Convergence Analysis of Classical Iterative Schemes

It is often thought that the convergence of Jacobi and Gauss-Seidel schemes goes together, i.e., one converges if the other does and that the latter method converges twice as fast. We dispel this notion by considering the following examples suggested by Ciarlet (1989). Consider, firstly, the set of three simultaneous linear algebraic equations given by

$$\begin{aligned}
x_1 + 2x_2 - 2x_3 &= -1 \\
x_1 + x_2 + x_3 &= 6 \\
2x_1 + 2x_2 + x_3 &= 9
\end{aligned} \tag{5.4.1}$$

It can be shown, using any direct method, that the correct solution is $x_i = \{1, 2, 3\}$. The solution obtained using Jacobi and Gauss-Seidel schemes with an initial guess of $x_i = \{0, 0, 0\}$ is shown in Fig. 5.3. It can be seen that the Jacobi method converges quickly while the Gauss-Seidel method diverges rapidly. Similar results are obtained for other initial guesses. The convergence behaviour of the SOR scheme with Jacobi and Gauss-Seidel schemes is not quite straightforward. Solutions with the SOR parameter values of $\omega = 1.25, 1.5, 1.75$ and 1.9 shows that the Gauss-Seidel-SOR method diverges rapidly in all the cases. The Jacobi-SOR method converges eventually in all cases although it goes through large negative and/ or positive values before converging in some cases. Also, increasing the value of the relaxation parameter appears to worsen convergence behaviour.

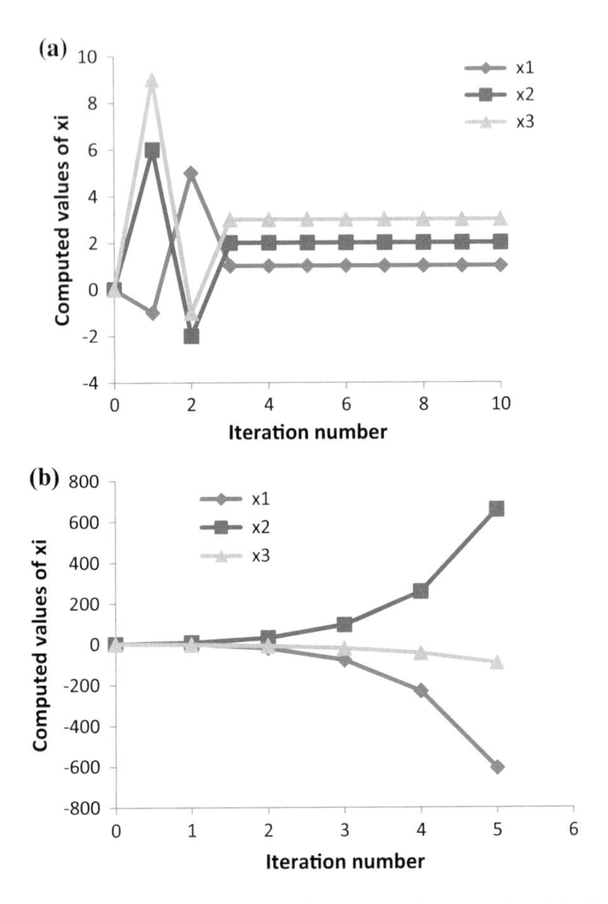

Fig. 5.3 Computed solutions of Eq. (5.4.1) with **a** the Jacobi method, and **b** the Gauss-Seidel method

Consider now a second example:

$$\begin{aligned} 2x_1 - x_2 + x_3 &= 3 \\ x_1 + x_2 + x_3 &= 6 \\ x_1 + x_2 - 2x_3 &= -3 \end{aligned} \tag{5.4.2}$$

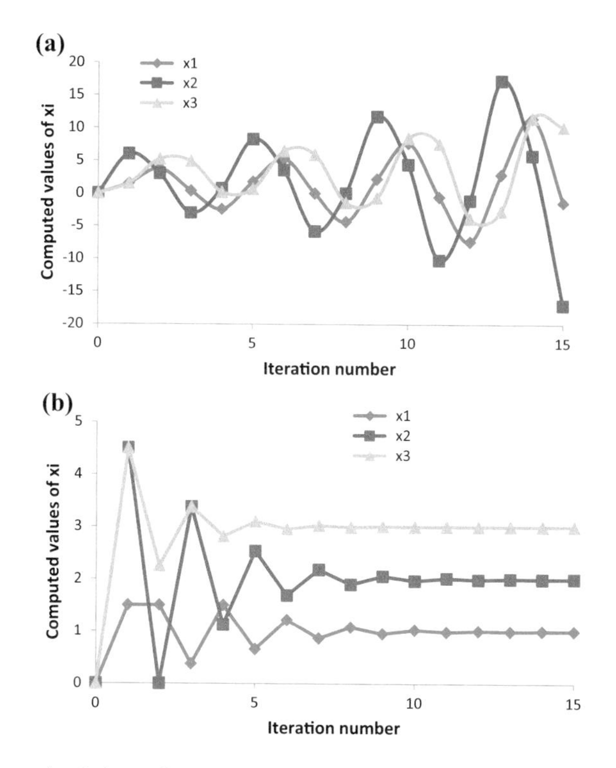

Fig. 5.4 Computed solutions of Eq. (5.4.2) with **a** the Jacobi method, and **b** the Gauss-Seidel method

Again, it can be shown that the correct solution is $x_i = \{1, 2, 3\}$. The solution obtained using the Jacobi and Gauss-Seidel schemes for this case is shown in Fig. 5.4 for an initial guess of $x_i = \{0, 0, 0\}$. For this case, the Jacobi scheme diverges but the Gauss-Seidel scheme converges. The Jacobi with SOR also diverges but the Gauss-Seidel scheme with SOR exhibits a more complicated behaviour; the computed values for relaxation parameter values of 0.75, 1.05, 1.10 and 1.14 are given in Table 5.1. The first value, corresponding to under-relaxation, appears to make the convergence faster while for the last value of 1.14, the scheme diverges.

Thus, the convergence behaviour of these classical schemes can be more complicated than what appears to be the case. A necessary and sufficient condition (Axelsson 1994) for convergence for an iterative scheme of the form given by Eq. (5.4.3)

Table 5.1 Successive values of x_1, x_2 and x_3 computed using the Gauss-Seidel SOR scheme for the set of equations given by Eq. (5.4.2) for different values of the SOR parameter ω

Iter no	$\omega = 0.75$			$\omega = 1.04$			$\omega = 1.1$			$\omega = 1.14$		
	x_1	x_2	x_3	x_1	x_2	x_3	x_1	x_2	x_3	x_1	x_2	x_3
0	0	0	0	0	0	0	0	0	0	0	0	0
1	−0.7500	4.5000	6.7500	1.5600	4.6176	4.7724	1.6500	4.7850	5.1893	1.7100	4.8906	5.4723
2	2.4375	1.1250	2.8125	1.4171	−0.3818	1.9075	1.2627	−0.9756	1.2890	1.1390	−1.3816	0.8056
3	2.3906	0.8438	2.1094	0.3129	3.9461	3.6984	0.2782	4.9737	4.4096	0.3038	5.7687	5.0586
4	1.7461	1.3359	2.4258	1.6763	0.4925	2.5398	1.9324	−0.8736	1.7914	2.0722	−2.0967	0.9878
5	1.3213	1.7051	2.7334	0.4283	3.1334	3.3105	−0.0090	4.7267	4.0656	−0.3383	6.3931	5.0229
6	1.1228	1.8853	2.8938	1.4508	1.1629	2.7867	2.0145	−0.5608	2.0430	2.5384	−2.6749	0.9290
7	1.0435	1.9589	2.9614	0.6576	2.6114	3.1484	0.0165	4.3907	3.8696	−0.6996	6.9530	5.1444
8	1.0146	1.9861	2.9868	1.2544	1.5566	2.8958	1.9349	−0.2241	2.2040	2.8388	−3.2343	0.7644
9	1.0047	1.9955	2.9957	0.8134	2.3201	3.0736	0.1211	4.0649	3.7319	−0.9667	7.5234	5.3403
10	1.0015	1.9986	2.9986	1.1357	1.7695	2.9478	1.8210	0.0853	2.3253	3.0897	−3.8235	0.5441
11	1.0005	1.9996	2.9996	0.9019	2.1656	3.0372	0.2359	3.7741	3.6230	−1.2121	8.1368	5.5809
12	1.0001	1.9999	2.9999	1.0707	1.8812	2.9735	1.7095	0.3568	2.4242	3.3365	−4.4651	0.2854
13	1.0000	2.0000	3.0000	0.9492	2.0852	3.0189	0.3420	3.5215	3.5325	−1.4649	8.8097	5.8566
14	1.0000	2.0000	3.0000	1.0365	1.9390	2.9865	1.6098	0.5914	2.5074	3.5984	−5.1720	−0.0069
15	1.0000	2.0000	3.0000	0.9738	2.0437	3.0097	0.4352	3.3040	3.4558	−1.7379	9.5531	6.1657

$$x^{k+1} = Px^k + q \quad k \geq 0 \tag{5.4.3}$$

for non-singular algebraic equations given by Eq. (5.4.4)

$$Ax = b \tag{5.4.4}$$

is that the spectral radius, ρ, of the iteration matrix, P, is less than unity, i.e.,

$$\rho(P) < 1 \text{ for convergence} \tag{5.4.5}$$

The spectral radius of a matrix P is the non-negative number defined by

$$\rho(P) = \max\{|\lambda_i|\}, \quad 1 \leq i \leq n \tag{5.4.6}$$

where λ_i are the eigenvalues of the square matix $P_{n \times n}$.
Defining the error, ε, in the solution of Eq. (5.4.3) as

$$\varepsilon = \tilde{x} - x \tag{5.4.7}$$

where $\tilde{x}$ is the exact solution of (5.4.4), the error propagation from iteration to iteration can be written as

$$\varepsilon^{k+1} = P\varepsilon^k \tag{5.4.8}$$

Thus, at the end of m recursive applications, we have

$$\varepsilon^m = P^m \varepsilon^0 \tag{5.4.9}$$

where ε^0 is in the initial error vector. If thc spectral radius of P is less than unity, then $\varepsilon^m \to 0$ as $m \to \infty$ leading to a converged solution. The rate at which the error decreases is governed, for large m, by the largest eigenvalue of P, i.e., by its spectral radius. Thus, the average convergence factor (per step for m steps) depends on the spectral radius; the smaller the spectral radius, the larger the convergence factor and the faster the rate of convergence. The asymptotic rate of convergence is proportional to – ln [ρ(P)]. This means that as $m \to \infty$, the error decreases by a factor of e for every 1/(− ln [ρ(P)]) number of iterations, or by one order (factor of ten) for every 2.3/(− ln [ρ(P)]) number of iterations.

Consider the simple case of the Laplace equation in the square [0, a] × [0, a] in the x- and y-directions. If this square is divided into a uniform mesh of M × M sub-rectangles, then for a five-point Jacobi scheme, the eigenvalues of the iteration matrix are given by

$$\lambda_{Jac} = 1 - \left[\left(\sin^2(\pi l/(2M)) + \sin^2(\pi m/(2M))\right]\right. = 1/2\left[\cos(\pi l/M) + \cos(\pi m/M)\right] \tag{5.4.10}$$

where l and m correspond to the grid indices in the x- and y-directions, respectively, and take values from 1 to (M − 1). The spectral radius of the Jacobi iteration matrix is given by the highest eigenvalue, that is, for l = m = 1 in Eq. (5.4.10) and is therefore

$$\rho(P_J) = \cos(\pi/M) \approx 1 - \pi^2/(2M^2) \quad \text{for large M} \tag{5.4.11}$$

The asymptotic convergence rate for this problem is therefore ($2\ M^2/\pi^2$). Thus, 0.466 n number of iterations are required to reduce the error by one order where n is the number of equations ($M \times M$) being solved simultaneously. Each equation of the Jacobi iteration requires five arithmetic operations, and thus the total number of arithmetic operations required to reduce error by four orders of magnitude for the Jacobi scheme is $9n^2$. This compares favourably with the estimate of $\frac{1}{3}n^3$ for the Gaussian elimination scheme discussed earlier for large values of n.

For the Gauss-Seidel method, it can be shown that the corresponding eigenvalues of the amplification matrix are equal to the square of the eigenvalues of the Jacobi method; thus,

$$\lambda_{GS} = 1/4\,[\cos(\pi l/M) + \cos(\pi m/M)]^2 \tag{5.4.12}$$

Hence the spectral radius of the iteration matrix for the Gauss-Seidel method for the Laplace equation is given by

$$\rho(P_{GS}) = \cos^2(\pi/M) \quad \approx 1 - \pi^2/M^2 \quad \text{for large M} \tag{5.4.13}$$

Thus, the Gauss-Seidel method converges twice as fast as the Jacobi method for this problem. For example, the total number of arithmetic operations required to reduce the error by four orders of magnitude will be $4.5n^2$.

The above results of the relative convergence of the Jacobi and Gauss-Seidel methods for the Laplace (and Poisson) equation can be generalized to the case where the coefficient matrix A is symmetric and positive definite. It can be shown (Hirsch 1988) that for such matrices, the SOR method would converge provided the overrelaxation factor satisfied the relation

$$0 < \omega < 2/[1 + \rho(P)] \tag{5.4.14}$$

where $\rho(P)$ is the spectral radius of the iteration matrix of the pure Jacobi or Gauss-Seidel method (i.e., without overrelaxation). The convergence rate for these methods will depend strongly on the overrelaxation factor and is greatly improved near the optimal relaxation factor, ω_{opt}, which corresponds to the minimum spectral radius of the iteration matrix. For the Jacobi method with SOR, the eigenvalues of the iteration matrix are related by (Axelsson 1994)

$$\lambda(P_{JSOR}) = (1 - \omega) + \omega\lambda(P_J) \tag{5.4.15}$$

where $\lambda(P_J)$ are the eigenvalues of the pure Jacobi iteration matrix. The optimum value of the overrelaxation parameter, ω, for the JSOR method is given by

$$\omega_{opt,JSOR} = 2/[2 - (\lambda_{min} + \lambda_{max})] \quad (5.4.16)$$

where λ_{min} and λ_{max} are the minimum and the maximum values of the iteration matrix corresponding to the pure Jacobi method. For the specific case of the Laplace equation with Dirichlet boundary conditions, the minimum and maximum eigenvalues are given by Eq. (5.4.10) as

$$\lambda_{max} = -\lambda_{min} = \cos(\pi/M)$$

and the optimal relaxation occurs for $\omega = 1$ corresponding to the Jacobi method without SOR itself!

For the point Gauss-Seidel method with SOR, the eigenvalues of the corresponding iteration matrix are related by

$$\lambda_{GSSOR} = 1 - \omega + \omega\, \lambda(P_{Jac})\, \lambda_{GSSOR}{}^{0.5} \quad (5.4.17)$$

The optimal relaxation factor, when A is symmetric and positive definite, is given by

$$\omega_{opt,\,GSSOR} = 2/\left[1 - \left\{1 - \rho^2\left(P_{Jac,opt}\right)\right\}^{0.5}\right] \quad (5.4.18)$$

The spectral radius for the optimal relaxation factor is given by

$$\rho_{GSSOR} = \omega_{opt} - 1 \quad (5.4.19)$$

For the case of the Laplace equation with Dirichlet boundary conditions, the spectral radius of the Jacobi iteration matrix is $\cos(\pi/M)$ and hence ω_{opt} for the GSSOR method is given by

$$\omega_{opt} = 2/[1 + \sin(\pi/M)] \approx 2(1 - \pi/M)$$

Hence the spectral radius at optimum relaxation is given by

$$\rho_{GSSOR} = 2\,(1 - \pi/M) - 1 = 1 - 2\pi/M \quad (5.4.20)$$

Comparing this with the corresponding expression without SOR (Eq. 5.4.13), we find that, for large M and in the asymptotic limit, the number of iterations required to reduce error by an order of magnitude varies as $n^{0.5}$ and that the total number of arithmetic operations required to reduce error by a decade varies as $n^{1.5}$, compared to n^2 variation for the pure GS method. This represents a considerably enhanced convergence rate for large values of n.

It is interesting to note from Eqs. (5.4.10), (5.4.12) and (5.4.20) that as the number of subdivisions (M) increases, the spectral radius approaches closer to unity and the convergence rate therefore decreases. Thus, for diffusion-dominated flows, convergence would be slower on finer grids and a larger number of iterations would be required to reduce the error by a given factor.

The above discussion is applicable in the asymptotic limit of large number of iterations and for large number of subdivisions (large M). In the initial stages of computation, the residual reduction may typically follow one of the two idealized curves shown in Fig. 5.5 depending on the spectrum of eigenvalues present in the iteration matrix and their damping characteristics. A residual history of the type of curve in dashed lines may be expected if the iterative scheme damps rapidly the high frequency errors but damps poorly the low frequency errors. A response of the type of curve shown as a solid line may be obtained if the low frequency errors are damped rapidly. A detailed discussion of this eigenvalue analysis of an iterative method is presented in Hirsch (1988). Such detailed analysis of the properties of iterative schemes is possible for symmetric matrices. For unsymmetric matrices, the convergence behaviour is more complicated and a non-monotone convergence may be expected, and the interested reader is referred to Axelsson (1994) for more details.

The above results on the conditions for convergence and the rate of convergence expressed in terms of the spectral radius of the iteration matrix are not very useful in practice because of the large size of matrices involved, typically of the order of $10^4 \times 10^4$ or more. The evaluation of the eigenvalues and the spectral radius for such cases may not be practically feasible. The following more restrictive condition,

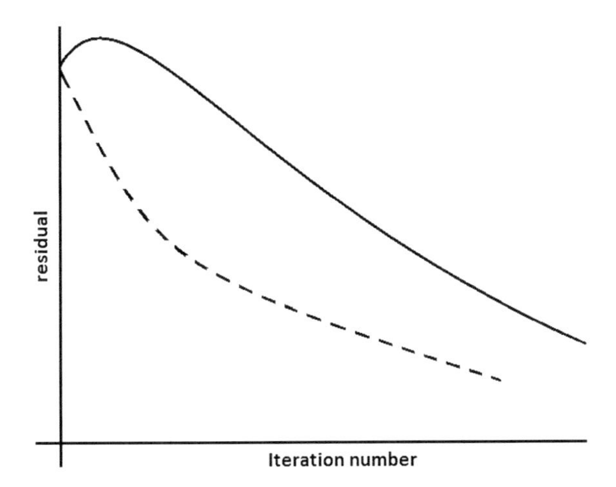

Fig. 5.5 Schematic idealized variations of residual with iteration number for a convergent iterative scheme: high frequency errors damped rapidly (*dashed line*) and damped slowly (*solid line*)

but one which is readily implementable, is often used as a guide for the convergence behaviour of iterative schemes:

A sufficient condition for convergence of an iterative scheme is that the matrix A (in Eq. 5.1.7) is irreducible and must have general diagonal dominance, i.e.,

$$\begin{aligned} |a_{ii}| &\geq \Sigma_{j\neq i}\left(|a_{ij}|\right) \text{ for all i} \\ |a_{ii}| &> \Sigma_{j\neq i}\left(|a_{ij}|\right) \text{ for at least one i} \end{aligned} \tag{5.4.21}$$

Both these conditions, namely, irreducibility and diagonal dominance, require some clarification. A reducible matrix does not require the simultaneous solution of all the equations; simultaneous solution of a reduced set of equations is possible. When this is not the case, the matrix A is said to be irreducible. Diagonal dominance of a matrix refers to the condition that

$$|a_{ii}| > \Sigma_{j\neq i}\left(|a_{ij}|\right) \quad \text{for all i} \tag{5.4.22}$$

A matrix which satisfies Eq. (5.4.22) is said to be strictly diagonally dominant. A matrix which satisfies the diagonal dominance condition for at least one row and satisfies the others with an equality sign is said to be generally diagonally dominant. A matrix satisfying the irreducibility and the general diagonal dominance condition is called irreducibly diagonally dominant matrix. It can be shown that (Ciarlet 1989) that under such conditions, the spectral radius of the iteration matrices corresponding to the Jacobi and Gauss-Seidel methods is less than unity and that the two methods converge for all initial vectors, x^0. Also, the SOR method would converge for $0 < \omega < 2$.

Let us summarize the results that we have obtained in this section. For the basic iterative methods, namely, the Jacobi method and the Gauss-Seidel method, the number of arithmetic operations required to reduce the residual by k orders of magnitude varies as $\sim kn^2$ where n is the number of equations being solved simultaneously. For the optimal SOR method, the number of arithmetic operations may vary as $\sim kn^{1.5}$, constituting a significant improvement for large n. However, the convergence rate decreases rather sharply for non-optimal (especially sub-optimal) values of the relaxation factor and, in a practical implementation of the iterative scheme, a search for an optimal value of ω should be incorporated for large n while taking account of the disparity of variation of the residual in the initial stages of the computation (Fig. 5.5).

The above results, which are valid for diffusion-dominated flows, compare very favourably with the best general purpose direct method for the solution of the same equation, namely, the Gaussian elimination method, which takes 1/3 n^3 number of arithmetic operations for the solution (to machine accuracy). The basic iterative methods are much simpler, easier to program, require less storage and take advantage of the structure of the matrix. However, these advantages are gained at the cost of a significant curtailment of applicability. While the Gaussian elimination method (with pivoting) would work for any non-singular matrix, the basic iterative methods are limited to diagonally dominant matrices. Also, they compare significantly less favourably with the best general purpose direct method for a diagonally dominant tridiagonal matrix, namely, the Thomas algorithm, for which the number of operations varies as $\sim n$. While SOR remains an effective means of enhancing the convergence rate of the basic methods, its efficacy is reduced for non-optimal values and may not work at all in some cases, as shown in Table 5.1 for GS-SOR with $\omega = 1.14$. (It may be noted that Eqs. (5.4.1) and (5.4.2) do not satisfy the diagonal dominance condition.) A number of methods and strategies have been developed to arrive at improving the convergence behaviour of the iterative methods. Some of these are discussed below.

5.5 Advanced Iterative Methods

A number of iterative methods have been devised to improve the convergence behaviour of the basic iterative schemes. These take the form of (i) improving the sensitivity of the SOR method to the choice of ω, e.g., Chebyshev iterative methods; (ii) taking advantage of the efficient Thomas algorithm for tridiagonal matrices, e.g., ADI methods; (iii) providing a better-conditioned splitting of the matrix A into M and N, e.g., strongly implicit methods; (iv) reformulation of the problem as a minimization problem, e.g. conjugate gradient methods; and (v) taking advantage of the error smoothening properties of some iterative methods, e.g., multigrid methods. An exhaustive review of these approaches is not possible and the salient points of some of these methods are discussed here to bring out the principles and concepts.

5.5.1 Chebyshev Iterative Methods

The basic iterative method given by Eq. (5.3.2) can be written in the following alternate form:

$$C\delta^{k+1} = -r^k, \quad x^{k+1} = x^k + \delta^{k+1}, \ k = 0, 1, 2, \ldots. \tag{5.5.1}$$

where r is the residual defined by

$$r = Ax - b \tag{5.5.2}$$

For matrices with real and positive eigenvalues, i.e., for diagonally dominant matrices, new and improved methods can be obtained by generalizing Eq. (5.5.1) and introducing some parameters. A first-order or one-step iterative method for the solution of Eq. (5.1.7) is defined by

$$C\delta^{k+1} = -\tau_k r^k, \quad x^{k+1} = x^k + \delta^{k+1}, \quad k = 0, 1, 2, \ldots \tag{5.5.3}$$

where τ_k is a sequence of parameters. Similarly, a second-order or two-step iterative method is defined by

$$Cs^k = r^k, \quad x^{k+1} = \alpha_k x^k + (1 - \alpha_k)x^{k-1} - \beta_k s^k, \quad k = 0, 1, 2, \ldots \tag{5.5.4}$$

where α_k and β_k are sequence of parameters with $\alpha_0 = 1$. When the parameters in Eqs. (5.5.3) and (5.5.4) are constant, i.e., when they do not change from iteration to iteration (that is, with k), these methods are known as first-order and second-order *stationary* iterative methods, respectively. A thorough discussion of the conditions of convergence and the convergence rates of these methods is given in Axelsson (1994), and only the main results are reproduced here. It can be shown that the first order stationary method with $\tau_k = \tau$ for all k converges if the matrix $C^{-1}A$ has real and positive eigenvalues $0 < \lambda_1 < \lambda_2 \ldots. < \lambda_n$ and τ is chosen such that

$$0 < \tau < \lambda_n \tag{5.5.5}$$

Similarly, the second-order stationary method with $\alpha_k = \alpha$ and $\beta_k = \beta$ and $\alpha_0 = 1$ converges for

$$0 < \alpha < 2 \text{ and } 0 < \beta < 2\alpha/\lambda_n \tag{5.5.6}$$

The optimum value of the parameters for best convergence for the first order method is given by

$$\tau = \tau_{opt} = 2/(\lambda_1 + \lambda_n) \tag{5.5.7a}$$

and the corresponding spectral radius is

$$\rho_0 = \{1 - \lambda_1/\lambda_n\}/\{1 + \lambda_1/\lambda_n\} \tag{5.5.7b}$$

For the second-order method, the optimum values of the parameters are

$$\begin{aligned} \beta_0 &= 2/\{\lambda_1 + \lambda_n\} \\ \beta &= 2\alpha/(\lambda_1 + \lambda_n) \\ \text{and} \quad \alpha &= 2/\{1 + (1 - \rho_0{}^2)^{0.5}\} \end{aligned} \tag{5.5.8}$$

where ρ_0 is given by Eq. (5.5.7b) and the spectral radius at optimum condition is given by

$$\begin{aligned}\rho_2 &= \left[\left\{1-\left(1-\rho_0^2\right)^{0.5}\right\}/\left\{1+\left(1-\rho_0^2\right)^{0.5}\right\}\right]^{0.5}\\ &= \left\{1-(\lambda_1/\lambda_n)^{0.5}\right\}/\left\{1+(\lambda_1/\lambda_n)^{0.5}\right\}\end{aligned} \tag{5.5.9}$$

where λ_1 and λ_n are the smallest and the largest eigenvalues of the matrix $C^{-1}A$.

The rates of convergence of these stationary methods can be compared with those of the Jacobi, Gauss-Seidel and the SOR schemes for the solution of the Laplace equation on a unit square with Dirichlet boundary conditions. For a division of each side into M segments, and taking the matrix C to be an identity matrix and M to be 50, the smallest and the largest eigenvalues of $C^{-1}A$ can be written as

$$\lambda_1 = 1-\cos(\pi/M) = 1-\cos(\pi/50) = 00.00197 \quad \text{and}$$
$$\lambda_n = 1+\cos(\pi/M) = 1.99803$$

Hence the smallest spectral radius for the optimum first-order stationary method, given by Eq. (5.5.7b), 0.99803 and it would take about 1150 iterations to reduce the residual by an order of magnitude. For the second-order method, the smallest spectral radius, given by Eq. (5.5.9), is equal to 0.93906 and it would take only about 37 iterations to reduce the error by an order of magnitude. Using the spectral radii formula given in Sect. 5.4 for the Jacobi, Gauss-Seidel and the GSSOR scheme, the number of iterations by each method to reduce the residual by a factor of ten can be computed to be 1165, 582 and 35, respectively. Thus, using the second-order stationary method, a convergence rate which is nearly as fast as that of the optimal SOR method can be achieved. Even faster rates of convergence can be achieved by *preconditioning* the matrix C which was taken in the above example as being equal to the identity matrix.

Proper application of the second-order stationary method requires a knowledge of the smallest and the largest eigenvalues of the matrix $C^{-1}A$. For some cases, as in the above example, these are known. In practice, these cannot be evaluated accurately using simple methods. When off-optimum values of the parameters are used (because λ_1 and λ_n are not known precisely), then the convergence rate is severely reduced as in the case of the SOR method. It is possible to make the convergence rate less sensitive to the estimates of the eigenvalues by making the parameters vary from iteration to iteration, i.e., by varying α_k and β_k in Eq. (5.5.4) with the iteration index k. The functional form of the optimum variation of α_k and β_k which would result in the minimum spectral radius is given by Chebyshev polynomials (Birkhoff 1972; Axelsson 1994) which can be used to obtain recursive formulae for the evaluation of the parameters of the non-stationary first- and second-order methods. The first-order Chebyshev method is potentially unstable and it is the stable second-order Chebyshev semi-iterative method which is stated below. If a and b, where $0 < a < \lambda_1$ and $b \geq \lambda_n$, are the estimates of the smallest and largest eigenvalues of the matrix $C^{-1}A$, respectively, then the second-order Chebyshev iterative method is given by

$$
\begin{aligned}
x^{k+1} &= \alpha_k x^k + (1-\alpha_k)x^{k-1} - \beta_k r^k \\
r^k &= C^{-1}\left(Ax^k - b\right) \\
x^1 &= x^0 - \beta_0 r^0/2 \\
\text{where}\quad \alpha_k &= \beta_k(a+b)/2;\ 1/\beta_k = (a+b)/2 - \beta_{k-1}(b-a)^2/4 \quad k = 1,2,\ldots \\
\text{and}\quad \beta_0 &= 4/(a+b)
\end{aligned}
\tag{5.5.10}
$$

The number of iterations, p, required for a relative error of ε is given, in the asymptotic limit, by

$$
p = {}^1\!/\!_2\,(b/a)^{0.5}\ln(2/\varepsilon) \tag{5.5.11}
$$

or about 1.5 $(b/a)^{0.5}$ to reduce the residual by a factor of ten. Note that here a and b are such that $0 < a \le \lambda_1$ and $b \ge \lambda_n$. If we take a = 0.001 and b = 1.999 in the above example for the Laplace equation in the unit square, then using the second order Chebyshev iterative method would require, according to Eq. (5.5.11), 67 iterations to reduce the residual by a factor of ten, which is still an order of magnitude improvement over the Gauss-Seidel method. If a is estimated to be 0.0001, i.e., an order of magnitude less, even then the number of iterations for an order of magnitude reduction in the residual would be 212. This shows the relative insensitivity of the method to be choice of the estimates of the lower and upper bounds for the eigenvalues. Note that the convergence rate is relatively more sensitive to the choice of the lower bound of the eigenvalues. Extensions of the Chebyshev iterative method to complex matrices with nearly real eigenvalues and to symmetric indefinite matrices, i.e., one which has both negative and positive eigenvalues is discussed in Axelsson (1994) where a thorough treatment of the method is given.

5.5.2 *ADI and Other Splitting Methods*

The alternating direction implicit (ADI) scheme was originally developed by Peaceman and Rachford (1955) for the iterative solution of elliptic equations. Consider the elliptic partial differential equation in Cartesian coordinates with Dirichlet boundary conditions:

$$
\partial/\partial x[A(x,y)\partial u/\partial x] + \partial/\partial y[C(x,y)\,\partial u/\partial y)] + G(x,y)\,u = S\,(x,y) \tag{5.5.12}
$$

where A and C are positive functions for the equation to be elliptic. Assuming G to be non-negative and considering a uniform mesh spacing of h and k in the x- and y-directions, respectively, Eq. (5.5.12) can be approximated (Birkhoff 1972) by the five-point difference equation as

$$(H+V+D)u = b \tag{5.5.13}$$

where

$$\begin{aligned}
&\quad\ Hu(x,y) = -a(x,y)\,u(x+h,y) + 2b(x,y)\,u(x,y) - c(x,y)\,u(x-h,\,y)\\
&\text{and}\ \ Vu(x,y) = -d(x,y)\,u(x,y+k) + 2e(x,y)\,u(x,y) - f(x,y)\,u(x,y-k)\\
&\text{with}\quad a = (k/h)A(x+h/2,\,y);\ c = (k/h)A(x-h/2,\,y);\ 2b = a+c\\
&\qquad\quad\ d = (h/k)C(x,\,y+k/2);\ e = (h/k)C(x,y-k/2);\ 2f = d+e
\end{aligned} \tag{5.5.14}$$

The matrix D in Eq. (5.5.13) is a diagonal matrix containing non-negative diagonal entries given by $hkG(x_i,\ y_i)$ at $(x_i,\ y_j)$. The vector b on the right hand consists of the term $hkS(x_i,\ y_j)$ along with additional source terms arising boundary points appearing in Eq. (5.5.14). For this two-dimensional case, matrices H and V are tridiagonal and contain difference approximations to the x-, and y-derivative terms appearing in Eq. (5.5.12). Peaceman and Rachford (1955) proposed a two-step solution of (5.5.13) for the case G = 0 and $D = \tau_i$:

$$(H+\tau_1^n I)u^{n+1/2} = b - (V - \tau_1^n I)u^n \tag{5.5.15a}$$

$$(V+\tau_2^n I)u^{n+1} = b - (H - \tau_2^n I)u^{n+1/2} \tag{5.5.15b}$$

where τ_1^n and τ_2^n are iteration parameters and I is the identity matrix. It can be shown (Birkhoff 1972; Axelsson 1994) that the iteration scheme is stable if these parameters are positive. The convergence rate of the iteration scheme depends strongly on the choice of the iteration parameters and, for the general case, it is equal to that of the optimum SOR method. However, it should be noted that each iteration step, that is, getting u^{n+1} from u^n, involves the solution of two tridiagonal systems given by Eqs. (5.5.15a) and (5.5.15b) consisting of $(N_i - 1)$ and $(N_j - 1)$ number of equations, respectively, where N_i and N_j are the number of divisions in the x- and y-directions, respectively. Thus, the solution of Eq. (5.5.12) using the ADI method is expected to be faster than the Gauss-Seidel method but not as fast as the optimum SOR method. However, if the matrices H, V and D commute, i.e., if

$$HV = VH;\quad HD = DH \quad \text{and}\ DV = VD \tag{5.5.16}$$

then the iteration parameters τ_1^n and τ_2^n can be chosen, in a non-stationary ADI method, in such a way (Birkhoff 1972; Axelsson 1994) that an asymptotic rate of convergence rate of $O(\log h^{-1})$, compared with that of $O(h^{-1})$ for the optimum SOR and of $O(h^{-2})$ with the Gauss-Seidel method, can be obtained with the ADI method. Commutativity of the operators requires that the coefficients A(x, y), C(x, y) and G (x, y) in Eq. (5.5.12) must be such that

$$A(x,y) = A(x); \quad C(x,y) = C(y) \quad \text{and } G(x,y) = G$$

In other words, the problem should be of separable-variables form in that the solution can be written in the form $u(x,\ y) = \varphi(x)\,\eta(y)$. For such problems the ADI scheme can prove to be a very powerful method.

The ADI scheme can be generalized as follows. If the coefficient matrix A in Eq. (5.1.7), i.e., Ax = b is decomposed into

$$A = A_1 + A_2, \tag{5.5.17}$$

then the ADI scheme can be written in the from

$$(I + \tau_1 A_1)\, x^{n+1/2} = (I - \tau_1 A_2)\, x^n + \tau_1 b \tag{5.5.18a}$$

$$(I + \tau_2 A_2)\, x^{n+1} = (I - \tau_2 A_1)\, x^{n+1/2} + \tau_1 b \tag{5.5.18b}$$

For stationary ADI, τ_1 and τ_2 do not change from iterations to iteration. It can be shown that if A_1 and A_2 are positive definite, the method would converge if τ_1 and τ_2 are positive. If A_1 and A_2 commute, then a sequence of τ_1^n and τ_2^n can be generated (Axelsson 1994) in such a way as to give a very fast convergence rate.

The ADI method can be extended to parabolic equations. Consider the unsteady diffusion problem in two dimensions:

$$\partial u/\partial t = \alpha\left(\partial^2 u/\partial x^2 + \partial^2 u/\partial y^2\right) \tag{5.5.19}$$

The equation can be readily solved using the FTCS-explicit scheme in a time-marching manner. However, the resulting scheme is only conditionally stable; it would be desirable to use an implicit scheme so as to be able use of large time steps, especially if one is interested only in the steady state solution. The difference approximation using FTCS-implicit for Eq. (5.5.19) can be written symbolically as

$$U^{n+1} - U^n = \Delta t\left(S_x + S_y\right)U^{n+1} \quad \text{or} \quad \left[1 - \Delta t\left(S_x + S_y\right)\right]U^{n+1} = U^n \tag{5.5.20a}$$

where

$$S_x U^{n+1} = \left(\alpha/\Delta x^2\right)\left(u_{i+1,j}^{n+1} - 2u_{i,j}^{n+1} + u_{i-1,j}^{n+1}\right) \tag{5.5.20b}$$

and

$$S_y U^{n+1} = \left(\alpha/\Delta y^2\right)\left(u_{i,j+1}^{n+1} - 2u_{i,j}^{n+1} + u_{i,j-1}^{n+1}\right) \tag{5.5.20c}$$

Direct solution of Eqs. (5.5.20a, b, c) requires a simultaneous solution of large number of equations at each time step due to the implicit nature of the equation. In the ADI method, the operator on the left hand side of Eqs. (5.5.20a, b, c) is split into one-dimensional operators and the resulting equation is solved in two (or three in case of three-dimensional problems) fractional times, each time using the efficient tridiagonal matrix algorithm. This is done as follows. The operator on the LHS of Eq. (5.5.20a) can be written as

$$\left[1-\Delta t\left(S_x+S_y\right)\right]U^{n+1}=(1-\Delta tS_x)\,(1-\Delta tS_y)\,U^{n+1}-\Delta t^2S_xS_yU^{n+1} \tag{5.5.21a}$$

Since the last term is of a higher order and proportional to Δt^2, it can be neglected in comparison because the time discretization is only first order accurate. Thus,

$$\left[1-\Delta t\left(S_x+S_y\right)\right]U^{n+1}\approx(1-\Delta tS_x)\,(1-\Delta tS_y)\,U^{n+1} \tag{5.5.21b}$$

Equation (5.2.20) can now be solved in two time steps:

$$(1-\Delta tS_x)U^{n+1/2}=(1+\Delta tS_y)U^n \tag{5.5.22a}$$

$$(1-\Delta tS_y)U^{n+1}=U^{n+1/2}-\Delta tS_yU^n \tag{5.5.22b}$$

Premultiplying Eq. (5.5.22b) with $(1-\Delta tS_x)$ and substituting (5.5.22a) into the resulting expression with the second order term neglected will lead to Eq. (5.5.20a). Using Eqs. (5.5.22a) and (5.5.22b) and using β_x and β_y to denote $\alpha\Delta t/\Delta x^2$ and $\alpha\Delta t/\Delta y^2$, respectively, the tridiagonal system of equations to be solved at each time step can be written as

$$u_{i,j}^{n+1/2}-\beta_x\left(u_{i+1,j}^{n+1/2}-2u_{i,j}^{n+1/2}+u_{i-1,j}^{n+1/2}\right)=u_{i,j}^n+\beta_y\left(u_{i,j+1}^n-2u_{i,j}^n+u_{i,j-1}^n\right) \tag{5.5.23a}$$

$$u_{i,j}^{n+1}-\beta_y\left(u_{i,j+1}^{n+1}-2u_{i,j}^{n+1}+u_{i,j-1}^{n+1}\right)=u_{i,j}^{n+1/2}-\beta_y\left(u_{i,j+1}^n-2u_{i,j}^n+u_{i,j-1}^n\right) \tag{5.5.23b}$$

Equation (5.5.23a) is solved, for example, using the TDMA scheme for each row, i.e., for each j by sweeping the grid from j = 1 to j = j_{max}. This gives the intermediate solution $u_{i,j}^{n+½}$ at all grid points. The final solution, $u_{i,j}^{n+1}$, at all grid points is obtained by solving the second equation for each column by sweeping the grid in the i-direction from i = 1 to i = i_{max}. In three dimensions, three sweeps, one in each direction, are required to obtain the solution at (n + 1)th time step from that at nth time step. It should be noted that appropriate boundary conditions must be incorporated into the intermediate solutions.

The factorization step can be generalized as follows. Consider the equation

$$\partial u/\partial T = Su + Q \tag{5.5.24}$$

where S is the space operator containing spatial derivatives and Q is not a function of u. The time derivative can be approximated by a generalized trapezoidal method as

$$U^{n+1} - U^{n} = \Delta t\left[\theta H^{n+1} + (1-\theta)H^{n}\right] \tag{5.5.25}$$

where H = SU + Q. For example, for the fully implicit scheme considered above, $\theta = 1$ and for the Crank-Nicolson scheme (resulting in a second order time accurate implicit scheme), $\theta = ½$. Substituting in Eq. (5.5.25) and rearranging, we obtain

$$\begin{aligned}&\left[1 - \theta\Delta t\left(S_x + S_y + S_z\right)\right]U^{n+1}\\ &= Q\Delta t + (1-\theta)\left(S_x + S_y + S_z\right)U^{n} + U^{n}\end{aligned} \tag{5.5.26}$$

The above equation can be approximately factored into the following form

$$\begin{aligned}&(1-\theta\tau\Delta tS_x)(1-\theta\tau\Delta tS_y)(1-\theta\tau\Delta tS_z)U^{n+1}\\ &= \tau\Delta tQ + [1+\tau(1-\theta)S_x][1+\tau(1-\theta)S_y][1+\tau(1-\theta)S_z]U^{n}\end{aligned} \tag{5.5.27}$$

where τ is a relaxation parameter which is equal to unity when time-accurate solution is required. If one is interested only in the steady state solution, τ can be chosen so as to accelerate the convergence towards the time-independent solution. Equation (5.5.27) is solved in steps as a succession of one-dimensional generalized trapezoidal schemes:

$$(1-\theta\Delta tS_x)U^{n+1/3} = [1+\tau\Delta t(1-\theta)\,S_x]U^{n} + \tau\Delta t\,Q \tag{5.5.28a}$$

$$(1-\theta\Delta tS_y)U^{n+2/3} = [1+\tau\Delta t(1-\theta)\,S_y]\,U^{n+1/3} \tag{5.5.28b}$$

$$(1-\theta\Delta tS_z)U^{n+1} = [1+\tau\Delta t(1-\theta)\,S_z]\,U^{n+2/3} \tag{5.5.28c}$$

For a three dimensional problem requiring time accurate solution ($\tau = 1$) with the Crank-Nicholson discretization scheme ($\theta = ½$), the following ADI scheme can be obtained from Eqs. (5.5.28a, b, c):

$$(1-½\Delta tS_x)U^{n+1/3} = (1+½\Delta tS_x)\,U^{n} + Q\Delta t \tag{5.5.29a}$$

$$(1-½\Delta tS_y)U^{n+2/3} = (1+½\Delta tS_y)\,U^{n+1/3} \tag{5.5.29b}$$

$$(1-½\Delta tS_z)U^{n+1} = (1+½\Delta tS_z)\,U^{n+2/3} \tag{5.5.29c}$$

The stability of these multi-step procedures can be evaluated using von Neumann type analysis. Hirsch (1988) shows that the ADI scheme for a transient diffusion problem is unconditionally stable. For the linear convection equation, it is stable in two-dimensions but unstable in three dimensions with the FTCS-implicit scheme. Thus, the approximate factorization step may introduce instability (note that the FTCS-implicit scheme is stable for the one-dimensional linear convection equation).

A further generalization of the fractional step method is the operator splitting method. Consider the initial value problem given by

$$\partial u/\partial t = Lu \tag{5.5.30}$$

where L is some operator which can be written as a linear sum of m pieces which are not necessarily linear. Thus,

$$Lu = L_1 u + L_2 u + L_3 u + \cdots + L_m u \tag{5.5.31}$$

In the operator splitting method, the evaluation of u_{n+1} from u_n is split into m steps. In each of these intermediate steps, the value of u is updated by evaluating only one of the m-terms. It is not necessary for the scheme for each of these intermediate steps be the same. If we denote by $U_i(n + 1, \Delta t)$ the scheme of updating the ith term of Eq. (5.5.31), then the operator splitting method proceeds from n to (n + 1)th time step in the following sequence of updates:

$$\begin{aligned}
u^{n+1/m} &= u_1(n, \Delta t) \\
u^{n+2/m} &= u_2(n + 1/m, \Delta t) \\
u^{n+3/m} &= u_3(n + 3/m, \Delta t) \\
&\cdot \\
&\cdot \\
u^{n+1} &= u_m(n + (m-1)/m, \Delta t)
\end{aligned} \tag{5.5.32}$$

The following example illustrates the nature of the operator splitting method and makes the distinction from the ADI scheme clear. Consider the one-dimensional unsteady advection-diffusion equation

$$\partial \varphi/\partial t + u\partial \varphi/\partial x = \Gamma \partial^2 \varphi/\partial x^2 \tag{5.5.32a}$$

Rearranging the terms on the right hand side, we can write the above equation as

$$\partial \varphi/\partial t = -u\partial \varphi/\partial x + \Gamma \partial^2 \varphi/\partial x^2 \tag{5.5.32b}$$

Using the operator splitting method, the above equation can be solved in two steps as a predictor-corrector sequence. The predictor step solves the equation

$$\partial\varphi/\partial t = -u\partial\varphi/\partial x \tag{5.5.33}$$

which has the analytical solution

$$\varphi(x,t) = \varphi(x - ut, t) \tag{5.5.34}$$

The corrector step solves the equation

$$\partial\varphi/\partial t = \Gamma\partial^2\varphi/\partial x^2 \tag{5.5.35}$$

which can be solved, for example, by using FTCS-implicit scheme as

$$\left(1 - 2\Gamma\Delta t/\Delta x^2\right)\varphi_i^{n+1} = \left(\Gamma\Delta t/\Delta x^2\right)\left(\varphi_{i+1}{}^{n+1} + \varphi_{i-1}{}^{n+1}\right) \tag{5.5.36}$$

Thus, one would get an intermediate solution at all i using Eq. (5.5.34) and then use Eq. (5.5.36) with intermediate values of φ_i to advance to φ_i^{n+1}. The use of this combination of the analytical solution for the convection term and a second order accurate scheme for the diffusion term can eliminate the excessive dissipation error associated with the use of the first-order upwind scheme for the first term while avoiding the numerical oscillations associated with the use of a central scheme at a steep convective gradient. Further discussion of the operator splitting methods can be found in Issa (1986), Issa et al. (1986) and Leonard and Niknafs (1991).

5.5.3 *Strongly Implicit Procedures*

In the basic iterative methods, the equation $A\varphi = b$ is solved as $M\varphi^{k+1} = N\varphi^k + b$ where M and N are constructed from A. As mentioned earlier, for an iterative method to work efficiently, the construction of M and N should be efficient, the evaluation of φ^{k+1} from φ^k should be efficient and the iterative scheme should converge fast. In the basic iterative methods discussed in Sect. 5.2, the first two tasks are done efficiently but the third condition is not easily satisfied. This results from the fact that in these methods, there is a significant component of $A\varphi$, namely, the part belonging to $-N\varphi$, which is evaluated at kth iteration level. Thus, only part of it, namely, $M\varphi$, is being evaluated implicitly, and the rest, i.e., $-N\varphi$, is being treated explicitly. If the quantum of this deferred correction can be reduced, then the convergence rate will be improved. In direct methods, for example, in the LU decomposition method, the coefficient matrix A is decomposed exactly into M, which is equal to LU and N = 0. Hence in these methods the whole of A is treated implicitly and the solution converges in one "iteration".

The problem with such exact decomposition is the computational cost. Exact LU decomposition requires $\sim n^3/3$ number of arithmetic operations in the general case. Strongly implicit procedures rely on making a decomposition which is relatively easy to perform but produces an M which is more like A and an N which contains less of A. The resulting solution of $M\varphi^{k+1} = N\varphi^k + b$ is therefore more strongly implicit than in basic iterative methods. One such possibility is the incomplete LU (ILU) decomposition. For a five-point computational molecule obtained from using central differences for a Poisson equation, there will be, upon lexicographic ordering, five non-zero diagonals in matrix A corresponding to points (i, j), (i − 1, j), (i + 1, j), (i, j − 1) and (i, j + 1). Since the exact decomposition of A into a product of L and U is costly, the trick is to find L and U approximately without requiring too many floating point operations. One such method is the incomplete Cholesky factorization developed for symmetric matrices (Axelsson 1994). This can be extended to asymmetric matrices resulting in the Incomplete LU (ILU) method. Here, an efficient but incomplete LU factorization of A is made possible by imposing the condition that every element that is zero in A is also zero in L and U. Since A has five non-zero diagonals, L and U have three non-zero element diagonals each, but their product, LU, which is equal to M, will contain seven non-zero diagonals. In the ILU method, the five non-zero diagonals of A are set equal to the corresponding ones of M. The two additional diagonals of the product of L and U are transferred to N such that M = LU = A + N. Algorithms can be developed for efficient and sequential evaluation of the elements of L and U which take advantage of the sparsity and known diagonal structure of A (Axelsson 1994). An iterative scheme can then devised as follows:

$$A\varphi = b \Rightarrow LU\varphi^{k+1} = b - N\varphi^k \tag{5.5.37}$$

which is then solved in two steps as

$$LY = b - N\varphi^k \text{ for } Y \text{ and } U\varphi^{k+1} = Y \tag{5.5.38}$$

The solution is not exact as $N\varphi$ is not evaluated implicitly and therefore iterative update is required. While the ILU method thus comes up with a method of efficient decomposition and efficient update of the iteration, the convergence rate of the iterative method is not fast enough to compensate for the additional computational effort required for the decomposition and the iterative update. In a variation of this method, Stone (1968) proposed an ILU decomposition which is more implicit than the conventional ILU method, i.e., there is more of $A\varphi$ in $M\varphi$. This is done in the following way. In the conventional ILU method, N contains only two non-zero diagonals corresponding to coefficients at nodes (i − 1, j + 1) and (i + 1, j − 1). Stone suggested that N can have non-zero values along all the seven diagonals of M (and not just the two as in the ILU method) but that these should be chosen such that $N\varphi \approx 0$. Thus, we seek a decomposition of $A\varphi$ into $M\varphi$ and $N\varphi$ such that, for point (i,j), both M and N have seven non-zero diagonals; these include the five non-zero diagonals of A corresponding to locations (i, j), (i − 1, j), (i + 1, j), (i, j − 1) and (i, j + 1) together with those corresponding to (i − 1, j + 1) and (i + 1, j − 1) which result from the multiplication of L and U. Thus, we seek

$$(M\varphi)_{i,j} = M_{i,j}\varphi_{i,j} + M_{i,j-1}\varphi_{i,j-1} + M_{i,j+1}\varphi_{i,j+1} + M_{i+1,j}\varphi_{i+1,j} + M_{i-1,j}\varphi_{i-1,j} + M_{i-1,j+1}\varphi_{i-1,j+1} + M_{i+1,j-1}\varphi_{i+1,j-1} \tag{5.5.39}$$

$$(N\varphi)_{i,j} = N_{i,j}\varphi_{i,j} + N_{i,j-1}\varphi_{i,j-1} + N_{i,j+1}\varphi_{i,j+1} + N_{i+1,j}\varphi_{i+1,j} + N_{i-1,j}\varphi_{i-1,j} + N_{i-1,j+1}\varphi_{i-1,j+1} + N_{i+1,j-1}\varphi_{i+1,j-1} \approx 0 \tag{5.5.40}$$

Since N contains the two extra diagonals of M, we have $N_{i-1,j+1} = M_{i-1,j+1}$ and $N_{i+1,j-1} = M_{i+1,j-1}$. In order to make $(N\varphi)_{i,j} \sim 0$, we choose the other coefficients of N such that Eq. (5.5.40) reduces to

$$M_{i-1,j+1}\left(\varphi_{i-1,j+1} - \varphi^*_{i-1,j+1}\right) + M_{i+1,j-1}\left(\varphi_{i+1,j-1} - \varphi^*_{i+1,j-1}\right) \approx 0 \tag{5.5.41}$$

Here $\varphi^*_{i-1,j+1}$ and $\varphi^*_{i+1,j-1}$ are approximations for $\varphi_{i-1,j+1}$ and $\varphi_{i+1,j-1}$ expressed in terms of $\varphi_{i,j}$, $\varphi_{i-1,j}$, $\varphi_{i+1,j}$, $\varphi_{i,j-1}$ and $\varphi_{i,j+1}$. Assuming smooth spatial variation of φ, which would be the case for the solution of a diffusion equation without strong local source terms, Stone proposed the following approximations:

$$\varphi^*_{i-1,j+1} = \alpha\left(\varphi_{i-1,j} + \varphi_{i,j+1} - \varphi_{i,j}\right) \quad \text{and} \quad \varphi^*_{i+1,j-1} = \alpha\left(\varphi_{i,j-1} + \varphi_{i+1,j} - \varphi_{i,j}\right) \tag{5.5.42}$$

where α should lie between 0 and 1 for stability. Substitution of expressions (5.5.42) into Eq. (5.5.41) would then yield the coefficients of $\varphi_{i,j}$, $\varphi_{i-1,j}$, $\varphi_{i+1,j}$, $\varphi_{i,j-1}$ and $\varphi_{i,j+1}$ in the N matrix in terms of $M_{i-1,j+1}$ and $M_{i+1,j-1}$ which themselves are known from the elements of L and U. This method is known as the Strongly Implicit Procedure (SIP) of Stone (1968). Schneider and Zedan (1981) extended it to an elliptic equation in two dimensions but on a nine-point molecule which involved the corner points also [e.g. (i + 1, j + 1), (i − 1, j − 1) etc.]. They derived the five-point molecule as a special case of the nine-point molecule, and reported that the modified SIP had even faster convergence rate while being easier to implement. Zedan and Schneider (1983) extended it to three dimensional flows wherein a procedure developed for a 19-point molecule was reduced to a 7-point molecule applicable, for example, for heat conduction problem in Cartesian coordinates. The algorithm for this procedure is explained below.

For an equation of AT = q, the 19-point molecule consists of the point (i, j, k) and the 18 nearest neighbours on either side in the three (i.e., x-, y- and z-) coordinate planes passing through the point (i, j, k). Thus, for point (i, j, k), the discretized equation can be written as

$$
\begin{aligned}
&A_1T_{i,j-1,k-1} + A_2T_{i-1,j,k-1} + A_3T_{i,j,k-1} + A_4T_{i+1,j,k-1} + A_5T_{i,j+1,k-1} \\
&\quad + A_6T_{i-1,j-1,k} + A_7T_{i,j-1,k} + A_8T_{i+1,j-1,k} + A_9T_{i-1,j,k} + A_{10}T_{i,j,k} \\
&\quad + A_{11}T_{i+1,j,k} + A_{12}T_{i-1,j+1,k} + A_{13}T_{i,j+1,k} + A_{14}T_{i+1,j+1,k} + A_{15}T_{i,j-1,k+1} \\
&\quad + A_{16}T_{i-1,j,k+1} + A_{17}T_{i,j,k-1} + A_{18}T_{i+1,j,k+1} + A_{19}T_{i,j+1,k+1} = q_{i,j,k}
\end{aligned}
\tag{5.5.43}
$$

We seek an approximate LU factorization of A such that [M] = [L] [U] = [A + N]. The L and U matrices are to have ten non-zero elements per row in the same locations as those in matrix A. Matrix M, which is the result of the multiplication of [L] and [U] will have the 19 non-zero elements of [A] and also 24 other non-zero elements resulting in a linear algebraic equation of the form

$$\sum_{m=1}^{10} A_m T_m + \sum_{n=1}^{24} \varphi_n T_n = q \tag{5.5.44}$$

Here, the elements of [L] and [U] can be determined by requiring that the original 19 elements of A remain unchanged in M. The new non-zero elements, φ_n, can then be determined from the product of L and U. This completes the ILU decomposition. An iterative scheme can now used to solve Eq. (5.5.44) by evaluating the terms involving the new elements $(\varphi_n T_n)$ at kth iteration while treating the other terms $(A_m T_m)$ implicitly at (k + 1)th iteration. Since this procedure is not very effective, Zedan and Schneider (1983) allowed [N] to have non-zero elements in all the 43 diagonals and tried to cancel the effect of $\varphi_n T_n$ by rewriting Eq. (5.5.44) as

$$\sum_{m=1}^{10} A_m T_m + \sum_{n=1}^{24} \varphi_n (T_n - \alpha D_n) = q \tag{5.5.45}$$

where α is a parameter (with a suggested value of 0.9) and D_n has values not only of the n-index but also of the m-index. They further deduced a 7-point formulation (see Fig. 5.6) for the discretized equation from the 19-point formulation of Eq. (5.5.43):

$$A^b T_{i,j,k-1} + A^s T_{i,j-1,k} + A^w T_{i-1,j,k} + A^p T_{i,j,k} + A^e T_{i+1,j,k} + A^n T_{i,j+1,k} + A^f T_{i,j,k+1} = q_{i,j,k} \tag{5.5.46}$$

They then devised a simpler LU decomposition which required fewer algebraic operations. With reference to Fig. 5.7 which shows the structures and notations for matrices [A], [L] and [U] for a 7-point computational molecule, the elements of [L] and [U] can be evaluated using the following set of equations for point (i, j, k):

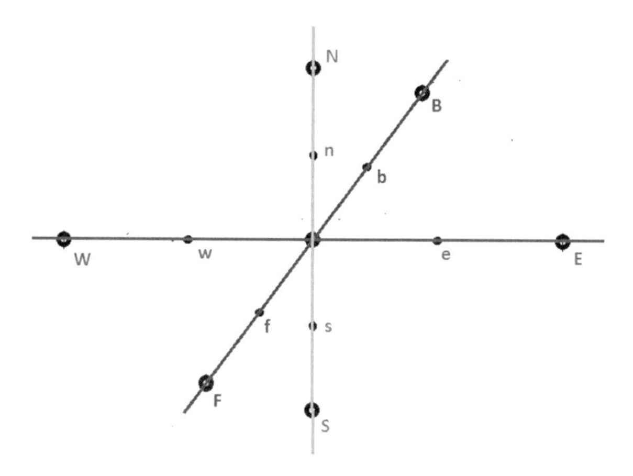

Fig. 5.6 Nomenclature for the seven-point computational molecule for the discretization (Eq. 5.5.46) of the diffusion equation in three dimensions

$$a_{i,j,k} = A^{b}_{i,j,k} / \left\{ 1 + \alpha \left[p_{i,j,k-1} - h_{i,j,k-1}\left(h_{i+1,j,k-1} + r_{i+1,j,k-1}\right) - \left(r_{i,j,k-1} - p_{i+1,j,k-1}h_{i,j,k-1}\right) \left(h_{i,j+1,k-1} + p_{i,j+1,k-1} + r_{i,j+1,k-1}\right) \right] \right\} \tag{5.5.47a}$$

$$b_{i,j,k} = -a_{i,j,k}h_{i,j,k-1} \tag{5.5.47b}$$

$$c_{i,j,k} = -a_{i,j,k}r_{i,j,k-1} - b_{i,j,k}p_{i+1,j,k-1} \tag{5.5.47c}$$

$$d_{i,j,k} = \left\{ A^{s}_{i,j,k} - a_{i,j,k}s_{i,j,k} + \alpha\left[\left(h_{i+1,j-1,k} + 2s_{i+1,j-1,k} + v_{i+1,j-1,k}\right)b_{i,j,k}s_{i+1,j,k-1} - s_{i-1,j,k} \left(A^{w}_{i,j,k} - a_{i,j,k}u_{i,j,k-1}\right)\right]\right\} / \left\{ 1 + \alpha\left[2s_{i,j-1,k} + u_{i,j-1,k} - s_{i-1,j,k}p_{i,j-1,k} - h_{i,j-1,k}\left(h_{i+1,j-1,k} + 2s_{i+1,j-1,k} + v_{i+1,j-1,k}\right)\right]\right\} \tag{5.5.47d}$$

$$e_{i,j,k} = -b_{i,j,k}s_{i+1,j,k-1} - d_{i,j,k}h_{i,j-1,k} \tag{5.5.47e}$$

$$f_{i,j,k} = \left[A^{w}_{i,j,k} - a_{i,j,k}u_{i,j,k-1} - d_{i,j,k}p_{i,j-1,k} - \alpha\left(a_{i,j,k}p_{i,j,k-1} + c_{i,j,k}p_{i,j+1,k-1} + d_{i,j,k}u_{i,j-1,k}\right)\right] / \left[1 + \alpha\left(2p_{i-1,j,k} + s_{i-1,j,k} + 2u_{i-1,j,k}\right)\right] \tag{5.5.47f}$$

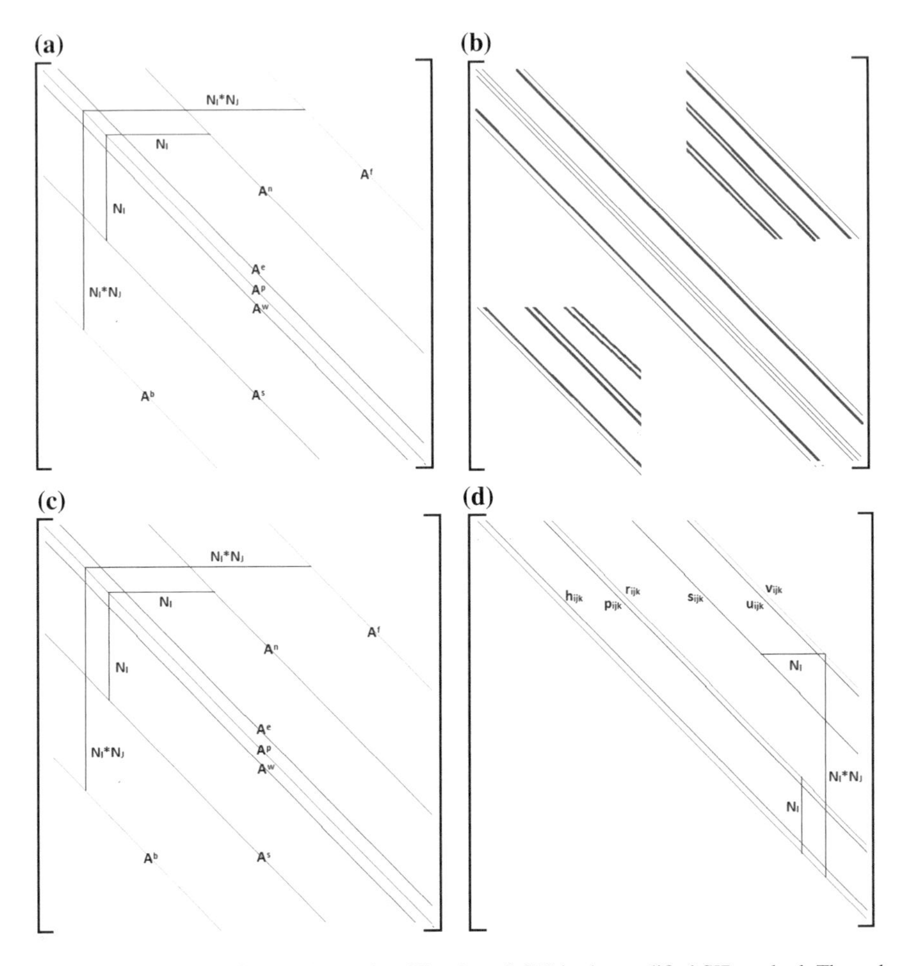

Fig. 5.7 Matrices **a** A, **b** M (= LU = A + N), **c** L and **d** U in the modified SIP method. The red diagonals in (**b**) are the twelve non-zero φ^m diagonals of M which are not present in A

$$
\begin{aligned}
g_{i,j,k} = {} & A^p_{i,j,k} - a_{i,j,k}v_{i,j,k-1} - b_{i,j,k}u_{i+1,j,k-1} - c_{i,j,k}s_{i,j+1,k-1} - d_{i,j,k}r_{i,j-1,k} \\
& - e_{i,j,k}p_{i+1,j-1,k} - f_{i,j,k}h_{i-1,j,k} + \alpha\Big[2\Big(\varphi^1_{i,j,k} + \varphi^2_{i,j,k} + \varphi^3_{i,j,k}\Big) + 3\varphi^4_{i,j,k} \\
& + 2(\varphi^5_{i,j,k} + \varphi^6_{i,j,k} + \varphi^7_{i,j,k} + \varphi^8_{i,j,k}) + 3\varphi^9_{i,j,k} + 2\Big(\varphi^{10}_{i,j,k} + \varphi^{11}_{i,j,k} + \varphi^{12}_{i,j,k}\Big)\Big]
\end{aligned}
\tag{5.5.47g}
$$

$$
h_{i,j,k} = \Big[A^e_{ijk} - b_{i,j,k}v_{i+1,j,k-1} - e_{i,j,k}r_{i+1,j-1,k} - \alpha(2\varphi^1_{i,j,k} + \varphi^3_{i,j,k} + 2\varphi^6_{i,j,k} + \varphi^9_{i,j,k} + \varphi^{11}_{i,j,k})\Big]/g_{i,j,k}
\tag{5.5.47h}
$$

$$
p_{i,j,k} = \big(-c_{i,j,k}u_{i,j+1,k-1} - f_{i,j,k}r_{i-1,j,k}\big)/g_{i,j,k}
\tag{5.5.47i}
$$

$$r_{i,j,k} = \left[A^{n}_{i,j,k} - c_{i,j,k}v_{i,j+1,k-1} - \alpha\left(\varphi^{2}_{i,j,k} + \varphi^{3}_{i,j,k} + 2\varphi^{4}_{i,j,k} + 2\varphi^{5}_{i,j,k} + \varphi^{7}_{i,j,k}\right)\right]/g_{i,j,k} \tag{5.5.47j}$$

$$s_{i,j,k} = \left(-d_{i,j,k}v_{i,j-1,k} - e_{i,j,k}u_{i+1,j-1,k}\right)/g_{i,j,k} \tag{5.5.47k}$$

$$u_{i,j,k} = \left(-f_{i,j,k}v_{i-1,j,k}\right)/g_{i,j,k} \tag{5.5.47l}$$

$$v_{i,j,k} = \left[A^{f}_{i,j,k} - \alpha\left(\varphi^{8}_{i,j,k} + \varphi^{9}_{i,j,k} + \varphi^{10}_{i,j,k} + \varphi^{11}_{i,j,k} + \varphi^{12}_{i,j,k}\right)\right]/g_{i,j,k} \tag{5.5.47m}$$

The elements of the twelve additional diagonals (shown in red in Fig. 5.7b, going from left to right) are obtained the multiplication rules of the product of [L] and [U]:

$$\varphi^{1}_{i,j,k} = b_{i,j,k}h_{i+1,j,k-1} \tag{5.5.48a}$$

$$\varphi^{2}_{i,j,k} = a_{i,j,k}p_{i,j,k-1} \tag{5.5.48b}$$

$$\varphi^{3}_{i,j,k} = b_{i,j,k}r_{i+1,j,k-1} + c_{i,j,k}h_{i,j+1,k-1} \tag{5.5.48c}$$

$$\varphi^{4}_{i,j,k} = c_{i,j,k}p_{i,j,k-1} \tag{5.5.48d}$$

$$\varphi^{5}_{i,j,k} = c_{i,j,k}r_{i,j+1,k-1} \tag{5.5.48e}$$

$$\varphi^{6}_{i,j,k} = e_{i,j,k}h_{i+1,j-1,k} \tag{5.5.48f}$$

$$\varphi^{7}_{i,j,k} = f_{i,j,k}p_{i-1,j,k} \tag{5.5.48g}$$

$$\varphi^{8}_{i,j,k} = d_{i,j,k}s_{i,j-1,k} \tag{5.5.48h}$$

$$\varphi^{9}_{i,j,k} = e_{i,j,k}s_{i+1,j-1,k} \tag{5.5.48i}$$

$$\varphi^{10}_{i,j,k} = d_{i,j,k}u_{i,j-1,k} + f_{i,j,k}s_{i-1,j,k} \tag{5.5.48j}$$

$$\varphi^{11}_{i,j,k} = e_{i,j,k}v_{i+1,j-1,k} \tag{5.5.48k}$$

$$\varphi^{12}_{i,j,k} = f_{i,j,k}u_{i-1,j,k} \tag{5.5.48l}$$

Equations (5.5.47a, b, c, d, e, f, g, h, i, j, k, l, m) and (5.5.48a, b, c, d, e, f, g, h, i, j, k, l) can be used to sequentially find the elements of [L], [U] and [N]. The solution of AT = q is done as per the iterative formula

$$[A+N]\,[T]^{k+1} = [A+N]\,[T]^{k} - \left\{[A][T]^{k} - [q]\right\} \qquad (5.5.49)$$

This can be rewritten in terms of the increment [δ] and the residual [ρ] as

$$[A\ +\ N][\delta]^{k+1} = [L][U][\delta]^{k+1} = [\rho]^{k} \qquad (5.5.50)$$

where $[\delta]^{k+1} = [T]^{k+1} - [T]^{k}$ and $[\rho]^{k} = [q] - [A][T]^{k}$. Equation (5.5.50) can be solved in two back and forward substitution processes:

$$[L][Y]^{k+1} = [\rho]^{k} \qquad (5.5.51a)$$

$$[U][\delta]^{k+1} = [Y]^{k+1} \qquad (5.5.51b)$$

It may be noted that the evaluation of [L] and [U] is done only once. Each iteration requires evaluation of the residual followed by solution of (5.5.51a) and (5.5.51b).

An extension of the modified SIP for unstructured meshes (see Chap. 6) is given by Leister and Peric (1994) for three-dimensional flows with seven non-zero diagonals.

5.5.4 Conjugate Gradient Methods

This class of methods embodies a principle of solving Ax = b which is very different from the methods that we have seen so far. Here, the problem is converted into a minimization of a quadratic functional, F, constructed from Ax = b such that the minimum of the functional corresponds to the solution of Ax = b. The problem is thus reduced to finding the minimum of a function. However, this is not as easy as it sounds. For a problem with n number of variables, the functional F is a surface in an n-dimensional hyperspace. For example, in a two-dimensional space described by the latitude and the longitude, the surface can be the profile of a mountain range. The problem will be to find the deepest valley in the mountain range. A well-known search method (see Sect. 7.2 for a more comprehensive discussion of search methods) for such a problem is the method of steepest descent. An initial guess x_0 for the solution can be used to evaluate the value of F_0 at x_0; this will then constitute a point on the mountain slope. The obvious thing to do would be to see in which direction the mountain is sloping down the most and take a large step in that direction. This search method has the well-known defect of not being very efficient in finding the bottom of a narrow valley in the topology of the functional space. In such a case, going in the steepest direction can lead to oscillation back and forth across the sides of the valley and many steps may be required to get to the bottom of the valley. The known cure for this problem is to make the successive search directions as different as possible. It is here that the conjugate gradient method makes its difference felt. The method is based on the discovery that it is possible to minimize the functional efficiently with respect to several directions simultaneously while searching in a new direction provided all the directions are mutually orthogonal, i.e., conjugate to each other. Thus, one starts with an initial guess x_0 for the minimum and constructs further improvements to it successively by adding one increment at a step in a direction which is orthogonal to all the previous directions and with a step size such that the functional is minimum. In the case of infinite precision arithmetic, the method will give the exact solution in n steps where n is the number of variables. In this sense, the conjugate gradient (CG) method is a direct method; however, since the problem is posed as one of minimization of the residual (see below), there is an approximate solution, x_k, available for $k < n$. In this sense, the method is more like an iterative method.

A well-constructed conjugate gradient has the additional feature that orders of magnitude decrease in the residual can be achieved for $k \ll n$, i.e., in far fewer number of steps than that required for the exact solution. Thus, though the CG method was originally proposed as a direct method (Hestenes and Stiefel 1952), it gained popularity after it was proposed as an iterative method by Reid (1971), Axelsson (1972, 1974) and Concus et al. (1976). While the original CG method was proposed for the case where the coefficient matrix A is symmetric and positive definite matrix, extensions to the non-symmetric case have also been developed (Young and Jea 1981; Axelsson 1994).

Let us examine how the CG method works by taking the simpler case of A being symmetric and positive definite (i.e., all its eigenvalues have positive real parts). For a linear system $Ax = b$, we construct a quadratic functional F as

$$F = \tfrac{1}{2} x^T A x - x^T b \tag{5.5.52}$$

where x^T is the transpose of x. The minimum of F occurs where the gradient,

$$\nabla F = Ax - b \tag{5.5.53}$$

is zero or $Ax = b$. One can readily see this for a 2×2 system. Consider a system with

$$A = \begin{bmatrix} a_{11} & a_{12} \\ a_{21} & a_{22} \end{bmatrix}, \; x = \begin{bmatrix} x_1 \\ x_2 \end{bmatrix}, \; b = \begin{bmatrix} b_1 \\ b_2 \end{bmatrix}, \; x^T = [x_1, x_2]^T \text{ and } a_{12} = a_{21} \text{ due to symmetry of A.}$$

Then, substituting these in Eq. (5.5.52), we get

$$F = \tfrac{1}{2}\left(a_{11}x_1^2 + a_{12}x_1x_2 + a_{21}x_1x_2 + a_{22}x_2^2\right) - \left(x_1b_1 + x_2b_2\right)$$

Now, $\nabla F = 0 \Rightarrow \partial F/\partial x_1 = 0$ and $\partial F/\partial x_2 = 0$. Thus,

$$\partial F/\partial x_1 = a_{11}x_1 + \tfrac{1}{2}\left(a_{12}x_2 + a_{21}x_2\right) - b_1 = a_{11}x_1 + a_{12}x_2 - b_1 \quad (\text{since } a_{12} = a_{21})$$
$$\partial F/\partial x_2 = \tfrac{1}{2}\left(a_{12}x_1 + a_{21}x_1\right) + a_{22}x_2 - b_2 = a_{21}x_1 + a_{22}x_2 - b_2$$

Setting $\partial F/\partial x_1 = 0$ and $\partial F/\partial x_2 = 0$ will give us the two equations that we seek to solve simultaneously, namely, $a_{11}\ x_1 + a_{12}\ x_2 = b_1$ and $a_{21}x_1 + a_{22}x_2 = b_2$

The principle of the conjugate gradient method is as follows. Starting with a guess value, x_0, for x which would minimize F, and p_0, the direction in which we search for a minimum, we seek to find an improved value of x which minimizes F, and also a new direction in which to search for the minimum. Thus, at the kth step of the iteration, we have a minimizer x^k and a direction vector p^{k-1}. We seek to find an improved value x^{k+1} by searching in a new direction p^k. We seek to do the minimization by seeking an α^k such that $F(x^{k+1})$, where $x^{k+1} = x^k + \alpha^k p^k$, is minimum. We also require at the same time that x^{k+1} is also the minimizer over the entire vector space $p^0, p^1, p^2, \ldots, p^k$. This is possible if $p^0, p^1, p^2, \ldots, p^k$ are mutually orthogonal. Using these two conditions, it is possible to find both the improved x^{k+1} and the new direction in which to search, p^k. We start the minimization with an arbitrary initial guess, x^0, and compute the corresponding residual r^0 as $r^0 = Ax^0 - b = (\nabla F)_0$, and take it as the search direction. Let us write the new guess for the minimum value as

$$x^1 = x^0 + \beta^0 r^0 \tag{5.5.54}$$

where β^0 is a parameter yet to be determined. Premultiplying Eq. (5.5.54) by A and subtracting b from both sides, we get

$$r^1 = r^0 + \beta^0 Ar^0 \tag{5.5.55}$$

We choose β^0 such that r^1 is orthogonal to r^0. Taking the dot (or scalar or inner) product of r^0 with Eq. (5.5.55) and setting the LHS to zero (because r^1 is orthogonal to r^0), we obtain β^0 as

$$\beta^0 = \left(r^0, r^0\right)/\left(r^0, Ar^0\right) \tag{5.5.56}$$

where (…) denotes the inner product. We seek an improved value of x using the general recursive relation involving two parameters, α^k and β^k for the kth iteration:

$$x^{k+1} = \alpha^k x^k + (1 - \alpha^k)x^{k-1} - \beta^k r^k, \quad k = 1,2,3,\ldots \tag{5.5.57}$$

Premultiplying (5.5.57) by A and subtracting b on both sides, the new residual, r^{k+1} can be calculated recursively as

$$r^{k+1} = \alpha^k r^k + (1 - \alpha^k) r^{k-1} - \beta^k A r^k, \quad k = 1,2,3,\ldots \tag{5.5.58}$$

Using (5.5.57) and (5.5.58), improved values of x^{k+1} and the residual r^{k+1} can be generated provided the parameters α^k and β^k are known. These can be determined uniquely for the case where A is symmetric and positive definite using the conjugacy conditions that r^{k+1} and r^k are mutually orthogonal and also that r^{k+1} and r^{k-1} are also mutually orthogonal, i.e., the inner products (r^{k+1}, r^k) and (r^{k+1}, r^{k-1}) are equal to zero. These conditions yield the following relations α^k and β^k (Axelsson 1994):

$$\alpha^k = \beta^k \mu^k \tag{5.5.59}$$

$$1/\beta^k = \mu^k - \left(\delta^k/\delta^{k-1}\right)/\beta^{k-1} \quad k = 1,2,3,\ldots \tag{5.5.60}$$

where

$$\mu^k = \left(r^k, A r^k\right)/\left(r^k, r^k\right) \quad \text{and} \quad \delta^k = \left(r^k, r^k\right) \tag{5.5.61}$$

Using Eqs. (5.5.57) to (5.5.61), x^{k+1} and r^{k+1} can be recursively evaluated provided r^k is not equal to zero for initial guesses for x^0 and β^0. It may be noted that when $r^k = 0$, then $Ax^k = b$ and the functional is minimized. Thus, the method is parameter-free. It can be further shown that the solution converges to the exact solution in exactly n steps. The short recurrence relation (involving only k and $k - 1$) keeps the computational time and memory requirements to acceptable limits. Despite this, for large n, the number multiplications required for the final solution varies as n^3 (Ciarlet 1989), and is thus less efficient than the Gaussian elimination method for full matrices. However, it has proved to be an efficient iterative method for sparse matrices with a high rate of convergence and often a solution to an acceptable accuracy can be achieved in far fewer number of steps than that required for the exact solution.

The rate of convergence of the CG method can be increased by preconditioning the matrix A so as to reduce its condition number. This is done by premultiplying the equation by a preconditioning matrix C. In order to preserve the symmetry of the matrix after preconditioning, this is done as follows:

$$C^{-1} A C^{-1} C x = C^{-1} b \tag{5.5.62}$$

The conjugate gradient method is applied to the preconditioned matrix $C^{-1}AC^{-1}$ (Golub and van Loan 1990). The choice of the preconditioner is crucial to the effectiveness of the overall solution. It should not only be good preconditioner in terms of reducing the condition number, it should also be readily invertible. The

LU-decomposable matrix arising out of incomplete Cholesky decomposition or SIP discussed in Sect. 5.5.3 have been found to be good preconditioners for the CG method.

The standard CG method described above has the restriction that it is applicable only when A is symmetric and positive definite (spd), which may not be the case, for example, when we discretize the convection-diffusion equation using upwind-central difference schemes. In such a case, symmetry can be brought into play by premultiplying $Ax = b$ by A^T:

$$A^T A x = A^T b \quad \text{or} \quad Px = Q \quad \text{where} \quad P = A^T A \text{ is spd.} \tag{5.5.63}$$

While this makes the conjugate gradient extendable to the case when A is not symmetric but positive definite, the condition number, i.e., the ratio of the maximum to the minimum eigenvalues, becomes very high and the CG method applied to $Px = Q$ loses the advantage of converging in a small number of steps to a sufficiently low value of the residual. A better method in such cases is the bi-conjugate gradient method which can be used even when A is neither symmetric nor positive definite (Meurant 1999). In this method, we generate four sequences of vectors, r^k, p^k, $\bar{r}^k$, $\bar{p}^k$, $k = 1, 2, \ldots$ Starting with r^1 and $\bar{r}^1$, and setting $p^1 = r^1$ and $\bar{p}^1 = \bar{r}^1$, the sequence of vectors is generated using the following recursive relations:

$$\begin{aligned}
\alpha^k &= \bar{r}^k \cdot r^k / (\bar{p}^k \cdot A \cdot p^k) \\
r^{k+1} &= r^k - \alpha^k A \cdot p^k \\
\bar{r}^{k+1} &= \bar{r}^k - \alpha^k A^T \cdot \bar{p}^k \\
\beta^k &= (\bar{r}^{k+1} \cdot r^{k+1}) / (\bar{r}^k \cdot r^k) \\
p^{k+1} &= r^k + \beta^k p^k \\
\bar{p}^{k+1} &= \bar{r}^k + \beta^k \cdot \bar{p}^k
\end{aligned} \tag{5.5.64}$$

The sequence of vectors satisfies the biorthogonality condition, namely, $\bar{r}^i \cdot r^j = r^i \cdot \bar{r}^j = 0,\ j<i$, and the biconjugacy condition, namely, $\bar{p}^i \cdot A \cdot p^j = p^i \cdot A^T \cdot \bar{p}^j = 0,\ j<i$ as well as the mutual orthogonality condition, namely, $\bar{r}^i \cdot p^j = r^j \cdot \bar{p}^j = 0,\ j<i$. The biconjugate gradient method requires twice the amount of effort per iteration as is required by the CG method but converges in about the same number of iterations. A number of variants of the CG method have been proposed such as the ORTHODIR, the ORTHOMIN and the ORTHORES methods (Young and Jea 1981; Axelsson 1994), the generalized minimal residual algorithm (GMRES) of Saad and Schultz (1986), the conjugate gradient squared (CGS) method of Sonneveld (1989), the stabilized method of CGS, namely, CGSTAB method of van den Vorst and Sonneveld (1990) and the Bi-CGSTAB method of van der Vorst (1992). The last four methods can be applied to non-symmetric matrices. While these methods have certain advantages (such as not requiring diagonal dominance), they require a large number of computations per step and are therefore not used for large CFD problems involving millions of grid

points. But they are making a comeback of sorts in combination with multigrid methods (Rasquin et al. 2010; Bolten et al. 2011).

5.5.5 *Multigrid Methods*

Recall the spatial decomposition of the error (x, t) in terms of Fourier components that was done in Sect. 3.3 and the subsequent analysis of dispersion and dissipation errors in Sect. 3.5 with respect to stability analysis of discretization scheme. An iterative solution scheme of the type given in Eq. (5.1.7) too can be considered in a similar mathematical framework. The error between the converged solution and the solution at kth iteration can be decomposed into Fourier components having wavelengths which are multiples of the grid spacing on a uniform grid. Multigrid methods (Brandt 1977; Hackbush 1985; Briggs 1987; Wesseling 1990) rely on the idea that some iterative methods such as the Gauss-Seidel method and the SIP method are good at damping out errors associated with high frequencies (short wavelengths) but are ineffective against low frequency (long wavelength) errors arising, for example, out of a Poisson equation solver. Here, the distinction of short and long wavelengths is with reference to the grid spacing. Errors of wavelength of the order of twice the grid spacing (the shortest wavelength possible) can be considered as short wavelength while errors having much larger wavelengths (say, five to ten times the grid spacing) should be considered as long wavelength errors. This is illustrated for a one-dimensional heat conduction case in Fig. 5.8. Here, the steady state solution varies linearly between T_1 and T_2. A simulated composite error consisting of short wavelength error (of wavelength of four times the grid spacing) and eight times the grid spacing is added to the steady state solution and is given as the initial guess (Fig. 5.8a). For clarity, the computed solution in a small region ($0.2 < x < 0.3$) is given for up to 200 iterations in Fig. 5.8b. It takes more than 100 iterations for the larger wavelength error to subside. Figure 5.8c shows the evolution of the error of the same wavelength on a grid which is four times as coarse as the one in Fig. 5.8b. One can see that the error, which has taken more than 100 iterations to die out on the fine grid, can be smoothed out in ten iterations on the coarse grid.

The key to reducing the error quickly therefore is to use a grid in which the error is of the short wavelength type. However, except in simulated cases such as that discussed above, typically, errors of all wavelengths will be present in the initial stages of an iterative solution. Also, for accuracy, we would like to have a fine grid and on this we can reduce only short wavelength errors quickly. In the multigrid technique, the equation is therefore solved on multiple grids, i.e., grids of different grid spacing. Consider the simplest case of two grids; a fine grid and a coarse grid. Starting with a small number of iterations on the fine grid, small wavelength errors are removed quickly and a smoother error is transferred to the coarse grid for further error reduction. After some iterations on the coarse grid, an estimated solution is then fed back to the fine grid which is then iterated further. This process of

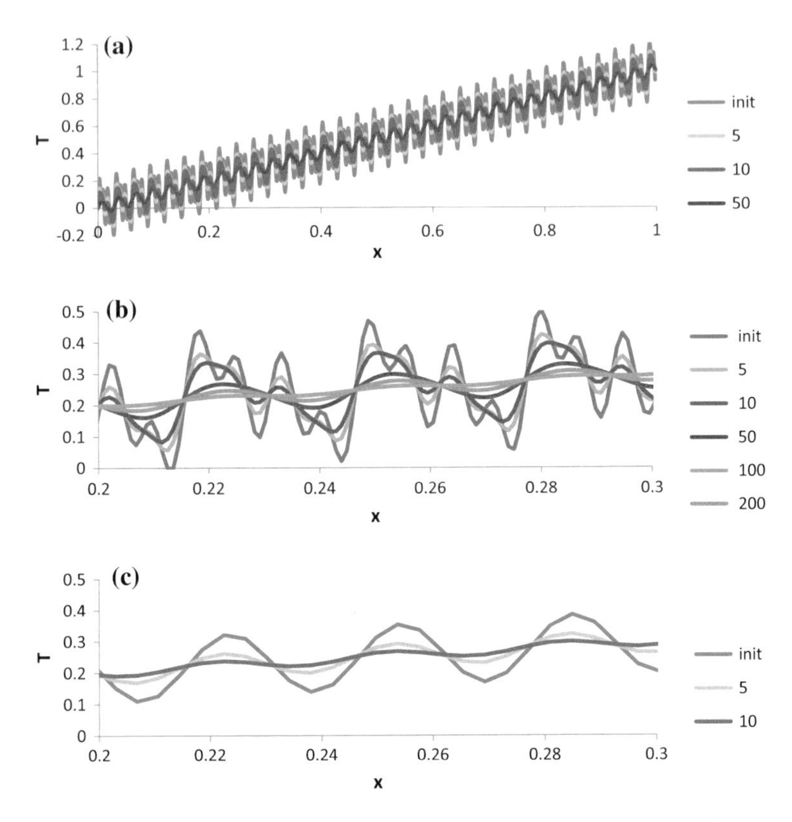

Fig. 5.8 Smoothing out of an initial error using Gauss-Seidel iterative method for the one-dimensional heat conduction: **a** the entire solution with $\Delta x = 1/1024$, **b** the solution over $0.2 < x < 0.3$, and **c** the solution obtained on a coarse grid with $\Delta X = 4/1024$

exchanging information from the fine grid to the coarse grid and back continues until the solution is sufficiently converged on the fine grid. Apart from faster error reduction, i.e., in a fewer number of iterations, this process has two additional advantages. Firstly, in elliptic problems, information of the boundary needs to be propagated to the interior for the converged solution to emerge. Since CFD calculations have a compact computational molecule, i.e., one which involves only the close neighbours, boundary information (such as the wall temperature) propagates into the interior by one grid point in one GS iteration. Since a fine grid has more number of points, the boundary information takes more number of iterations to reach the centre than with a coarse grid. By doing some computations on a coarse grid, the boundary information gets propagated more quickly in a multigrid method. The second advantage stems from the fact that the number of arithmetical operations required to sweep through the domain in one iteration will be fewer (significantly so as the grid size increases) on a coarse grid. Taken together, these three advantages give a significant speed-up when a multigrid method is used.

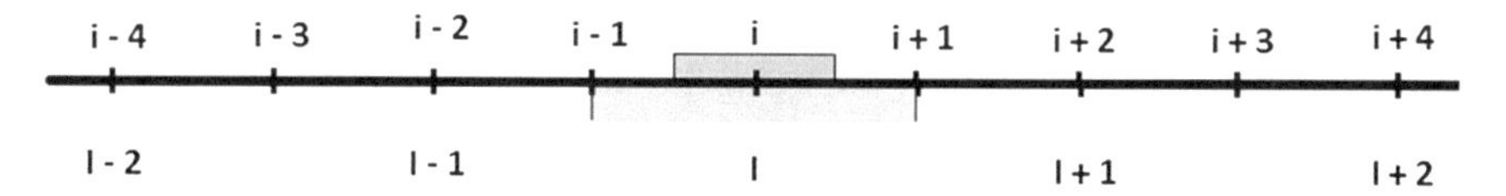

Fig. 5.9 A two-grid discretization with uniform spacing of a one-dimensional domain

Let us look at how the multigrid calculation proceeds for a simple one-dimensional steady state heat conduction case. Consider the two-grid case shown in Fig. 5.9 where the coarse grid spacing (ΔX) is twice that of the fine grid spacing (Δx). The fine grid nodes are denoted by small letters (i, i +1, etc.) while the coarse grid nodes are denoted using capital letters. The governing equation, namely,

$$d^2T/dx^2 = f(x), \quad T|_{x=0} = T_0 \quad \text{and } T|_{x=L} = T_L \tag{5.5.65}$$

can be discretized using a central differencing approximation. For point i, this can be written as

$$(T_{i-1} - 2T_i + T_{i+1})/\Delta x^2 = f_i \tag{5.5.66}$$

We do a small number of iterations of Eq. (5.5.66) on the fine grid with given initial guess and reduce the high frequency component of the error. After k iterations on the fine grid, the solution satisfies Eq. (5.5.66) with a residual ρ^k:

$$\left(T_{i-1}^k - 2T_i^k + T_{i+1}^k\right)/\Delta x^2 = f_i - \rho_i^k \tag{5.5.67a}$$

The error at node i at iteration k, ε_i^k, is defined as

$$\varepsilon_i^k = (\tilde{T}_i - T_i^k), \text{ where } \tilde{T}_i \text{ is the exact solution.} \tag{5.5.67b}$$

It evolves as per the following equation

$$(\varepsilon_{i-1}^k - 2\varepsilon_i^k + \varepsilon_{i+1}^k)/\Delta x^2 = \rho_i^k \tag{5.5.68}$$

Since it is already smoothened to some extent by GS iterations for k steps, it can be solved further on a coarse grid without significant loss of accuracy. In order to export this information on to the coarse grid, we note from Fig. 5.9 that the domain of interest of node I on the coarse grid spans half of the domain of node (i − 1), the full domain of node i and half the domain of node (i + 1) on the fine grid. Thus, we write the equivalent of Eq. (5.5.68) for nodes (i − 1) and (i + 1) and derive an error evolution equation for grid node I on the coarse grid. We thus have

$$\begin{aligned}&[1/2\,(\varepsilon_{i-2}{}^k - 2\varepsilon_{i-1}{}^k + \varepsilon_i{}^k) + (\varepsilon_{i-1}{}^k - 2\varepsilon_i{}^k + \varepsilon_{i+1}{}^k) + 1/2\,(\varepsilon_i{}^k - 2\varepsilon_{i+1}{}^k + \varepsilon_{i+2}{}^k)]/\Delta x^2\\ &= (1/2\,\rho_{i-1}{}^k + \rho_i{}^k + 1/2\,\rho_{i+1}{}^k) \equiv \rho_I{}^k\end{aligned} \tag{5.5.69a}$$

In cases where the grid spacing is not uniform, spacing-weighted average (face area- and cell volume-weighted averaging in two and three-dimensions, respectively) can be done to derive the above equation. Simplifying Eq. (5.5.69a) and using the node notation on the coarse grid, where applicable, one finds that the error evolution equation on the coarse grid is given by

$$(\varepsilon_{I-1}{}^{k} - 2\varepsilon_{I}{}^{k} + \varepsilon_{I+1}{}^{k})/\Delta X^2 = \rho_{I}{}^{k} \tag{5.5.69b}$$

It may be noted that while the LHS of Eq. (5.5.69b) can be obtained from writing a difference approximation for the second derivative directly on the coarse grid, the RHS is obtained only by weighted average of the fine grid nodes falling in the domain of the coarse grid node. Equation (5.5.69b) is now solved until sufficient reduction of residual has occurred, say, for K number of iterations, on the coarse grid. At this stage, an estimated value of error on the coarse grid, denoted as ε_I^{k+K} is available at all grid nodes. From Eq. (5.5.67b), we can now get an improved estimate of the variable value at I, denoted here as T_I^{k+K}, i.e., its value after k iterations on the fine grid and K iterations on the coarse grid:

$$T_I^{k+K} = T_I^{k} + \varepsilon_I^{k+K} \tag{5.5.70}$$

We can now use these coarse grid values of T_I^{k+K} to interpolate improved estimate of the solution on the fine grid. On the simple grid shown in Fig. 5.9, one can use $T_I^{k+K} = T_i^{k+K}$ where the fine and the coarse grid nodes coincide. When the fine grid node is in between coarse grid nodes (e.g. i − 1), then its value is interpolated. Thus,

$$T_{i-1}{}^{k+K} = \tfrac{1}{2}\left(T_{I}{}^{k+K} + T_{I-1}{}^{k+K}\right);\ T_{i}{}^{k+K} = T_{I}{}^{k+K};\ T_{i+1}{}^{k+K} = \tfrac{1}{2}\left(T_{I}{}^{k+K} + T_{I+1}{}^{k+K}\right) \tag{5.5.71}$$

With these new values T on the fine grid as the initial guess, the G-S iterations are continued on the fine grid using Eq. (5.5.66) and the results are extrapolated back to the coarse grid. This alternating process is repeated until sufficient reduction of residual occurs on the fine grid.

In the multigrid literature, the extrapolation of results from the fine grid to the coarse grid is called prolongation, and the interpolation of coarse grid results on to the fine grid is called restriction. Also, the error evolution equation (Eq. 5.5.68) is usually called the residual equation or the defect correction equation. Thus, the multigrid method is characterized by discretized equation on the fine grid (Eq. 5.5.66), a prolongation scheme (Eq. 5.5.69a), a residual equation on the coarse grid (Eq. 5.5.69b) and a restriction scheme (Eq. 5.5.71). It may be noted that the discretized equation is solved on the fine grid while only the defect correction equation is solved on the coarse grid. The method can be readily extended to more number of grids. In such cases, only the defect correction is solved on all but the finest grid. Indeed, the advantage of the multigrid method becomes more predominant as the number of grid levels is increased. However, this depends on the grid size and also on the problem.

Let us now examine the method more formally. Let us consider a linear problem defined by

$$A\phi = b \tag{5.5.72a}$$

which is converted into an iterative scheme of the form

$$M\phi^{n+1} = b - N\phi^{n} \quad \text{where } A = M + N \tag{5.5.72b}$$

Subtracting $M\phi^n$ from both sides, one can rewrite Eq. (5.5.72b) in the residual form as

$$M\Delta\phi^{n} = b - A\phi^{n} = r^{n} \tag{5.5.73a}$$

where

$$\Delta\phi^{n} = \left(\phi^{n+1} - \phi^{n}\right) \tag{5.5.73b}$$

and r^n is the residual. In the multigrid method, we solve Eqs. (5.5.72a, b) on the finest grid and Eqs. (5.5.73a, b) on all coarse grids. Considering a two-grid scheme with representative grid spacings of h and H on the fine and the coarse meshes, respectively, we solve, on the fine grid,

$$M_h\phi_h^{n+1} = b_h - N_h\phi_h^{n} \tag{5.5.74}$$

or its equivalent form

$$M_h\Delta\phi_h^{n} = r_h \tag{5.5.75}$$

It may be noted that $r_h \to 0$ as $n \to \infty$. However, within a few iterations on the fine grid, the high frequency components of the error would be damped, and it would be no longer necessary to employ a fine grid. One would therefore like to solve the residual form of the equation on a coarse grid:

$$M_H\Delta\phi_H^{n} = r_H \tag{5.5.76}$$

However, r_H is not known on the coarse grid and must be deduced from the residual on the fine grid. To this end, we define a restriction operator I_h^H such that

$$r_H = I_h^H r_h \tag{5.5.77}$$

For a one-dimensional case with uniform grid spacing and with H = 2 h, r_H and r_h are column matrices containing n/2 and n number of elements where n is the number of unknowns. In this case, I_h^H would thus be a matrix of size n/2 × n. Since every coarse grid node corresponds to a fine grid node on a uniform grid with $H = 2\,h$, one may take r_H to be the corresponding values of r_h at the common nodes. This is called injection of the residual, and I_h^H would of the form

$$
\begin{array}{lll}
 & & \begin{array}{ccccccc} & & i\text{-}1 & i & i+1 & & \\ & & \downarrow & \downarrow & \downarrow & & \end{array} \\
I_h^H = & \begin{array}{r} \\ \\ I\text{-}1 \longrightarrow \\ I \longrightarrow \\ I+1 \longrightarrow \end{array} & \left\{\begin{array}{ccccccc}
. & . & . & . & . & . & . \\
. & . & . & . & . & . & . \\
. & 0 & 1 & 0 & . & . & . \\
. & . & 0 & 1 & 0 & . & . \\
. & . & . & 0 & 1 & 0 & .
\end{array}\right\}
\end{array}
\tag{5.5.78a}
$$

For the case given by Eq. (5.5.69a), the restriction operator would take the form

$$
\begin{array}{lll}
 & & \begin{array}{ccccccc} & & i\text{-}1 & i & i+1 & & \\ & & \downarrow & \downarrow & \downarrow & & \end{array} \\
I_h^H = & \begin{array}{r} \\ \\ I\text{-}1 \longrightarrow \\ I \longrightarrow \\ I+1 \longrightarrow \end{array} & \left\{\begin{array}{ccccccc}
. & . & . & . & . & . & . \\
. & . & . & . & . & . & . \\
. & \frac{1}{4} & \frac{1}{2} & \frac{1}{4} & . & . & . \\
. & . & \frac{1}{4} & \frac{1}{2} & \frac{1}{4} & . & . \\
. & . & . & \frac{1}{4} & \frac{1}{2} & \frac{1}{4} & .
\end{array}\right\}
\end{array}
\tag{5.5.78b}
$$

It may be noted that in both cases, the sum of weights in a row equals unity. After carrying out calculations on the coarse grid, one would get an improved estimate of $\Delta\phi_H$. This needs to be taken back to the fine grid for further error reduction. In order to do this interpolation, we introduce the prolongation operator, I_H^h such that

$$\Delta\phi_h = I_H^h \Delta\phi_H \tag{5.5.79}$$

Here, I_H^h is an an n × n/2 matrix. For the simple one-dimensional case described above, I_H^h can take the following form

$$I_h{}^H = \begin{array}{c} \\ \\ \\ i\text{-}1 \longrightarrow \\ i \longrightarrow \\ i+1 \longrightarrow \end{array} \begin{array}{c} \quad\quad I\text{-}1 \;\; I \;\; I+1 \\ \quad\quad \downarrow \;\; \downarrow \;\; \downarrow \\ \left\{ \begin{array}{ccccccc} . & . & . & . & . & . & . \\ . & . & . & . & . & . & . \\ . & 0 & \frac{1}{2} & \frac{1}{2} & 0 & . & . \\ . & . & 0 & 1 & 0 & . & . \\ . & . & 0 & \frac{1}{2} & \frac{1}{2} & 0 & . \end{array} \right\} \end{array} \tag{5.5.80}$$

The relative weights for the restriction and prolongation operators for a two-dimensional case are shown in Figs. 5.10a and 5.10b, respectively. In the restriction operator, a simple injection is possible but the scheme shown in Fig. 5.10a includes the effect of the residual at the neighbouring nodes, and is to be preferred for a smooth variation. Here, the residual at the coarse grid point P has contributions from the fine grid point at that location itself (denoted by point A) and from the surrounding neighbours B to I. In the case of prolongation (Fig. 5.10b), the fine grid node may coincide with the coarse grid (such as points A and P), or it can be at the middle of two coarse grid nodes (either vertically, e.g., point D or horizontally, e.g., point B) or it can be surrounded by four coarse grid nodes at the four immediate corners, e.g., point C. In the first case, a simple injection can be used (i.e., assigning $\Delta\phi_h$ to be equal to $\Delta\phi_H$ at that node). For the other two cases, average value of the two or four nearest coarse grid neighbours can be used to compute $\Delta\phi_h$ as indicated in Fig. 5.10b.

With this formalism, one can investigate the stability of the two-grid strategy in the following way. Assume that we do some iterations on the fine grid and perform further computations on the coarse grid to come up with an improved value of $\Delta\phi_H$, and thus of $\Delta\phi_h$ and ϕ_h at the next step. After k such steps on the fine grid, we have

$$\begin{aligned} \Delta\phi_h^{k+1} = I_H^h \Delta\phi_H^k &= I_H^h M_H^{-1} r_H^k \\ &= I_H^h M_H^{-1} I_h^H r_h^k \\ &= I_H^h M_H^{-1} I_h^H \left(b - A\phi^k\right) \end{aligned} \tag{5.5.81}$$

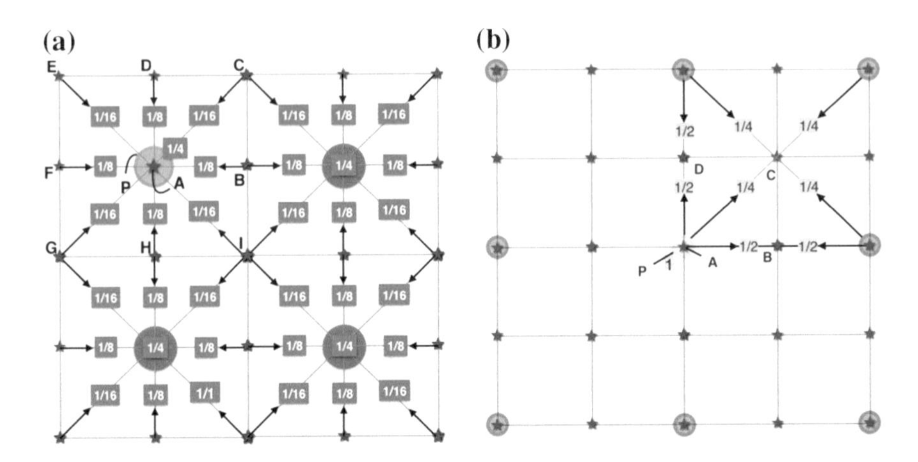

Fig. 5.10 Restriction (**a**) and prolongation (**b**) operations for a two-dimensional case. The numbers indicate the relative weight for the contribution neighbouring points. In **a**, the residual at the coarse grid point P is determined by the fine grid point at the same location as well as by the neighbouring points B to I. In **b**, three cases, represented by point A, points B or D, and point C are distinguished separately

Or,

$$\phi_h^{k+1} = \left(1 - I_H^h M_H^{-1} I_h^H A\right)\phi^k + I_H^h M_H^{-1} I_h^H b \tag{5.5.82}$$

In the case where the coarse grid residual equation is solved exactly, then, M_H^{-1} can be replaced by A_H^{-1} in the above equation, which then can be written as

$$\phi_h^{k+1} = G_{CCG}\,\phi^k + \; I_H^h A_H^{-1} I_h^H b \tag{5.5.83a}$$

where G_{CCG} is the amplification factor associated with the coarse grid correction (CCG), and is given by

$$G_{CCG} = \left(1 - I_H^h A_H^{-1} I_h^H A_h\right) \tag{5.5.83b}$$

where a distinction is made between the coefficient matrix on the fine grid (A_h) and on the coarse grid (A_H). If n_1 fine grid iterations are made, each with an amplification factor of G_S, for each coarse grid correction, then the effective amplification factor of the two-grid scheme becomes

$$G_h^H = G_S^{n_1} \cdot G_{CCG} \tag{5.5.84}$$

The scheme would be convergent if the spectral radius of amplification factor is less than unity, i.e., if $\rho(G_h^H) < 1$. Analysis of schemes with various combinations smoothers and restriction and prolongation operators for elliptic problems shows that the convergence rate is independent of the mesh size (Hackbush and

Trottenberg 1982) unlike in the case of standard schemes such as the Gauss-Seidel scheme where the asymptotic convergence rate is proportional to the number of unknowns, or the SOR scheme where it is proportional to the square root of the number of unknowns under optimal conditions. Therefore the major computational effort in the multi-grid scheme is associated with the smoothing computations on the fine grid. This effort varies linearly with the number of grid points for standard smoothers like the Gauss-Seidel scheme for elliptic problems. Therefore the overall computational effort scales nearly linearly for the multigrid method.

The two-grid scheme can be readily extended to multiple grids. Let us denote the finest grid level by h, the next coarser mesh by $h - 1$, the next by $h - 2$, and so on up to $h - m$. The discretized form of the governing Eq. (5.5.72a) on the h-grid (the finest grid) is given by

$$A_h\{\phi_h\} = b_h \tag{5.5.85}$$

After this is solved for a few iterations, we have the residual given by

$$A_h\{\phi_h\} = b_h - r_h \tag{5.5.86}$$

This is prolongated to the next coarser grid, and the following discretized equation is solved on the $(h - 1)$ grid:

$$A_{h-1}\{\Delta\phi_{h-1}\} = I_h^{h-1} r_h \tag{5.5.87}$$

This is further prolongated to the next coarser grid and the following discretized equation is solved on the $(h - 2)$ grid:

$$A_{h-2}\{\Delta\phi_{h-2}\} = I_{h-1}^{h-2} r_{h-1} \tag{5.5.88}$$

This process is continued until on the coarsest (h–m) grid, the following equation is solved:

$$A_{h-m}\{\Delta\phi_{h-m}\} = I_{h-m-1}^{h-m} r_{h-m-1} \tag{5.5.89}$$

The results from the coarsest grid are then restricted to find course grid correction for the next finer grid, i.e.,

$$\{\Delta\phi_{h-m-1}\} = I_{h-m}^{h-m-1}\{\Delta\phi_{h-m}\} \tag{5.5.90}$$

followed by

$$\{\Delta\phi_{h-m-2}\} = I_{h-m-1}^{h-m-2}\{\Delta\phi_{h-m-1}\} \tag{5.5.91}$$

$$\begin{gathered} \cdots\cdots\cdots\cdots \\ \{\Delta\phi_{h-1}\} = I_{h-2}^{h-1}\{\Delta\phi_{h-2}\} \end{gathered} \tag{5.5.92}$$

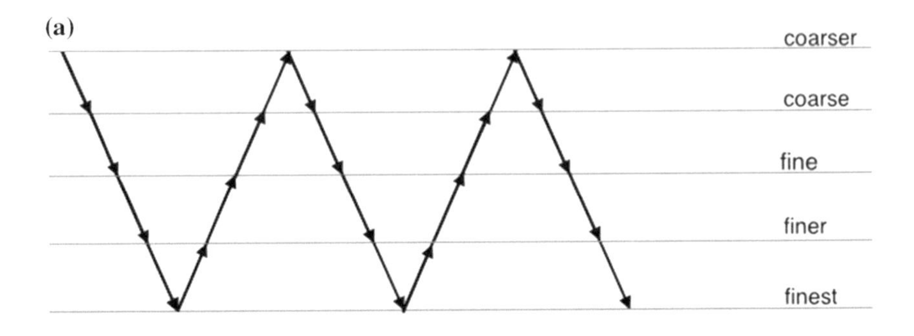

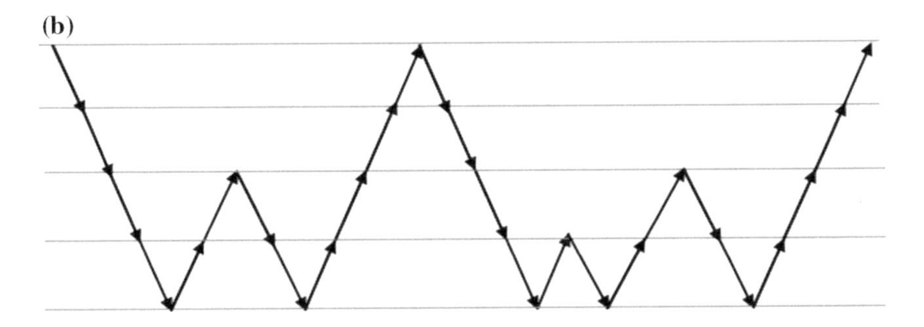

Fig. 5.11 Illustration of **a** the V-cycle and **b** the W-cycle in a multigrid calculation scheme

and finally

$$\{\Delta\phi_h\} = I_{h-1}^{h}\{\Delta\phi_{h-1}\} \tag{5.5.93}$$

from which we obtain a new value of ϕ on the finest grid:

$$\{\phi_h^{new}\} = \{\phi_h^{old}\} + \{\Delta\phi_h\} \tag{5.5.94}$$

The above method constitutes what is known as the V-cycle in which the errors are eliminated successively in each of the grids (see Fig. 5.11). Other cycles, such as the W-cycle and the F-cycle, can also be used. The W-cycle requires more number of computations to be done at fine grid levels. This may be necessary in strongly coupled, non-linear problems.

The method can be readily extended to two and three dimensional problems, to problems involving non-uniform meshes and unstructured meshes and to linear and non-linear problems. When more complicated situations than the simple one studied above are considered, several variations of multigrid emerge. In the full multigrid method (FMG) , the calculations are not started at the finest grid level but at the coarsest grid level where the discretized equation itself is solved. The solution then is used to interpolate the solution on the next finer grid where again the discretized equation is solved and the resulting solution is transferred to the next fine grid level.

This is continued until the finest grid level is achieved. Once this is done, the regular multigrid method approach is used wherein the residual equation only is solved in all but the finest grid level. The V-cycle FMG scheme used by Shyy and Sun (1993) is shown in Fig. 5.12. Note that residual reduction of a particular level is sought at intermediates steps (open circles in the figure). The FMG method is expected to improve the convergence rate while incurring little additional computational effort. The effectiveness of the multigrid method for the benchmark case of the lid-driven cavity in 3-D using the SIMPLE algorithm on a collocated mesh is discussed by Lilek et al. (1997).

For non-linear problems, the prolongation operation (Eq. 5.5.69a) is no longer correct because of the non-linearity of the prolongation operator. For the system Ax = b, we derive the residual equation by writing

$$A\tilde{x} - Ax^k = A(\tilde{x} - x^k) = A\varepsilon^k = b - \rho^k \tag{5.5.95}$$

where $\tilde{x}$ is the exact solution. For a non-linear problem, $(A\tilde{x} - Ax^k) \neq A(\tilde{x} - x^k)$ and therefore the residual equation cannot be derived as for a linear problem. Three options are possible. One can linearize the non-linear terms by Picard substitution. One can also do a global linearization of the problem: one can write $A\tilde{x} - Ax^k = J\varepsilon^k$ where J is the Jacobian of A, i.e., $\nabla^T A$. In this Newton-multigrid method, the residual equation is written as

$$J\varepsilon^k = b - \rho^k \tag{5.5.96}$$

This method has excellent convergence characteristics but does not treat the non-linearity using the multigrid concept. This is done in the Full Approximation Scheme (FAS) approach (Henson 2003). Here, the error is not computed at coarse grid level; instead an approximate solution is sought by solving the discrctized equation with prolongated solution from the fine grid.

Another distinct development in the multigrid method is the algebraic multigrid (AMG) approach as opposed to geometric multigrid (Stüben 2001, 2003). In

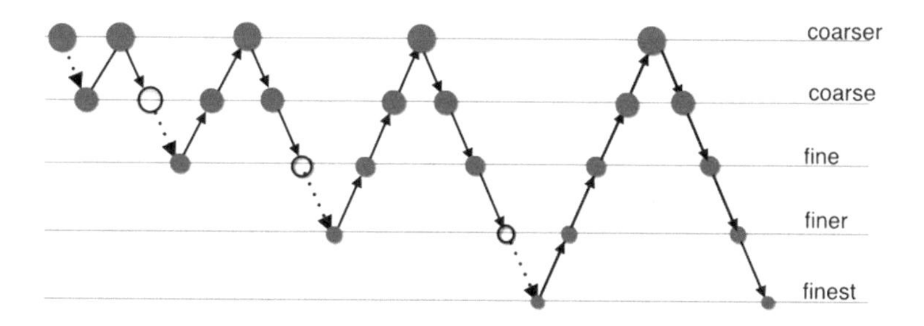

Fig. 5.12 The V-cycle full multigrid (FMG) scheme used by Shyy and Sun (1993). The open circle indicates a solution that has been carried out until the required level of residual reduction is achieved at that level. A closed circle indicates a small number of smoother calculations at the grid level. The dashed line indicates an interpolation step

geometric multigrid method, grid coarsening strategies are defined a priori based on geometric considerations (such as Eq. (5.5.69a)). The success of the solution depends in this case strongly on the error smoothening properties of the matrix equation solver. In strongly anisotropic media, such fixed solution strategies may not be effective. In AMG, using a simple smoother such as the Gauss-Seidel scheme, coarser levels and interpolation strategies are dynamically and automatically evolved without reference to the physical geometry. The fine-level variables are divided into two disjoint sets, the fine-level (F^h) and the coarse level (C^H), the latter being a small proportion of the whole set. On the fine grid level, the equation solved is the same as in the geometric grid approach, namely,

$$A^h x^h = b^h \quad \text{or} \quad \sum_{j \in F^h} a_{ij}^j x_j^h = b_i^h \tag{5.5.97}$$

In AMG, an interpolation operator, I_H^h is constructed such that the error on the coarse level can be interpolated to the fine level and the residual on the fine level can be prolongated to the coarse level using $(I_H^h)^T$. Using this interpolation operator, the coarse level operator A^H is defined as

$$A^H = I_h^H A^h I_H^h \text{ with } I_h^H = \left(I_H^h\right)^T \tag{5.5.98}$$

The fine level solution is updated after coarse level smoothing as

$$x^{k+K} = x^k + I_H^h \varepsilon^H \tag{5.5.99}$$

where the error ε^H is the solution of

$$A^H \varepsilon^H = \rho^H \quad \text{or} \quad \sum_{j \in C^H} a_{ij}^H \varepsilon_j^H = \rho_i^H \tag{5.5.100}$$

where the residual on the coarse level, ρ^H, is prolongated as

$$\rho^{Hk} = I_h^H\left(\rho^{hk}\right) \quad \text{where} \quad \rho^{hk} = b - A^h x^{hk} \tag{5.5.101}$$

The construction of the interpolation operator and the splitting of fine level variables into F^h and C^H are done through heuristic algorithms. Although AMG has been developed for scalar elliptic PDEs, it has been used for a wide range of problems in CFD (Stüben 2003).

There is continued interest in the development of multigrid methods. Their robustness and applicability have increased to such an extent that they are considered as the most efficient solvers for CFD applications that are currently available for general flow problems. While best results are obtained for elliptic problems, their efficacy decreases for convection-dominated cases, for example, at high Reynolds numbers (Shyy and Sun 1993); see also Sect. 4.6 where other literature is cited regarding convergence rate.

5.6 Summary

In this chapter, we have dealt with the important aspect of efficient solution of linear algebraic equations. Given that for the general fluid flow problem we have to deal with non-linear and coupled set of partial differential equations, we often resort to implicit discretization and sequential solution of the equations. This results in sets of large sets of algebraic equations. With increasing computational power, problems of increasing physical and geometrical complexity are being addressed. This increases the demand on the computational efficiency of the linear equations solvers. We have seen in this chapter some of the basic methods and some advanced methods that are available at the disposal of the CFD engineer to tackle the problem. We see that increasing sophistication of the approach can have tremendous gains in the computing power. This is one of the key developments that have contributed to the exponential growth of CFD in the engineering community in recent decades.

Problems

1. For the rectangular duct problem in Chap. 1, assemble the coefficient matrix for 10 grid points in each coordinate direction. Check for banded structure, sparsity and diagonal dominance.
2. Repeat Problem #1 for 40 grid points in each coordinate direction.
3. Repeat Problem #1 for a quarter-duct case, i.e., where you take advantage of symmetry and do the calculations only for one quarter of the duct.
4. Consider the problem of finding an interpolating polynomial passing through a given set of points. If there are n + 1 points, then it is possible to find the coefficients of an nth order polynomial of the form

$$P(x) = a_0 + a_1 x + a_2 x^2 + a_3 x^3 + \ldots + a_n x^n \qquad (Q5.1)$$

If $(x_0, y_0), (x_1, y_1), (x_2, y_2), \ldots, (x_n, y_n)$ are the n + 1 points through which the polynomial needs to pass, then we have

$$P(x_0) = a_0 + a_1 x_0 + a_2 x_0^2 + a_3 x_0^3 + \cdots + a_n x_0^n = y_0 \qquad (Q5.2a)$$
$$P(x_1) = a_0 + a_1 x_1 + a_2 x_1^2 + a_3 x_1^3 + \cdots + a_n x_1^n = y_1 \qquad (Q5.2b)$$
$$P(x_2) = a_2 + a_1 x_2 + a_2 x_2^2 + a_3 x_2^3 + \cdots + a_n x_2^n = y_2 \qquad (Q5.2c)$$
$$P(x_0) = a_3 + a_1 x_3 + a_2 x_3^2 + a_3 x_3^3 + \cdots + a_n x_3^n = y_3 \qquad (Q5.2d)$$
$$\cdot \quad \cdot \quad \cdot \quad \cdot \quad \cdot \quad \cdot \quad \cdot$$
$$\cdot \quad \cdot \quad \cdot \quad \cdot \quad \cdot \quad \cdot \quad \cdot$$
$$P(x_0) = a_0 + a_1 x_n + a_2 x_n^2 + a_3 x_n^3 + \cdots + a_n x_n^n = y_n \qquad (Q5.2e)$$

The above (n + 1) equations algebraic equations with coefficients, $a_0, a_1, a_2, \ldots$ a_n as the variables, can be put in the form of

$$Ax = b \tag{Q5.3}$$

The coefficient matrix in this case, namely, A, is known as the Vandermonde matrix.
For the specific case of the six points (x_i, y_i) given in the following table, assemble the Vandermonde matrix:

	1	2	3	4	5	6
x	0.1	0.3	0.6	0.9	1.2	1.5
y	0.29552	0.78333	0.97385	0.42738	−0.44252	−0.97753

5. Verify that the Vandermonde matrix of Problem #4 is not diagonally dominant.
6. Solve eqn. (Q5.3) of Problem #4 using Gaussian elimination.
7. Repeat Problem #6 using LU decomposition.
8. Repeat Problem #6 using an appropriate conjugate gradient technique.
9. Write a computer program to solve the set of algebraic equations for Problems #2 and #3 using Gaussian elimination.
10. Repeat Problem #9 using LU decomposition method.
11. Repeat Problem #9 using (a) Jacobi and (b) Gauss-Seidel method for a residual reduction factor 10^4. Find the asymptotic convergence rate in each case.
12. Repeat Problem #9 using the successive over-relaxation method and explore the influence of the SOR parameter ω on the rate of convergence.
13. Repeat Problem #12 for 20, 40 and 80 grid points in each coordinate direction and find the optimal value of the SOR parameter in each case. Find the asymptotic rate of convergence in each case.
14. Repeat Problem #9 using the ADI scheme and compare the rate of convergence with that obtained in Problems #11, #12 and #13.
15. Repeat Problem # 14 using Chebyshev iterative method.
16. Repeat Problem #14 using the strongly implicit procedure.
17. Repeat Problem #14 using the conjugate gradient technique.
18. Repeat Problem #9 for an 80×80 and a 160×160 grid at the finest level using the multigrid technique using (a) 3, (b) 4 and (c) 5 levels of fineness of grids.
19. A rectangular block of dimensions 3 m height × 1 m width × 2 m length is suspended by a slender thread and is allowed to cool in air by natural circulation. If the initial temperature of the block is 800 deg C throughout, if the ambient temperature is 30 deg C, find T(x, y, x, t). Assume that the natural convective heat transfer coefficient is 5 W/m^2K on the bottom wall, 10 W/m^2K on the side (vertical) walls and 15 W/m^2K on the top wall. Assume that the density, thermal conductivity and specific heat of steel are 8000 kg/m^3, 16 W/mK and 500 J/kgK, respectively. Solve the 3-D unsteady heat conduction problem using the ADI scheme. Use the Crank-Nicolson scheme for discretization of the equation.
20. Repeat Problem #19 using the Gauss-Seidel method and the multigrid method.

Chapter 6
Dealing with Irregular Flow Domains and Complex Physical Phenomena

The methodology discussed in Chaps. 2–5 is applicable for simple cases of fluid flow. Practical flow problems are often more complicated and require additional measures to get to a solution. If one leaves aside the very difficult and case-specific question of the problem formulation itself (e.g. what flow to simulate, what data is available, why one would want to use CFD as opposed to other modelling approaches, what exactly one would like to get from the stimulation, etc.), and assume that there is a clear case to be made for a CFD simulation, then the problems associated with practical flow situations can be divided, especially from CFD solution point of view, into two categories: those associated with irregular and complicated flow domain, and those associated with complicated physics. In the former, the simplistic approach to grid generation adopted in earlier chapters is not applicable. In the latter case, the equations derived in Chap. 2 are inadequate to describe the physical phenomena. In this chapter, we discuss these two cases separately for pedagogical convenience. As may be expected in complicated cases, there is often more than one approach. Which one of these one selects may depend on several parameters including such factors as assured possibility of getting a solution, known goodness of the approach in some benchmark cases, familiarity with the approximations or concepts involved or the ready availability of only one particular approach!

6.1 Dealing with Irregular Geometries

We have seen in Chap. 1 that the first step in the CFD approach is finding the grid points at which the solution is to be determined. For simple cases in which the flow domain can be fitted along lines of constant orthogonal coordinate planes or for regular polygons, the identification of grid points is rather straightforward as illustrated in Figs. 1.1 and 1.2. However, practical flow systems have irregular shapes; one such air ducting system in a power plant is shown in Fig. 6.1. Here, the duct, which carries the flue gas from the furnace has a cross-section that is

S. Jayanti, *Computational Fluid Dynamics for Engineers and Scientists*,
https://doi.org/10.1007/978-94-024-1217-8_6

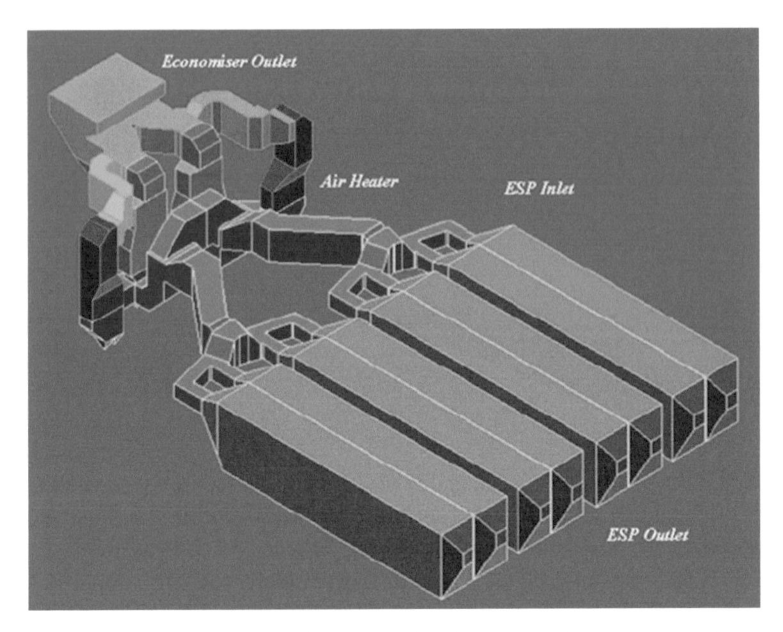

Fig. 6.1 Typical air ducting in a coal-fired power plant (Avvari and Jayanti 2016). ESP stands for electrostatic precipitator which is used to remove the dust (fine ash particles) contained in the flue gas. Reprinted with permission from Elsevier

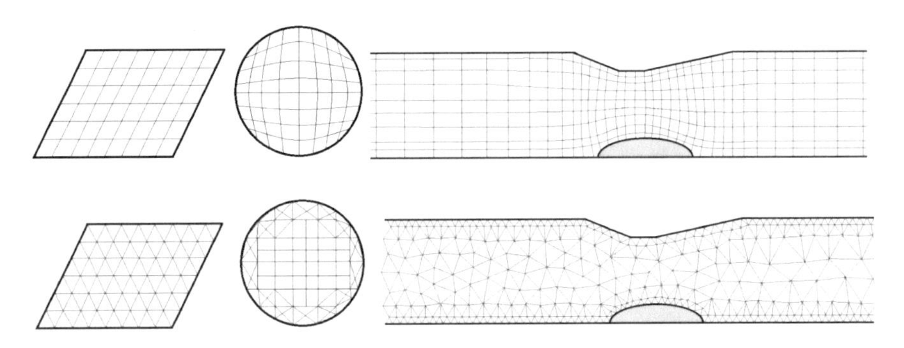

Fig. 6.2 Schematic sketches of structured (top) and unstructured (bottom) grids for some simple 2-d domains

constantly changing. The flow also splits into two or more streams which may merge later. Abrupt changes of direction and elevation are also present. These are needed to circumvent obstacles, fittings that are present in the flow path, or to accommodate thermal or chemical interaction with other equipment, or for operational reasons such as facilitating maintenance schedules, meeting new (for example, environmental) regulations and demands, or even for the sake of reliability. In dealing with computations of flow through such flow domains, there are two principal approaches that one can adapt:

- in the first approach, we distort a Cartesian mesh to fit the shape of the flow domain and retain the "structured grid" characteristic in which the grid points are still at the intersection of lines of constant coordinate (but not necessarily mutually orthogonal) directions. This approach is known as the body-fitted grid approach and one can use finite difference or finite volume or finite element techniques to discretize the governing equations.
- in the second approach, we abandon the comfort of the structured grid and divide the flow domain into tiles or bricks of polygonal (but not necessarily exactly of the same shape or size) control volumes, which when put together make up the entire flow domain. The flow variables are sought at a point (usually the centroid) associated with each control volume. In this "unstructured grid" approach, one needs to use finite volume or finite element techniques to discretize the governing equations.

Structured and unstructured grids for some simple flow domains have already been shown in Figs. 3.1 and 3.2, respectively. Some more are contrasted in Fig. 6.2 for some 2-dimensional cases. In a structured grid, the information of the structure, such as the number, location and orientation of the neighbouring elements, is readily available, and can often be evaluated as and when it becomes necessary. Implementation of high order accuracy discretization schemes (the higher the accuracy, the more the number of successive grid points that need to be used to approximate a derivative; see Sect. 3.1) is therefore relatively straightforward. Another important advantage of structured grids is that the grid elements can be elongated in regions where the flow is primarily one-dimensional (such as in the boundary layers) without significantly affecting the accuracy of the solution. This results in a smaller number of grid points to discretize a given domain, and leads to faster solution. Finally, the coefficient matrices arising from the use of a structured grid retain diagonal structure which helps in implementing advanced linear algebraic equation solvers.

In the unstructured grid approach, the information of the number, location and orientation of neighbouring grid points needs to be stored separately at the time of grid generation, and is not inherently present in the structure of the grid. However, the possibility of accommodating arbitrarily-shaped control volumes in the computation is an advantage in tackling irregular geometries as well as in grid refinement, grid coarsening and grid adaptation. Local grid refinement is possible in the unstructured grid by inserting more points in the region where refinement is needed. In Fig. 6.3a, the unstructured grid is able to do grid refinement only in the two stagnation point regions of flow over a semi-elliptical object. In the structured grid approach, grid refinement in these regions also leads to partial grid refinement, i.e., only along one axis, in other regions too, as illustrated in Fig. 6.3b. Also, as we will see later, local grid refinement or coarsening can be more easily done in the unstructured grid case; in the structured grid case, this would mean going through the entire grid generation procedure once again.

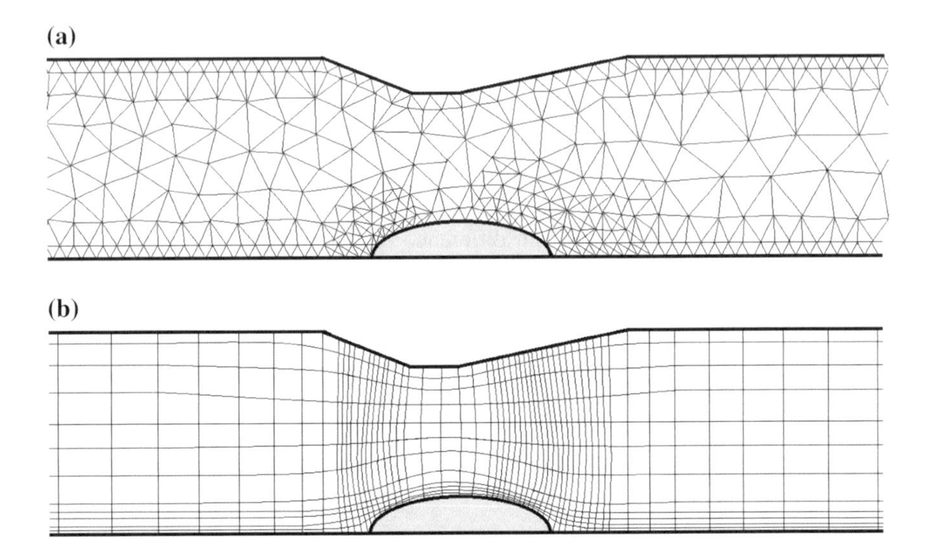

Fig. 6.3 Possibility of local grid refinement with (**a**) an unstructured grid and (**b**) a structured grid

There is considerable detail that needs to be developed in using either approach. The body-fitted grid approach is discussed below in Sect. 6.2. The unstructured grid approach is discussed in Sect. 6.3.

6.2 The Body-Fitted Grid Approach

The major departure in the body-fitted grid approach from the simple approach discussed in Chaps. 3–5 is in the use of a non-orthogonal grid. This brings in three changes to the calculation procedure:

- Firstly, one makes a distinction between the grid in the physical plane and that in the computational plane. As shown schematically for a two-dimensional case in Fig. 6.4, in the physical (x-y) plane, the coordinate lines of the body-fitted grid are expressed in terms of (ξ, η) which are curvilinear lines, non-orthogonal and wrap around internal surfaces, if any. The external boundaries are located along lines of constant ξ or η. Thus, the boundary A′B′ in the computational plane corresponds to the horizontal line $\eta = 0$; in the physical plane, it corresponds to the curvilinear boundary AB. Similarly, boundary BC corresponds to curvilinear line $\xi = 1$ in the physical plane and to the straight vertical line B′C′ given by $\xi = 1$ in the computational plane. In the same way, boundaries CD and DA correspond to curvilinear lines $\eta = 1$ and $\xi = 0$, respectively in the physical plane whereas in the computational plane, they are the straight lines C′D′ and D′A′, respectively.
- Secondly, the conservation equations are discretized in the computational plane where the grid is rectangular and uniform, but are expressed in terms of the new

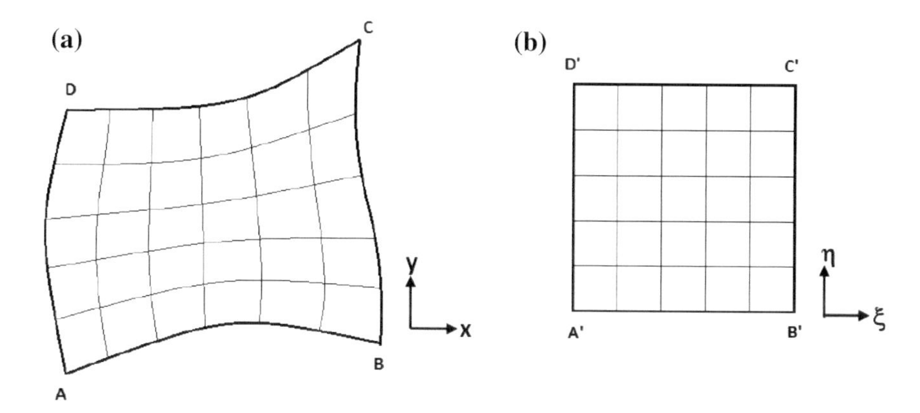

Fig. 6.4 Schematic diagram of domain ABCD in **a** the physical (x, y) plane and **b** the computational (ξ-η) plane

coordinate variables ξ and η. Therefore the governing equations describing changes in terms of physical space coordinates need to be transformed from the x-y coordinate system to the ξ-η coordinate system.

- Thirdly, although the mathematical nature (i.e., hyperbolic, parabolic or elliptic nature) of the governing equations does not change, more terms will appear in the transformed equations, including cross-derivative terms and terms associated with coordinate transformation. These will have a profound influence on the way we solve these equations.

These aspects are elaborated below.

6.2.1 Transformation of the Governing Equations

Consider the simple case of two-dimensional, steady, constant-property heat conduction equation with a source term S(x, y) where S is an algebraic function of x, y and other variables but does not contain any derivatives:

$$\partial^2 T/\partial x^2 + \partial^2 T/\partial y^2 = S(x, y) \tag{6.2.1}$$

This equation is valid within the domain shown in Fig. 6.4a and is subjected to Dirichlet boundary conditions. Given the non-rectangular cross-section of the flow domain, we wish to work in the rectangular computational plane (ξ, η) shown in Fig. 6.4b with the assurance that there is a one-to-one mapping between (x, y) and (ξ, η) and that lines of constant ξ and η represent the physical boundaries. We seek to solve for $T(\xi_i, \eta_j)$, and use the mapping to obtain the corresponding $T(x_i, y_j)$. We seek therefore to transform Eq. (6.2.1) from (x, y) coordinates to (ξ, η). This is done as follows. We can write

$$\partial T/\partial x = (\partial T/\partial \xi)(\partial \xi/\partial x) + (\partial T/\partial \eta)(\partial \eta/\partial x) \tag{6.2.2}$$

and

$$\begin{aligned}\partial^2 T/\partial x^2 &= \partial/\partial x(\partial T/\partial x) = \partial/\partial x[(\partial T/\partial \xi)(\partial \xi/\partial x) + (\partial T/\partial \eta)(\partial \eta/\partial x)] \\ &= \partial/\partial x[(\partial T/\partial \xi)(\partial \xi/\partial x)] + \partial/\partial x[(\partial T/\partial \eta)(\partial \eta/\partial x)] \\ &= (\partial \xi/\partial x)^2(\partial^2 T/\partial \xi^2) + 2(\partial^2 T/\partial \xi \partial \eta)(\partial \xi/\partial x)(\partial \eta/\partial x) + (\partial T/\partial \xi)(\partial^2 \xi/\partial x^2) \\ &\quad + (\partial \eta/\partial x)^2(\partial^2 T/\partial \eta^2) + (\partial T/\partial \eta)\partial^2 \eta/\partial x^2 \\ &= \xi_x^2 T_{\xi\xi} + \eta_x^2 T_{\eta\eta} + 2\xi_x \eta_x T_{\xi\eta} + \xi_{xx} T_\xi + \eta_{xx} T_\eta\end{aligned} \tag{6.2.3}$$

where the subscripts indicate differentiation. Similarly, it can be shown that

$$\partial^2 T/\partial y^2 = T_{yy} = \xi_y^2 T_{\xi\xi} + \eta_y^2 T_{\eta\eta} + 2\xi_y \eta_y T_{\xi\eta} + \xi_{yy} T_\xi + \eta_{yy} T_\eta \tag{6.2.4}$$

Thus, Eq. (6.2.1) becomes

$$\left(\xi_x^2 + \xi_y^2\right)T_{\xi\xi} + \left(\eta_x^2 + \eta_y^2\right)T_{\eta\eta} + 2(\xi_x \eta_x + \xi_y \eta_y)T_{\xi\eta} + (\xi_{xx} + \xi_{yy})T_\xi + (\eta_{xx} + \eta_{yy})T_\eta = S(\xi, \eta) \tag{6.2.5}$$

It can be seen that the transformed equation is much more complicated than the original Poisson equation, although it can be shown that the equation remains elliptic. The complications arise from several factors. Firstly, the coefficients of the various terms are functions of the metrics of the coordinate transformation (see below) and vary from location to location. This is in contrast to Eq. (6.2.1) which has constant coefficients for the derivatives. Secondly, some of the coefficients are squares of the derivatives and are therefore sensitive to the smoothness of the grid. Due to the squaring, discontinuities in the grid lines get amplified. Thirdly, the second-order cross-derivative term ($\partial^2 T/\partial \xi \partial \eta$) appears; discretization of this term using central differences may lead to loss of diagonal dominance of the resulting set of algebraic equations. Finally, first derivative terms also appear with variable coefficients.

The coefficients of the temperature derivatives appearing in Eq. (6.2.5) involve what are known as the metrics of the mapping between the physical plane and the computational plane. These can be obtained as follows. Consider the two-dimensional mapping between (x, y) and (ξ, η). Here, we have

$$\xi = \xi(x, y) \quad \text{and} \quad \eta = \eta(x, y) \tag{6.2.6}$$

$$x = x(\xi, \eta) \quad \text{and} \quad y = y(\xi, \eta) \tag{6.2.7}$$

Thus,

$$d\xi = (\partial\xi/\partial x)dx + (\partial\xi/\partial y)dy \quad \text{and} \quad d\eta = (\partial\eta/\partial x)dx + (\partial\eta/\partial y)dy \tag{6.2.8}$$

$$dx = (\partial x/\partial\xi)d\xi + (\partial x/\partial\eta)d\eta \quad \text{and} \quad dy = (\partial y/\partial\xi)d\xi + (\partial y/\partial\eta)d\eta \tag{6.2.9}$$

One can thus write

$$\begin{bmatrix} d\xi \\ d\eta \end{bmatrix} = \begin{pmatrix} \xi_x & \xi_y \\ \eta_x & \eta_y \end{pmatrix} \begin{bmatrix} dx \\ dy \end{bmatrix} = \begin{pmatrix} \xi_x & \xi_y \\ \eta_x & \eta_y \end{pmatrix} \begin{pmatrix} x_\xi & x_\eta \\ y_\xi & y_\eta \end{pmatrix} \begin{bmatrix} d\xi \\ d\eta \end{bmatrix} \tag{6.2.10}$$

or

$$\begin{pmatrix} \xi_x & \xi_y \\ \eta_x & \eta_y \end{pmatrix} = \begin{pmatrix} x_\xi & x_\eta \\ y_\xi & y_\eta \end{pmatrix}^{-1} = J^{-1} \begin{pmatrix} y_\eta & -x_\eta \\ -y_\xi & x_\xi \end{pmatrix} \tag{6.2.11}$$

where $J = (\xi_x\eta_y - \xi_y\eta_x)$ is the Jacobian which is the determinant of the matrix $\partial(\xi, \eta)/\partial(x, y)$, i.e.,

$$J = \begin{vmatrix} \xi_x & \xi_y \\ \eta_x & \eta_y \end{vmatrix} \tag{6.2.12}$$

The inverse of J is the area (in 2-d and volume in 3-d) scaling factor between the physical and the computational planes.

These metrics can therefore be determined from the functional relationship, if known, of the mapping between the physical and the computational plane. Alternatively, if the coordinates of the grid points are known in the two planes, then the metrics can be evaluated numerically. Determination of the coordinates of the grid points, both on the boundary and the interior of the flow domain, is the objective of grid generation. This is discussed in detail in Sect. 6.2.2.

Transformations involving vector quantities such as velocity can be done in more than one way. The vector velocity field can be described in terms of its Cartesian, covariant or contravariant velocity components. These velocity components are illustrated for the two-dimensional case in Fig. 6.5. The Cartesian velocity components (u, v) are aligned along fixed x- and y-directions. Covariant velocity components (u_ξ, v_η), denoted by convention with subscripts (these are not to be confused with derivatives), are oriented parallel to the curvilinear coordinate lines (ξ, η) while the contravariant velocity components (u^ξ, v^η), which are denoted by convention by superscripts, are oriented along the normals to the curvilinear coordinate lines. The three velocity components in the three systems coincide with each other when the curvilinear coordinate system is orthogonal.

The unit vectors $(\hat{\mathbf{e}}_\xi, \hat{\mathbf{e}}_\eta)$ in the covariant coordinate system can be obtained by considering arc lengths along the curvilinear coordinates. Since only one coordinate direction changes along the curvilinear coordinate line, one can write

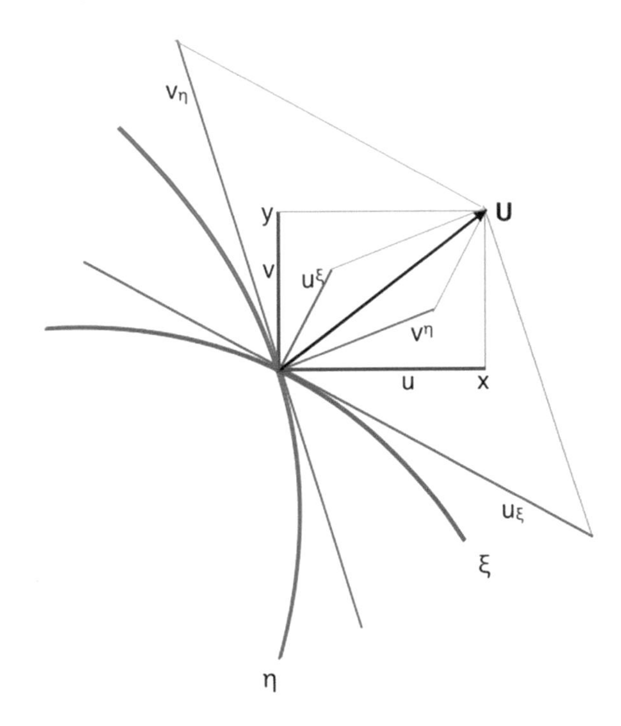

Fig. 6.5 Cartesian (u, v), covariant (u_ξ, v_η) and contravariant (u^ξ, v^η) components of a vector in a curvilinear coordinate system

$$dl_\xi \hat{\mathbf{e}}_\xi = x_\xi d\xi \hat{\mathbf{i}} + y_\xi d\xi \hat{\mathbf{j}} \quad \text{and} \quad dl_\eta \hat{\mathbf{e}}_\eta = x_\eta d\eta \hat{\mathbf{i}} + y_\eta d\eta \hat{\mathbf{j}} \tag{6.2.13}$$

and

$$\hat{\mathbf{e}}_\xi = \frac{dl_\xi \hat{\mathbf{e}}_\xi}{dl_\xi} = \frac{x_\xi \hat{\mathbf{i}} + y_\xi \hat{\mathbf{j}}}{\sqrt{x_\xi^2 + y_\xi^2}} \quad \hat{\mathbf{e}}_\eta = \frac{dl_\eta \hat{\mathbf{e}}_\eta}{dl_\eta} = \frac{x_\eta \hat{\mathbf{i}} + y_\eta \hat{\mathbf{j}}}{\sqrt{x_\eta^2 + y_\eta^2}} \tag{6.2.14}$$

The covariant velocity components are then given by

$$(u_\xi, u_\eta) = (\mathbf{u} \cdot \hat{\mathbf{e}}_\xi, \mathbf{u} \cdot \hat{\mathbf{e}}_\eta) = \left(\frac{ux_\xi + vy_\xi}{\sqrt{x_\xi^2 + y_\xi^2}}, \frac{ux_\eta + vy_\eta}{\sqrt{x_\eta^2 + y_\eta^2}} \right) \tag{6.2.15}$$

The unit vectors of the contravariant coordinate system are defined as

$$\hat{\mathbf{e}}^\xi = \frac{\nabla \xi}{|\nabla \xi|} = \frac{\xi_x \hat{\mathbf{i}} + \xi_y \hat{\mathbf{j}}}{\sqrt{\xi_x^2 + \xi_y^2}} \quad \hat{\mathbf{e}}^\eta = \frac{\nabla \eta}{|\nabla \eta|} = \frac{\eta_x \hat{\mathbf{i}} + \eta_y \hat{\mathbf{j}}}{\sqrt{\eta_x^2 + \eta_y^2}} \tag{6.2.16}$$

The covariant velocity components are then given by

$$\left(u^{\xi}, u^{\eta}\right) = \left(\mathbf{u} \cdot \hat{\mathbf{e}}^{\xi}, \mathbf{u} \cdot \hat{\mathbf{e}}^{\eta}\right) = \left(\frac{u\xi_x + v\xi_y}{\sqrt{\xi_x^2 + \xi_y^2}}, \frac{u\eta_x + v\eta_y}{\sqrt{\eta_x^2 + \eta_y^2}}\right) \tag{6.2.17}$$

The conservation equations can be expressed in terms of Cartesian, covariant or contravariant coordinate systems. Consider the continuity equation for incompressible flow, which can be stated as

$$\nabla \cdot \mathbf{U} = 0. \tag{6.2.18}$$

In terms of Cartesian velocity components (u, v), this equation can be transformed to the (ξ, η) plane as

$$\nabla \cdot \mathbf{U} = \xi_x \partial u/\partial \xi + \eta_x \partial u/\partial \eta + \xi_y \partial u/\partial \xi + \eta_y \partial u/\partial \eta = 0 \tag{6.2.19}$$

In terms of the covariant velocity components (u_ξ, u_η), the continuity equation becomes

$$\nabla \cdot \mathbf{U} = \frac{1}{J}\left\{\partial/\partial\xi\left(y_\eta u_\xi - x_\eta u_\eta\right) + \partial/\partial\eta\left(-y_\xi u_\xi + x_\xi u_\eta\right)\right\} = 0 \tag{6.2.20}$$

In terms of the contravariant velocity components (u_ξ, u_η), the continuity equation becomes

$$\nabla \cdot \mathbf{U} = \frac{1}{J}\left\{\partial/\partial\xi\left(|J|u^{\xi}\right) + \partial/\partial\eta\left(|J|u^{\eta}\right)\right\} = 0 \tag{6.2.21}$$

Mathematically all the forms are equivalent. Indeed, numerical solution frameworks have been built using all the three components (Melaaen 1992). Two of the known disadvantages with the covariant or contravariant velocity components-based formulations are related to the solution at discrete level. Strong conservation form, in which all the flux and source terms can be expressed in terms of divergence of a vector or tensor, is a desirable feature of the governing equation in flows involving strong discontinuities. A strong conservation form of the momentum conservation equation is possible only when velocity components in fixed coordinates are used. A velocity component associated with curvilinear coordinates changes direction and will require an apparent force to make this happen. For example, for flow through a bend, centrifugal and Coriolis force terms appear when the equations are written in cylindrical coordinates; these force terms are not in divergence form. The second major disadvantage of curvilinear coordinates is the appearance of terms involving local curvature which can introduce some error if the mapping from (x, y) to (ξ, η) is not analytical, which is most often the case. Numerical evaluation of the curvature terms can introduce errors in regions of large changes in grid spacing. On the other hand, curvilinear coordinates can preserve the reduced dimensionality of the flow better. As a trivial example, one can consider the fully-developed flow in a circular pipe which is one-dimensional if cylindrical coordinates are used to describe the problem and two-dimensional if Cartesian coordinates are used. Similarly, the flow through a 90° bend will have only one dominant velocity component if curvilinear coordinate components are used. Using Cartesian velocity components will have, say, u-velocity as the principal velocity upstream and v-velocity component as the principal velocity downstream of the bend. Despite these advantages, curvilinear velocity components are used only in special cases and Cartesian velocity components in fixed directions are most generally used. The compressible Navier-Stokes equations in fixed Cartesian velocity components are given below in conservation form after a discussion on the transformation to a three-dimensional curvilinear coordinate system.

Consider the three-dimensional mapping between (x, y, z) and (ξ, η, ζ). It is assumed that a one-to-one mapping exists and that it is unique so that the Jacobian of the transformation, J, is non-zero. For this generalized transformation, one can write

$$\begin{aligned}\begin{pmatrix}\xi_x & \xi_y & \xi_z\\ \eta_x & \eta_y & \eta_z\\ \zeta_x & \zeta_y & \zeta_z\end{pmatrix} &= \begin{pmatrix}x_\xi & x_\eta & x_\zeta\\ y_\xi & y_\eta & y_\zeta\\ z_\xi & z_\eta & z_\zeta\end{pmatrix}^{-1}\\ &= J^{-1}\begin{pmatrix} y_\eta z_\zeta - y_\zeta z_\eta & -(x_\eta z_\zeta - x_\zeta z_\eta) & x_\eta y_\xi - x_\xi y_\eta\\ -(y_\xi z_\zeta - y_\zeta z_\xi) & x_\xi z_\zeta - x_\zeta z_\xi & -(x_\xi y_\zeta - y_\xi x_\zeta)\\ y_\xi z_\eta - y_\eta z_\xi & -(x_\xi z_\eta - x_\eta z_\xi) & y_\xi y_\eta - x_\eta y_\xi\end{pmatrix}\end{aligned} \tag{6.2.22}$$

Here J is the determinant of the matrix ∂(ξ, η, ζ)/∂(x, y, z) and is given by

$$\begin{aligned}J &= \xi_x(\eta_y\zeta_z - \eta_z\zeta_y) - \xi_y(\eta_x\zeta_z - \eta_z\zeta_x) + \xi_z(\eta_x\zeta_y - \eta_y\zeta_x)\\ &= 1/\left[x_\xi(y_\eta z_\zeta - y_\zeta z_\eta) - x_\eta(y_\xi z_\xi - y_\zeta - z_\xi) + x_\zeta(y_\xi z_\eta - y_\eta z_\xi)\right]\end{aligned} \tag{6.2.23}$$

The equations of conservation of mass, momentum and energy can be written in conservative, vector form as

$$\partial U_1/\partial t + \partial E_1/\partial\xi + \partial F_1/\partial\eta + \partial G_1/\zeta = 0 \tag{6.2.24}$$

where

$$\begin{aligned}U_1 &= U/J\\ E_1 &= (E\xi_x + F\xi_y + G\xi_z)/J\\ F_1 &= (E\eta_x + F\eta_y + G\eta_z)/J\\ G_1 &= (E\zeta_x + F\zeta_y + G\zeta_z)/J\end{aligned} \tag{6.2.25}$$

where

$$
\begin{aligned}
U &= \{\rho \quad \rho u \quad \rho v \quad \rho w \quad \rho e_t\}^T \\
E &= \{\rho u \quad (\rho u^2 + p - \tau_{xx}) \quad (\rho uv - \tau_{yx}) \quad (\rho uw - \tau_{zx}) \quad (\rho u e_t + pu - u\tau_{xx} - v\tau_{xy} - w\tau_{xz} + q_x)\}^T \\
F &= \left\{\rho v \quad (\rho uv - \tau_{xy}) \quad (\rho v^2 + p - \tau_{yy}) \quad (\rho vw - \tau_{zy}) \quad \left(\rho v e_t + pv - u\tau_{yx} - v\tau_{yy} - w\tau_{tz} + q_y\right)\right\}^T \\
G &= \{\rho w \quad (\rho uw - \tau_{xz}) \quad (\rho vw - \tau_{yz}) \quad (\rho w^2 + p - \tau_{zz}) \quad (\rho w e_t + pw - u\tau_{zx} - v\tau_{zy} - w\tau_{zz} + q_z)\}^T
\end{aligned}
\tag{6.2.26}
$$

The stresses appearing in Eq. (6.2.26) need to be transformed to (ξ, η, ζ) plane; this is done by writing, for example, the τ_{yz} component as

$$
\begin{aligned}
\tau_{yz} &= \mu(\partial w/\partial y + \partial v/\partial x) \\
&= \mu\left(\xi_y \partial w/\partial \xi + \eta_y \partial w/\partial \eta + \zeta_y \partial w/\partial \zeta + \xi_z \partial v/\partial \xi + \eta_z \partial v/\partial \eta + \zeta_z \partial v/\partial \zeta\right)
\end{aligned}
\tag{6.2.27}
$$

The above equations can be discretized using finite differences and solved as before. For this, the metrics of the transformation matrix are needed. This is done as part of the grid generation which is explained in the following section.

6.2.2 *Structured Grid Generation*

The objective of grid generation is to find the mapping between (x, y, z) and (ξ, η, ζ) over the entire flow domain including on the boundary. We wish to make a structured mesh, one which is topologically rectangular in two-dimensions and cuboid in three-dimensions. Each interior grid point is surrounded by four immediate neighbours in 2-d and six in 3-d. The grid points are located at the intersection of lines of constant coordinate values. In the physical plane, the coordinate lines are curvilinear, and the external and internal boundaries are located on lines or surfaces of constant ξ or η or ζ. In the computational plane, we require them to be mutually orthogonal and equi-spaced (Fig. 6.4b). Given that the transformed conservation equations contain the metrics of the mapping as coefficients, we require the mapping to be smooth so that the first and the second derivatives of the mapping exist.

In some simple cases, the mapping may be obvious and can be done analytically. In 2-d cases, conformal mapping techniques may be used to generate orthogonal grids. The interested reader is referred to Ives (1982) and Duraiswami and Prosperetti (1992). For general three-dimensional domains, one needs generic algorithms to do grid generation. Two classes of methods exist; these are algebraic methods of grid generation and differential equation-based methods.

One successful algebraic method is based on the concept of transfinite interpolation, originally introduced by Gordon and Hall (1973). Consider the domain shown in Fig. 6.6a containing a four-sided domain with the outer edges being represented by curved lines c_1 to c_4. We wish to map it on to a unit square in the computational domain. Each curved boundary is mapped on to a corresponding vertical or horizontal boundary in the computational plane. In order to obtain a structured grid, it is necessary that opposite sides have equal number of grid points. In the computational plane, this can be done readily by dividing each side into a number of uniform length segments. In order to ensure a one-to-one mapping, the corresponding boundary in the physical plane must also be divided into the same number of linear segments; these do not need to be equally spaced in the physical plane. For the sake of illustration, the constant-ξ boundaries (the left and the right boundaries) in Fig. 6.6a are divided into four segments and the constant-η boundaries (the bottom and the top boundaries) are divided into five segments. It is not necessary to have uniform segment length in the physical plane although this is so in the computational plane for ease of writing difference approximations for derivatives. In the computational plane, where constant-ξ and constant-η curves are linear (besides being orthogonal), the interior points can be generated by joining the corresponding points on the boundary (Fig. 6.6b). In the physical plane, these are curvilinear and a linear interpolation of the coordinates of the interior points from known boundary points is not unique. Consider point P corresponding to (ξ, η) of (0.4, 0.5) in the computational plane (Fig. 6.6b). Its x- and y-coordinates can be obtained by linear interpolation between the left and right boundary points corresponding to $\eta = 0.5$ (E and F) or using the top and the bottom boundary points corresponding to $\xi = 0.4$ (G and H). In the general case, both give different (x, y) for point P. In fact, using one-sided linear interpolation is not likely to be useful because some of the internal points may actually lie outside the flow domain, as illustrated in Fig. 6.6c. The boundary points chosen may also not correspond to those calculated using one-sided linear interpolation. In transfinite interpolation, these errors are eliminated by adjusting the coordinates of P such that the boundary points coincide. This is done as follows.

For point P (i, j), let the coordinates in the computational plane be (ξ_i, η_j). These are known because we choose the computational plane to be orthogonal and equi-spaced between 0 and 1 in both directions. An estimate of the corresponding coordinates of point P (i, j) in the physical plane can be made by interpolating between the two boundary points at $\eta = \eta_j$. Thus,

$$\begin{aligned} x_P^*(i,j) &= (1-\xi_i)x_L(\eta_j) + \xi_i x_R(\eta_j) \\ y_P^*(i,j) &= (1-\xi_i)y_L(\eta_j) + \xi_i y_R(\eta_j) \end{aligned} \tag{6.2.28}$$

Here, x_L and x_R are the x-coordinates of the left and the right boundary points corresponding to the curvilinear coordinate line η_j. Similar y_L and y_R are the y-coordinates of the same boundary points. When this is done for all the boundary points from $\eta = 0$ to $\eta = 1$, we find that the (x_P^*, y_P^*) of the boundary points at $\eta = 0$ and $\eta = 1$ are not the same as those of the points on the boundaries. We therefore introduce boundary point corrections at the bottom (Δx_0 and Δy_0) and the top (Δx_1 and Δy_1), see Fig. 6.6d, e, respectively. These are defined as follows:

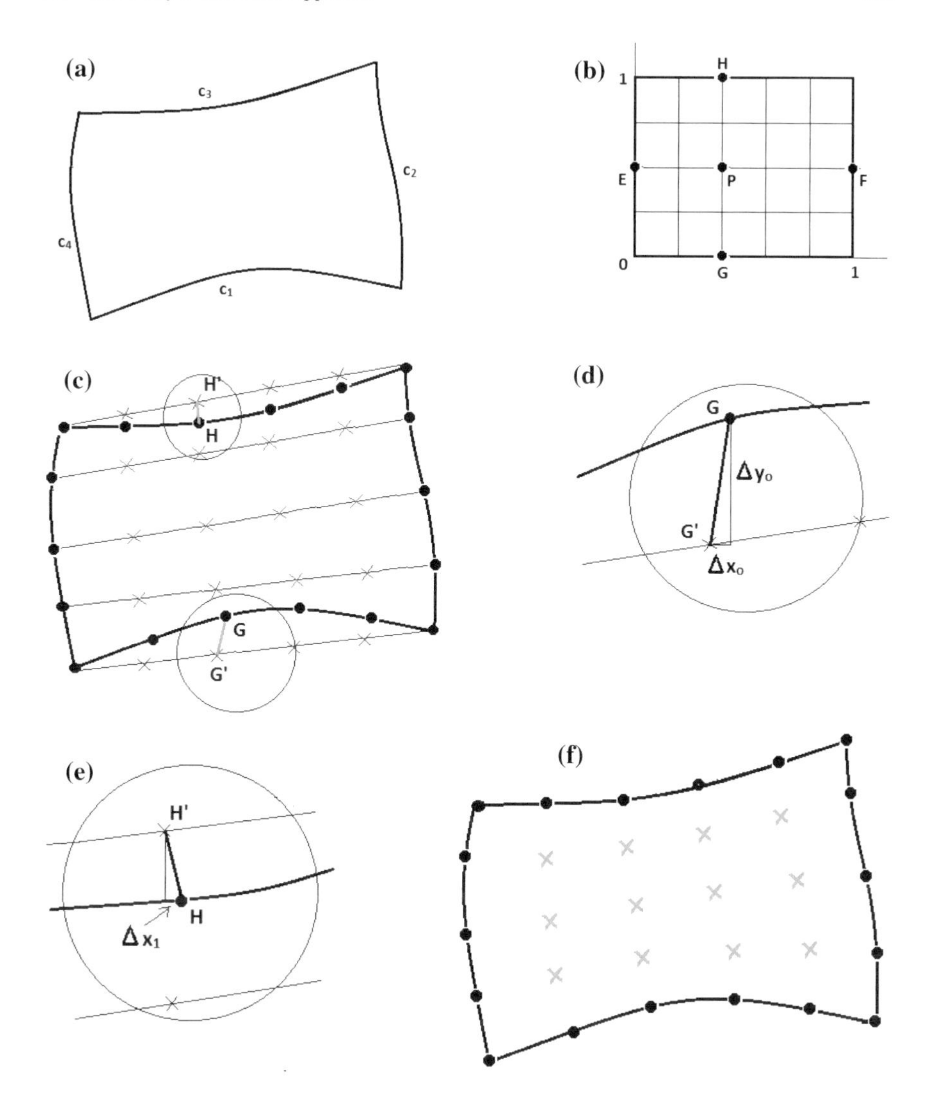

Fig. 6.6 Algebraic grid generation using transfinite interpolation: **a** the physical domain bounded by four curves, **b** grid in the computational domain, **c** determination of the interior and boundary points through interpolation between curves c_2 and c_4 showing points G and H on the boundaries and their respective interpolated points G′ and H′, **d** and **e** displacement corrections originating from boundary points G and H, and **f** adjusted, final positions of the internal points

$$\begin{aligned}
\Delta x_0(i) &= x_P^*(i,1) - x_P(i,1) = (x_{G'} - x_G)\\
\Delta y_0(i) &= y_P^*(i,1) - y_P(i,1) = (y_{G'} - y_G)\\
\Delta x_1(i) &= x_P^*(i,NJ) - x_P(i,NJ) = (x_{H'} - x_H)\\
\Delta y_1(i) &= y_P^*(i,NJ) - x_P(i,NJ) = (y_{H'} - y_H)
\end{aligned} \tag{6.2.29}$$

We then find the correct x and y coordinates for the entire domain using the following relations:

$$x_P(i,j) = x_P^*(i,j) + (1 - \eta_j)\Delta x_0(i) + \eta_j \Delta x_1(i) \quad 0 \le \eta \le 1 \tag{6.2.30a}$$

$$y_P(i,j) = y_P^*(i,j) + (1 - \eta_j)\Delta y_0(i) + \eta_j \Delta y_1(i) \quad 0 \le \eta \le 1 \tag{6.2.30b}$$

The method can be readily extended to three dimensional grids. Using other interpolating functions, it is possible to introduce orthogonality requirement near boundaries and clustering of the grid lines in anticipated regions of large gradients (Chung 2002) to some extent. One of the major advantages of the method is its simplicity, especially when compared to the methods discussed below. However, it carries slope discontinuities from the boundary into the domain which may affect the accuracy of estimation of the metrics that appear in the transformed equations.

Due to the boundary adjustments (Δx_1, Δy_1 etc.,) made in the transfinite interpolation method, the interior grid points are no longer on a straight interpolating line. A line passing through all the points lying on a η = constant coordinate line, where $0 < \eta < 1$, would be like a zigzag line due to the minor displacements arising from one-sided interpolation. This line would effectively be represented by a polynomial function of very high degree (if there are m points on the coordinate line, then a polynomial function of degree $(m - 1)$ can be constructed that would pass through all the grid points). The metrics, for example, η_x, η_y and the second derivatives such as η_{xx}, that appear as coefficients in the transformed equation (Eq. 6.2.5) will exhibit unduly large variation which may have an impact on the accuracy of the solution. Smoothness of the coordinate lines is therefore essential.

In the differential equations approach, which was pioneered by Thompson and co-workers (see Thompson et al. 1974, 1985), the grid is generated by solving partial differential equations. The idea of using partial differential equations can be understood intuitively by considering the curvilinear coordinate lines in the physical space to be streamlines. Along the streamline, the streamfunction, ϕ, remains constant. Streamlines are smooth and do not intersect themselves or one another. Together with isopotential lines, they constitute a "flow net" (White 1986) which can serve as a grid for fluid flow calculations. Both the streamfunction and the potential function satisfy the Laplace equation, namely,

$$\partial^2\phi/\partial x^2 + \partial^2\phi/\partial y^2 = 0 \tag{6.2.31}$$

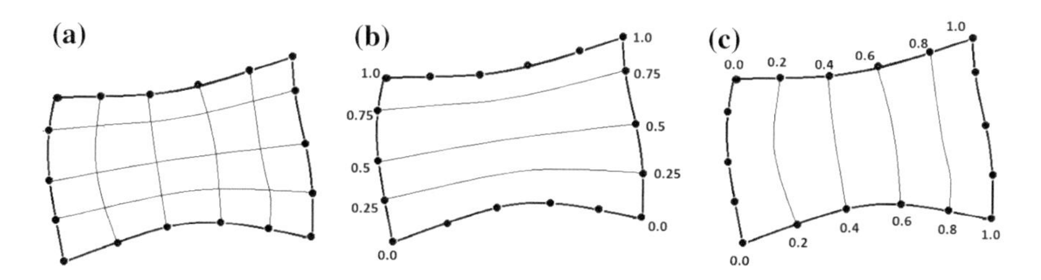

Fig. 6.7 **a** Lines of constant ξ and constant η in the physical plane. **b** Boundary conditions for the constant η-lines in the physical plane given by Eq. (6.2.29). **c** Boundary conditions for the constant ξ-lines in the physical plane given by Eq. (6.2.28)

The flow net over a flow domain can be visualized as grid for computations. Consider the physical domain shown in Fig. 6.7a and the expected constant-ξ and constant-η lines in it schematically drawn for a coarse grid. It is assumed here that the boundary points on the four curved sides are located algorithmically at fairly equal distances along the curved line. The smoothly-drawn constant ξ and η lines look like streamlines and isopotential lines; one can therefore analogously seek that ξ and η satisfy Laplace-type equations within the flow domain, i.e.,

$$\partial^2\xi/\partial x^2 + \partial^2\xi/\partial y^2 = 0; \quad \xi \text{ specificied on boundaries} \tag{6.2.32}$$

$$\partial^2\eta/\partial x^2 + \partial^2\eta/\partial y^2 = 0; \quad \eta \text{ specified on boundaries} \tag{6.2.33}$$

Thus, η and ξ lines can be generated subject to boundary conditions shown in Fig. 6.7b, c, respectively. These boundary conditions take the form of specified values of η (in Fig. 6.7b) and ξ (in Fig. 6.7c) at the grid nodes. Thus, Eqs. (6.2.32) and (6.2.33) appear to be well-posed mathematical problems (though devoid of any physical meaning; we will see that grid lines can also be generated using other types of partial differential equations). However, in this form, the equations are not very useful in determining the interior grid points. Discretization of Eqs. (6.2.32) or (6.2.33) would require knowledge of the coordinates of the interior grid points in the physical plane, which is what we are trying to determine!

The problem can be resolved by posing it in an inverse form, where the derivatives are with respect to ξ and η. It would be possible to solve the equations numerically as the grid in the computational plane is known, orthogonal and equi-spaced. We therefore transform Eqs. (6.2.32) and (6.2.33) from (x, y) plane to (ξ, η). This yields a system of two elliptic equations:

$$a x_{\xi\xi} - 2b x_{\xi\eta} + c x_{\eta\eta} = 0 \tag{6.2.34}$$

$$a y_{\xi\xi} - 2b y_{\xi\eta} + c y_{\eta\eta} = 0 \tag{6.2.35}$$

where the subscripts indicate differentiation and where a, b and c are given by

$$a = \left(x_\eta^2 + y_\eta^2\right); \quad b = \left(x_\xi x_\eta + y_\xi y_\eta\right); \quad c = \left(x_\xi^2 + y_\xi^2\right) \tag{6.2.36}$$

With this, the problem of structured grid generation is reduced to one of solving Eqs. (6.2.34) and (6.2.35) subject to the Dirichlet boundary condition of specified values of the x- and y-coordinates of the grid points over the entire boundary in the physical plane (Fig. 6.8a, b, respectively). In the computational plane, (ξ_i, η_j) are known for all points, including those on the boundary.

We begin the solution by fixing the (x_i, y_j) in the physical plane for every (i, j) on the boundary in the computational plane. These then form the Dirichlet boundary conditions for the solution of the two pde's. It may be noted that the equations are non-linear and the coefficients, a, b and c, are not known a priori. One can use the transfinite interpolation method to identify the provisional locations of the interior grid points. These can then be used to estimate numerically the coefficients a, b and c using Eq. (6.2.36). Using these metrics, and using the boundary values of x (i, j), Eq. (6.2.34) is discretized using finite difference approximations and the resulting set of Ax = b is solved to obtain x (i, j) for all interior points. Similarly, Eq. (6.2.35) is solved with y (i, j) on the boundary to obtain y (i, j) for all the interior points. This then completes one iteration of determining (x_i, y_j) for all interior points. These (x_i, y_j) have been obtained using estimated values of coefficients a, b and c using the old values of (x_i, y_j). We therefore use the new coordinates to again to estimate the metrics required to calculate the coefficients and these are again used to sequentially evaluate x(i, j) and y(i, j). This process is carried on iteratively until convergence. At this point, we will have obtained the x- and y-coordinates for all the interior points which are located along constant ξ and constant η lines. These can be used finally to discterize the transformed governing equation in the computational plane to get T (i, j).

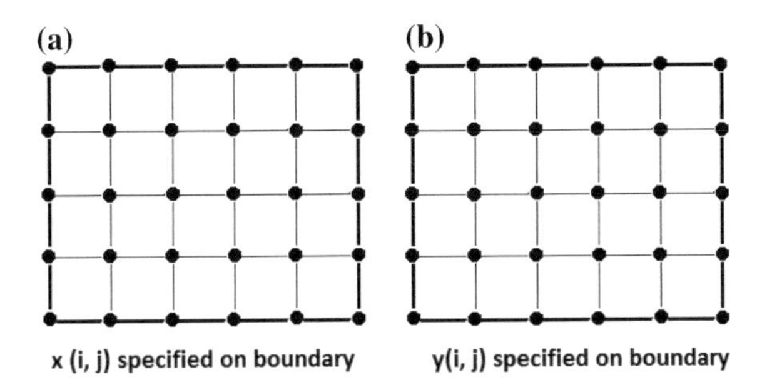

Fig. 6.8 The coordinates of x at interior grid nodes are given by Eqs. (6.2.30a, b) and specified x-values on the boundaries as in (**a**). The coordinates of y at interior grid nodes are given by Eq. (6.2.31) and specified y-values on the boundaries as in (**b**)

One can see that the elliptic equation-based grid generation method requires considerable computational effort to generate the mesh. The resulting interior grid will be smooth and the metrics of the transformation will be well-defined over the computational domain.

The partial differential equations used to generate the grid are by no means unique. In fact, in order to exercise grid control such as clustering of points in regions of anticipated large gradients, one often uses a Poisson equation of the form:

$$\partial^2\xi/\partial x^2 + \partial^2\xi/\partial y^2 = P(\xi, \eta) \qquad (6.2.37)$$

$$\partial^2\eta/\partial x^2 + \partial^2\eta/\partial y^2 = Q(\xi, \eta) \qquad (6.2.38)$$

Here, P and Q are functions which can either attract or repel grid lines at selected points in the physical plane, as shown schematically in Fig. 6.9. These can be likened to finding the steady state temperature distribution with heat sources or sinks. A lumped heat source can repel temperature contours from the source location while a heat sink can pull the iso-temperature towards itself. It may be noted that the coordinate transformation of Eqs. (6.2.37) and (6.2.38) leads to the following equations:

$$ax_{\xi\xi} - 2bx_{\xi\eta} + cx_{\eta\eta} = -J^2\{Px_{\xi} + Qx_{\eta}) \qquad (6.2.39)$$

$$ay_{\xi\xi} - 2by_{\xi\eta} + cy_{\eta\eta} = -J^2\{Py_{\xi} + Qy_{\eta}) \qquad (6.2.40)$$

where a, b and c are given by Eq. (6.2.36) and $J = x_{\xi}y_{\eta} - x_{\eta}y_{\xi}$ is the determinant of the transformation. Solution of these equations will give the coordinates of the interior points.

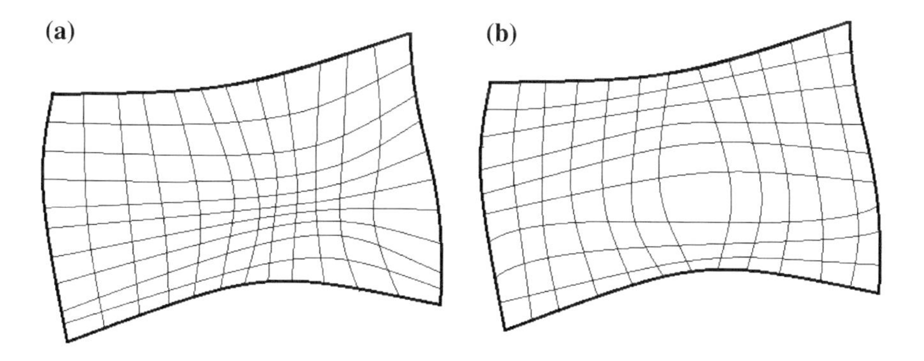

Fig. 6.9 **a** Attraction towards and **b** repulsion away from a grid point of grid lines using local source terms P and Q in Eqs. (6.2.39) and (6.2.40)

Other types of pde's can also be used. For example, in order to generate a grid for an open domain, for example, for flow over an aerofoil, one may use a hyperbolic equation-based method (Steger and Sorenson 1980). One equation is provided by specifying the area scaling factor, namely, J:

$$x_\xi y_\eta - x_\eta y_\xi = J \tag{6.2.41}$$

Yet another is provided by imposing an orthogonality condition:

$$x_\xi x_\eta + y_\xi y_\eta = 0. \tag{6.2.42}$$

The above two equations are non-linear and need to be solved iteratively. This is done by restating them as a system of hyperbolic equations. To begin with, the equations are linearized as follows. Using the superscript k to indicate the iteration number, the product *fg*, where both *f* and *g* are being determined iteratively, can be written as

$$fg \approx f^{k+1} g^k + g^{k+1} f^k - f^k g^k \tag{6.2.43}$$

Using this linearization for each product term, Eqs. (6.2.41) and (6.2.42) can be written as

$$\begin{aligned} x_\xi^{k+1} y_\eta^k + x_\xi^k y_\eta^{k+1} + x_\xi^k y_\eta^k - x_\eta^{k+1} y_\xi^k - x_\eta^k y_\xi^{k+1} + x_\eta^k y_\xi^k &= J \\ x_\xi^{k+1} x_\eta^k + x_\xi^k x_\eta^{k+1} - x_\xi^k x_\eta^k + y_\xi^{k+1} y_\eta^k + y_\xi^k y_\eta^{k+1} - y_\xi^k y_\eta^k &= 0 \end{aligned}$$

Noting that $x_\xi^k x_\eta^k + y_\xi^k y_\eta^k = 0$ and $x_\xi^k y_\eta^k - x_\eta^k y_\xi^k = J^k$, the above equations can be recast as

$$[P]U_\xi + [Q]U_\eta = [R] \tag{6.2.44}$$

where

$$[P] = \begin{bmatrix} x_\eta^k & y_\eta^k \\ y_\eta^k & -x_\eta^k \end{bmatrix} \quad [Q] = \begin{bmatrix} x_\xi^k & y_\xi^k \\ -y_\xi^k & x_\xi^k \end{bmatrix} \quad [R] = \begin{bmatrix} 0 \\ J + J^k \end{bmatrix} \quad [Q] = \begin{bmatrix} x^{k+1} \\ y^{k+1} \end{bmatrix} \tag{6.2.45}$$

In hyperbolic grid generation, which is generally used for external flows over a body, the grid starts from the surface of the body and spreads out into the open space. We assume that the body is mapped on to the ξ-direction in computational space and that it evolves in the η-direction. For this hyperbolic nature of evolution in the η-direction to be represented by Eq. (6.2.44), the eigenvalues of $[P]^{-1}[Q]$ must be real. These eigenvalues are given by

$$\lambda_{1,2} = \pm \sqrt{\frac{\left(x_{\eta}^{k}\right)^{2} + \left(y_{\eta}^{k}\right)^{2}}{\left(x_{\xi}^{k}\right)^{2} + \left(y_{\xi}^{k}\right)^{2}}} \qquad (6.2.46)$$

Equation (6.2.44) can be solved as a marching forward type problem starting with an initial distribution of points on the boundary corresponding to $\eta = 0$. This enables a faster solution to the grid generation problem than in the elliptic grid generation method but boundary discontinuities tend to propagate into the interior of the flow domain. Grid generation methods based on the solution of parabolic partial differential equations, which allow some diffusive correction of boundary discontinuities while still offering the possibility of using a marching forward type of solution approach have also been proposed for structured grid generation.

In many practical cases, multiply-connected calculation domains are encountered. These domains have internal elements within the computational domain (for examples, tubes on the shell side of a shell-and-tube heat exchanger or the rows of blades in a steam turbine) such that when the flow domain is squeezed from all sides, it will not collapse to a domain of zero area in 2-d or zero volume in 3-d. It would not be possible in such cases to traverse the entire boundary without jumping over from one closed curve to another closed curve. A structured grid can be created for such flow domains by creating imaginary boundary lines (or planes) within the flow domain. Figure 6.10 shows several possible grids of a doubly connected domain obtained by different mappings of the physical domain boundaries on to the computational plane.

6.2.3 *Solution of Navier-Stokes Equations Using the Structured Grid Approach*

In the computational plane, the solution of the N-S equations is not very different from the methods discussed in Chaps. 3–5 except for a few details which are worth recalling:

- The governing equations need to be transformed into the computational plane. While this does not change the mathematical nature of the pde's, a number of additional terms appear. For example, a simple Laplace equation, when transformed, acquires cross-derivative terms and first derivative terms in the general case. Also the coefficients of the various terms in the scalar transport equations are dependent on the metrics of the transformation and vary from location to location.
- While the mapping between the physical and the computational planes may be performed analytically in some simple cases, it is done numerically in most cases and the metrics of the transformation are evaluated numerically.

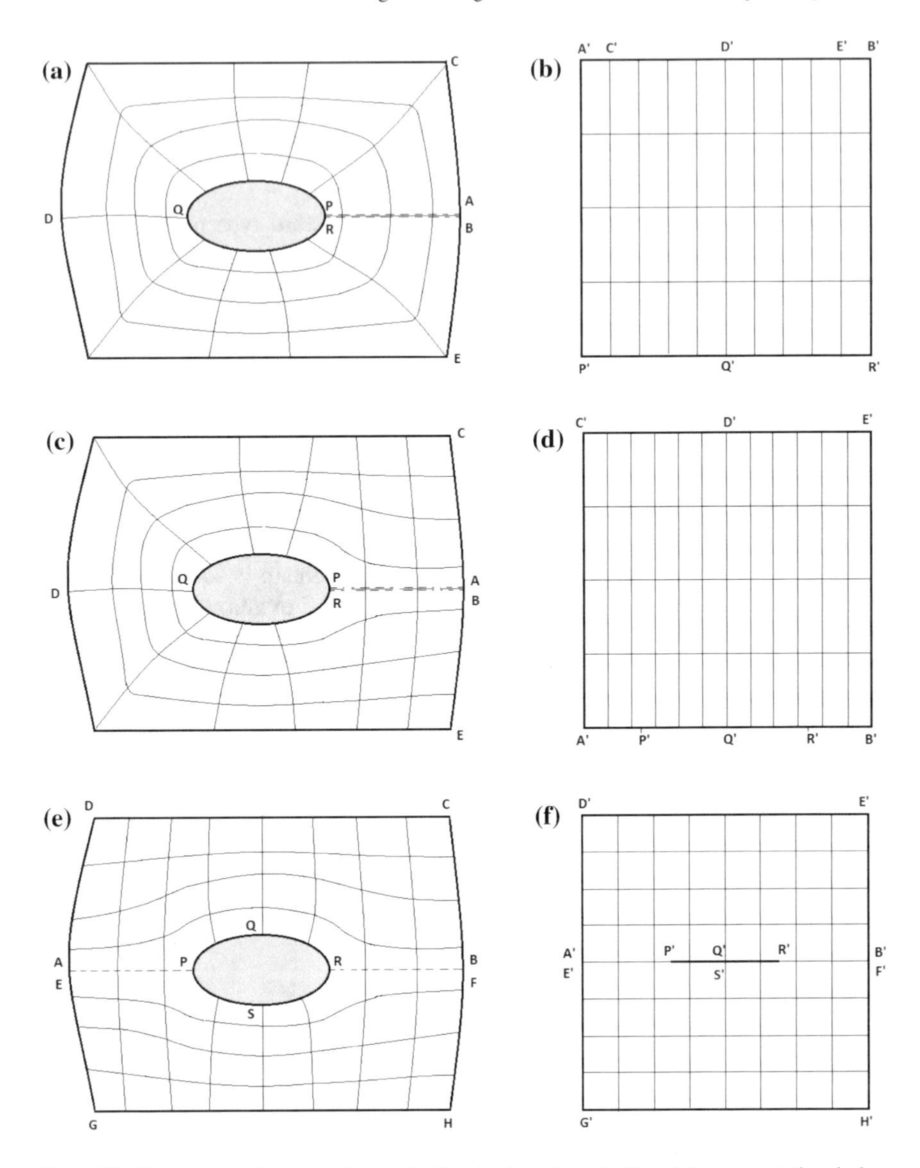

Fig. 6.10 Three types of structured grids in the physical plane (left) and the computational plane (right): **a** and **b** O-type grid, **c** and **d** C-type grid, and **e** and **f** H-type grid

- The grid in the physical plane is non-uniform and non-orthogonal in the general case while it is, by choice, orthogonal and equi-spaced in the computational space.

Thus, it makes sense to discretize the transformed governing equations in the computational plane, solve them to find the variables at the grid nodes and carry them over to the corresponding points in the physical plane. In this case, the cross-derivative terms and the first derivative terms, when discretized, may lead to loss of diagonal dominance of the set of linearized algebraic equations. In such a case, these additional terms need to be included in a "deferred correction approach", i.e., should be evaluated explicitly in terms of the values at the previous iteration. In highly skewed grids or in case of grids with discontinuities in gradients, these terms may become dominant and may lead to convergence difficulties. Since coefficients of the principal terms are now highly dependent on the grid smoothness etc., simplistic criteria based on uniform grid spacing etc., may not be applicable and the stability limits need to be evaluated afresh. Better stability may be expected if implicit schemes are used. The structured nature of the grids preserves the diagonal structure of the coefficient matrix although the coefficients themselves can vary widely due to non-uniform coefficients.

Apart from these complications, the structured grid formulation enables us to make use of the fundamental concepts discussed in Chaps. 3–5 in a relatively straightforward way to obtain the solution for the case of a geometrically complex flow domain.

6.3 The Unstructured Grid Approach

We have seen in Sect. 1.2 a brief exposition of the finite volume method (FVM)-based evaluation of the flow field for flow through a duct of triangular cross-section using an unstructured grid. Here, we dwell on it at length to see how it can be applied to the general case where the flow domain is irregular and the equations consist of the full Navier-Stokes equations. The general formulation of the problem is discussed in Sect. 6.3.1. Algorithms for unstructured grid generation are discussed in Sect. 6.3.2 and some aspects of solution are discussed in Sect. 6.3.3.

6.3.1 Formulation of the Finite Volume Method

Recall that the equations governing fluid flow can be put in the form of a scalar transport equation:

$$\partial(\rho\phi)/\partial t + \partial(\rho u_j\phi)/\partial x_j = \partial\{\Gamma_\phi(\partial\phi/\partial x_j)\}\partial x_j + S_\phi \tag{6.3.1}$$

For different values of ϕ, Γ_ϕ and S_ϕ, we can recover the continuity, momentum balance and energy balance equations. It may be noted that Cartesian velocity components in fixed directions are used to describe the velocity field; these velocity components can therefore be treated as scalars. The second term on the left hand side represents the net outflux of ϕ due to flow while the first term on the right hand side represents the net influx of ϕ through diffusion. Equation (6.3.1) can be rewritten in vector form as

$$\partial(\rho\phi)/\partial t + \nabla \cdot (\rho \mathbf{u} \phi) = \nabla \cdot (\Gamma_\phi \nabla \phi) + S_\phi \quad (6.3.2)$$

The above form of the governing equation is said to be in conservation form as both the flux terms are expressed as divergence terms. We can bring the two fluxes together and write:

$$\partial(\rho\phi)/\partial t + \nabla \cdot \mathbf{F} = S_\phi \quad (6.3.3)$$

where

$$\mathbf{F} = \mathbf{F}_{conv} + \mathbf{F}_{diff} = (\rho \mathbf{u} \phi) - (\Gamma_\phi \nabla \phi) \quad (6.3.4)$$

In the finite volume method, the flow domain is divided into a large number of contiguous cells which, when put together, will fill the entire control volume without overlapping. These cells are tetrahedral or hexahedral in 3-D and triangular or quadrilateral (or polygonal) in 2-D. Curved surfaces (curved lines in 2-D) are approximated by planar surfaces (line segments in 2-D). Equation (6.3.3) is integrated over each such cell:

$$\int_{CV} [\partial(\rho\phi)/\partial t] dV + \int_{CV} [\nabla \cdot \mathbf{F}] dV = \int_{CV} S_\phi dV \quad (6.3.5)$$

where CV denotes the cell volume. We assume, in finite volume method, that the value of the variable is constant over the whole cell block although it varies from cell to cell. Thus, the terms $[\partial(\rho\phi)/\partial t]$ and S_ϕ can be evaluated appropriately at the cell centroid. Using Gauss' divergence theorem, the volume integral in the second term can be converted to an integral over the cell surface. This reduces Eq. (6.3.5) to

$$[\partial(\rho\phi)/\partial t] V_{cell} + \int_{CS} \mathbf{F} \cdot \mathbf{dS} = S_\phi V_{cell} \quad (6.3.6)$$

The cell surface is composed of a number of planar surfaces through which convective and diffusive fluxes occur. These depend on the local values of velocity components, gradients of the scalar variables and diffusivities (see Eq. 6.3.4). Therefore, the surface integral in Eq. (6.3.6) is evaluated over each planar surface separately and is then summed. Thus, the discretized form of the governing equation can be written as

$$[\partial(\rho\phi)/\partial t]V_{cell} + \sum_i \left(\int_A \mathbf{F}\cdot\mathbf{dS}\right)_i = S_\phi V_{cell} \tag{6.3.7a}$$

or

$$[\partial(\rho\phi)/\partial t]V_{cell} + \sum_i (\mathbf{F}_{av}\cdot\mathbf{A})_i = S_\phi V_{cell} \tag{6.3.7b}$$

where $\mathbf{F}_{av}$ is the average flux evaluated at the centroid of each planar surface A_i.

The definition of the cell over which Eq. (6.3.3) is integrated is arbitrary except that one assumes that the value of the scalar ϕ is constant over that cell. Consider the tetrahedral cell j shown in Fig. 6.11a bounded by vertices P, A, B and C. For this cell, Eqs. (6.3.7a, b) takes the form

$$[\partial(\rho\phi)/\partial t]_j V_j + (\mathbf{F}\cdot\mathbf{A})_{PAB} + (\mathbf{F}\cdot\mathbf{A})_{PBC} + (\mathbf{F}\cdot\mathbf{A})_{PCA} + (\mathbf{F}\cdot\mathbf{A})_{ABC} = (S_\phi)_j V_j \tag{6.3.8}$$

Evaluation of Eq. (6.3.8) requires an evaluation of the volume of cell j, the areas and the direction cosines of the outward normal vector of each face, and the flux vector on each face. Here, one must exercise care to ensure that the area and flux terms are done consistently for each cell so that the flux leaving cell j through a surface is equal to the flux that is entering a neighbouring cell through the same surface. If this is not done, there can be imbalance of fluxes, and spurious, unphysical or unintended sources of ϕ may appear. Consistent evaluation of areas and fluxes is possible if the same variables are used to evaluate the terms. Let us first consider the evaluation of surface areas and volumes of cells.

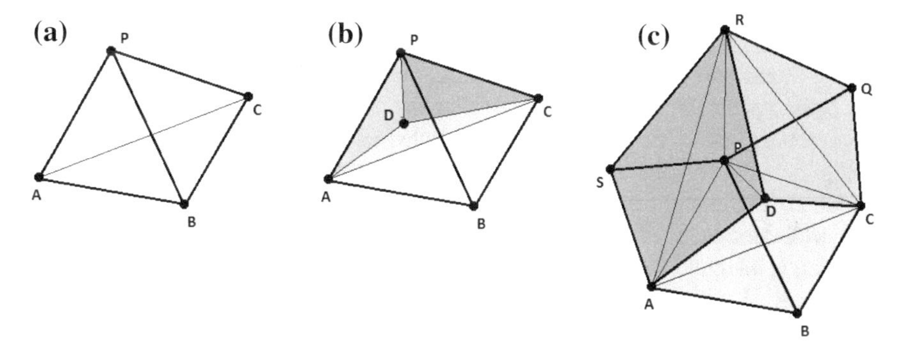

Fig. 6.11 Three types of 3-d cells: **a** a tetrahedral cell PABC, **b** a pyramidal cell PABCD, and **c** a hexahedral cell ABCDPQRS

The surface vector $\mathbf{S}_{ABC}$ can be evaluated solely in terms of the coordinates of its vertices:

$$\mathbf{S}_{ABC} = {}^1\!/_2(\mathbf{x}_{AB} \times \mathbf{x}_{BC}) = {}^1\!/_2(\mathbf{x}_{BC} \times \mathbf{x}_{CA}) = {}^1\!/_2(\mathbf{x}_{CA} \times \mathbf{x}_{AB}) \tag{6.3.9}$$

Here, the vector $\mathbf{x}_{AB}$ is defined as $x_B - x_A$. Similarly, the volume of the tetrahedral cell PABC can be obtained as

$$V_{PABC} = {}^1\!/_3 \sum_{faces} (\mathbf{x} \cdot \mathbf{S})_{faces} = {}^1\!/_6[\mathbf{x}_{PA} \cdot (\mathbf{x}_{AB} \times \mathbf{x}_{BC})] = {}^1\!/_6[\mathbf{x}_{PA} \cdot (\mathbf{x}_{BC} \times \mathbf{x}_{CA})] \tag{6.3.10a}$$

or

$$V_{PABC} = \frac{1}{6} \begin{vmatrix} x_P & y_P & z_P & 1 \\ x_A & y_A & z_A & 1 \\ x_B & y_B & z_B & 1 \\ x_C & y_C & z_C & 1 \end{vmatrix} \tag{6.3.10b}$$

The above formulae form the basis for the evaluation of areas and volumes of other shapes. The area of the quadrilateral ABCD can be evaluated as the sum of the areas of the triangles ABC and ADC:

$$S_{ABCD} = {}^1\!/_2[(\mathbf{x}_{AB} \times \mathbf{x}_{BC}) + (\mathbf{x}_{CD} \times \mathbf{x}_{DA})]$$

The above formula can be applied to a quadrilateral even when all the four verticess are not in the same plane. Similarly, a pyramid consisting of one vertex (P) and a quadrilateral base (ABCD, see Fig. 6.11b) can be decomposed into two tetrahedrals PABC and PACD. The hexahedral made up of eight vertices (see Fig. 6.11c) can be decomposed into three pyramids, each of which can be decomposed into two tetrahedrals. Consider the hexadral ABCDPQRS shown in Fig. 6.11c. Vertex P is contact with three of the six bounding surfaces. Therefore, one pyramid each can be made with P as vertex and each of the other three non-contacting quadrilaterals as the base, namely, PABCD, PADRS and PCQRD.

The fluxes vary locally with position and are usually a function of ϕ. Here, there can be two possibilities depending on where the solution variable is evaluated. In a cell-centred method, the value of ϕ is associated with a cell, for example, ϕ_j and ϕ_m in Fig. 6.12a. The value is not precisely defined at a location and is assumed to be uniform over the entire cell. In a cell-vertex scheme (see Fig. 6.12b), ϕ is evaluated at a well-defined node but the definition of the cell over which the governing equation is integrated is variable. For face ABC which is uniquely determined by its vertices A, B and C, one consistent way of evaluating the flux through the face would be to evaluate it at its vertices and take the average, i.e.,

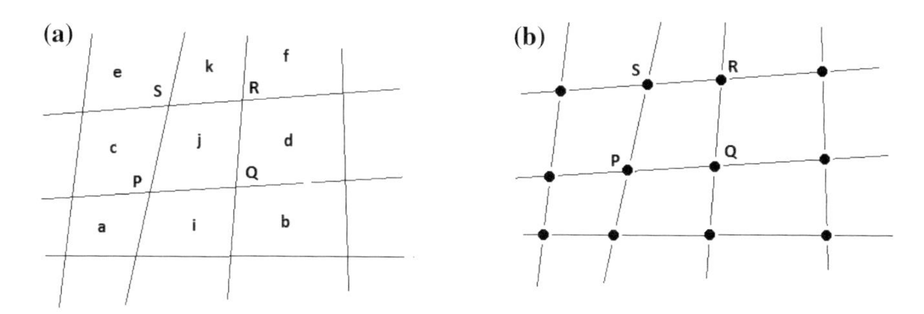

Fig. 6.12 **a** Cell-centred and **b** cell-vertex schemes of finite volume formulation

$$\mathbf{F}_{ABC} = 1/3[\mathbf{F}(\phi_A) + \mathbf{F}(\phi_B) + \mathbf{F}(\phi_C)] \tag{6.3.11}$$

Since this would involve three evaluations of the flux, one can alternately evaluate it as

$$\mathbf{F}_{ABC} = \mathbf{F}[(\phi_A + \phi_B + \phi_C)/3] \tag{6.3.12}$$

While the second expression appears to be simple, it will be accurate only if the flux is a function of ϕ alone. In turbulent flows, the diffusivity itself may be a function of the flow, and may therefore vary from location to location. In this case, Eq. (6.3.11) may give a more correct estimation of the flux.

Another important consideration in evaluating the fluxes is the nature of evaluation of the two flux components, namely, the convective flux and the diffusive flux. Evaluation of the diffusive flux requires an estimate of the diffusivity and of the gradient of ϕ. In a cell-centred method, the gradient can be evaluated to a second order accuracy using the ϕ-values of the two cells on either side of the face, for example, cell j and cell k for the surface RS in Fig. 6.12a.

$$\begin{aligned}[\mathbf{F}_{\mathbf{diff}} \cdot \mathbf{S}]_{RS} &= [-D_{eff} \nabla\phi \cdot \mathbf{S}]_{RS} \\ &= -[D_{eff}(\phi_k - \phi_j)/(x_k - x_j)]S_{RSx} - [D_{eff}(\phi_k - \phi_j)/(y_k - y_j)]S_{RSy}\end{aligned} \tag{6.3.13}$$

Here D_{eff} is the effective diffusivity which is evaluated based on a weighted average of the diffusivity for cell j and cell k.

The convective flux is usually evaluated using an upwind scheme; thus,

$$[\mathbf{F}_{conv} \cdot \mathbf{S}]_{RS} = [\rho \mathbf{u} \phi_j \cdot \mathbf{S}]_{RS} \quad \text{if } (\mathbf{u} \cdot \mathbf{S})_{RS} > 0 \tag{6.3.14a}$$

$$= [\rho \mathbf{u} \phi_k \cdot \mathbf{S}]_{RS} \quad \text{if } (\mathbf{u} \cdot \mathbf{S})_{RS} < 0 \tag{6.3.14b}$$

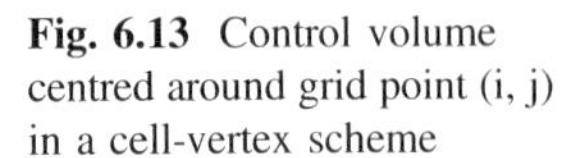
Fig. 6.13 Control volume centred around grid point (i, j) in a cell-vertex scheme

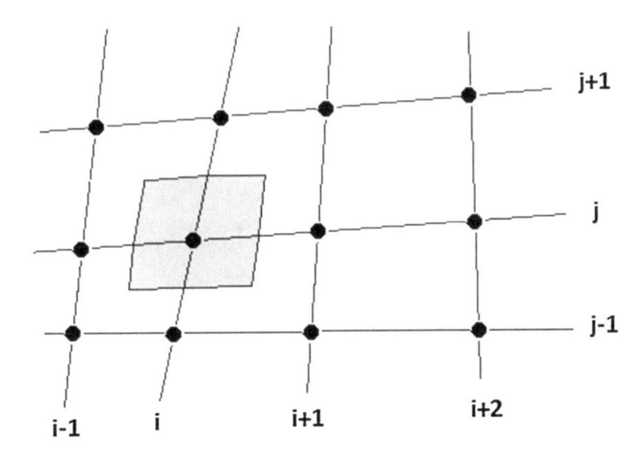

Here, **u** may vary from cell j to cell k. A weighted average of the two should be taken to evaluate the value of u at the cell face. Note that Eqs. (6.3.14a, b) is only first-order accurate; evaluation to second order accuracy requires reconstruction of the local velocity gradient vector (Murthy and Mathur 1997).

A cell-vertex scheme is often found with structured body-fitted grids which have been discussed in Sect. 6.2. On such a grid, a cell can be thought to be extending half grid distance in the two (or three) coordinate directions (Fig. 6.13). The integration is carried out as before (Eqs. 6.3.7a, b). The fluxes at the face locations can be evaluated using Eqs. (6.3.11) and (6.3.12) while the diffusive and convective fluxes can be evaluated using Eqs. (6.3.13) and (6.3.14a, b). Implementation of higher order schemes for the convective fluxes is relatively straightforward on a structured mesh. When the cell-vertex scheme does not use a structured mesh, a local reconstruction of the local velocity gradient vector would be necessary to implement schemes with second or higher order accuracy.

6.3.2 Unstructured Grid Generation

Grid generation has the objective of distributing grid points throughout the flow domain. The values of the flow variables are evaluated at these grid points or at locations closely associated with them. In structured grids, these grid points lie at the intersection of linear or curvilinear coordinate lines. In unstructured grids, they do not necessarily do so and are placed algorithmically throughout the computational domain. These grid points are used to divide the flow domain into polygonal tiles in 2-d or polyhedral or prismatic blocks in 3-d over each of which the governing equations are integrated. This is done in two steps. In the first step, grid points are distributed throughout the flow domain. In the second step, these are joined together to form polygons or polyhedra. While these steps are fairly simple for regular geometries, an algorithmic approach is needed for irregular flow domains because the number of grid points in practical cases is very large (and may run into millions), and the joining these to make polyhedra that do not intersect each other is non-trivial. In this section, we shall look at algorithms specifically developed for grid generation in two-dimensions. The first step of distribution of nodes in the interior is called domain nodalization and the second step of generating triangular mesh elements is called triangulation.

Consider the doubly-connected domain shown in Fig. 6.14a the boundaries of which are closed polygons (a curved surface can be approximated as a many-sided polygon). The domain external to the inner polygon and internal to the external polygon needs to be meshed using triangles of a nominal size of l_S. This size may be arrived at based on the number of grid points that one would like to populate the domain with. To begin with, we divide the external boundary and the internal boundary into edges of size l_s starting at a convenient point on each boundary. In

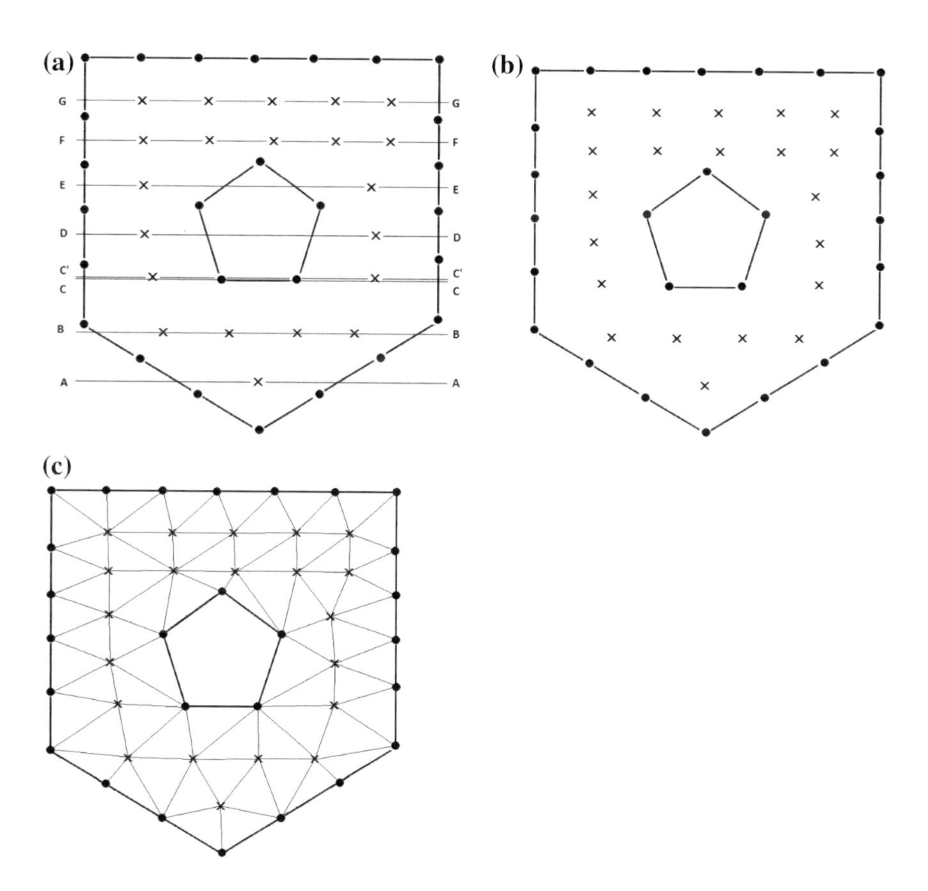

Fig. 6.14 Nodalization of the domain: **a** locating the interior nodes on horizontal lines, **b** the boundary and internal nodes, and **c** a non-algorithmic, free-hand triangulation illustrating creation of mostly of similar size

order to find the internal nodes, we draw, starting from the lowest node, horizontal lines separated by the distance l_s. In the absence of the internal domain, these horizontal lines, for example, lines AA and BB, would intersect two boundary edges. Starting with, say, the left intersection point, we can put along each horizontal line grid nodes which are approximately distance l_s apart. Some horizontal lines, such as lines DD and line EE, will intersect the external and the internal boundary edges. Along each horizontal segment within the computational domain and bounded by two boundary edges, we again place nodes which are roughly distance l_s apart. There may be lines which coincide with part of the horizontal boundary, or lie very close to it (such as line CC). Placing grid nodes along these lines may result in some grid points which are very close to each other; such lines can be moved up or down slightly, say, by $\pm 1/10\ l_s$, and then be subdivided into segment of l_s each, for example, see line C′C′. This procedure results in internal grid nodes distributed throughout the domain with a separation distance of about l_s (see Fig. 6.14b). Joining these nodes results in a domain which is composed of triangles of fairly equal size, see Fig. 6.14c. In some cases, one may wish to place very small nodes near the surface in order to resolve the boundary layer. Usually, these boundary layers are in the form of quadrilaterals (or hexahedrals in 3-d). In such a case, these need to be formed first by projecting the boundary into the flow domain by a small distance in a direction normal to the surface and dividing this small distance into small nodes. Joining these by lines that are nearly parallel to the boundary will give rise to quadrilateral elements. The rest of the domain can be nodalized as explained earlier. The quadrilateral boundary elements can also be subdivided into small triangles to form small triangular elements in the boundary layer and progressively larger triangles as we move away from the walls. Both these possibilities are shown in Fig. 6.15.

Joining the internal nodes and the boundary nodes to form triangles is not a trivial task. Consider the four adjacent points that make up a quadrilateral ABCD in Fig. 6.16. This quadrilateral can be divided into two triangles in two ways as shown in Fig. 6.16 by joining either the diagonal AC or the diagonal BD. Out of these two

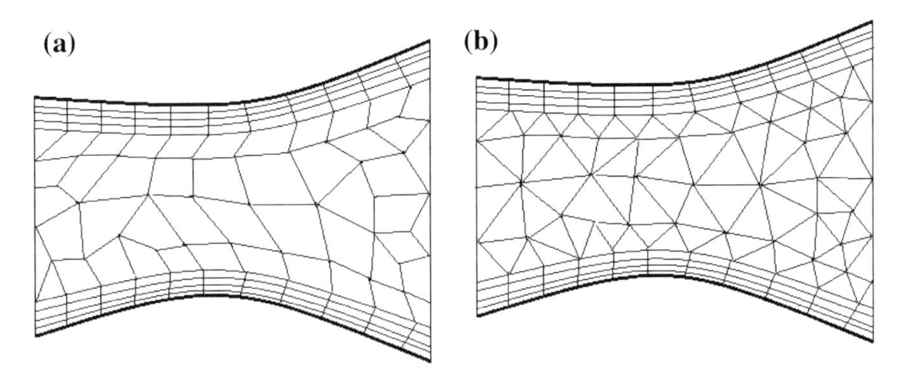

Fig. 6.15 Unstructured grid with boundary layer type of gridnear the boundaries and **a** triangular elements and **b** quadrilateral cells in the interior

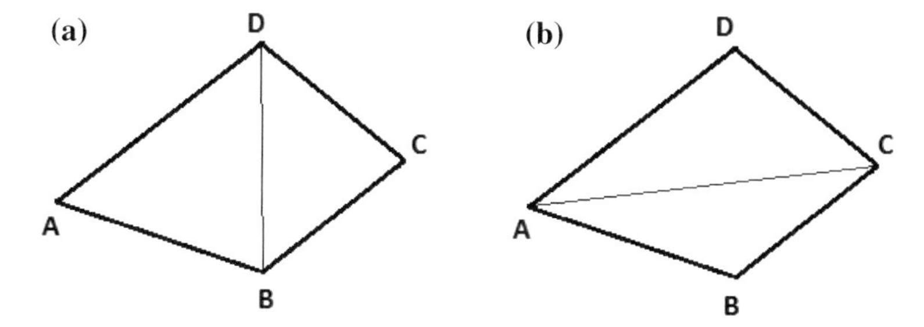

Fig. 6.16 Two ways of triangulating the quadrilateral ABCD: **a** the less distorted option: ∠A + ∠C < 180°, and **b** the more distorted option, ∠B + ∠D > 180°

possibilities, assuming that ABCD is not a rectangle, one of the possibilities produces two triangles that are less distorted than with the other combination. This happens when the sum of the angles opposite the common diagonal is less than 180°. Also, in this case, the fourth point of the quadrilateral lies outside the circumcircle passing through the three vertices used to form the triangles. Both these conditions are satisfied by the triangulation of ABCD into ΔABD and ΔBCD (Fig. 6.17a) and not by the triangulation into ΔABC and ΔACD (Fig. 6.17b).

Since a more distorted triangle leads to more errors when the governing equation is discretized (Eqs. 6.3.7a, b), it is preferable to have less distorted triangles. This can be done in the course of grid generation as follows. Consider the vertices A to F in Fig. 6.18a which have been joined to form four triangles in the course of progressive grid generation (which is discussed below). Now a new grid node P needs to be included or inserted into the grid. Since the new point P falls in ΔBDF, one possibility is to create three new triangles by joining node P to the three vertices of ΔBDF. This is shown in Fig. 6.18a. Now we can see if there is a better way of triangulation which results in less distorted set of triangles. This can be done by drawing circumscribing circles on each of the new triangles and verifying that the fourth node of the quadrilateral falls outside the circle. This is the case for the two

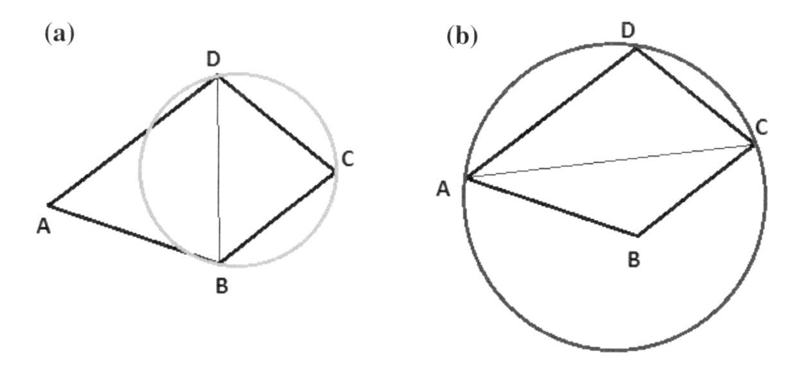

Fig. 6.17 Triangulation of a new grid node P: **a** the straightforward choice of joining P with the three vertices of ΔBDF into which it falls, and **b** taking the less distorted triangles option by drawing circumscribing circles of the three new triangles

new triangles ΔBPF and ΔDPF but not for ΔBPD. Therefore the quadrilateral BPDC is triangulated in the alterative way, i.e., into ΔBPC and ΔDPC as shown in Fig. 6.18b. This process does not change the number of triangles that are formed but makes the overall set of triangles less distorted, which is a desirable feature. This feature, known as Delaunay triangulation scheme, is incorporated in the Bowyer-Watson algorithm (Bowyer 1981; Watson 1981) for triangulation by successive insertion of grid points that need to be triangulated. This is described below.

The algorithm starts with a convex hull, in the form of a super triangle, which contains all the points (including the boundary points) that need to be triangulated. A point from this set of grid nodes is introduced into the triangle. The circumcircle of the convex hull obviously contains the new point. Therefore the original triangle is removed and three new triangles are created by joining the new point with the vertices. At this point, there is only one way of joining these vertices to make non-overlapping triangles. If another point is now introduced, we get into the situation described in Fig. 6.18. We first find into which of the existing triangles the new grid point falls. This can be done by ensuring that the new grid point is always on the left side as we move along from one node of a triangle to another node in the counterclockwise direction. Consider the two adjacent triangles ΔBAF and ΔBDF in Fig. 6.18a. Point P is the new point. Consider the three edges of ΔBAF. As we move along the edges in the counterclockwise direction, we note that point P is to the left of edge FA and edge AB but not of edge BF. For ΔBDF, we find that point P is to the left of edges BD, DF and FB as we move along the edges in the counterclockwise direction. An algorithmic way of finding if point P lies to the left of, say, edge AB is as follows. We construct vectors AB and AP, noting that AB is the path from node A to node B in the counterclockwise direction, and take the cross product of the two vectors. If the product is positive, then P is to the left of vector B; if it is negative, it is to the right side (if the cross product is zero, then vertices A, B and P cannot be joined to form a triangle). Thus, for P to lie the left of the vector AB,

$$\mathbf{x}_{AB} \times \mathbf{x}_{AP} > 0 \text{ or } \mathbf{x}_{AB}\, \mathbf{x}_{AP} \sin\theta > 0 \tag{6.3.15a}$$

where

$$\mathbf{x}_{AB} = (x_B - x_A)\mathbf{i} + (y_B - y_A)\mathbf{j} \tag{6.3.15b}$$

Once the triangle containing the new point is identified, three new triangles can be formed by joining the new point with the vertices of the triangle, as in Fig. 6.18a. We now draw the circumcircles over the new triangles and go through the process of identifying the best possible way of triangulating the neighbouring quadrilaterals, as shown in Fig. 6.18b. This completes the insertion of a new point. This results in the creation of three new triangles; we remove the original triangle which contained the new point from the list of triangles leaving a net addition of two triangles per insertion. This process is continued until there are no more points to be inserted.

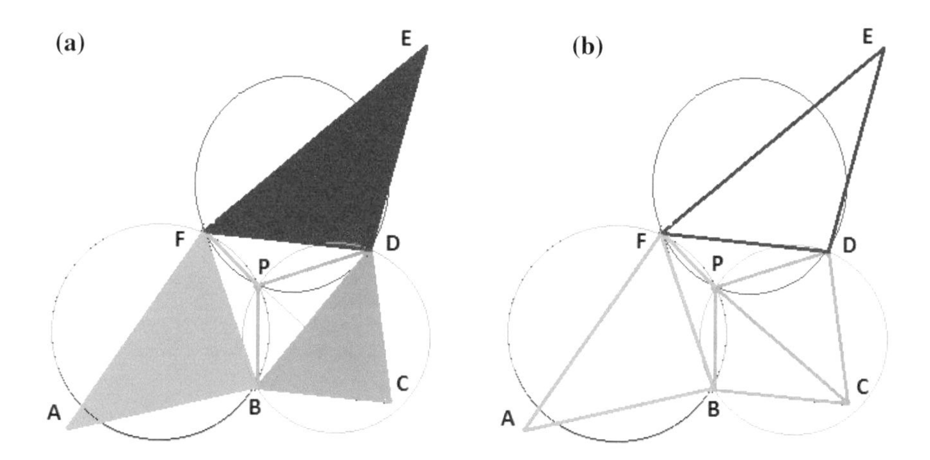

Fig. 6.18 **a** The domain with 11 nodes and **b** the triangulated grid

The entire process of triangulation is illustrated with the help of Figs. 6.19 and 6.20. The domain consists of a square with a chamfered edge at the right hand bottom and contains four interior points. The boundary nodes have been fixed as shown in Fig. 6.20a; there are thus a total of 11 nodes, denoted as A to K, that need to be triangulated. The final triangulated grid, which is obtained through successive insertions of points, is shown for later reference in Fig. 6.20b.

The 12-step process of triangulation which leads to the final grid is summarized in Fig. 6.19. We start with a super circumscribing triangle PQR which contains all the nodes, as shown in Fig. 6.19a. Then the eleven nodes are introduced one by one in no particular order. The sequence of introduction of nodes shown in Fig. 6.19 is as follows: I, F, J, K, A, E, B, H, G, C and D. In each case, the triangle in which the insertion point falls is determined and three new triangles are created. The less-distorted option is chosen in each case using the circumscribing circle criterion. This results in changes during the insertions of nodes K, A, E, B, H, G, C and D in which more than one change is required for nodes A, C and D, as indicated in Fig. 6.19. The final set of triangles obtained at the end of insertion are shown in Fig. 6.20a; deleting the nodes of the super triangle PQR and its links with the grid nodes leaves the set of triangles shown in Fig. 6.20. Close scrutiny shows that the triangulation is such all quadrilaterals made up of two adjacent triangles are triangulated in the less distorted way.

Another approach to triangulation is the advancing front method (Nelson 1978; Lo 1985; Merriam 1991) which guarantees that no area that lies outside the boundaries is included in the triangulated computational domain. This may happen with the Delaunay triangulation scheme in regions that are highly concave. In the advancing front method (AFM), the triangulation starts from the boundary and progresses inwards. The boundaries of the domain, both internal and external, are segmented into edges which are numbered sequentially in such a way that the internal boundary is in clockwise direction and the external boundary nodes are

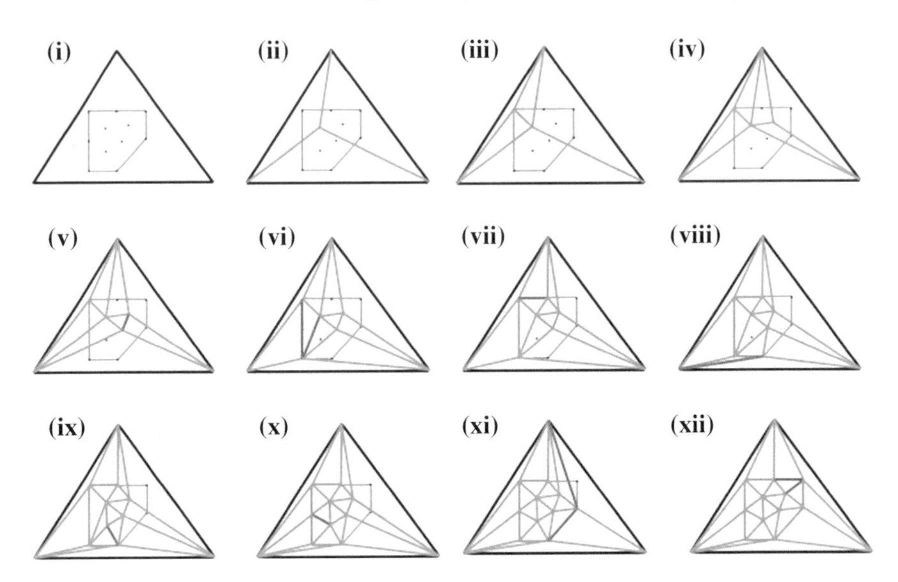

Fig. 6.19 The 12-step process of triangulation of the eleven nodes, chosen (randomly) in the sequence I-F-J-K-A-E-B-H-G-C-D, shown in Fig. 6.20a. The sides in red colour indicated where the circumscribing circle criterion (Eqs. 6.3.15a, b) was used to arrive at less distorted triangulation

arranged in the anticlockwise direction. At the beginning of the triangulation, the front consists of the boundary edges. Starting with an arbitrary edge on the boundary, one finds all the interior and boundary points which lie to the left of this edge and selects the one which lies closest to form a triangle with. At this point, the edge which has been joined to the new point is removed from the front and a new front consisting of the two new edges linked in the counterclockwise manner is created. The process continues with the last of the new edges. All the points to the left of new edge on the front are checked to see which of them fall to the left side and the closest one is added to make a new triangle. The front is again redefined and is advanced. If, at any point, a dead end is reached, that is, when there are no more points that lie to the left, a search is started from the preceding edge. This process is continued until all the nodes get interconnected.

The triangulation of the chamfered square geometry shown in Fig. 6.20a using the advancing front method is illustrated in Fig. 6.21. At the beginning, the closed boundary ABCDEFGA constitutes the advancing front. Starting with edge GA, which is the last segment of the advancing front, we find that nodes I, H, B, E, K, J, C and D to be the nodes that are to the left of edge GA and that node H is the one closest to it. We therefore join nodes G and H and nodes H and A to arrive at a first triangle GAH and a new advancing front consisting of the closed boundary ABCDEFGHA. This time, starting with HA, we find node B to be the closest node to the left of edge HA and form the second triangle HAB and a new advancing front BCDEFGHB. We continue this process and join successively nodes K, C, J, D, E, I,

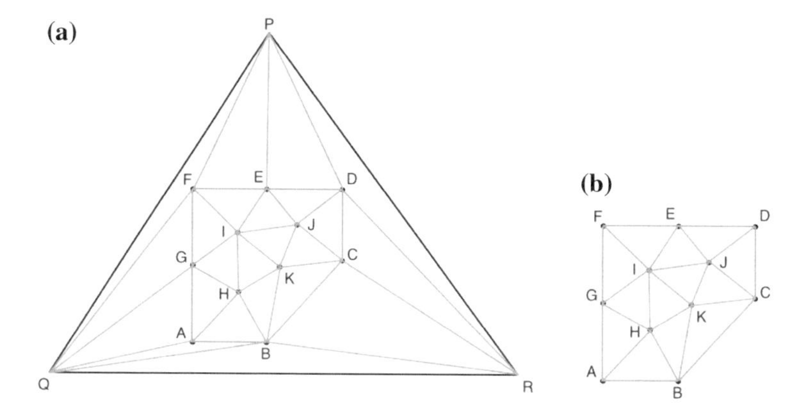

Fig. 6.20 **a** The final set of triangles after insertion of all the grid points showing that only the external nodes are connected to the vertices of the super triangle PQR, and **b** the Delaunay-triangulated grid in which all quadrilaterals are triangulated in the less distorted way

F, G, H, K and finally the node J to the complete set of 13 triangles. The sequence of creation of triangles is given in the last sub-figure. For the sake of clarity, the newly created triangle is shown in blue colour and those already created are shown in green colour. One advantage of the AFM is that only internal and boundary points are joined to form triangles and no area external to the flow domain is included in the calculation. However, the triangles are not always optimally joined (although not in the case shown in Fig. 6.21; here the final triangulation is the same as in Fig. 6.20); an alternative joining algorithm using the Delaunay triangulation condition would give grids with lower aspect ratio. A number of methods for unstructured grid generation have been reviewed by Mavriplis (1995).

6.3.3 Solution of Navier-Stokes Equations on an Unstructured Grid

Compared to the structured grid generation methods, unstructured grid generation is more complicated from an algorithmic point of view, and requires considerable skills in programming to develop general purpose grid generators. However, once the grid nodes are identified and the connectivity with the neighbouring cells established, solving the governing equations on an unstructured grid does not pose any special problems. Over each cell, Eqs. (6.3.7a, b) is discretized by evaluating the fluxes and the surface vectors. The discretization is simpler than for a structured body-fitted grid because cross-derivatives are not present and grid-mapping related coefficients are also not present. Unstructured grid is invariably used with a collocated (non-staggered) grid approach in pressure-based methods (such as the SIMPLE family of methods or the MAC scheme (Harlow and Welch 1965)) for the solution of Navier-Stokes equations. The face velocity information that is needed for

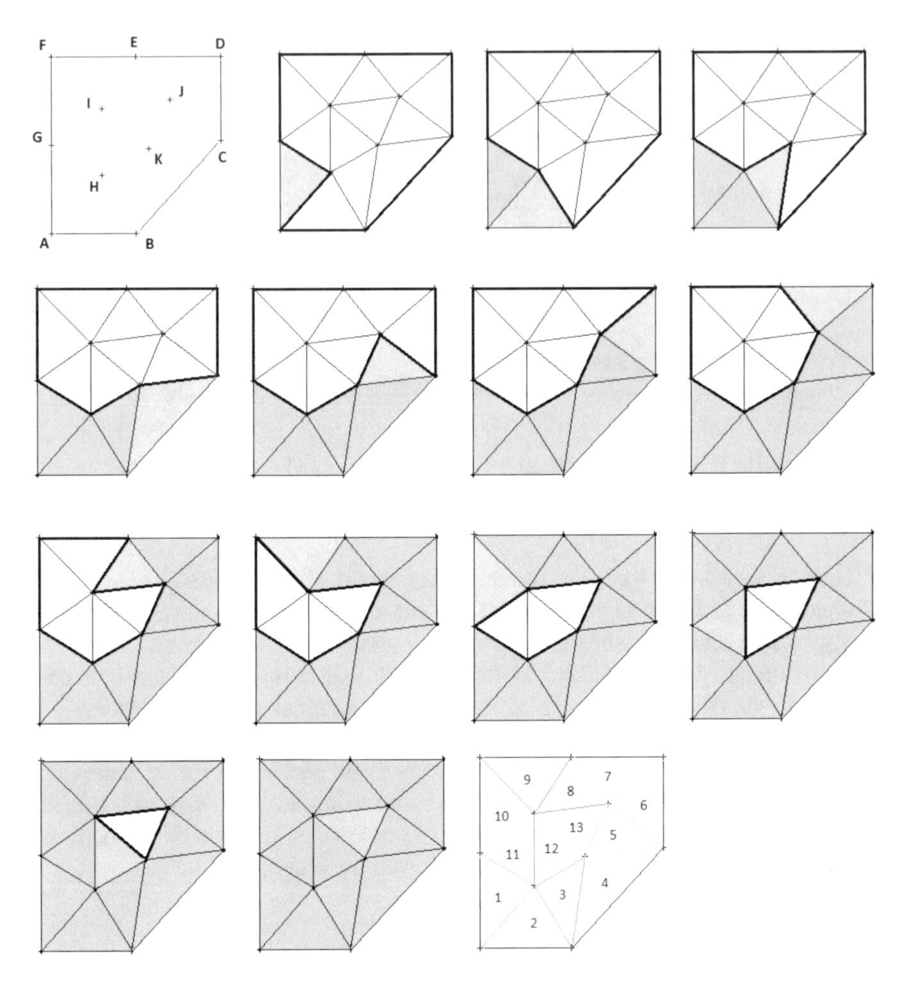

Fig. 6.21 Triangulation of the nodes A to K using the advancing front method. The step-wise progression of the method is shown from left to right. The sequence of creation of triangles is given in the last sub-figure. For the sake of clarity, the newly created triangle is shown in blue colour and those already created are shown in green colour

formulating the pressure correction equation (see Sect. 4.4.4) is obtained through the Rhie-Chow interpolation scheme (Rhie and Chow 1983); here, it can be obtained as

$$u_f = \bar{u}_f + \bar{D}_f\left[(\nabla p)_f - \left(\overline{\nabla p}\right)_f\right] \tag{6.3.16}$$

where the overbar indicates an interpolated quantity and the subscript f stands for the face location. The first pressure gradient term on the right hand side, namely, $(\nabla p)_f$, is evaluated using the pressure at cell nodes on either side of the face f. This brings in the desired influence of the cell pressure into the pressure correction equation and avoids the chequerboard oscillations in pressure and velocity that are otherwise associated with the use of non-staggered grids.

Integration of the governing partial differential equation (pde) in one cell results in one algebraic equation per cell; doing this over all the cells gives rise to one system of algebraic equations for one partial differential equations. For fluid flow problems, there are four or more coupled partial differential equations that need to be solved together, and therefore the systems of algebraic equations are coupled and non-linear. Using Picard's substitution or Newton's linearization, these algebraic equations can be linearized so as to reduce each set to the form Ax = b. In general, the coefficient matrix A is not diagonal although it is sparse. Basic iterative methods such as the Gauss-Seidel method and the SOR method can be readily applied to solve these. While some diagonal structure-based methods such as the ADI and the SIP cannot be used for unstructured grids, some advanced linear algebraic equation solvers such as the multigrid method can be used for unstructured grid formulations. The reader is referred to references given in Chap. 5.

Due to its attractive features such as ease grid generation, grid adaptation, ability to tackle large, irregular geometries, the unstructured grid-based solution has become the choice by default over the past two decades for the solution of Navier-Stokes equations for flow in irregular geometries has gained popularity quickly in the process engineering community where flows through complicated flow domains are encountered.

6.4 Dealing with Complex Physics

While Navier-Stokes equations are at the core of fluid flow and CFD, there are many cases when they are insufficient to represent the phenomena of interest and need to be supplemented by additional equations. This situation is discussed briefly in this section.

6.4.1 Why Modelling Is Necessary

The fundamental equations governing the motion of a fluid are the equations describing the conservation of mass, momentum and energy. For an isothermal, incompressible flow of a Newtonian fluid, the Navier-Stokes equations and the continuity equation along with appropriate boundary conditions completely describe the flow. For compressible flows, the energy conservation equation and an equation of state for the fluid will be required additionally. Since the equations have been derived for a vanishingly small control volume and represent fundamental conservation laws and statements of laws of thermodynamics, they will be valid for all flows irrespective of whether or not the flow is steady or turbulent or compressible. However, two assumptions have been made in deriving these equations, viz., the continuum assumption so that the equations may not be applicable for very rarified gases, and a Newtonian fluid assumption expressing a linear relationship between stress and strain rate for an isotropic fluid. In addition, some effects such as those related to electromagnetism, nuclear reactions etc. have not been considered. An inertial coordinate system has been assumed throughout although extension to a non-inertial coordinate system is fairly straightforward (see, for example, Hirsch 1988; White 1991; Warsi 2006). Apart from these restrictions, these equations have universal validity. However, they have to be augmented with additional mathematical relations—algebraic or ordinary or partial differential equations which are, in many cases, empirical and are not as widely applicable as the above laws—in a number of practical cases to represent more "physics" of the flow and related phenomena. Such mathematical modelling is necessary in the following cases:

- When additional effects, not considered hitherto, are present. Examples of these are the presence of chemical reactions and radiative heat transfer, both of which may contribute to the energy equation. The simulation of a mixing problem, a common problem in the process industries, requires additional model equations (Jayanti 2001).
- When relations incorporated therein are not applicable. Here, specific mention can be made of the Newtonian fluid assumption which breaks down for a number of fluids which exhibit a more complicated rheological behaviour than that exhibited by a Newtonian fluid.
- When the temporal or spatial complexity of an actual fluid flow need not be resolved to the smallest scale found in practice, for example, in turbulent flow where flow variables exhibit high frequency, temporal fluctuations; in flow through porous media where the flow domain consists of millions of minute pores; in gas-solid reactions involving finely dispersed catalysts. In all these cases, one is not interested in the fine details of the temporal or spatial structure of the flow (though these are important for the physics or chemistry of the problem), and one would be happy with a macro-level description of the flow. This requires replacing the original equations with equivalent relations representing the overall effect at grid-resolved scales.
- When more than one phase is present. Here, although the fundamental equations are valid for each phase, the equations for a mixture of the two phases are different because, say, as in the case of flow boiling and fluidized beds, interfacial phenomena may be occurring at physical and temporal scales that cannot be resolved in practical applications.
- When electrochemical phenomena, such as those occurring in fuel cells and flow batteries, are involved. Here, the reaction rate is influenced also by the electric field which needs to be resolved along with flow, concentration and temperature fields.

In all these cases, additional terms and equations should be added to the set of basic and fundamental equations mentioned above. Given the complexity of the phenomena, often, these additional terms describe a simplified "mathematical model" of the actual phenomena. Typically, these mathematical models require some empirical information (in fact, many models are data-driven and need to be calibrated to ensure accuracy in a narrow range of parameters) to close the system of equations. These closure relations have a limited range of applicability and one should be aware of their limitations in applying these models. A thorough discussion of these models is beyond the scope of the present book; the basic modeling approaches in dealing with turbulent reacting flows are briefly discussed below.

6.4.2 Turbulence Modelling

Turbulence is a three-dimensional unsteady viscous flow phenomenon that occurs at high Reynolds numbers. It is characterized by rapid and highly localized fluctuations in flow parameters such as velocity components, pressure, temperature, species concentration etc. These fluctuations are generated in regions of high shear (such as near a wall or in a mixing layer) by an internal instability mechanism that is characteristic of the flow. The fluctuations are thus generated, sustained and regulated by the flow itself. They persist as long as the flow is maintained. Although the fluctuations are small in magnitude (typical amplitude of the velocity fluctuations is in the range of 1–10% of the mean velocity), their effect on the flow can be enormous. Flow-dependent quantities such as the pressure drop and heat transfer rate can be an order of magnitude higher in turbulent flow. Transport properties such as friction factor, heat and mass transfer coefficients and reaction rates can change by orders of magnitude compared to the value that would prevail if the flow remained laminar at the same Reynolds number. In many cases, the extent of enhancement (or suppression in some rare cases such as the drag on a sphere around transition Reynolds numbers) of the transport coefficient depends on the Reynolds number itself in a non-trivial way. It is therefore important to get an accurate representation of these turbulent fluctuations into flow analysis.

Turbulent fluctuations of different velocity components are not entirely random; a persistent feature of turbulent flow (when viewed appropriately, see Schlichting (1960), Lesieur (1986), Frisch (1995)) is the presence of three-dimensional eddies of different sizes that are convected by the mean flow. These eddies are present everywhere in the flow domain on average and randomly appear, persist for some time and slowly disappear. Thus, unlike in the case of steady laminar flow, a fluid particle or any other particle with sufficiently small inertia does not follow the same path through apparently steady turbulent flow. If ten such particles are released from the same location in ten identical experiments, one may see ten different paths. The particle need not be small; an aeroplane flying through a turbulent air pocket will also experience a fluctuating amount of drag and cause the passengers a lot of discomfort! A simplistic view of turbulence is that these eddies are created by flow instabilities in regions of high mean shear (velocity gradients) such as near walls and in mixing layers, and are then ejected into the bulk motion where they break up or die away through interaction with the mean flow and with each other. One important consequence of the presence of eddies is the occurrence of rapid convective (rather than diffusive) mixing of different layers of the fluid.

One can get an idea of the intensity of turbulence by decomposing the instantaneous velocity, u, (Fig. 6.22a) into a time-averaged velocity, U, and a fluctuating component, u′ (see Fig. 6.22b). When the time-averaged velocity does not change with time, then the turbulent flow can be assumed to be steady even though the instantaneous flow variables exhibit temporal fluctuations. From this decomposition, one can define turbulent kinetic energy, k, as $k = \frac{1}{2}\overline{u'_i u'_i}$ where u'_i is fluctuating part of the ith velocity component. Since the index i is repeated, it implies summation over the repeated index i.

The level of turbulent kinetic energy associated with an eddy size exhibits a characteristic variation which is shown schematically in Fig. 6.23. Here, the variables are expressed in the spectral domain. Using fast Fourier transform techniques, u′(t) is converted into frequency domain and the "energy" associated with each wave component (which is proportional to the square of the amplitude of that particular wave component) is plotted in the figure as a function of the wave number. The wave number is inversely proportional to the wavelength of the wave component, and can be loosely associated with the inverse of eddy size. The spectral energy associated with the wave number can be a measure of the square of the velocity fluctuation. This turbulent energy spectrum represents an energy cascade which is anchored at the low wave number end by the large "energy containing eddies", typically of the size of L/6 where L is the characteristic length scale of the flow, and at the high wave number range by the Kolmogorov length scale ~ $(\nu^3/\varepsilon)^{1/4}$ where ν is the kinematic viscosity and ε is the energy dissipation rate (see below for a quantitative definition). In steady turbulent flow, turbulent kinetic energy is drawn from the mean flow, cascaded down the eddy sizes and is dissipated in the smallest eddies. The mechanism of turbulent kinetic energy generation by the large eddies is dependent on the mean velocity gradients, and is therefore specific to the type of flow (for example, flow through a pipe, wake flow behind a bluff body, boundary layer flow over an aerofoil etc.) and to the characteristic velocity and length scales of the flow. The mechanism of turbulent energy dissipation rate, which occurs predominantly in very small eddies, is expected to be universal in nature and said to depend only on the local properties. A thorough discussion of the scales and structure of turbulence can be found in books such as those of Tennekes and Lumley (1981), Reynolds (1987), Pope (2000) and Davidson (2004).

There are three principal approaches to the prediction of turbulent flows based on what aspects of the turbulence structure are resolved. In the direct numerical simulation (DNS) approach, the unsteady Navier-Stokes equations are solved using highly accurate numerical schemes on a very fine spatial grid so that all characteristic time and length scales are resolved. Though fundamentally correct, the computing effort required for this approach, and the restrictions on the geometry of the flow domain and the boundary conditions that go with the methods used, restrict its applicability mainly to academic research. In the large eddy simulation (LES) simulation approach, the large scale structures of turbulence are fully resolved and computed while the Kolmogorov-scale structures are modeled using sub-grid

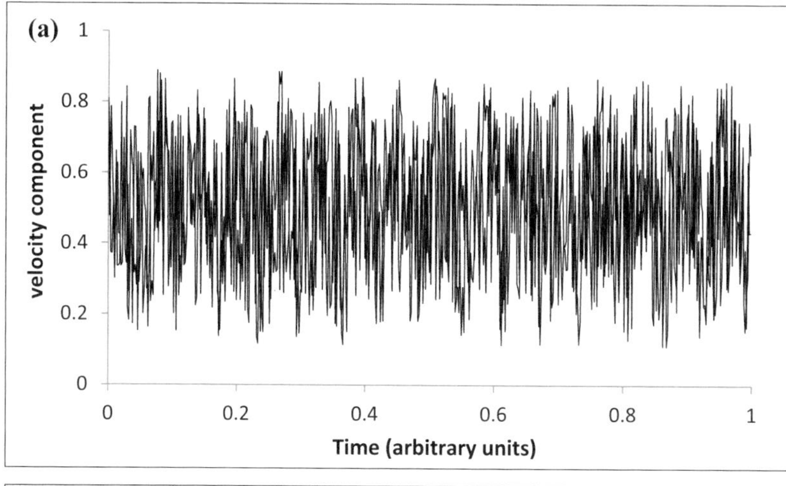

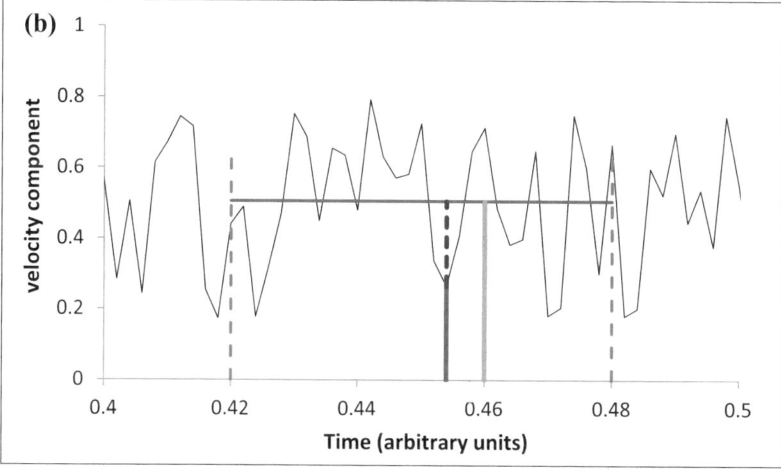

Fig. 6.22 **a** Typical variation of a velocity component, u(t), in turbulent flow, and **b** time-averaging to decompose it into a mean (U) and a fluctuating component, u′(t)

closure relations. Since only large eddies are resolved, LES works with relatively coarse grids compared to DNS but shares its disadvantage of requiring three-dimensional, unsteady computations even when these are not warranted from an application point of view. The Reynolds-averaged Navier Stokes equations (RANS) approach (or its equivalent of Favre-averaged Navier-Stokes equations approach for compressible flows) is the most widely used of the three approaches used when dealing with industrial flow problems.

The RANS approach is based on the solution of time-averaged conservation equations that obey the macro-dimensionality of a problem in the sense that if the

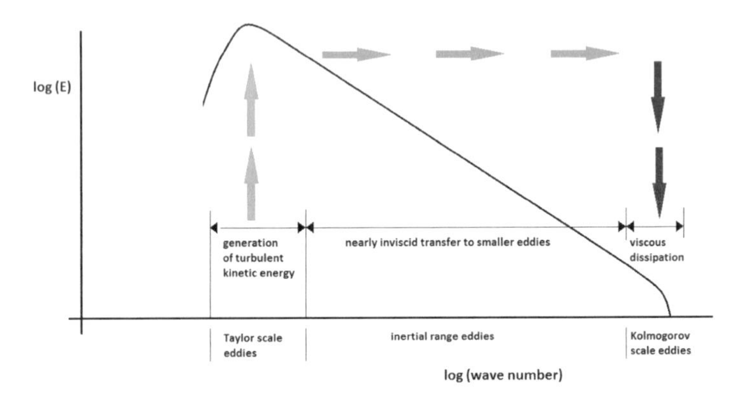

Fig. 6.23 Energy spectrum in turbulent flow: turbulent kinetic energy (E) is produced by Taylore-scale instability mechanisms and is cascaded down the inertial range to be dissipated in the Kolmogorov-scale eddies

flow is essentially steady (because the driving force and the boundary conditions are time-invariant) and, say, two-dimensional (because there is no driving force in the third dimension), then the flow variables of interest are obtained by solving the steady-state, two-dimensional form of the time-averaged governing equations. Since time-averaging leads to a loss of information about the fluctuations and the interplay between the mean and the fluctuating components that is encapsulated in the energy cascade shown in Fig. 6.23, the solution obtained from the RANS approach is not inherently accurate and needs to be bench-marked and calibrated.

In turbulent flow, an instantaneous scalar variable can be decomposed into a mean component (which is indicated by an overbar over the variable) and a fluctuating component:

$$f = \bar{f} + f' \quad \text{where } \bar{f}(t) = \frac{1}{T} \int_{t-T/2}^{t+T/2} f(t)dt \tag{6.4.1}$$

This is called Reynolds decomposition. It is easy to show from the definition of the time averaging operator, that if *f* and *g* are two scalars, then the following time-averaging rules apply:

$$\overline{f+g} = \bar{f} + \bar{g} \tag{6.4.2a}$$

$$\overline{\partial f/\partial x_i} = \partial \bar{f}/\partial x_i \tag{6.4.2b}$$

$$\overline{\partial f/\partial t} = \partial \bar{f}/\partial t \tag{6.4.2c}$$

$$\overline{fg} = \bar{f}\bar{g} + \overline{f'g'} \tag{6.4.2d}$$

Brief comments are necessary regarding the last two relations. Equation (6.4.2c) is applicable for transient variations with characteristic time periods of variation that are much larger than those associated with the slowest turbulent fluctuations, which typically correspond to the turnover time of the largest eddies. These eddies have a characteristic eddy velocity of ~5% of the mean flow velocity, and a large eddy size of ~L/6. For water flow through a pipe of 0.05 m diameter at a Reynolds number of 100,000, the mean flow velocity is 2 m/s. Thus, the characteristic large eddy life time is given by eddy size/ eddy velocity or (0.05/6)/(0.05 * 2) or ~0.08 s! This is a very short time duration, and we can expect Eq. (6.4.2c) to be applicable for typical system transients. The second term on the RHS of Eq. (6.4.2d) will be close to zero if the fluctuations of f and g are statistically independent. If the two fluctuations are somehow related and thus not entirely random, then $\overline{f'g'}$ can be either positive or negative. For example, in near-wall fluid flows, the quantity $\overline{u'v'}$ is non-zero and negative.

Using the above rules of time-averaging, the continuity and the Navier-Stokes equations (assuming incompressible flow with constant thermophysical properties) can be time-averaged. The resulting equations are

$$\partial \bar{u}_m / \partial x_m = 0 \tag{6.4.3}$$

$$\partial \bar{u}_i / \partial t + \partial \left(\bar{u}_i \bar{u}_m + \overline{u'_i u'_m} \right) / \partial x_m = -\frac{1}{\rho} \partial \bar{p} / \partial x_i + \nu \partial^2 \bar{u}_i / \partial x_m \partial x_m \tag{6.4.4}$$

These are called Reynolds-Averaged Navier-Stokes (RANS) equations. They have the same general form as the instantaneous Navier-Stokes equations, with the velocities and other solution variables now representing time-averaged variables. However, the LHS of Eq. (6.4.4) contains an additional term, namely, $\overline{u'_i u'_m}$, which in fact represents six terms, namely, $\overline{u'u'}$, $\overline{v'v'}$, $\overline{w'w'}$, $\overline{u'v'}$, $\overline{u'w'}$ and $\overline{v'w'}$. In the general case, these are non-zero and vary spatially and with time in a non-trivial manner. Thus, time-averaging of the four governing equations containing four unknowns leads to a four time-averaged equations containing a total of ten unknowns. This is not surprising because, by filtering out the rapid fluctuations, the information on how these influence each other and the mean flow is lost and this has to be brought back into the equations somehow. No amount of further manipulation of the time-averaged equations will fill this gap. This is known as turbulence closure problem. It is necessary to introduce some "turbulence modelling" that goes beyond these fundamental equations in order to "close" the problem.

An insight into the nature of the additional variables can be gained by rewriting Eq. (6.4.4) as follows:

$$\partial \bar{u}_i/\partial t + \partial(\bar{u}_i\bar{u}_m)/\partial x_m = -\frac{1}{\rho}\partial\bar{p}/\partial x_i + \partial/\partial x_m\left(\nu\partial\bar{u}_i/\partial x_m - \overline{u'_i u'_m}\right) \qquad (6.4.5)$$

The fully-expanded x-momentum balance equation thus reads as

$$\begin{aligned}\partial\bar{u}/\partial t + \partial\left(\bar{u}^2\right)/\partial x + \partial(\bar{u}\bar{v})/\partial y + \partial(\bar{u}\bar{w})/\partial z = &-\frac{1}{\rho}\partial\bar{p}/\partial x + \partial/\partial x\left(\nu\partial\bar{u}/\partial x - \overline{u'u'}\right) \\ &+ \partial/\partial y\left(\nu\partial\bar{u}/\partial y - \overline{u'v'}\right) + \partial/\partial z\left(\nu\partial\bar{u}/\partial z - \overline{u'w'}\right)\end{aligned} \qquad (6.4.6)$$

This juxtaposition of the $\overline{u'_i u'_m}$ terms with the viscous stress terms in the ith momentum equation highlights the role of these terms as equivalent "turbulent" stresses which contribute to the transport of ith-direction momentum in the mth direction. A simplistic but effective modeling of these terms is therefore through the Boussinesq hypothesis, namely,

$$-\overline{u'_i u'_m} = \nu_t(\partial\overline{u_i}/\partial x_m + \partial\overline{u_m}/\partial x_i) - \frac{1}{3}\delta_{im}\overline{u'_n u'_n} \qquad (6.4.7)$$

where δ_{im} is the Kronecker delta function. For example, we have, from Eq. (6.4.7),

$$-\overline{u'u'} = 2\nu_t\left(\frac{\partial\bar{u}}{\partial x}\right) - \frac{2}{3}k \quad \text{where } k = \frac{1}{2}\overline{u'_n u'_n} = \frac{1}{2}\left(\overline{u'u'} + \overline{v'v'} + \overline{w'w'}\right) \qquad (6.4.8a)$$

$$-\overline{u'v'} = \nu_t\left(\frac{\partial\bar{v}}{\partial x} + \frac{\partial\bar{u}}{\partial y}\right) \qquad (6.4.8b)$$

The last term in Eq. (6.4.7) is needed so as to ensure that the sum of the normal stresses, namely, $\overline{u'_n u'_n}$, is not zero. Here ν_t is the turbulent or eddy kinematic viscosity and needs to be determined. A constant and large value (compared to the viscosity of the fluid itself to account for the much larger pressure drop in turbulent flow) of ν_t would not be correct as this would make the time-averaged turbulent flow equations mathematically identical to the laminar flow equations leading to similar velocity profiles. For example, for the specific case of steady, fully-developed one-dimensional flow between two plates, Eq. (6.4.7), together with Eq. (6.4.8b), reduces to

$$0 = -\frac{1}{\rho}\frac{d\bar{p}}{dx} + \frac{d}{dy}\left(\nu\frac{d\bar{u}}{dy} - \overline{u'v'}\right) = -\frac{1}{\rho}\frac{d\bar{p}}{dx} + \frac{d}{dy}\left(\nu_{eff}\frac{d\bar{u}}{dy}\right) \qquad (6.4.9)$$

where $\nu_{eff} = \nu + \nu_t$. Solution of Eq. (6.4.9) for flow between parallel plates would give a laminar-like velocity profile with a maximum to average velocity ratio of 1.5 instead of ~ 1.15 that is expected in turbulent flow.

One of the early successes in turbulence modeling is the mixing length model of Prandtl (1925). For the one-dimensional flow case, the Reynolds stress term in Eq. (6.4.9) is modeled as

$$-\overline{u'v'} = l_m^2 \left|\frac{d\bar{u}}{dy}\right| \left(\frac{d\bar{u}}{dy}\right) \tag{6.4.10a}$$

or

$$\nu_t = l_m^2 |d\bar{u}/dy| \quad \text{where } l_m = \kappa y \text{ and } \kappa \approx 0.4 \tag{6.4.10b}$$

Here, y is the normal distance from the wall and l_m is the so-called mixing length. This expression is found to be valid in the near-wall region in many flows. Here, the argument goes that a positive v' brings in a fluid particle from y to $y + \Delta y$ where the velocity is higher. This mixing of fluid particles of different velocities causes a momentary velocity defect, u', which can be estimated as $u' \approx \bar{u}(y) - \bar{u}(y + \Delta y) \approx -(d\bar{u}/dy)\Delta y$. Since in turbulent flow, $u' \sim v'$ in terms of magnitude, and assuming that $\overline{u'v'} \sim C \,\|u'\|\,\|v'\|$, where C is a constant, one can write

$$-\overline{u'v'} \sim C\left(\frac{\partial \bar{u}}{\partial y}\Delta y\right)\left(\frac{\partial \bar{u}}{\partial y}\Delta y\right) \equiv l_m^2 \left|\frac{d\bar{u}}{dy}\right| \left(\frac{\partial \bar{u}}{\partial y}\right)$$

Similar expressions for the mixing length have been proposed for other flows such as for jets, wakes, fully developed flows in circular pipes etc. The primary disadvantage with this model is the need to specify the mixing length, the meaning of which becomes uncertain in three-dimensional flows. Also, the model is incapable of accounting for advection and diffusion of turbulence. It presumes that all turbulence is locally generated whereas it may have been generated elsewhere and brought to the point of interest by the fluid dynamic processes of advection and diffusion.

The simplest level of turbulence modeling that addresses both these shortcomings is the two-equation model of turbulence. There is a plethora of these (see Wilcox 1993); the earliest successful version is the *k*-ε model in which the turbulent viscosity is modeled as

$$\nu_t = C_\mu k^2/\varepsilon \tag{6.4.11}$$

where *k* and ε are the turbulent kinetic energy and its dissipation rate, respectively, and are defined as follows:

$$k = \tfrac{1}{2}\overline{u'_m u'_m} \tag{6.4.12a}$$

$$\varepsilon = \overline{\nu \partial u'_i/\partial x_m \partial u'_i/\partial x_m} \tag{6.4.12b}$$

Both k and ε are scalar field variables and can vary with position and time. They are flow properties and are non-zero in turbulent flow. From the definition, one can infer that the contribution to *k* comes primarily from the energy containing eddies of the turbulent energy spectrum while the contribution to ε comes primarily from the Kolmogorov length scale eddies where much of the dissipation takes place. However, these eddies are not explicitly resolved in the k-ε turbulence model. Conservation equations in the form of a transport equation can be developed for both k and ε (see Warsi 2006); however, a number of new variables involving products of fluctuating components, their derivatives and the mean flow variables are introduced in the process. Through further approximations and modeling, these equations are reduced to the following transport equation form:

$$\begin{aligned}\partial k/\partial t+\partial(\bar{u}_m k)/\partial x_m &= \partial/\partial x_m\left[\left(\nu+\frac{\nu_t}{\sigma_k}\right)\partial k/\partial x_m\right]\\&\quad+\left[\nu_t(\partial\overline{u_i}/\partial x_m+\partial\overline{u_m}/\partial x_i)-\frac{2}{3}k\delta_{im}\right]\partial\overline{u_i}/\partial x_m-\varepsilon\end{aligned}\tag{6.4.13}$$

$$\begin{aligned}\partial\varepsilon/\partial t+\partial(\bar{u}_m\varepsilon)/\partial x_m &= \partial/\partial x_m\left[\left(\nu+\frac{\nu_t}{\sigma_\varepsilon}\right)\partial\varepsilon/\partial x_m\right]\\&\quad+C_{1\varepsilon}\frac{\varepsilon}{k}\left[\nu_t(\partial\overline{u_i}/\partial x_m+\partial\overline{u_m}/\partial x_i)-\frac{2}{3}k\delta_{im}\right]\partial\overline{u_i}/\partial x_m-C_{2\varepsilon}\frac{\varepsilon^2}{k}\end{aligned}\tag{6.4.14}$$

The constants appearing in the model have been determined primarily through calibration of model predictions for a variety of cases. To some extent, therefore these constants vary from researcher to researcher but one set of values is as follows: $C_{1\varepsilon} = 1.44$; $C_{2\varepsilon} = 1.92$; $C_\mu = 0.09$; $\sigma_k = 1.0$; $\sigma_\varepsilon = 1.3$. The last two constants can be viewed as turbulent "Prandtl" numbers expressing the turbulent diffusion of *k* and ε, respectively, relative to turbulent momentum diffusion.

This "standard" k-ε model is applicable only in the highly turbulent region of the flow. It does not account for the suppression of turbulence in the region immediately adjacent to a wall. Therefore this model has to be used with wall functions which can be adjusted to account for wall roughness (Jayanti et al. 1990). Other variants of these have been proposed. These include a low Reynolds number k-ε turbulence model (Jones and Launder 1972) which accounts for relaminarization in regions of high shear, the renormalization group theory (RNG)-based k-ε model (Yakhot and Orszag 1986) and the k-ω model (Wilcox 1988, 1993) where ω, the specific dissipation rate, defined as dissipation rate (ε) per unit turbulent kinetic energy or ε/k, is used as the resolved scalar in place of ε. The low Reynolds number k-ε model and the k-ω model do not use wall functions and can be used, with a very fine grid, to calculate the velocity field right up to the wall. An entirely different approach is the Reynolds stress modelling (Launder 1989) in which transport equation-type conservation equations are derived for the Reynolds stresses without having to model these using the Boussinesq hypothesis (Eq. 6.4.7). Despite the modelling assumptions and simplifications made to deal with a large number of higher moment closure terms that arise in the process of derivation of the conservation equations, this approach is more fundamentally correct in dealing with the effect of external force fields on turbulence than the k-ε family of turbulence models. However, it requires the solution of seven partial differential equations that are coupled tightly leading sometimes to convergence problems during numerical solution.

Thus, turbulence requires considerable additional modelling effort. From a solution point of view, a generic turbulence model such as the k-ε model requires the solution of two additional partial differential equations along with the continuity and the momentum balance equations. These are handled in a relatively straightforward way in a sequential solution approach like that of the SIMPLE family of methods discussed in Sect. 4.3.4.

6.4.3 *Modelling of Reacting Flows*

A chemical reaction involves exchange or sharing of electrons between atoms and molecules leading to the formation or destruction of chemical species. We note that for a chemical reaction to occur, the reacting species will have to be brought into intimate contact. This is possible only if at least one of the reactants is a fluid. If the effects of chemical reactions are important to be resolved—either because they have a profound effect on the properties of the fluids or because it is the formation of the products that is important–then the laws governing fluid flow, namely, the continuity and the Navier-Stokes equations, need to be supplemented with additional equations which determine the concentration of the species that are participating in the chemical reactions, and the rate and extent of the chemical reactions. Since chemical reactions often involve release or absoption of heat, the energy balance equation for the fluid also gets affected. In this section, we look at these effects, and see how these can be accounted for mathematically.

The change in the concentration of a chemical species can be obtained from a mass conservation equation for that species. For a fluid containing a mixture of species, the mass conservation equation for species *i* can be written as

$$\partial \rho_i / \partial t + \nabla \cdot (\rho_i \mathbf{u}_i) = S_i \qquad (6.4.15)$$

where ρ_i is the density of the ith species, $\mathbf{u}_i$ is *its* velocity and S_i is the rate (in kg/m^{3}s) of generation of species i during the chemical reaction. This is similar to the continuity equation for a fluid except that the variables pertain to a specific species. In this form, the equation is not very useful because one needs to determine the velocity of each species. A mixture formulation would enable the individual species to be differentiated while the mixture as a whole moves with a single velocity. To this end, we define the following properties of the mixture:

$$\text{mass density} = \rho = \sum_{\alpha} \rho_{\alpha} \tag{6.4.16a}$$

$$\text{mass fraction of species } \alpha = \mathrm{Y}_{\alpha} = \rho_{\alpha}/\rho \tag{6.4.16b}$$

$$\text{mass velocity} = \mathbf{u} = \sum_{\alpha} (\rho_{\alpha}\mathbf{u}_{\alpha})/\rho \tag{6.4.16c}$$

Using the Fick's law of diffusion (Bird et al. 2002) which states that

$$\mathbf{J}_{\mathrm{i}} = \rho_{\mathrm{i}}(\mathbf{u}_{\mathrm{i}} - \mathbf{u}) = -\Gamma\nabla \mathrm{Y}_{\mathrm{i}} \tag{6.4.17}$$

where $\mathbf{J}$ is the diffusive flux of species i, and Γ is the mass diffusivity of the species, one can derive the more useful form of species balance equation in terms of mixture properties:

$$\partial\rho_{\mathrm{i}}/\partial \mathrm{t} + \nabla\cdot(\rho_{\mathrm{i}}\mathbf{u}) = \nabla\cdot(\Gamma\nabla \mathrm{Y}_{\mathrm{i}}) + \mathrm{S}_{\mathrm{i}} \tag{6.4.18}$$

or

$$\partial(\rho \mathrm{Y}_{\mathrm{i}})/\partial \mathrm{t} + \nabla\cdot(\rho \mathrm{Y}_{\mathrm{i}}\mathbf{u}) = \nabla\cdot(\Gamma\nabla \mathrm{Y}_{\mathrm{i}}) + \mathrm{S}_{\mathrm{i}} \tag{6.4.19}$$

The evaluation of diffusivity for a binary mixture, i.e., one which contains only two species, is relatively straightforward. For gases, the diffusivity depends only the pressure, temperature and the species involved. When there are three or more species, the diffusivity also depends on the concentrations, and a multicomponent mixture formulation (Bird et al. 2002) is required to evaluate the diffusivity that appears in Eq. (6.4.17). For a liquid mixture, simple theoretical estimates can be made for dilute mixtures; the diffusivity is often estimated empirically and is required as input to the CFD solution.

One may note that Eq. (6.4.19) is in terms of the species mass fractions and the mixture density. Summing Eq. (6.4.18) over all species and noting that $\Sigma_{\alpha}\mathbf{J}_{\alpha} = \Sigma_{\alpha}\mathrm{S}_{\alpha} = 0$, the mixture continuity equation can be written in terms of mixture density and mass velocity as

$$\partial\rho/\partial \mathrm{t} + \nabla\cdot(\rho\mathbf{u}) = 0 \tag{6.4.20}$$

This equation is the same as that for a single fluid with the same density and velocity. For a mixture of gases containing N_{s} species, the mass or species conservation requires the solution of Eq. (6.4.20) for the mixture density and $(\mathrm{N}_{\mathrm{s}} - 1)$ number of species mass conservation equations of the type of Eq. (6.4.19).

The momentum balance equation for the mixture as a whole remains the same as that for a single fluid except that the viscosity should be that of the mixture. It may be noted that significant density and velocity changes may arise due to chemical reactions involving concentration changes even in hydrodynamically incompressible flows. As a result, the density and diffusivities cannot be assumed to be constant even if they are constant for a given species. The mixture momentum equation for a reacting should therefore be written, even for incompressible flows with constant mixture thermophysical properties, as

$$\partial(\rho\mathbf{u})/\partial t + \nabla\cdot(\rho\mathbf{u}\mathbf{u}) = -\nabla p + \nabla\cdot\{\mu(\nabla\mathbf{u} + (\nabla\mathbf{u})^{T})\} + \rho\mathbf{g} \tag{6.4.21}$$

The energy balance equation for the mixture needs to be modified to account for the heat liberated or absorbed during the chemical reaction:

$$\partial(\rho h_t)/\partial t + \nabla\cdot(\rho\mathbf{u}h_t) = \partial p/\partial t + \nabla\cdot(\lambda\nabla T) + \nabla\cdot\{\mu[\nabla\mathbf{u} + (\nabla\mathbf{u})]^{T}\}\mathbf{u} + Q_R \tag{6.4.22}$$

where h_t is the total enthalpy defined as $h_t = h + ½\,\mathbf{u}\cdot\mathbf{u}$, λ is the mixture thermal conductivity of the fluid and Q_R is the rate of heat release from chemical reaction. The static enthalpy is evaluated in terms of the specific heat for a mixture of gases assuming that the constituent species are thermally perfect, i.e., their static enthalpies are functions only of temperature. If this is not the case, then the enthalpies of the individual species need to be determined and the enthalpy of the mixture is evaluated as $h = \sum_{\alpha}(Y_{\alpha}\,h_{\alpha})$.

The evaluation of the rate of generation of species, S_i, and the heat source term Q_R due to chemical reaction can be described in the context of a generic chemical reaction scheme involving N_S number of species in N_R number of reactions. A typical chemical reaction may involve a number of intermediate reactions, and a particular chemical species may participate as a reactant in one reaction and appear as a product in a different reaction. The jth-reaction in this scheme can be written as

$$a_{r1j}X_1 + a_{r2j}X_2 + \ldots \Leftrightarrow a_{p1j}X_1 + a_{p2j}X_2 + \ldots \quad \text{where } i = 1, 2, \ldots N_S \tag{6.4.23}$$

where a_{rij} and a_{pij} are the stoichiometric coefficients of reactant species X_i. The reaction rate R_j for the jth reaction is modeled as a function of the molar concentrations of the reactants and the products as

$$R_j = k_{fj}\prod_i [X_i]^{\alpha_{ij}} - k_{bj}\prod_i [X_i]^{\beta_{ij}} \tag{6.4.24}$$

where $[X_i]$ is the molar concentration of species i, k_{fj} and k_{bj} are the forward and backward jth reaction rate constants and α_{ij} and β_{ij} are the forward and the backward rate exponents for species i in the jth reaction. The rate constants are evaluated in terms of an Arrhenius law involving activation energy and a temperature-dependent pre-exponential factor (Levenspiel 1972). In some cases, an overall, irreversible, single-step reaction with a specified rate may also be given in a different format. It is expected that the stoichiometry of the chemical reaction and the rate of reaction are known.

In a reacting fluid, the reactants get consumed as a result of the reaction and these need to be supplemented by a steady flux of reactants in order to keep the reaction going. In the limiting case where the reactant species are intimately pre-mixed, the reaction front will propagate into the fluid continuum at a steady rate, which, in case of gaseous combustion, is known as laminar flame velocity, which among other parameters is determined by the diffusivity of the species involved. Another limiting case is that of a steady reaction front in which the reactant species are fed into the reactor at a steady rate (which is known as the diffusion or non-premixed flame case in gaseous combustion) . Here, the rate of reaction is also limited by the rate at which the reactant species can diffuse to the flame front. Thus, in practical cases, the reaction rate may be influenced by both chemical kinetic considerations and reactant or product diffusion rates. In asymptotic conditions, one of these two effects dominates. For example, at very low temperatures, it is the chemical kinetics and activation of the reaction that determines the effective reaction rate. At sufficiently high temperatures, no further acceleration of the reaction rate is possible by increasing the temperature, and the rate may be limited only by diffusion of the active species. Increasing the concentration may then help in increasing the reaction rate. In-between these two asymptotic cases, one may have both effects contributing to the reaction rate. If k_c is the chemical kinetics-limited estimate of the reaction rate and k_d is the diffusion-limited reaction rate, then the effective reaction rate may be determined as $k_{eff} = (1/k_c + 1/k_d)^{-1}$.

With this formulation, the source term S_A for species A in Eq. (6.4.15) can be written for the entire set of chemical reactions as

$$S_{A} = M_A \sum_j (n_{Aj} R_j) \quad j = 1, 2, \ldots, N_R \tag{6.4.25}$$

where M_A is the molecular weight of species A and $n_{Aj} = (a_{pAj} - a_{rAj})$ is the overall stoichiometric coefficient for species A and in reaction j. The source term Q_R due to chemical reaction in the energy equation is given in terms of heat of jth reaction (ΔH_{Rj}) as

$$Q_R = -\sum_j (R_j \, \Delta H_{Rj}) \quad j = 1, 2, \ldots, N_R \tag{6.4.26a}$$

with

$$\Delta H_{Rj} = \sum_i \left(a_{pij} h_i - a_{rij} h_i \right) \tag{6.4.26b}$$

Thus, when dealing with a reacting flow of a fluid containing a mixture of N_S constituent species, one would need to solve one mixture continuity equation, one mixture momentum equation, one energy equation and (N_S − 1) species mass conservation equations. These would need to be supplemented with the equation of state for the mixture and the reaction scheme along with the necessary chemical kinetic parameters. Thus, the number of equations solved increases as the number of active species increases. In addition, a tighter coupling of the equations ensues because the properties of the fluid change with changing composition. Finally, the highly non-linear rate expressions and the large amount of heat exchanges can compromise the convergence of the iterative process.

6.4.4 Turbulent Combustion

Combustion of a hydrocarbon fuel is a very rapid reaction resulting in very high temperatures. Due to the high temperatures involved, the reactants and products can be present in dissociated or meta-stable states, and therefore such reactions typically involve hundreds of interconnected intermediate chemical reactions. In principle, the corresponding species transport equations can be solved using the generalized finite rate formulation discussed above. However, in practice, coupling the full chemistry with the flow equations for three-dimensional flows is extremely difficult. Turbulent flow with its fluctuations brings in additional difficulties:

- Firstly, since the chemical kinetic reaction rate is a strong, non-linear function of temperature, fluctuations in temperature lead to fluctuations in the instantaneous reaction rate. If the reaction time scale is significantly less than that of the eddy, then the reaction rate will also show wild oscillations which may be even more amplified than those of temperature. Therefore, the reaction rate evaluated using the mean (time-averaged) temperature may be significantly different from the time average of the instantaneous reaction rate.
- The second major difficulty arises from the concentration fluctuations. For a chemical reaction to occur in a binary reaction, both the species must be present at the same location at a given time for sufficient period of time for the reaction to progress. The presence of eddies in turbulent flow will tend to break up pockets of the reacting species and mix them rapidly. Hence, a much faster rate of reaction can be supported than in purely laminar flow where reactant species will have to arrive at the reaction site by the much slower mechanism of molecular diffusion. This again means that the rate of reaction cannot be obtained purely from the time-averaged description of the species concentration. The role of turbulent fluctuations in bringing together reacting species is to be modeled for a truer estimate of the reaction rate in turbulent flow.

Given the above complicating factors, a number of assumptions and simplifications are made in treating turbulent combusting flows resulting in a hierarchy of models of varying degrees of sophistication. A full treatment of these is not possible here but the main conceptual frameworks are discussed below.

In the simplest of gaseous combustion models, the overall reaction is described in terms of a single-step, irreversible chemical reaction involving fuel and oxidants leading to the formation of combustion products. For example, for the combustion of methane, this is written as

$$CH_4 + 2O_2 \rightarrow CO_2 + 2H_2O \tag{6.4.27}$$

In the general case, this is written as

$$1\,\mathrm{kg\ of\ fuel} + \mathrm{i\,kg\ of\ oxidant} \rightarrow (1 + \mathrm{i})\mathrm{kg\ of\ products} \tag{6.4.28}$$

where i is the stoichiometric ratio (the formulation can be extended to equivalence ratios other than unity). Two additional species conservation equations are solved for the fuel and the oxidant. The heat of the reaction per mole of fuel can be determined either from the chemical composition of the fuel or experimentally; this can then be used to evaluate the heat release rate in the energy equation.

Regarding the reaction rate, two broad approaches are used. The simplest approach assumes that the reaction rate is infinitely fast so that the fuel and the oxidant cannot co-exist and the reaction proceeds until one (or both) of these is completely exhausted to form the products. In this case, the three species conservation equations can be combined into one equation for the conservation of the mixture fraction, which, in the case of a diffusion flame with separate inlets for the fuel and the oxidant, is defined as

$$f = (\chi - \chi_{OX})/(\chi_{FU} - \chi_{OX}) \tag{6.4.29}$$

Here $\chi = Y_{FU} - Y_{OX}/i$ where i is the stoichiometric ratio, Y_{FU} and Y_{OX} are the mass fractions of the fuel and the oxidant, respectively, and the subscripts FU and OX refer to the conditions at the fuel inlet and the oxidant inlet. The mixture fraction is a field variable within the computational domain and the mass fractions of the fuel, the oxidant and the products can be computed uniquely as a function of f:

$$Y_{FU} = \frac{f - f_{st}}{1 - f_{st}}; \quad Y_{OX} = 0; \quad Y_{PR} = 1 - Y_{FU} \quad \text{if } f \geq f_{st} \tag{6.4.30}$$

and

$$Y_{FU} = 0; \quad Y_{OX} = \frac{1 - f}{f_{st}}; \quad Y_{PR} = 1 - Y_{OX} \quad \text{if } f < f_{st} \tag{6.4.31}$$

where f_{st} = mixture fraction at stoichiometric conditions = $1/(1 + i)$. This is known as the flame sheet approximation. The variation of f within the computational domain is given by the following transport equation:

$$\partial(\rho f)/\partial t + \partial(\rho u_j f)/\partial x_j = \partial\{(\mu \partial f/\partial x_j)\}/\partial x_j \tag{6.4.32}$$

Once the composition of the mixture of gases is known from Eqs. (6.4.30)–(6.4.32), the heat of the reaction can be calculated from the heats of formation of the species. Thus, this flame sheet approximation of the reaction as a single-step, irreversible reaction occurring at infinitely fast rate requires the solution of only one additional partial differential equation (Eq. 6.4.29) to represent combustion. If the flow is turbulent, the following time-averaged form of Eq. (6.4.29) can be used:

$$\frac{\partial(\rho\bar{f})}{\partial t} + \frac{\partial(\rho\bar{u}_i\bar{f})}{\partial x_i} = \frac{\partial}{\partial x_i}\left(\frac{\mu_t}{\sigma_t}\frac{\partial\bar{f}}{\partial x_i}\right) \tag{6.4.33}$$

where μ_t is the turbulent (dynamic) viscosity, σ_t is the turbulent "Prandtl" number for the mixture fraction having a value of around unity, and an overbar indicates time-averaged variable.

In a more realistic approach, a finite rate of the combustion reaction (Eq. 6.4.29) can be considered. Here, there are several possibilities:

- One may neglect the effect of turbulence and specify an effective chemical kinetics-limited reaction rate in the form of Arrhenius parameters. This requires the solution of the mixture continuity, momentum and energy conservation equations along with two additional mass conservation equations for two of the three active species, for example, fuel and oxidant.
- Another possibility is to assume that the reaction rate is determined by the rate at which turbulent mixing of the reacting species occurs. Spalding (1971) proposed the eddy break-up (EBU) model in which the reaction rate is assumed to be very fast, and that it is limited only by the rate at which pockets of fuel or oxidant or products are broken up into eddies and mixed well. Calculation of the reaction rate therefore requires two estimates, namely, the size of the fuel pocket and the time scale of its break-up. Spalding suggested the former would be related to the variance of fuel concentration fluctuations while the latter would be proportional to the characteristic time scale of turbulence, namely, the eddy life-time given by k/ε. Thus, according to this model, the reaction rate is given by

$$\bar{w}_F = \rho C_{EBU}\frac{\varepsilon}{k}\left(\overline{Y_F'^2}\right)^{1/2} \tag{6.4.34}$$

where C_{EBU} is a constant. The variance of the fuel mass fraction fluctuations is to be obtained by solving a transport equation for the variance of the mixture fraction:

$$\frac{\partial\left(\rho\overline{f'^2}\right)}{\partial t} + \frac{\partial\left(\rho\bar{u}_i\overline{f'^2}\right)}{\partial x_i} = \frac{\partial}{\partial x_i}\left(\frac{\mu_t}{\sigma_t}\frac{\partial\overline{f'^2}}{\partial x_i}\right) + C_g\mu_t\left(\frac{\partial\bar{f}}{\partial x_i}\right)^2 - C_d\rho\frac{\varepsilon}{k}\overline{f'^2} \tag{6.4.35}$$

where the constants σ_t, C_g and C_d take the values of 0.7, 2.86 and 2.0, respectively.

- A variant of the eddy break-up model, which does not require the solution of an additional transport equation (for the variance of the mixture fraction), was proposed by Magnussen and Hjertager (1977). In this "eddy dissipation (ED)" model, the reaction rate is assumed to be governed by the mean species concentration rather than by the variance of the concentration fluctuations. Accordingly, the reaction rate is given by

$$\bar{w}_F = -\bar{\rho} A \frac{\varepsilon}{k} \min\left[\bar{Y}_F, \frac{\bar{Y}_{O_2}}{\upsilon}\right] \quad \bar{w}_P = \bar{\rho} A \cdot B \frac{\varepsilon}{k}\left[\frac{\bar{Y}_P}{1+\upsilon}\right] \tag{6.4.36}$$

The rate thus depends on the minimum of the three species that are present. A feature of both the EBU and the ED models is that the reaction does not require to be ignited as there is no chemical activation of the reaction. Wherever turbulence is present, the reaction takes place and proceeds until one of the species gets exhausted.

- Another approach used to take account of concentration fluctuations is to model them using probability density functions. For turbulent combusting flows involving chemical reactions, estimation of the reaction rate using time-averaged concentrations can lead to erroneous results. For example, no reaction can take place at a position if the mixture is composed of fuel and products at one instant and of oxidant and products at another instant, even though on a time-average basis, the concentrations of both the fuel and the oxidant are non-zero. One approach to modeling concentration fluctuations (see, for example, Pope 2000; Warnatz et al. 1995) is to model the statistics of fluctuation of the mixture fraction, f, using a probability density function, p(f), such that

$$0 \leq p(f) \text{ and } \int_0^1 p(f)df = 1 \tag{6.4.37}$$

The time average of any variable ξ, such as Y_F, which is a function of f, is then given by

$$\bar{\xi} = \int_0^1 \xi f\, p(f)\, df \tag{6.4.38}$$

In general, one does not know the probability density function (pdf); this is expected to vary with flow and within the flow. While the evolution of the pdf itself can be modeled through further transport equations (Pope 2000), a simpler approach is the presumed pdf approach. Here, one uses simple, easily integrable probability density functions such as the double delta function, the beta function and the clipped Gaussian function. For example, the double delta function can be formulated as follows:

$$p(f) = A\delta(f - \bar{f}_{+}) + B\delta(f - \bar{f}_{-}) \quad \text{with } \bar{f}_{+} = \bar{f} + \alpha,\ \bar{f}_{-} = \bar{f} - \alpha \qquad (6.4.39)$$

where $\bar{f}$ is the time-averaged mixture fraction. The three model constants, namely, A, B, α are determined from the conditions that the distribution functions must satisfy the normalization integral, give the correct mean of f and the correct mean square of f.

Similarly, the beta function is given by

$$p(f) = \frac{f^{\alpha-1}(1-f)^{\beta-1}}{\int f^{\alpha-1}(1-f)^{\beta-1}df} \qquad (6.4.40)$$

where

$$\alpha = \bar{f}\left[\frac{\bar{f}(1-\bar{f})}{\overline{f'^2}} - 1\right] \quad \text{and} \quad \beta = (1-\bar{f})\left[\frac{\bar{f}(1-\bar{f})}{\overline{f'^2}} - 1\right] \qquad (6.4.41)$$

The mean and mean square values of f, namely, $\bar{f}$ and $\overline{f'^2}$, respectively, that appear in these equations are obtained by solving Eqs. (6.4.33) and (6.4.35), respectively. From known local values of these quantities, the local values of A, B and α are fixed. The local p(f) is then given by Eq. (6.4.39) and the mean concentrations of the chemical species can be calculated using Eq. (6.4.38).

- While the above approaches are robust turbulent combustion models in the sense that they give a solution, they are based on a single-step, irreversible reaction. Such treatment is not sufficient in dealing with some practical issues related to combustion such as flame lift-off, stability, extinction and pollutant formation for which the prediction of formation of important intermediate radicals is important (Warnatz et al. 1995). The possibility of dealing with reaction chemistry in sufficient detail must be included in the analysis. This can be done through more advanced models such as the eddy dissipation concept model (Magnussen and Hjertager 1977) and flamelet models (Peters 1984) which allow treatment of detailed chemistry together with turbulence under some specialized conditions. These are not discussed here.

In the case of solid or liquid fuels, additional modelling is required to model the release rate of volatile gaseous fuels from these during heating. In case of some solid fuels such as coal and biomass, part of the combustion occurs in the form of a heterogeneous reaction involving hot gas and porous char (Smoot and Smith 1985; Lawn 1987; Warnatz et al. 1995). A framework for the tracking of these liquid droplets or solid particles through the combustion chamber is also necessary and is briefly discussed below.

6.4.5 Multiphase Flows

Multiphase flow involves flow of two or more fluid phases; these may be of the same fluid, as in steam-water flow through a boiler tube, or of different fluids, as in the case of droplet entrainment by wave action in atmospheric air, air bubble entrapment by the surface waters of rivers, the flow of two immiscible liquids in a solvent extraction column. Multiphase flow also encompasses the cases of gas-solid or liquid-solid flows where only one phase is in the fluid phase. These flows occur in many industrial processes as well as in natural phenomena. A fundamental understanding of the nature of these flows is obviously of interest. However, these flows are physically very complex and a single calculation methodology cannot be used for all types of flows. Traditionally, multiphase flow modelling in the framework of CFD-based simulations has taken three distinct routes:

- Category A cases in which the principal objective is to track the interface location. The flow field in the liquid phase is usually calculated by taking proper account of kinematic and dynamic boundary conditions at the interface (see Sect. 2.3.3). The "dam break" problem and the sloshing tank problem are classic examples that are often used as benchmarking exercises. More challenging applications include dynamics of nucleate boiling (Dhir 2006) and droplet or bubble dynamics on impact or release at or from a gas-liquid interface (Sussman et al. 1994; Scardovelli and Zaleski 1999). These cases are usually dealt with using the volume of fluid (VOF) model (Hirt and Nichols 1982) or the level-set method (Osher and Sethian 1988).
- Category B cases in which one phase is continuous while the other is in dispersed form in very dilute concentrations. In this case, we expect there to be little interaction among the particles of the dispersed phase, and the particle-continuous phase interaction can be taken into account through particle trajectory calculations. This approach, known as the Eulerian (for continuous phase)-Lagrangian (for the particulate phase) approach, can be used, for example, in dealing with the combustion of liquid or solid fuels.
- Category C cases in which one phase is continuous and the other phase is dispersed but is in high volumetric concentration. In this case, the typical approach is to use an Eulerian-Eulerian formulation for both phases and include an "interpenetrating continua" model to account for the partial (on a time- and space-averaged basis) presence of both the phases at any point in the computational domain.

Examples of these three types of multiphase flows are given in Fig. 6.24. The shape of a gas bubble during formation and rise at a submerged nozzle is shown in Fig. 6.24a at different times. This is an example of Category A type of flow. These calculations have been done using the VOF method, see below. Figure 6.24b shows the computed trajectories of coal particles in a pulverized coal-fired boiler. This Category B flow computations have been carried out using an Eulerian-Lagrangian type of approach which is discussed later. The trajectories of coal particles (of 70

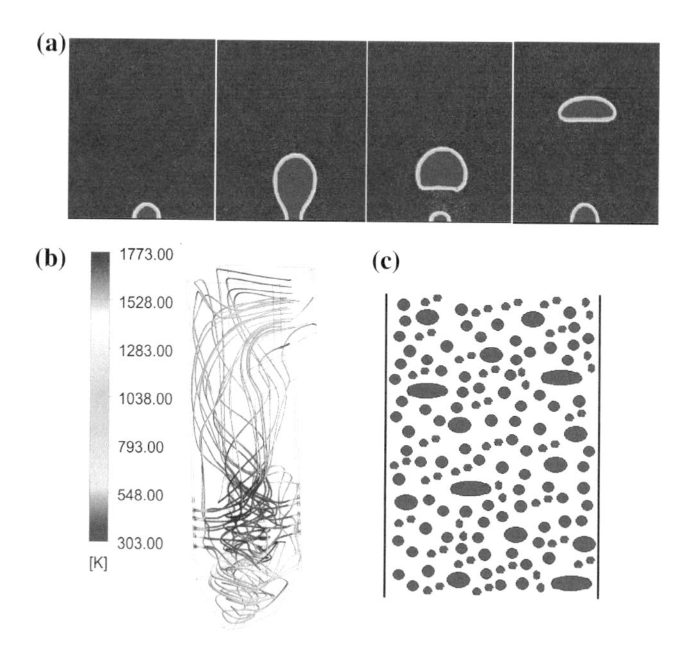

Fig. 6.24 **a** Bubble shape during formation and release predicted the VOF method at t = 0.02, 0.263, 0.28 and 0.31 s (Nowneswara Reddy and Jayanti 2013), reprinted with permission from Elsevier. **b** Tracking of burning coal particles in a furnace, and **c** dispersed bubble flow in gas-liquid flow through a vertical tube

micron diameter) are coloured by temperature. A number of physico-chemical processes such as heating, devolatilization and combustion happen as the coal particle traverses through the furnace. Figure 6.24c shows schematically dispersed bubble flow in air-water flow in a vertical tube. Here, the volume fraction of the dispersed phase can be high (5–30%) which necessitates an Eulerian-Eulerian description of the flow. A number of complications, such as bubbles of different sizes, non-spherical bubbles, turbulence modulation by the dispersed phase, etc., may be present in real systems. Another example of high volume fraction flows is the fluidized bed coal combustor in which relatively large coal particles (typically 5–10 mm in size) are burned together with limestone which captures SO_2 formed during combustion. There is intimate coupling between flow of fluids, movement of solids–which are present in large volume fractions—and combustion and flue gas desulphurization reactions that occur within the combustor. Category C type of flow calculations are carried out using Eulerian-Eulerian multiphase flow calculations.

There are special cases where a two-phase flow is treated essentially as a single phase flow. The presence of the other phase is acknowledged only in the shape of the flow domain with, usually, a shear-free boundary condition imposed at the gas-liquid interface. For example, in the case of a falling liquid film flowing downwards over a plate, the flow field and dynamics of the gas phase may not be resolved but the interfacial wave structure and its influence on the liquid flow field can be resolved by solving the Navier-Stokes equations in the liquid phase alone. There may also be some specials cases, for example, stratified gas-liquid flow through a horizontal pipe, in which both phases are continuous and are separated by a distinct and continuous interface across which interfacial transfer of mass, momentum and heat may be occurring. Here, one can solve the individual phasic conservation equations in each domain with the interfacial jump conditions acting as boundary conditions for each phase. In such cases, the flow governing equations remain unchanged from their single phase flow counterparts. The computational domain may dynamically change to accommodate the interfacial boundary conditions. For multiphase flow cases of Categories A, B and C, additional or modified governing equations need to be used to formulate the problem. The salient aspects of these cases are discussed briefly below without going into the details.

The volume-of-fluid (VOF) model developed by Hirt and Nichols (1982) can track the interface between two phases, and has been used widely in analyzing various two-phase flow systems such as stratified flow, free surface flow, bubble flow etc. A specific example is that of Gerlach et al. (2007) who used it to study the formation and dynamics of a single bubble in a quiescent liquid at atmospheric pressures and elevated pressures. In the VOF model, the calculations are made for a single fluid having density and viscosity weight-averaged by the volume fraction of each phase:

$$\rho = \alpha_l \rho_l + (1 - \alpha_l)\rho_g \tag{6.4.42}$$

$$\mu = \alpha_l \mu_l + (1 - \alpha_l)\mu_g \tag{6.4.43}$$

where α_l is volume fraction of liquid, α_g volume fraction of gas, ρ_l and ρ_g are densities of liquid and gas respectively. The velocity is also determined based on a weighted-average basis as

$$\mathbf{u} = \frac{\alpha_l \rho_l \mathbf{u}_l + \alpha_g \rho_g \mathbf{u}_g}{\rho} \tag{6.4.44}$$

Since the VOF model is usually used for separated flows, this enables exact treatment of each phase in the individual domains. The flow and fluid properties are smeared only at the interface. In order to maintain a sharp interface, the interface is reconstructed after every time step. The location of the interface is obtained from the volume fraction equation for phase k:

$$\partial(\alpha_k)/\partial t + \nabla \cdot (\alpha_k \mathbf{u}) = 0 \tag{6.4.45}$$

The primary phase volume fraction is obtained from the constraint that $\sum_{k=1}^{n} \alpha_k = 1$. The distinguishing feature of the VOF model is that no specific interfacial conditions are specified. However, interface-specific terms can be added. For example, the effect of interfacial tension is often added, where needed (for example, while simulating bubble dynamics but not for the dam break problem), as a volumetric source term at the interface using the model of Brackbill et al. (1992).

There are essentially two ways of calculating the flow field in dispersed flow. In the Eulerian-Eulerian approach, both the phases are assumed to be so thoroughly mixed that they can be regarded as interpenetrating continua. In any small volume within the flow domain ("cell"), a fraction of the volume is occupied by one phase, the rest being occupied by the other phase (a generalization to arbitrary number of phases is also possible). Conservation equations for each phase are then written (Drew and Lahey 1979) introducing a new variable, phase fraction, to represent the fact that only part of the volume is being occupied by that phase. The equations for the two phases are coupled by the phase fraction and interaction terms representing the exchange of mass, momentum and heat between the two phases. The generic scalar transport equation for phase a can be written as follows:

$$\begin{aligned}&\partial(\alpha_a \rho_a \phi_a)/\partial t + \nabla \cdot (\alpha_a \rho_a u_a \phi_a - \alpha_a \Gamma_a \nabla \phi_a) \\ &= \alpha_a S_a + \sum_{b=1}^{N_P} c_{ab}(\phi_b - \phi_a) + \sum_{b=1}^{N_P} (\dot{m}_{ab}\phi_b - \dot{m}_{ba}\phi_a)\end{aligned} \tag{6.4.46}$$

Here ϕ is the scalar and c_{ab} is the drag coefficient-like interaction term between the phases that depends on the difference between the scalar variable values across the interface. The last terms represents exchange of scalar associated with phase change processes such as boiling and condensation. Closure relations for these exchange processes are needed to completely the equations that need to be solved (Drew and Lahey 1979). A distinct off-shoot of the generic multiphase flow model is the approach based on kinetic theory of granular flows for the simulation of gas-solid flows and fluiduzed beds. The interested reader is referred to the paper of Gidaspow et al. (2003) as a starting point.

In the Eulerian-Lagrangian approach, the motion of the two phases is calculated separately and alternately without explicitly accounting for the volumetric fraction occupied by the other phase. The continuous phase flow field is calculated first using the traditional Eulerian approach. The motion of the particulate phase is then calculated using a Lagrangian approach in which the trajectories of individual particles are calculated through the flow domain by solving the following equations for the velocity and position of the particle:

$$d(m\mathbf{u}_P)/dt = \sum(\mathbf{F}_{ext}) \quad \text{and} \quad d(\mathbf{x}_P)/dt = \mathbf{u}_P \tag{6.4.47}$$

Here m is the mass of the particle, $\mathbf{u}_p$ is its velocity and $\mathbf{x}_p$ is its position. The external forces include drag force, gravitational force and other forces such as electrostatic force, thermophoretic force etc., as appropriate. Heat and mass transfer effects as well as turbulence dispersion of particles can also be included while doing the particle trajectory calculations. Several such particle histories are required to obtain a representative description of particle motion. The continuous phase affects the particle motion through drag, buoyancy and other forces. The effect of the particulate phase on the continuous phase is represented through source/sink terms appearing in the conservation equations for the continuous phase. Alternating computation of each phase iteratively will eventually lead to the correct evaluation of the coupling terms.

From a computational point of view, all multiphase flow simulations require higher computational resources than the corresponding single phase flow simulations. This is not only due to the number of additional partial differential equations that need to be solved but also due to the complexity of closure relations that need to be satisfied. Of the three approaches, the VOF-based route requires transient solution with severe restriction on the allowable time step. For a two-phase flow, the Eulerian-Eulerian approach has twice the number of equations that need to be solved together as compared to single phase flow. The inter-phase and interface exchange terms (Eq. 6.4.45) make them tightly coupled and render them more difficult to solve. The Eulerian-Lagrangian approach is the most robust of the three in terms of generating a solution without too much difficult, but is limited to rather small range of application in terms of volume fraction of the dispersed phase.

Multiphase flow CFD is still a developing field and robust and generic algorithms are not yet available for the general case. The accuracy of predictions depends greatly on the goodness and validity of the closure relations that are used. Advances have been made in some niche applications and the reader is advised to consult current literature in the relevant area of application for an accurate assessment of the capabilities of CFD for multiphase flow applications.

6.4.6 Other Phenomena Requiring Modelling

In a combusting environment where high temperatures are produced, radiative heat transfer is the dominant mode of heat transfer. All hot surfaces emit radiation as per Planck's law (Hottel and Sarofim 1967) which then passes through the surrounding medium, where it may get modulated in a number of ways (see below) and reach other hot surfaces where it may be absorbed, transmitted or reflected. The two hot surfaces can thus exchange heat radiatively, even in vacuum. This radiative exchange of heat does not follow the usual transport equation type of law. The attenuation or modulation of the emitted beam as it passes through a medium is given by the radiative transfer equation (RTE) which is an integro-partial differential equation for a given beam. The spectral intensity of radiation, I_λ, of the beam is defined as the radiant energy per unit time, per unit wavelength interval about wavelength λ, per unit surface area perpendicular to the direction of travel, per unit solid angle centred around Ω. In the general case,

$$I_\lambda = I_\lambda(x, y, z, \theta, \phi, \lambda) \tag{6.4.48}$$

Here, x, y and z define the point source of the radiation, and θ and ϕ are the mutually orthogonal angular coordinates in a spherical coordinate system. The total intensity is obtained by integrating I_λ over all λ, and the net radiant energy flux, q_R, is obtained by integrating I_λ over all wavelengths and all directions. Thus,

$$I = \int_0^\infty I_\lambda(x, y, z, \theta, \phi, \lambda)d\lambda \quad (6.4.49)$$

and

$$q_R = \int_0^{4\pi} I \cos\theta \, d\Omega = \int_0^{2\pi} I(\theta, \phi)\cos\theta \sin\theta \, d\theta \, d\phi \quad (6.4.50)$$

The hemispherical flux is obtained by integrating over a hemisphere of 2π steradians. When radiative heat transfer is present, the divergence of the radiant energy flux, i.e., $\nabla \cdot q_R$, appears as a source term on the right hand side of the energy balance equation. Thus, Eq. (2.2.38) representing the energy balance in terms of internal energy gets modified as

$$\partial/\partial t(\rho e) + \nabla \cdot (\rho u e) = \nabla \cdot (k\nabla T) + \tau : \nabla u - p\nabla \cdot u + \nabla \cdot q_R \quad (6.4.51)$$

Estimation of q_R is thus important for determining the temperature distribution. This requires the calculation of the radiant energy exchange between hot and cold surfaces within the computational domain. Some of the basic concepts related to this are discussed below.

The emissive power of a black body, $E_{\lambda b}$, which is equal to $\pi I_{\lambda b}$, into a medium with a refractive index (the ratio of speed of light in vacuum to that in the medium) is given by the Planck's law:

$$E_{\lambda b}(\lambda) = \pi I_{\lambda b}(\lambda) = \frac{2\pi c_1}{n^2 \lambda^5 (e^{c_2/\lambda T} - 1)} \quad (6.4.52)$$

Here, n is the refractive index, $c_1 = hc_0^2$ and $c_2 = hc_0/\mathcal{K}$ where h is Planck's constant ($= 6.62560 \times 10^{-34}$ Js), c_0 is the speed of light in vacuum ($= 2.998 \times 10^8$ m/s) and $\mathcal{K}$ is the Boltzmann constant ($= 1.3806 \times 10^{-23}$ J/K). The hemispherical total emissive power of a black surface into vacuum (where n = 1) is given by

$$E_b = \int_0^\infty E_{\lambda b}(\lambda)d\lambda = \int_0^\infty \pi I_{\lambda b}(\lambda)d\lambda = \sigma T^4 \quad (6.4.53)$$

Here σ is the Stefan-Boltzmann constant and has a value of 5.6696×10^{-8} W/ (m^2K^4) and T is the absolute temperature in degree Kelvin (K). The emissivity, ε, of a real surface is less than that of a black body and is defined such that the emissive power $E = \varepsilon E_b$ where $0 \leq \varepsilon \leq 1$. When a beam of intensity I_λ enters an absorbing, emitting and scattering medium (see Fig. 6.25), its intensity is attenuated. The variation of the spectral intensity along the path is given by the radiative transfer equation:

$$\frac{dI_\lambda(s)}{ds} + (a_\lambda + \sigma_{s\lambda})I_\lambda(s) = a_\lambda n^2 \frac{\sigma T^4}{\pi} + \frac{\sigma_{s\lambda}}{4\pi} \int_{\Omega'=0}^{4\pi} I_\lambda(s)\, \Phi(\lambda, \Omega, \Omega')d\Omega' \quad (6.4.54)$$

Here, s is the distance along the path in the direction of the solid angle Ω, Ω' is the in-scattering direction vector, a_λ is the spectral absorptivity, $\sigma_{s\lambda}$ is the spectral out-scattering coefficient, and Φ is the phase function representing the probability that radiation of wavelength λ propagating in the direction of Ω' and confined within the solid angle $d\Omega'$ is scattered into the direction Ω in the solid angle interval of $d\Omega$ and wavelength interval of $d\lambda$. The first term on the left hand side of Eq. (6.4.54) accounts for the net change in intensity per unit path length. The second term accounts for the loss in intensity by absorption and out-scattering. The first term on the right hand side is the gain in intensity by emission from the surrounding medium and the second term indicates the gain in intensity due to in-scattering.

The solution of the above integro-differential equation is very complicated when scattering needs to be considered and consumes enormous amount of computational time. Several approximate methods have been developed to deal with radiative heat transfer in practical combusting systems; these have been reviewed in detail by Viskanta and Menguc (1987). While simple models such as the Rosseland model

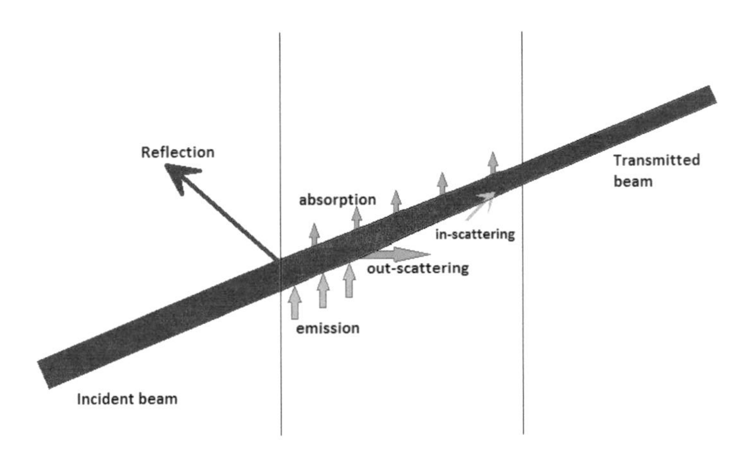

Fig. 6.25 Interactions of a beam of radiation with a medium

and the spherical harmonics model can be developed for optically thick media, i.e., where the beam gets attenuated rapidly, many practical cases of typical combustion applications deal with optically thin media. For such applications, ray-tracing methods are normally used. The accuracy of the model is limited mainly by the number of sources considered, the number of rays traced per source and the computational grid. Practical problems involving radiation are compounded in difficulty by the spectral dependence of the coefficients of absorption, scattering etc. The concentration of gaseous species, for example, in a combustor environment, the presence of fine particulates such as soot, the surface characteristics of the emitting and the receiving surfaces also play an important role in radiative heat transfer. The interested reader is referred to books dedicated radiative heat transfer (Hottel and Sarofim 1967; Rhine and Tucker 1991; Siegel and Howell 2002) for a thorough discussion of these issues.

Modeling of electrochemical phenomena is important in such devices such as fuel cells and flow batteries where there can be significant flow-related influence on the performance of the device. In these cases, some of the reactions involve charge transfer and the rate of these reactions is influenced not only by the chemical kinetics and diffusion of species but also by the local electric potential (O'Hayre et al. 2009). Thus, for true rates of reactions to be computed, the electric potential in the media needs to be resolved along with species concentrations, flow velocity and thermal energy exchanges. The kinetics of the electrochemical reaction rate are often expressed in the form of a largely empirical Butler-Volmer equation, which is a highly non-linear function of the overpotential and concentrations of the reactant, and product species. The modeling of fuel cells is further complicated by the multiscale nature of the phenomena that are involved and the layered structure of the media. The smallest scales of interest are in the nanometer range (~ 10 nm) and correspond to the size of the catalyst particles. The surface areas and pore volumes available at this scale can have a strong influence on the overall rate of the electrochemical reactions. The catalyst layer has a thickness that is nearly 1000 times larger and is of the order of 10–20 μm. The gas diffusion layer which is made of electrically conducting material about 10 times thicker (about 100–200 μm); the flow field passages through which the reacting gases are circulated are another 10 times larger and are about ~ 1 mm in depth. All these layers have a length and breadth that are 10–100 times larger. In a practical fuel cell power unit, 50–100 individual cells may be stacked together. Different phenomena are important at different scales. Heat transfer and flow distribution aspects are important at stack level, reactant transport is important at cell level and charge transport and electrochemical aspects are important at sub-micron level. There is strong coupling among all the levels. Many of these phenomena are expressed in the form of additional conservation equations and constitutive equations. The interested reader is referred to O'Hayre et al. (2009) for fundamental concepts related to fuel cells and research papers for more information.

6.5 Summary

This chapter brings out the difficulties involved in dealing with practical CFD applications. Two types of complications—those arising from having to deal with irregular shapes of the flow domains of interest and those arising from having to deal with complicated physics—have been discussed at length. When dealing with irregular flow geometries, two approaches, namely, the body-fitted grid approach and the unstructured grid approach, can be used to obtain solutions of the governing partial differential equations. When dealing with complicated physiscs, additional variables, conservation equations and constitutive models are needed to take account of the physics that is not captured in the set of governing equations derived in Chap. 2. Of the two types of complexities, satisfactory computational frameworks have been developed to tackle the difficulties related to irregular geometries, and it can be claimed confidently that single phase flow calculations of sufficient accuracy can be generated in flow domains of arbitrary complexity as long as the physics is relatively simple. While computational frameworks for tackling complex physics have been developed on a case-to-case basis, a large degree of empiricism is brought into play in order to obtain a realizable solution. Flow problems involving complex physics thus require case-specific validation, calibration and sensitivity analysis to generate some degree of confidence in the predictions of CFD simulations.

Problems

1. Consider the 2-d Euler equations given by

$$\partial U/\partial t + \partial E/\partial x + \partial F/\partial y = 0 \qquad \text{(Q6.1)}$$

where

$$U = \begin{pmatrix} \rho \\ \rho u \\ \rho v \\ \rho e_t \end{pmatrix}; \quad E = \begin{bmatrix} \rho u \\ \rho u^2 + p \\ \rho uv \\ (\rho e_t + p)u \end{bmatrix}; \quad F = \begin{bmatrix} \rho v \\ \rho vu \\ \rho v^2 + p \\ (\rho e_t + p)v \end{bmatrix}$$

Show that the transformation into a curvilinear coordinate system (ξ, η) yields the following equations:

$$\partial(U/J)/\partial\tau + \partial\{(\xi_t U + \xi_x E + \xi_y F)/J\}/\partial\xi + \partial\{(\eta_t U + \eta_x E + \eta_y F)/J\}/\partial\eta = 0 \qquad \text{(Q6.2)}$$

where J is the Jacobian of the transformation given by

$$J = \partial(\xi, \eta)/\partial(x, y) = \begin{vmatrix} \xi_x & \xi_y \\ \eta_x & \eta_y \end{vmatrix} \qquad \text{(Q6.3)}$$

2. Show that transformation of the two-dimensional incompressible flow equations in Cartesian coordinates into a curvilinear coordinate system (ξ, η) yields the following equations:

$$\partial U/\partial \tau + \partial E/\partial \xi + \partial F/\partial \eta = S \tag{Q6.4}$$

where

$$U = \begin{pmatrix} 0 \\ \rho u \\ \rho v \end{pmatrix}; \quad E = \begin{bmatrix} y_\eta u - x_\eta v \\ \rho(y_\eta u^2 - x_\eta uv) + y_\eta p - y_\eta H_1 + x_\eta H_2 \\ \rho(y_\eta uv - x_\eta v^2) - x_\eta p - y_\eta H_2 + x_\eta H_3 \end{bmatrix}$$

$$F = \begin{bmatrix} -y_\xi u + x_\xi v \\ \rho(-y_\xi u^2 + x_\xi uv) - y_\xi p + y_\xi H_1 - x_\xi H_2 \\ \rho(-y_\xi uv + x_\xi v^2) + x_\xi p + y_\xi H_2 - x_\xi H_3 \end{bmatrix};$$

$$S = \begin{bmatrix} 0 \\ \rho\dot{x}\left\{(y_\eta u)_\xi - (y_\xi u)_\eta\right\} + \rho\dot{y}\left\{-(x_\eta u)_\xi + (x_\xi u)_\eta\right\} \\ \rho\dot{x}\left\{(y_\eta v)_\xi - (y_\xi v)_\eta\right\} + \rho\dot{y}\left\{-(x_\eta v)_\xi + (x_\xi v)_\eta\right\} \end{bmatrix}$$

where

$$H_1 = \frac{2\mu}{J}\left\{(y_\eta u)_\xi - (y_\xi u)_\eta\right\}; \; H_2 = \frac{\mu}{J}\left\{-(x_\eta u)_\xi + (x_\xi u)_\eta + (y_\eta v)_\xi - (y_\xi v)_\eta\right\}$$

$$H_3 = \frac{2\mu}{J}\left\{-(x_\eta v)_\xi + (x_\xi v)_\eta\right\}; \quad \dot{x} = x_\tau \quad \dot{y} = y_\tau$$

Here $\dot{x}$ and $\dot{y}$ are grid speeds and zero if the physical mesh is stationary and J is the Jacobian of the transformation and is given by Eq. (Q6.3).

3. Show (Vinokur 1974) that the 3-d Navier-Stokes equations (in compressible form and with the Stokes' hypothesis the second coefficient of viscosity is given by $\lambda = -2\mu/3$) in generalized curvilinear coordinates can be put in the following strongly conservative form:

$$\partial U_1/\partial \tau + \partial E_1/\partial \xi + \partial F_1/\partial \eta + \partial G_1/\partial \zeta = 0 \tag{Q6.5}$$

where

$$U_1 = U/J; \quad E_1 = (E\xi_x + F\xi_y + G\xi_z)/J; \quad F_1 = (E\eta_x + F\eta_y + G\eta_z)/J$$

$$G_1 = (E\zeta_x + F\zeta_y + G\zeta_z)/J \quad \text{where } J = \frac{\partial(\xi,\eta,\zeta)}{\partial(x,y,z)} = \begin{vmatrix} \xi_x & \xi_y & \xi_z \\ \eta_x & \eta_y & \eta_z \\ \zeta_x & \zeta_y & \zeta_z \end{vmatrix}$$

$$U = \begin{bmatrix} \rho \\ \rho u \\ \rho v \\ \rho w \\ \rho e_t \end{bmatrix}; \quad E = \begin{bmatrix} \rho u \\ \rho u^2 + p - \tau_{xx} \\ \rho uv - \tau_{xy} \\ \rho uw - \tau_{xz} \\ \rho e_t u - u\tau_{xx} - v\tau_{xy} - w\tau_{xz} + q_x \end{bmatrix};$$

$$F = \begin{bmatrix} \rho v \\ \rho uv - \tau_{xy} \\ \rho v^2 + p - \tau_{yy} \\ \rho vw - \tau_{yz} \\ \rho e_t v - v\tau_{xy} - v\tau_{yy} - w\tau_{yz} + q_y \end{bmatrix};$$

$$G = \begin{bmatrix} \rho w \\ \rho uw - \tau_{xz} \\ \rho vw - \tau_{yz} \\ \rho w^2 + p - \tau_{zz} \\ \rho e_t w - u\tau_{xz} - v\tau_{yz} - w\tau_{zz} + q_z \end{bmatrix};$$

$$\tau_{xx} = \frac{2}{3}\mu\{2(\xi_x u_\xi + \eta_x u_\eta + \zeta_x u_\zeta) - (\xi_y v_\xi + \eta_y v_\eta + \zeta_y v_\zeta) - (\xi_z w_\xi + \eta_z w_\eta + \zeta_z w_\zeta)\}$$

$$\tau_{yy} = \frac{2}{3}\mu\{2(\xi_y v_\xi + \eta_y v_\eta + \zeta_y v_\zeta) - (\xi_x u_\xi + \eta_x u_\eta + \zeta_x u_\zeta) - (\xi_z w_\xi + \eta_z w_\eta + \zeta_z w_\zeta)\}$$

$$\tau_{zz} = \frac{2}{3}\mu\{2(\xi_z w_\xi + \eta_z w_\eta + \zeta_z w_\zeta) - (\xi_x u_\xi + \eta_x u_\eta + \zeta_x u_\zeta) - (\xi_y v_\xi + \eta_y v_\eta + \zeta_y v_\zeta)\}$$

$$\tau_{xy} = \mu(\xi_y u_\xi + \eta_y u_\eta + \zeta_y u_\zeta + \xi_x v_\xi + \eta_x v_\eta + \zeta_x v_\zeta)$$

$$\tau_{xz} = \mu(\xi_z u_\xi + \eta_z u_\eta + \zeta_z u_\zeta + \xi_x w_\xi + \eta_x w_\eta + \zeta_x w_\zeta)$$

$$\tau_{yz} = \mu(\xi_y w_\xi + \eta_y w_\eta + \zeta_y w_\zeta + \xi_z v_\xi + \eta_z v_\eta + \zeta_z v_\zeta)$$

$$q_x = -k(\xi_x T_\xi + \eta_x T_\eta + \zeta_x T_\zeta);$$

$$q_y = -k(\xi_y T_\xi + \eta_y T_\eta + \zeta_y T_\zeta);$$

$$q_z = -k(\xi_z T_\xi + \eta_z T_\eta + \zeta_z T_\zeta).$$

4. “Parabolized” form of compressible flow N-S (PNS) equations permit efficient solution of supersonic flows because they allow space-marching techniques to be employed, and are valid in both the inviscid and the viscous portions of the flow field. In these equations, it is assumed that streamwise viscous term derivatives and those associated with heat flux in the streamwise direction are

negligible compared to the corresponding cross-wise and normal components. Derive the PNS equations in curvilinear coordinates by dropping, from Eq. (Q6.5) above, the time derivative terms and the viscous (and heat flux) term derivatives in the streamwise (ξ-) direction. See Rubin and Tannehill (1992) and Tannehill et al. (1997) for a discussion of the PNS equations and the restrictions and advantages of their use.

5. You wish to simulate the flow of air over a symmetric aerofoil (see Fig Q6.1); the shape of the upper half of the aerofoil is given by the curve

$$y = x^{0.75}(1 - x)^{1.75}, \quad 0 \leq x \leq 1$$

   Construct a structured grid of (a) C-type, (b) O-type and (c) H-type using the transfinite interpolation approach. Which one would you use when?
6. Repeat Problem #4 using the elliptic/parabolic/hyperbolic grid generation method, as appropriate.
7. You wish to simulate the air flow over a cylinder (Fig. Q6.2). Knowing that there can be flow separation and shedding of vortices at even moderate Reynolds numbers, and knowing that the effect of these vortices can be felt quite a long distance downstream, you wish to create an H-type grid for it using the elliptic grid generation approach.
8. Consider the benchmark case of compressible flow through a converging-diverging nozzle of the type shown in Fig. Q6.3. Generate a curvilinear structured grid for it using the algebraic transfinite interpolation method.
9. You wish to find the flow distribution in a trapezoidal lid-driven cavity of the shape given in Fig. Q6.4. Generate a structured grid for it using transfinite interpolation and find the metrics of the transformation that are required in the transformed equations (see Problem #2) at each grid point.
10. Two electronic chip components generate a lot of heat which needs to be taken away effectively so as to maintain a reasonably low operating temperate for the chip. You therefore propose to use heat diffuser plate of the shape shown in Fig. Q6.5 where the circles of 40 and 50 mm diameter represent the two electronic components which generate heat of 1000 W/m along the perimeter. This heat needs to be carried away by conduction along the diffuser plate and by convective heat transfer from the edges of the diffuser plate. The convective heat transfer coefficient is 25 W/m^2K and the ambient is at 25 °C. You wish to find the temperature distribution over the plate. Generate a structured 2-d grid for the plate using elliptic grid generation method and solve the steady 2-d heat conduction equation to find the temperature distribution. Under what conditions is the 2-d formulation of the problem reasonably correct?
11. Triangulate the grid shown in Fig. Q6.6 using the advancing front method.
12. Repeat Problem #11 using the Delaunay triangulation scheme.
13. Triangulate the grid shown in Fig. Q6.7 using the advancing front method. Is it possible for a triangulation of the type shown with a dashed line, in which area outside the domain is included in triangulation, using this method? Justify your answer.

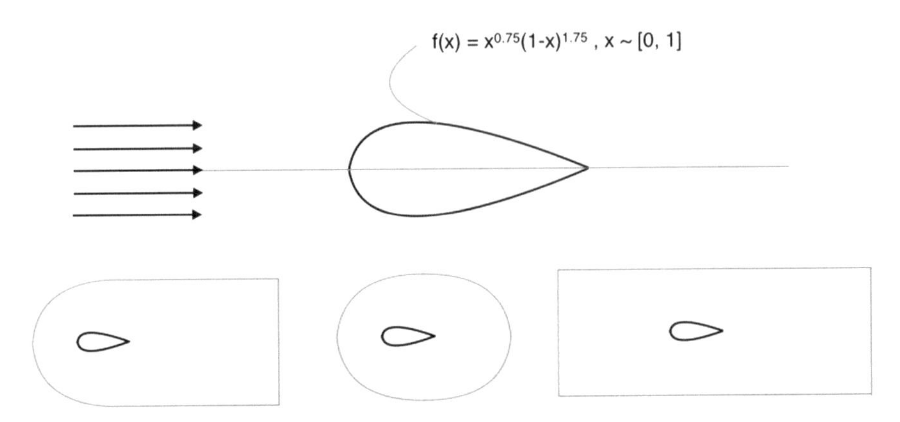

Fig. Q6.1 Steady, 2-dimensional flow over a symmetric aerofoil (Problem #5). The origin is at the upstream stagnation point. Construct C-type, O-type and H-type structured grid for the flow domain indicated in the bottom set of figures

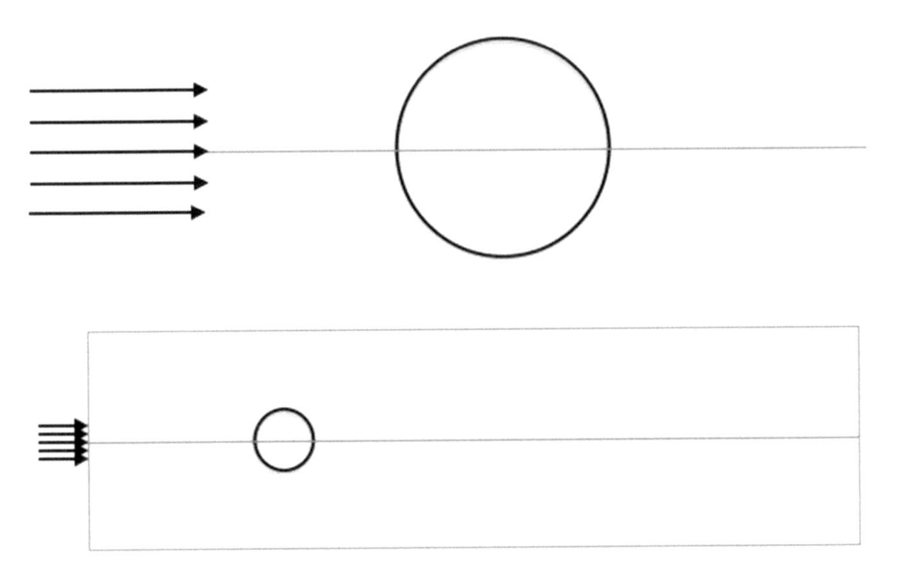

Fig. Q6.2 Unsteady, 2-dimensional flow over a circular, stationary cylinder (Problem #7)

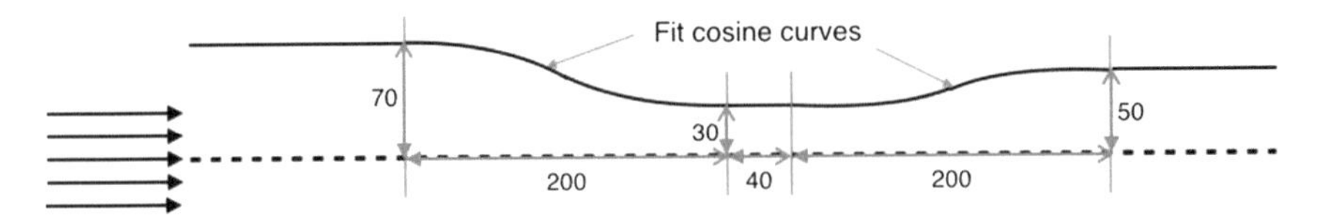

Fig. Q6.3 Converging-diverging nozzle geometry for Problem #8. All dimensions are in mm

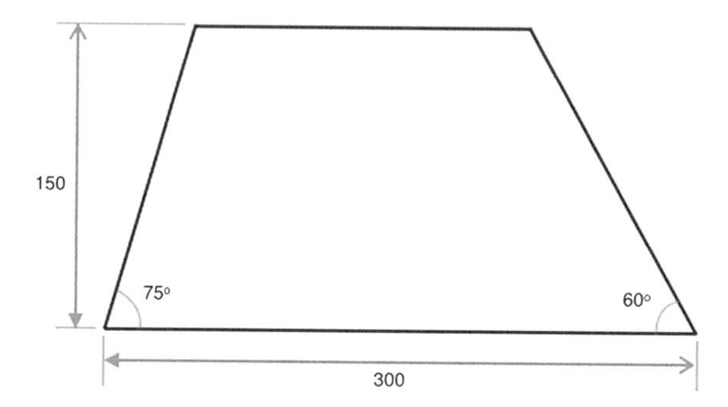

Fig. Q6.4 Geometry of the asymmetric trapezoidal cavity for Problem #9. All dimensions are in mm

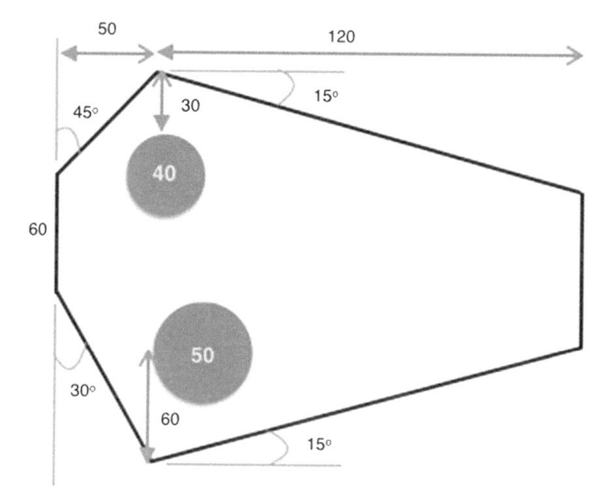

Fig. Q6.5 Geometry for the 2-d, steady-state heat conduction case discussed in Problem #10 All dimensions are in mm

14. Repeat Problem #13 using the Delaunay triangulation scheme. Is it possible for the triangle with the dashed line to be triangulated as part of this method? Justify your answer.
15. Write a computer program to generate a triangular mesh to the domain shown in Fig. Q6.8 with a target side length of 5 mm. Use (a) the advancing front method, and (b) the Delaunay triangulation scheme for triangulating the domain.
16. Generate an unstructured mesh for flow over an aerofoil of the shape given in Fig. Q6.1 using advancing front method.
17. Generate unstructured mesh for flow over the cylinder (Fig. Q6.2) using Delaunay triangulation.

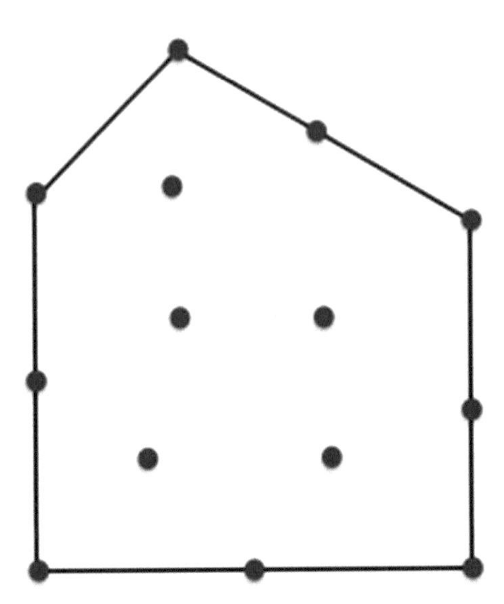

Fig. Q6.6 Geometry for Problem #11

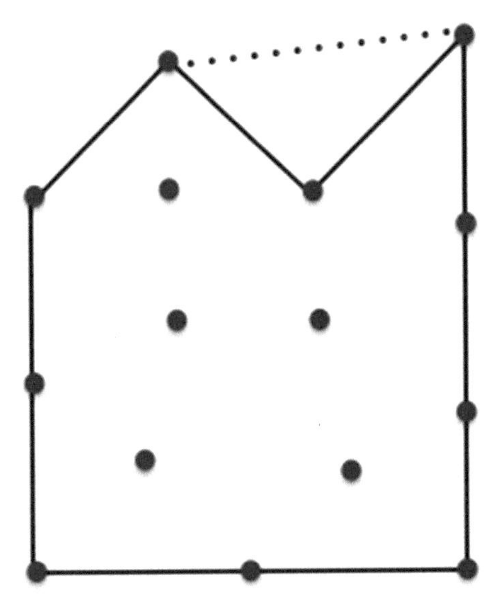

Fig. Q6.7 Geometry for Problem #13

18. Generate unstructured meshes using triangular cells for the computational domain shown in Figs. Q6.3 and Q6.4 using the advancing front method for grid generation.

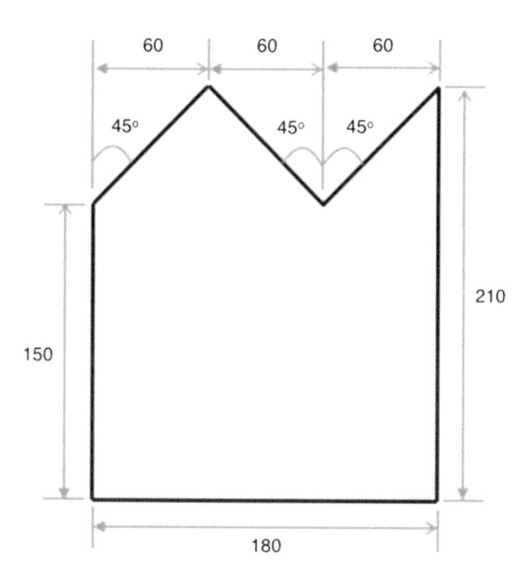

Fig. Q6.8 Geometry for Problem #15. All dimensions are in mm

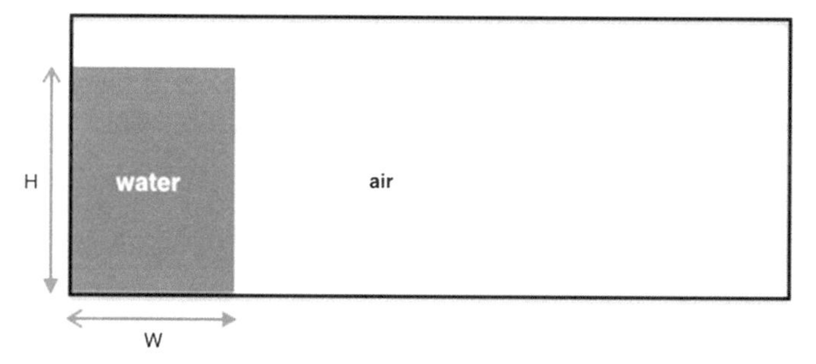

Fig. Q6.9 Geometry and initial conditions for the dam break problem

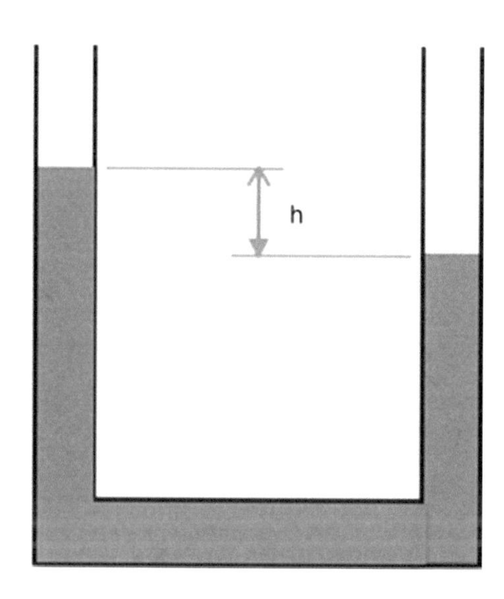

Fig. Q6.10 Geometry and initial conditions for the case of oscillating manometer

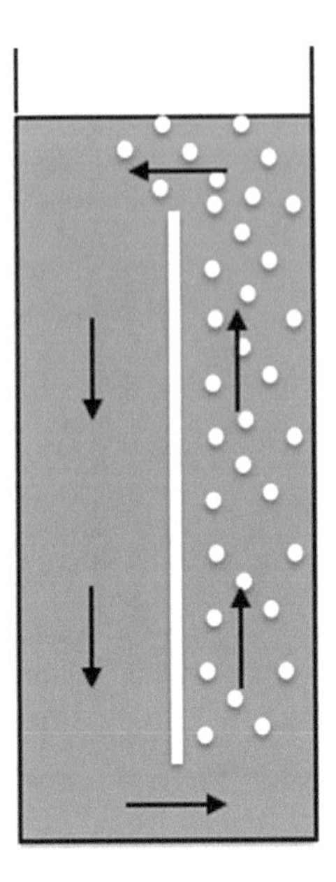

Fig. Q6.11 Geometry and formulation of the internal air loop reactor problem

Fig. Q6.12 Geometry and formulation of the two-dimensional bubble column problem

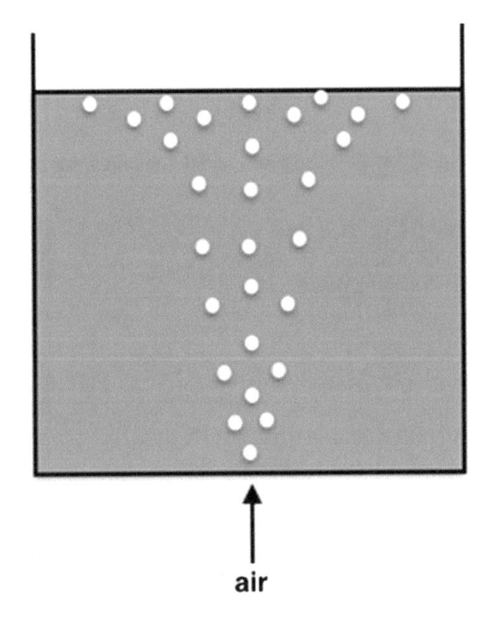

19. Repeat Problem #18 using the Delaunay triangulation method.
20. Use the advancing front method for the generation of a triangulated 2-d unstructured mesh for the computational domain shown in Fig. Q6.5.
21. Repeat Problem #20 using the Delaunay triangulation algorithm.
22. Consider fully developed turbulent flow between two parallel plates separated by a height H in the y-direction. For this one-dimensional flow, the pressure gradient in the flow direction, dp/dx, will be constant and x-component of the velocity, u, varies only in the y-direction. For the case of air flow, find u(y) using Prandtl's mixing length model. Take $H = 0.05$ m, $dp/dx = -1$ N/m^3, density of air = 1.2 kg/m^3, kinematic viscosity = 1.5 m^2/s.
23. Repeat Problem #22 using the k-ε model of turbulence.
24. Repeat Problem #23 for the case of developing flow between two plates of length 2 m. Take the inlet profiles to be uniform for quantities of interest, and assume that the flow is fully developed at the end of 2 m.
25. Consider the case of flow over a backward facing step (see Problem #20i of Chap. 2). Set up the problem for an inlet Reynolds number of 100,000 and step size equal to half the height of the upstream channel. Calculate the flow field using the k-ε model of turbulence and the SIMPLE scheme.
26. Use the flow field computed in Problem #24 to compute the trajectory of a particle of size 10/20/50/100/200/500/1000 μm and density of 2500 kg/m^3. Assume that gravity is acting in the direction of flow, i.e., $g_x = 9.806$ m/s^2 and $g_y = 0$. The inlet velocity of the particles may be taken to be zero. One would like to see at distance the particle attains its 95% of its terminal velocity. Assume that the drag coefficient is 0.44. Use Lagrangian particle tracking approach to find the trajectory of each particle.
27. Use the flow field computed in Problem #25 for flow over a backward facing step and the Lagrangian particle tracking approach to determine the particle size that will be entrained in the eddy created downstream of the step.
28. Consider the classic two-phase flow test case of the two-dimensional case dam break problem sketched in Fig. Q6.9. The initial height and width of water column at $t = 0$ are H and W, respectively. The water column collapses by gravity, a wave forms and propagates to the right, hits the wall, reverses and continues moving to the left. Try to capture the interface position using the volume of fluid (VOF) model. See the early work of Harlow and Welch (1965) who predicted the flow field and the interface evolution using the marker and cell (MAC) scheme.
29. Consider another simple illustration of the interface dynamics in the form of the oscillating manometer case (see Fig. Q6.10). The two-dimensional Cartesian computational domain consists of a U-tube of a width of w which is open at both ends. At $t = 0$, the U-tube is filled with a liquid in such a way that the height in the left leg is higher than that in the right leg. As a result of gravity, the water in the left column rushes down, that in the right column rushes up and overshoots the equilibrium position (of equal heights), and the reverses the direction. This happens several times before it reaches the excess initial potential energy is dissipated by viscous friction. Use the volume of fluid model

to capture the oscillating interface position. See Suresan and Jayanti (2004) for a description of the problem and some numerical and experimental results.

30. Consider the idealized case of an internal airlift loop reactor shown schematically in Fig. Q6.11. A two-dimensional rectangular domain is divided into two compartments, and air bubbles are introduced into one side, the right side compartment in this case. The bubbles rise up and escape through the pressure boundary (corresponding physically to the gas-liquid interface) at the top. This induces a circulation of the liquid which circulates in the loop reactor. Simulate this flow field using the Eulerian-Eulerian two-phase flow model. See Mudde and Van den Akker (2001) for a discussion of this case.
31. A problem similar to the one above is the 2-d bubble column reactor shown in Fig. Q6.12. Here, there is no separator plate; gas is introduced in the centre of the reactor. As it funnels up, it creates an internal circulation which affects the way the gas bubbles spread out. Simulate this steady state case using the Eulerian-Eulerian two-phase flow model. Typical air bubbles in water are in the diameter range of 3–7 mm and have a rise velocity of the order of 0.2 m/s.
32. A catalytic chemical reaction occurs on the surface of the catalyst, and is thus a heterogeneous reaction. Consider the catalytic steam reforming of methanol over a catalyst. Methanol is vapourized and mixed with steam and is made to pass over a $Cu/ZnO/Al_2O_3$ catalyst so that the following reaction takes place:

$$CH_3OH + H_2O \rightarrow CO_2 + 3H_2$$

Experiments (Agrell et al. 2002) show that the rate of reaction of steam reforming of methanol, r_{SRM}, is given by the following kinetic expression under certain conditions:

$$r_{SRM} = k\, p_{CH_3OH}^{0.26}\, p_{H_2O}^{0.03} \quad \text{where } k = A \exp\left(\frac{-E}{RT}\right)$$

with $A = 1.9 \times 10^{12}$ and $E = -100.9$ kJ/mol.

Assume that we have the gaseous mixture flowing between two plates of infinite width, length L which are separated by a constant gap of height H. Assume that the catalyst is embedded on the walls so that the reaction takes place at the walls of the 2-d domain (see Fig. Q6.13). Assume that a 50:50 mol % mixture of methanol vapour and steam enter the reactor at the inlet You wish to find out how much of methanol gets converted over the given reactor length L, under steady operating conditions. Formulate the problem, i.e., write the governing equations in their simplified for this reacting flow and suggest suitable boundary conditions. Assume laminar flow.
33. Consider the well-studied reaction of burning of methane gas in air, which is a homogeneous combustion reaction resulting in a flame. While the overall reaction is rather simple, an engineer is often interested in fine details such as the ignition delay and stability of the flame. These phenomena can be understood only if the formation of intermediate radicals are also modeled. One such

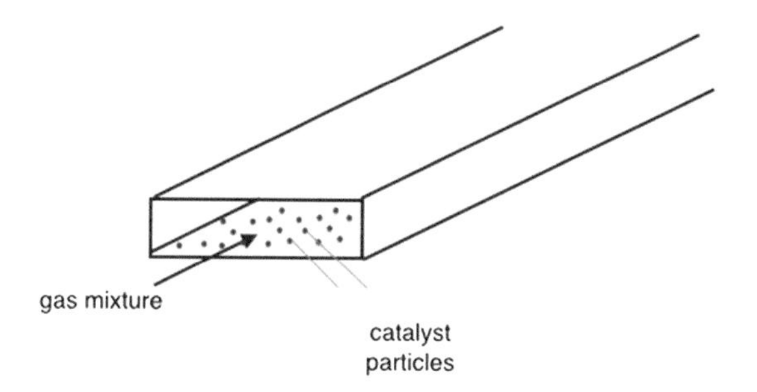

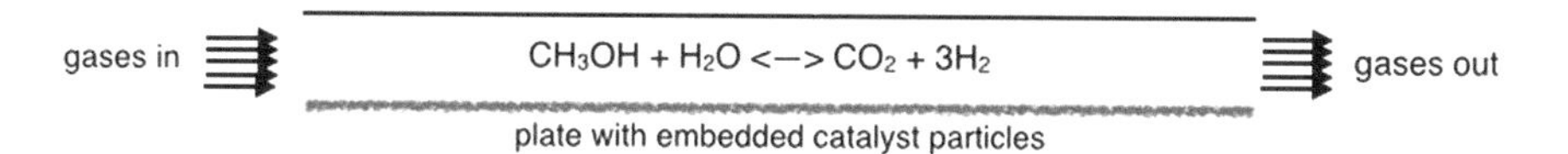

Fig. Q6.13 Geometry and formulation of the problem of catalytic steam reforming of methane

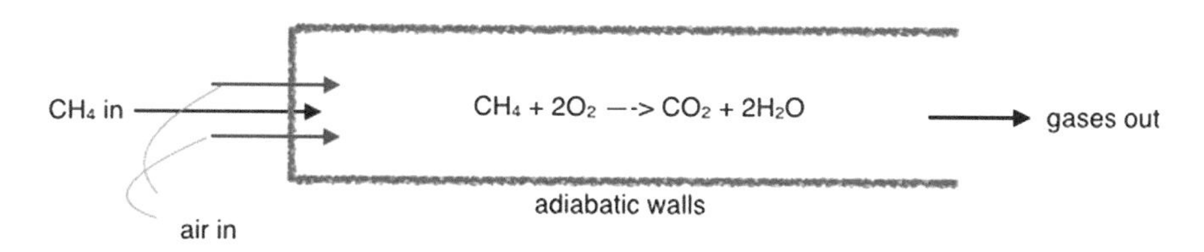

Fig. Q6.14 Geometry and formulation of the water-gas shift reactor problem

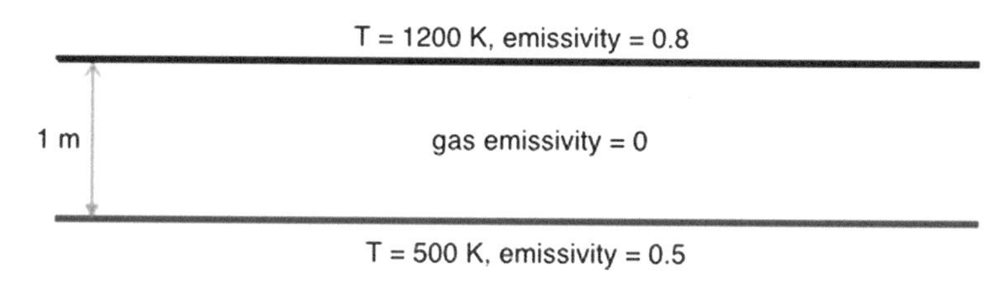

Fig. Q6.15 Radiative heat exchange between two infinitely long and wide parallel plates separated by 1 m

—highly simplified—reaction scheme for methane combustion involves a four-step reaction (Jones and Linstedt 1988) which includes the formation of hydrogen and carbon monoxide:

(a) $CH_4 + 1/2O_2 \rightarrow CO + 2H_2$
(b) $CH_4 + H_2O \rightarrow CO + 3H_2$
(c) $H_2 + 1/2O_2 \Leftrightarrow H_2O$
(d) $CO + H_2O \Leftrightarrow CO_2 + H_2$

It may be noted that reactions (a) and (b) are uni-directional whereas are reactions (c) and (d) are reversible. The reaction rates for these have been given by Jones and Linstedt (1988) as follows:

$$r^f_{(a)} = \{0.44 \times 10^{12} \exp(-30{,}000/RT)\} [CH_4]^{0.5} [O_2]^{1.25}$$
$$r^f_{(b)} = \{0.30 \times 10^{9} \exp(-30{,}000/RT)\} [CH_4][H_2O]$$
$$r^f_{(c)} = \{0.25 \times 10^{17}\, T^{-1} \exp(-40{,}000/RT)\} [H_2]^{0.5} [O_2]^{2.25} [H_2O]^{-1}$$
$$r^f_{(d)} = \{0.275 \times 10^{10} \exp(-20{,}000/RT)\} [CO][H_2O]$$

Here, the concentration is in terms of kmol and the activation energy is in terms of calories and not Joule. The reverse reaction rates can be calculated based on the equilibrium constant K which is a function of pressure and temperature and can be obtained from the report by Gordon and McBride (1971) and its more recent versions. Consider the same computational domain as for Problem #32 and formulate the problem for this gaseous mixture flow with a 4-step homogeneous reaction mechanism.

34. Consider the problem of turbulent combustion of methane in a 2-dimensional channel as shown in Fig. Q6.14. The fuel, CH_4, enters through the central slot and the oxidant, in this case, air, enters through two slots, one above and one below the central port. Write down the equations to be solved and the boundary conditions for this turbulent combustion problem. Use the k-ε model for turbulence and the mixture fraction-based infinite rate chemistry model for combustion.
35. Repeat Problem #34 using the eddy break-up model to deal with turbulent combustion.
36. Repeat Problem #34 if you wish to the probability density function route to modeling turbulence-chemistry interaction. Use the double delta function for the presumed pdf. Work out the details of implementation of this model by listing the equations to be solved.
37. Calculate the radiative heat exchange between two infinitely long and wide parallel plates separated by a distance of 1 m (see Fig. Q6.15) at temperatures 1200 and 500 K and having emissivities 0.8 and 0.5, respectively.
38. Repeat Problem #37 if the gas between the plates is a radiatively participating medium with an emissivity of 0.4.
39. Consider the transient heat conduction case discussed in Problem #20j of Chap. 2. Imagine that it is dropped out of a geo-stationary satellite and is getting cooled simultaneously by radiating heat into the outer universe. If the block, of dimensions 3 m × 2 m × 1 m, is initially at a uniform temperature of 1500 K, find its temperature variation with time. Take the emissivity of the block to be 0.8 and the outer space to be black body at a temperature of 3 K.
40. Find the steady-state temperature distribution within the block if its 3 m × 2 m is facing earth all the time. Under steady conditions, the radiative heat received

from this face is radiated into the outer space (which is at 3 K) by all the other five faces of the block. Take the earth to be a sphere with a radius of 6000 km emitting radiation at a temperature of 300 K with an emissivity of 0.5 and that the object is at a distance of 36,000 km from the earth.

Chapter 7
CFD and Flow Optimization

We have seen over the preceding several chapters how the equations governing flow can be solved for even turbulent, reacting flows through irregular geometries of the kind that are often encountered in industrial applications. In the form of user-friendly software running on desktop computers, CFD has been adopted by researchers and engineers from a range of disciplines from aerospace to ocean engineering. Despite these developments, CFD has remained primarily a flow analysis tool capable of simulating realistic engineering flow situations. In its conventional form, it is best used as a "what-if" analysis tool. Since a CFD solution requires the specification of the flow domain and geometrical details inside, its applicability as a design tool is limited. For example, one can use CFD to find the drag and lift forces over a wind turbine blade of a given shape rotating at a constant speed, but one cannot use it to find the shape and size for a given drag/lift factor. One would have to try different shapes, find the drag and lift forces for each case, and then use this information somehow to arrive at the desired shape. Similarly, the heat transfer enhancement over a dimpled surface can be calculated using CFD for a given size, shape and distribution of dimples but the "design" of the dimpled surface to meet a required heat transfer enhancement cannot be done readily. Yet, since CFD enables a fairly accurate simulation of flow, being able to use it as a design tool would be highly attractive and even necessary to realize its full benefits. Recent efforts have focused on using optimization techniques to make this happen, and this is discussed in this chapter. In Sect. 7.1, we will look at the formulation of optimization problem in the general context. In Sect. 7.2, we will discuss several methods of finding the optimal solution. In Sect. 7.3, we will discuss a couple of case studies which illustrate the application of these concepts to CFD-based optimization.

S. Jayanti, *Computational Fluid Dynamics for Engineers and Scientists*,
https://doi.org/10.1007/978-94-024-1217-8_7

7.1 Formulation of the Optimization Problem

Before we formally approach the problem formulation, let us consider a simple example of design optimization related to fluid flow. Let us say that we have to find a suitable pipe diameter that would carry a fluid at a volumetric flow rate of Q over a distance L. As the diameter increases, the flow velocity decreases, and both the size of pump required and the power required to keep pumping the fluid decrease leading to reducing lifetime cost of the pipeline. However, as the pipe size increases, the cost of the pipeline and that of the pipe fittings that go with it will increase. This situation is illustrated schematically in Fig. 7.1. The total cost is a sum of these two components. One can see that there is an optimal intermediate diameter at which the total cost is minimum, and, on cost consideration, this should be our design choice. A similar situation arises when we consider heat exchanger design. Here, one would try to minimize the surface area required for heat transfer as this would immediately lead to savings in material costs and lead to better utilization of space in a industrial setting. One of the ways of achieving a reduction in the surface area requirement is to enhance the heat transfer coefficient. This can be done by increasing the surface roughness or by adding specially shaped inserts. Both these would often lead to increase in pressure drop which would increase the cost of operation. Both measures would also tend to increase the cost of fabrication of the heat exchanger compared to one without either measure. Therefore, there will be an optimum value of the surface roughness (or the number of inserts per unit length) at which the net benefit from increasing the heat transfer coefficient will be the maximum and that will be our design value.

In practical cases, a number of complications will often arise. One may have limited number choices to choose from. In the first example, pipes of certain sizes only may be available commercially. One may have a range of sizes that can be

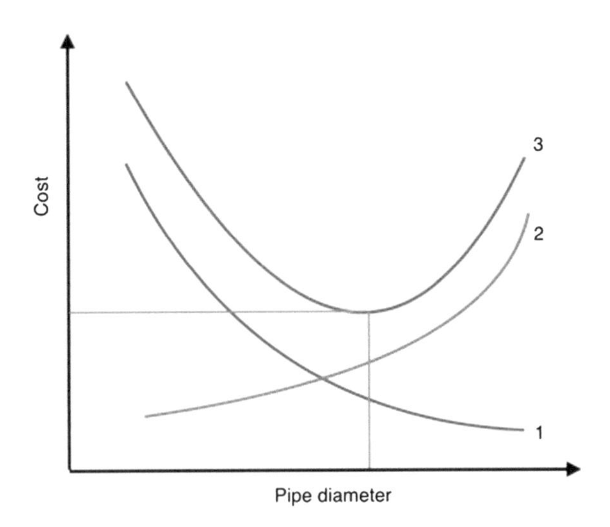

Fig. 7.1 Formulation of the optimization problem for the pipe diameter carrying a given flow rate. *Note* 1. decreasing cost of pump with increasing diameter, 2. increasing cost of pipeline, and 3. the overall cost of the pipeline system

practically accommodated. For instance, in the second example, there can be a limit on the maximum length of the heat exchanger that can be accommodated. There can also be a number of choices for the shape of the inserts that are used to increase the heat transfer coefficient. These shapes belong to categories rather than being rational numbers. There can be a number of other factors influencing the cost. There can be multiple criteria (e.g. pressure drop not too high but good heat transfer coefficient) on which the selection needs to be based. Keeping these factors in mind, the optimization problem is mathematically formulated in terms of an objective function (f) that needs to be either minimized or maximized, a design vector (**x**) which contains the design variables that influence the objective function in a known way, and the equality and non-equality constraints on the design variables that need to be satisfied in the process of optimization. A constrained optimization problem can thus be stated as follows:

Find the design variable vector x which minimizes the objective function

$$f(x) = f(x_1,\ x_2,\ x_3, \cdots x_n) \tag{7.1.1a}$$

subject to m equality constraints

$$g_i(x) = 0 \quad i = 1, 2, 3, \cdots m \tag{7.1.1b}$$

and p inequality constraints

$$h_j(x) \leq 0 \quad j = 1, 2, 3, \cdots p \tag{7.1.1c}$$

It may be noted that while there are no restrictions on the number of inequality constraints, the number of equality constraints cannot be greater the number of design variables, i.e., $m \leq n$. In fact, if $m = n$, then the solution is already fixed and there is no leeway or discretion to optimize the design variables. If $m > n$, the formulation is inconsistent. The set of feasible designs, i.e., those values of design variables that satisfy the constraints, is called the constraint set. The optimum solution must lie within this feasible set.

If there are no constraints, then the optimization problem becomes unconstrained optimization. A constrained optimization problem can be converted into an unconstrained one using the penalty method whereby the objective function is penalized for constraint violations. Similarly, a problem requiring maximization of F(x) can be converted into a minization problem by defining $f(x) = -F(x)$. An inequality criterion of form $H_j(x) \geq 0$ can be converted into the form of Eq. (7.1.1c) by defining $h_j(x) = -H_j(x)$. Further, a design variable can be a rational number or discrete (where its value needs to be selected from a finite set of values) or integer (e.g., the number of blades in a wind turbine). The functions f(x), g(x) and h(x) may be linear or non-linear functions of the design variables. Thus the formulation given by Eqs. (7.1.1a, b and c) is quite generic.

The solution to the optimization problem corresponds to a minimum of the objective function $f(\mathbf{x})$ within the feasible set. If the minimum occurs at $\mathbf{x}^*$, then the minimum is global minimum if $f(\mathbf{x}^*) \leq f(\mathbf{x})$ for all $\mathbf{x}$ in the feasible region. If the condition that $f(\mathbf{x}^*) \leq f(\mathbf{x})$ holds good only in the neighbourhood of $\mathbf{x}^*$, then the solution is considered as a local minimum. In either case, the condition for a minimum requires the first derivative of $f(\mathbf{x})$ at $\mathbf{x}^*$ to be zero and the second derivative at $\mathbf{x}^*$ to be positive. In the general case where there are n design variables, then $f(\mathbf{x})$ is a function of several variables. In such a case, one can expand $f(\mathbf{x})$ about $\mathbf{x}^*$ in Taylor series as

$$f(\mathbf{x}) = f(\mathbf{x}^*) + \nabla f(\mathbf{x}^*)^T \mathbf{d} + \tfrac{1}{2}\mathbf{d}^T H(\mathbf{x}^*)\mathbf{d} + R \tag{7.1.2a}$$

where

$$\mathbf{d} = \mathbf{x} - \mathbf{x}^* \tag{7.1.2b}$$

$$\nabla f(x^*) = \{\, \partial f(\mathbf{x}^*)/\partial x_1 \quad \partial f(\mathbf{x}^*)/\partial x_2 \quad \partial f(\mathbf{x}^*)/\partial x_3 \quad \cdots \quad \partial f(\mathbf{x}^*)/\partial x_n \,\} \tag{7.1.2c}$$

and H is the Hessian matrix defined as

$$H = \begin{pmatrix} \frac{\partial^2 f(\mathbf{x}^*)}{\partial x_1 \partial x_1} & \frac{\partial^2 f(\mathbf{x}^*)}{\partial x_1 \partial x_2} & \frac{\partial^2 f(\mathbf{x}^*)}{\partial x_1 \partial x_3} & \cdots & \frac{\partial^2 f(\mathbf{x}^*)}{\partial x_1 \partial x_n} \\ \frac{\partial^2 f(\mathbf{x}^*)}{\partial x_2 \partial x_1} & \frac{\partial^2 f(\mathbf{x}^*)}{\partial x_2 \partial x_2} & \frac{\partial^2 f(\mathbf{x}^*)}{\partial x_2 \partial x_3} & \cdots & \frac{\partial^2 f(\mathbf{x}^*)}{\partial x_2 \partial x_n} \\ & \cdots & & \cdots & \cdot \\ & \cdots & & \cdots & \cdot \\ \frac{\partial^2 f(\mathbf{x}^*)}{\partial x_n \partial x_1} & \frac{\partial^2 f(\mathbf{x}^*)}{\partial x_n \partial x_2} & \frac{\partial^2 f(\mathbf{x}^*)}{\partial x_n \partial x_3} & \cdots & \frac{\partial^2 f(\mathbf{x}^*)}{\partial x_n \partial x_n} \end{pmatrix} \tag{7.1.2d}$$

and R is the truncation error.

The condition for optimality can then be written as

$$\nabla f(\mathbf{x}^*) = 0 \tag{7.1.3a}$$

and

$$\mathbf{d}^T H(\mathbf{x}^*)\mathbf{d} > 0 \tag{7.1.3b}$$

or that the Hessian is positive definite, i.e., all its eigenvalues are strictly positive. For a maximization problem, the Hessian should be negative definite.

For simple problems, the optimal solution can be found graphically. For practical problems, there are principally two types of approaches for finding the optimal solution: the *indirect* method which is based on satisfying the optimality criteria given by Eqs. (7.1.3a and b), and the *direct* method, which is based on an algorithmic, iterative search of the feasible design space. We will briefly illustrate the application of the indirect method to two simple constrained optimization problems and discuss the direct method in the next section.

Consider the problem of finding the maximum volume of a box that can be inscribed in a sphere of radius R. Take the origin to coincide with the centre of the sphere and let the sides of the box be 2X, 2Y and 2Z in the three coordinate directions. The volume of the box is then given by 8XYZ. Since the vertices of the box lie on the sphere, we have a constraint, namely,

$$X^2 + Y^2 + Z^2 = R^2 \tag{7.1.4}$$

We therefore seek to maximize the volume of the box V:

$$\text{maximize } V(X, Y, Z) = 8XYZ \tag{7.1.5}$$

subject to the constraint given by Eq. (7.1.4). In order to find the solution, we first eliminate Z from Eq. (7.1.5) using the constraint (7.1.4) by writing Z as

$$Z = \left(R^2 - X^2 - Y^2\right)^{1/2} \tag{7.1.6}$$

so that the objective function becomes

$$V(X, Y) = 8XY\left(R^2 - X^2 - Y^2\right)^{1/2} \tag{7.1.7}$$

If X*, Y* are the coordinates of the extremum point where V(X,Y) is maximum (or minimum), then

$$\partial V/\partial X|_{X^*,Y^*} = 8Y^*\{(R^2 - X^{*2} - Y^{*2})^{1/2} - X^{*2}/(R^2 - X^{*2} - Y^{*2})^{1/2}\} = 0 \tag{7.1.8}$$

and

$$\partial V/\partial Y|_{X^*,Y^*} = 8X^*\{(R^2 - X^{*2} - Y^{*2})^{1/2} - Y^{*2}/(R^2 - X^{*2} - Y^{*2})^{1/2}\} = 0 \tag{7.1.9}$$

From these, we obtain $2X^{*2} + Y^{*2} = R^2$ and $X^{*2} + 2Y^{*2} = R^2$ Solving these and using Eq. (7.1.6), we obtain

$$X^* = Y^* = Z^* = (R/3)^{1/2} \tag{7.1.10}$$

It is left to the reader to show that the Hessian at the extremum, H(X*, Y*, Z*), of V(X, Y, Z) is negative-definite indicating that the volume is maximum.

In theory, it is possible that a problem with only equality constraints can be reduced to a constraint-free optimization problem by eliminating some of the design variables. For a problem involving n variables and m equality constraints, a new objective function can be defined in terms of (n-m) variables by eliminating m variables using the equality constraints. However, in practice, this may not be feasible or it may not render the problem any less tractable. Another way of dealing with a constrained optimization problem with both equality and inequality constraints is to render it into an unconstrained problem using the Lagrange multiplier technique, and using the Kuhn-Tucker conditions (Rao 2009) to identify stationary points that may be potential optimal points. An inequality condition such as the one given by Eq. (7.1.1c) is said to be active if it is satisfied with an equality sign at the stationary point, i.e., $h_j(x^*) = 0$. A point x^* is said to be a regular point of the design space if $f(x)$ is differentiable at x^* and the gradient vectors of all equality and active inequality constraints are linearly independent. One can then construct a Lagrange function $L(\mathbf{x}, \mathbf{u}, \mathbf{v}, \mathbf{s})$ which adds the equality and the inequality constraints to the objective function:

$$L(\mathbf{x}, \mathbf{v}, \mathbf{u}, \mathbf{s}) = f(\mathbf{x}) + \sum_{i=1}^{m} v_i g_i(\mathbf{x}) + \sum_{i=1}^{p} u_i \left(h_i(\mathbf{x}) + s_i^2\right) \tag{7.1.11}$$

The stationary points of the Lagrange function can be found using the Kuhn-Tucker conditions:

$$\partial L/\partial x_j = \partial f/\partial x_j + \sum_{i=1}^{m} v_i^* \partial g_i/\partial x_j + \sum_{i=1}^{p} u_i^* \partial h_i/\partial x_j = 0,\ j = 1, 2, \cdots, n \tag{7.1.12}$$

$$\begin{array}{ll} g_i(\mathbf{x}^*) = 0, & i = 1, 2, \cdots, m \\ h_i(\mathbf{x}^*) + s_i^2 = 0, & i = 1, 2, \cdots, p \\ u_i^* s_i = 0, & s_i^2 \geq 0,\ u_i^2 \geq 0,\ i = 1, 2, \cdots, p \end{array}$$

These equations for (n + m + 2p) equations that are needed to solve n values of x^*, m values of v^*, p values each of u^* and s^* that occur in Eq. (7.1.11).

Not all stationary points are optimal points, as the Kuhn-Tucker conditions are only necessary conditions and not sufficient conditions. However, if the objective function and the constraint functions are all convex (see Rao 2009), then the minimum obtained using K-T conditions will be the global minimum. However, it is very difficult to determine if a given optimization problem is convex. Several other direct methods for special classes of optimization problems are discussed in Rao (2009) and Deb (2004), among others. For most engineering problems, the objective function and/or the constraints cannot be expressed as explicit functions of the design variables. For such cases, direct methods cannot be used and indirect methods, based on iterative search algorithms, are more useful. These are discussed in the next section.

7.2 Iterative Search Methods for Optimization Problems

In these methods, algorithmic, step-by-step calculation procedures are used repetitively to arrive at the optimal solution. As in the case of iterative methods for the solution of linear algebraic equations discussed in Sect. 5.3, the optimal solution is approached by taking incremental steps towards it in an algorithmic way. Classical search methods can be classified into two major groups: direct methods that use only the objective function value, and gradient-based methods that require the information on the gradient of the objective function. Bracketing methods and golden section methods, often illustrated in the context of single-variable optimization techniques, are examples of the former. The Newton-Raphson method, the secant method are examples of gradient-based methods. Most often, engineering problems require multivariable optimization. Extensions of these methods are possible. A number of non-traditional methods employing newer concepts have also been developed recently. Some of these methods are discussed briefly below; the interested reader is referred to standard textbooks on optimization such as those of Deb (2004), Nocedal and Wright (2006), Thévenin and Jániga (2008), Rao (2009) and Mohammadi and Pironneau (2010).

7.2.1 Direct Iterative Search Methods

Let us consider the simpler case of single-variable optimization problem, namely, minimize f(x) over an interval $a < x < b$. In bracketing methods, the minimum is found in two stages. In the first stage, assuming unimodal distribution of f(x), we search in the direction of decreasing f(x) for an interval (characterized by a lower bound x_l and an upper bound x_u) in which the minimum lies. This can be done by finding three successive of f(x), namely, $f(x - h_1)$, f(x) and $f(x + h_2)$ such that $f(x - h_1) > f(x) < f(x + h_2)$ where h_1 and h_2 are step sizes, which can be equal or unequal, as per the algorithm. This would then give $x_l = x - h_1$ and $x_u = x + h_2$. Once the minimum is bracketed, a more accurate estimate of the minimum is obtained in the second stage by employing a region elimination method such as the interval halving method or the golden section method. In the latter, the bracketed interval of $x_l \leq x_{min} \leq x_u$ is projected to on a [0, 1] interval and a function evaluation is made at two intermediate locations, i.e., at $x_{\tau 1} = x_l + \tau(x_u - x_l)$ and $x_{\tau 2} = x_u - \tau(x_u - x_l)$, where τ is the golden fraction having a numerical value of 0.618. If $f(x_l) > f(x_{\tau 1}) < f(x_{\tau 2})$, then the minimum lies in the interval $[x_l, x_{\tau 2}]$, otherwise, it will lie in the interval $[x_{\tau 1}, x_u]$. Now, this interval is projected on to [0, 1], and the function evaluation is done at the two intermediate points. Actually, the function value at one of these two points will already have been evaluated, and only one more function value will be needed. Using the latest set of function values, we again determine the sub-interval in which the minimum would lie. Thus, in the golden section method, the interval in which the minimum would lie is reduced to 61.8% of the previous value. The search would be continued until the interval became sufficiently small. This method is illustrated schematically in Fig. 7.2.

In case of single-variable optimization, the search direction for the minimum is fixed; it corresponds to the positive or negative x direction, depending on whether f(x) is decreasing or increasing with x. In the case of multivariable optimization involving N decision variables, there can be multiple directions for search. It is possible to extend the single-variable approach of bracketing to multiple directions in sequence. In the univariate method, we change one variable at a time by fixing the values of all the other variables at the starting point. This makes it a single-variable optimization problem. We vary this variable along, for example, the first coordinate direction to find the best approximation of minimum. We continue

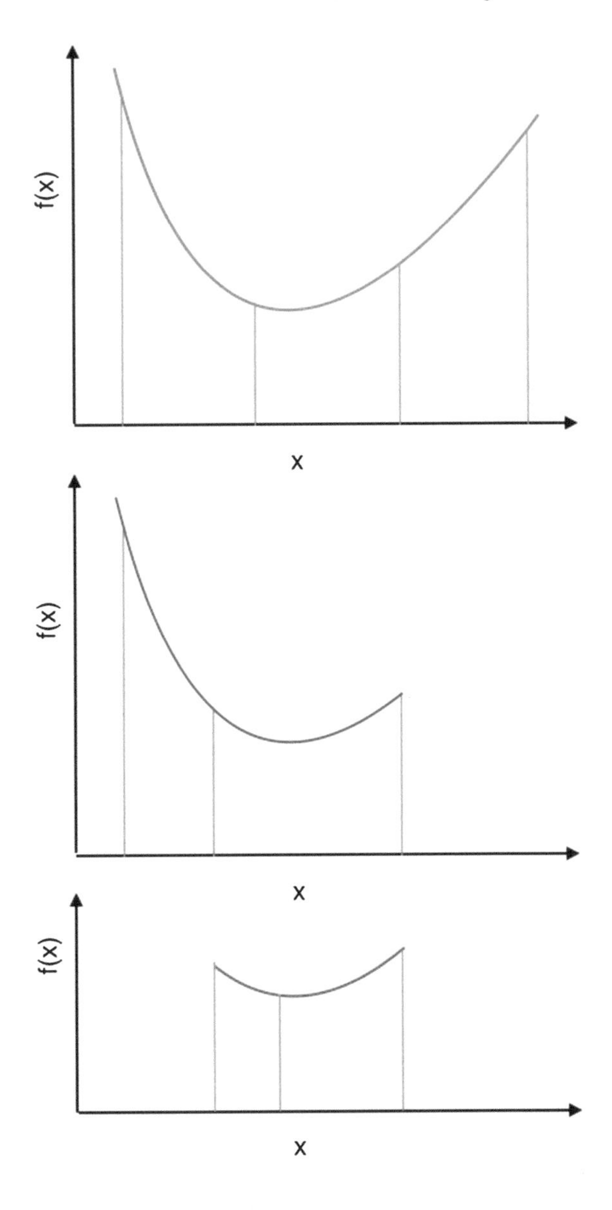

Fig. 7.2 The golden section method for the minimization of a unimodal function

from this approximate minimum point by searching in a new direction, for example, in the second coordinate direction. We continue this until all coordinate directions are exhausted. Then, from the latest approximation of the minimum point, we start the process of sequential minimization by searching along the coordinate directions until no further improvement is possible. Other search methods such as the random search method or the grid search method can also be employed (Rao 2009). The most popular among the direct search methods is the conjugate directions method of Powell (1964). This method uses the information of previous search directions to create a new search direction that is linearly independent of (or conjugate to) the previous search directions. Minimization is carried out in this direction and a new conjugate direction is then determined to proceed with the search further. For problems where the objective function is quadratic in N-dimensional space, for example,

$$Q(\mathbf{X}) = \tfrac{1}{2}\mathbf{X}^T\mathbf{A}\mathbf{X} + \mathbf{B}^T\mathbf{X} + \mathbf{C}, \tag{7.2.1}$$

if $Q(\mathbf{X})$ is minimized sequentially each time in mutually conjugate directions, then the method is guaranteed to converge in N or fewer steps. For functions higher than quadratic, there is no formal proof of convergence; however, it may still be expected to work as around a minimum, the function may be approximated by a quadratic. The method has been successfully used for many non-quadratic functions.

It can be shown that for a quadratic function Q, if $\mathbf{X}_{m1}$ and $\mathbf{X}_{m2}$ are the minima obtained by searching along the same direction S from two starting points $\mathbf{X}_1$ and $\mathbf{X}_2$ (see Fig. 7.3a), then the line joining $\mathbf{X}_{m1}$ and $\mathbf{X}_{m2}$, i.e., the line $(\mathbf{X}_{m1} - \mathbf{X}_{m2})$ will be conjugate to the search direction S. Thus, for a two-dimensional case, Powell's conjugate directions algorithm starts with a starting point 1 (see Fig. 7.3b) and a search direction along coordinate direction X_2. We reach the minimum point 2 by minimizing along the direction 1-A. From point 2, we search along coordinate direction X_1 to reach the minimum point 3 by searching along direction 2-B. From point 3, we again search in the coordinate direction X_2 to reach point 4 in the direction 3-C. Now, the line joining 2 and 4 is the new, conjugate search direction. We search along this direction (along line 2-4-D) to reach the minimum point 5. From point 5, we once again search along coordinate direction X_2 to reach point 6 by minimizing along line 5-E. We now search along the direction 2-4 but starting from point 6, i.e., along line 6-F to reach the new minimum 7. The search now continues along the new conjugate direction 5-7 to reach point 8 by minimizing along direction 7-G. From 8, we search in the direction parallel to 5-6 but starting from point 8 to find the next minimum (which is not shown in Fig. 7.3b), and continue further until we reach the minimum.

A totally different approach, but still using only objective function values and not its gradients, is the simplex method developed in the early 1960s (Spendley et al. 1962; Nelder and Mead 1965). A simplex in an N-dimensional space is a geometric figure formed by (N + 1) points. In two dimensions, it would be a triangle, and in three dimensions (or in a three-decision variable problem), it would be a tetrahedron. To begin with, the objective function is evaluated at (N + 1) feasible points. The aim of the method is to replace the worst point, i.e., the one having the highest objective function value (for a minimization problem) with one having a lower value. The coordinates of the "better" point are found by reflecting the worst point about the centroid of the rest of the N points, i.e., in the direction of lower objective function values. If $\mathbf{X}_w$ are the coordinates of the worst point, $\mathbf{X}_c$ are those of the centroid of the rest of the vertices of the simplex, and $\mathbf{X}_r$ are those of the reflected point, then

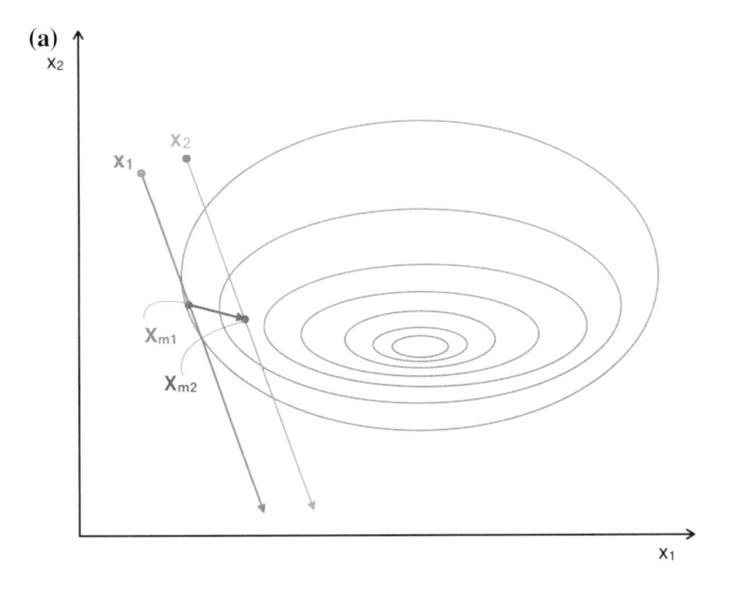

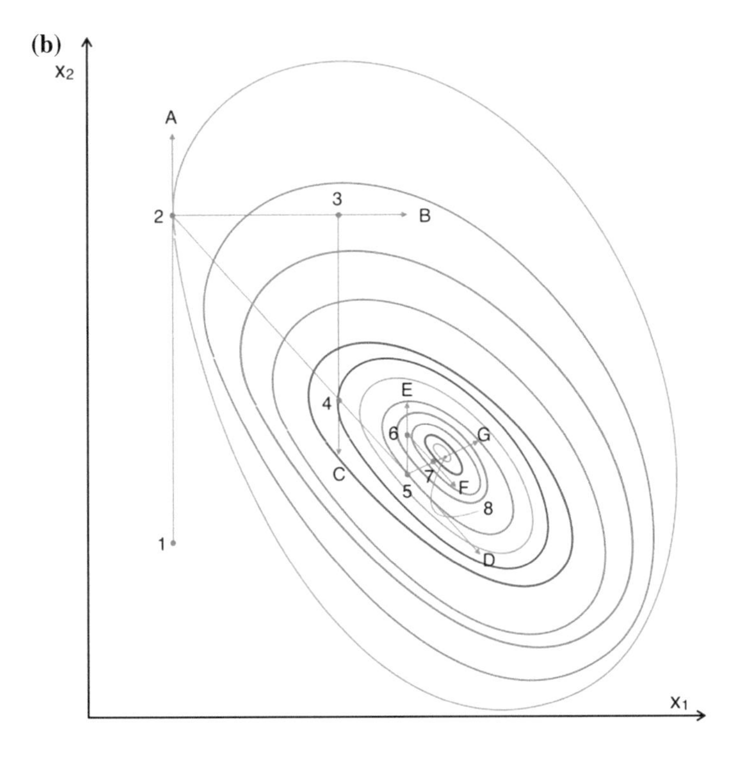

Fig. 7.3 Powell's conjugate directions method for minimization of a function: **a** illustration of the conjugate direction, and **b** sequential progression towards the minimum using the conjugate directions algorithm

$$\mathbf{X_C} = \frac{1}{N}\sum_{i=1}^{N+1} \mathbf{X_i},\ i \neq w \tag{7.2.2a}$$

$$\mathbf{X_r} = (1+\alpha)\mathbf{X_C} - \alpha\mathbf{X_W} \tag{7.2.2b}$$

Here α is the reflection factor; it determines the distance to which the new point is reflected. It is defined as the ratio of the distance between $\mathbf{X}_r$ and $\mathbf{X}_c$ to that between $\mathbf{X}_w$ and $\mathbf{X}_c$. The function value at the reflected point, i.e., $f(\mathbf{X}_r)$ is evaluated. If $f(\mathbf{X}_r) < f(\mathbf{X}_w)$, then the worst point is replaced in the simplex by the reflected point. If $f(\mathbf{X}_r) < f(\mathbf{X}_b)$ where $\mathbf{X}_b$ is the best point in the simplex, i.e., the one having the lowest objective function value, then we continue further by expanding in the same direction. A new point, $\mathbf{X}_e$, is found such that

$$\mathbf{X_r} = (1-\beta)\mathbf{X_c} - \beta\mathbf{X_r} \quad 0 \leq \beta \leq 1 \tag{7.2.3}$$

where β is the coefficient of expansion. If, on the other hand, $f(\mathbf{X}_r) > f(\mathbf{X}_w)$, we find a new $\mathbf{X}_r$ with a lesser value of α in Eq. (7.2.2b) so that the point is retracted towards the centroid. This process is carried out until a better point than the current worst is found. If a better point than $\mathbf{X}_r$ is found, then the simplex shrinks. The process is continued until the simplex shrinks sufficiently. At this point, the centroid of the simplex is taken as the minimum. If despite repeated contraction, a new $\mathbf{X}_r$ is not found such that $f(\mathbf{X}_r) < f(\mathbf{X}_w)$, then the process may be started with the second worst point of the simplex. A modified form of this method, namely, the complex method of Box (1965), used for constrained multivariable optimization problems, is used in the case study discussed in Sect. 7.3.3. Further details of the method are discussed in that section.

7.2.2 Gradient-Based Search Methods

The methods discussed in the above section use only the objective function value. Given that the optimal point x^* requires that $\partial f/\partial x|_{x^*} = 0$, methods used for finding the roots of non-linear algebraic equations, namely, the Newton-Raphson method, the quasi-Newton method and the secant method, can be used to derive formulae for search directions and step sizes. Expanding f(x) at $x = x_i$, and retaining the first three terms in the Taylor series expansion, we get

$$f(x) = f(x_i) + f'(x_i)(x - x_i) + {}^1\!/\!_2 f''(x_i)(x - x_i)^2 \tag{7.2.4}$$

Taking the derivative of Eq. (7.2.2a and b) with respect to x, we obtain

$$f'(x) = f'(x_i) + f''(x_i)(x - x_i) \tag{7.2.5}$$

Evaluating equation at the optimal point x^* and noting that $f'(x^*) = 0$, we get

$$f'(x^*) = f'(x_i) + f''(x_i)(x^* - x_i) = 0$$

or

$$x^* = x_i - f'(x_i)/f''(x_i) \tag{7.2.6}$$

In Newton's method, starting with an estimated optimal point x_i, the first and second derivatives of f(x) at x_i are evaluated, and an improved guess value of the optimal point is obtained using

$$x_{i+1} = x_i - f'(x_i)/f''(x_i) \tag{7.2.7}$$

In the quasi-Newton method, which can be used when f(x) is not easy to differentiate or is not available in closed form, the derivatives are evaluated numerically using finite differences. In this case, the improved value of x, namely, x_{i+1}, is given by

$$x_{i+1} = x_i - \frac{\Delta x\{f(x_i + \Delta x) - f(x_i - \Delta x)\}}{2\{f(x_i + \Delta x) - 2f(x_i) + f(x_i - \Delta x)\}} \tag{7.2.8}$$

where Δx is a small step size. A similar expression can be derived for the estimation of x_{i+1} using the secant method where the second derivative is estimated in terms of finite difference approximations of the first derivatives.

Gradient-based methods have far superior convergence rates compared to the bracketing or simplex methods, especially in the neighbourhood of the optimum point. This advantage is gained at the cost of determination of the first and the second derivatives at every new point. Where an analytical expression for the objective function is not available or when it is not easily differentiable, then these derivatives need to be evaluated numerically. These difficulties are more pronounced in multi-dimensional problems. Here, a distinction is drawn therefore between those methods that use only the first derivative, such as the steepest descent method of Cauchy (1847) and the conjugate gradients method of Hestenes and Stiefel (1952) and Fletcher and Reeves (1964), and those that use the second derivative too, for example, the Newton method and the Marquardt (1963) method. Quasi-Newton methods in which the second derivative, or the Hessian matrix in case of multidimensional problems, is approximated by another matrix which is iteratively updated have also been developed (Fletcher and Powell 1963; Broyden 1967; Shanno 1970). These methods are briefly reviewed here. For detailed algorithms, the reader is referred to specialized books and current literature.

In the steepest descent method, the search starts in the direction of $-\nabla f|_0$ where $\nabla f|_0$ is the gradient of the objective function evaluated at the starting point $\mathbf{X}_0$. The step size λ_0 is obtained by trying to minimize f ($\mathbf{X}_0 - \lambda_0 \nabla f_0$). The new minimum, $\mathbf{X}_1$, is given by $\mathbf{X}_1 = \mathbf{X}_0 - \lambda_0(\nabla f)|_0$. Now, the gradient is evaluated afresh at $\mathbf{X}_1$, and the search is continued in the $-\nabla f|_{X1}$ direction. In the conjugate gradient method, proposed originally by Hestenes and Stiefel (1952) for the solution of a system of linear algebraic equations, successive search directions are determined such that they are mutually orthogonal to the previous search directions. This method has a quadratic convergence rate for quadratic functions, and has the guarantee of convergence in N steps or fewer for an N-dimensional problem if infinite precision arithmetic is used. In a numerical computation with finite precision arithmetic and round-off errors, the minimization at every univariate optimization step is not exact, and this will have a spill-over effect on the determination of subsequent minima. As a result, the solution may not be obtained in N steps. The round-off errors will also accumulate with repetitive calculations. Fletcher and Reeves (1964) proposed the method as a variant of the steepest descent method to obtain quadratic convergence.

Consider the quadratic function Q(**X**) given by Eq. (7.2.1) above. Starting from an arbitrary point X_0, and using the steepest-descent search direction

$$\mathbf{S}_0 = -\nabla Q_0 \tag{7.2.9a}$$

and the step length from

$$\mathbf{X}_1 - \mathbf{X}_0 = \lambda_0 \mathbf{S}_0 \tag{7.2.9b}$$

where λ_0 is obtained from the minimization of Q(**X**) along direction $\mathbf{S}_0$ as

$$\lambda_0^* = -\left(\mathbf{S}_0^T \nabla Q_0\right) / \left(\mathbf{S}_0^T \mathbf{A} \mathbf{S}_0\right). \tag{7.2.9c}$$

The next search direction, $\mathbf{S}_1$, is found as a linear combination of $\mathbf{S}_0$ and $-\nabla Q_1$:

$$\mathbf{S}_1 = -\nabla Q_1 + \beta_1 \mathbf{S}_0 \tag{7.2.10}$$

Here, the constant β_1 is to be determined such $\mathbf{S}_0$ and $\mathbf{S}_1$ are conjugate with respect to Q. This condition gives

$$\beta_1 = -[(\nabla Q_1)^T(\nabla Q_1)]/[(\nabla Q_0)^T \mathbf{S}_0] = [(\nabla Q_1)^T(\nabla Q_1)]/[(\nabla Q_0)^T(\nabla Q_0)] \tag{7.2.11}$$

Equations (7.2.10) and (7.2.11) can be generalized to obtain recursive relations for $\mathbf{S}_i$ and β_i. The Fletcher-Reeves method is based on these relations and can be stated as follows. One starts with an arbitrary point $\mathbf{X}_0$ and the search direction $\mathbf{S}_0$ given by Eq. (7.2.9a), and determines the step length and the new point $\mathbf{X}_1$ using Eqs. (7.2.9b) and (7.2.9c). This is then continued recursively. For iteration i, we find

$$\nabla Q_i = \nabla Q(\mathbf{X}_i) \tag{7.2.12a}$$

$$\mathbf{S}_i = -\nabla Q_i + \{|\nabla Q_i|^2/(|\nabla Q_{i-1}|^2)\} \mathbf{S}_{i-1} \tag{7.2.12b}$$

$$\mathbf{X}_{i+1} = \mathbf{X}_i + \lambda_i^* \mathbf{S}_i \tag{7.2.12c}$$

where λ_i^* is obtained from the minimization of $Q(\mathbf{X}_i + \lambda_i \mathbf{S}_i)$. If the solution does not converge in (N + 1) steps, then the search is continued by setting

$$S_{N+2} = -\nabla Q_{N+2} \tag{7.2.13}$$

This step is equivalent to the steepest descent search step but with a starting point of $\mathbf{X}_{N+2}$ instead of $\mathbf{X}_0$. This "reset" is used to reduce the influence of round-off errors which would have accumulated over the previous N + 1 steps.

For multidimensional optimization problems, the equivalent form of Eq. (7.2.7) in Newton's method takes the form

$$\mathbf{X}_{i+1} = \mathbf{X}_i - [\mathbf{H}_i]^{-1} \nabla f_i \tag{7.2.14}$$

where $\mathbf{H}_i$ is the Hessian matrix evaluated at $\mathbf{X}_i$. This method can be used for even non-quadratic objective functions and has quadratic convergence rate close to the minimum. But it is costly from the point of computational requirements. Apart from the need for additional memory to store the Hessian, it requires the computation of the Hessian matrix, its inversion and multiplication with the gradient vector at every iteration.

The Marquardt method (Marquardt 1963) seeks to make the Newton method more robust by proposing to write Eq. (7.2.14) as

$$\mathbf{X}_{i+1} = \mathbf{X}_i - \left[\widetilde{\mathbf{H}}_i\right]^{-1} \nabla f_i \tag{7.2.15}$$

Here $\widetilde{\mathbf{H}}_i$ is written as

$$\widetilde{\mathbf{H}}_i = \mathbf{H}_i + \alpha_i \mathbf{I} \tag{7.2.16}$$

where α_i is a positive constant. For large values of α, Eq. (7.2.15) reduces to the steepest descent method while for small values, it reverts to the Newton method. In the Marquardt method, large values of α are used in the beginning of the iteration, thus effectively starting as the steepest descent method. As the iteration progresses, the value of α is reduced by writing it as

$$\alpha_{i+1} = c_1 \alpha_i \text{ where } 0 < c_1 < 1 \tag{7.2.17a}$$

The algorithm starts with α_0 of 10^4 and keeps on reducing the value of α_{i+1} as per Eq. (7.2.17a) as long as $f_{i+1} < f_i$. If $f_{i+1} > f_i$, then α_{i+1} is increased by writing it as

$$\alpha_{i+1} = c_2 \alpha_i, \; c_2 > 1 \tag{7.2.17b}$$

While the Marquardt method attempts to increase the robustness of the Newton method, the family of methods known as quasi-Newton methods attempt to reduce the computational effort involved in Newton method by seeking to approximate $[\mathbf{H}_i]^{-1}$ as $\mathbf{B}_i$ and writing Eq. (7.2.14) as

$$\mathbf{X}_{i+1} = \mathbf{X}_i - \lambda_i^* \mathbf{B}_i \nabla f_i \tag{7.2.18}$$

An estimate for $\mathbf{B}$ can be obtained as follows. Expanding $\nabla f(\mathbf{X})$ about $\mathbf{X}_0$, one can write

$$\nabla f(\mathbf{X}) = \nabla f(\mathbf{X}_0) + \mathbf{H}(\mathbf{X}_0)\,(\mathbf{X} - \mathbf{X}_0) \tag{7.2.19}$$

For two successive points $\mathbf{X}_i$ and $\mathbf{X}_{i+1}$, one can therefore write

$$\nabla f_i = \nabla f_0 + \mathbf{H}_0\left(\mathbf{X}_i - \mathbf{X}_0\right) \tag{7.2.20a}$$

and

$$\nabla f_{i+1} = \nabla f_0 + \mathbf{H}_0(\mathbf{X}_{i+1} - \mathbf{X}_0) \tag{7.2.20b}$$

Subtracting (7.2.20a) from (7.2.20b), one can obtain

$$\mathbf{H}_0\left(\mathbf{X}_{i+1} - \mathbf{X}_i\right) = \nabla f_{i+1} - \nabla f_i$$

or

$$\left(\mathbf{X}_{i+1} - \mathbf{X}_i\right) = [\mathbf{H}_0]^{-1}(\nabla f_{i+1} - \nabla f_i) = \mathbf{B}_0(\nabla f_{i+1} - \nabla f_i) \tag{7.2.21}$$

Equation (7.2.21) represents N equations for the determination of the N^2 elements of $\mathbf{B}$, which means that $\mathbf{B}_i$ cannot be uniquely determined from successive values of $\mathbf{X}_i$ and $\mathbf{X}_{i+1}$ and the corresponding values of the function and its first derivative. A number of attempts have been made to determine $\mathbf{B}_i$ recursively such that it is a good approximation of $[\mathbf{H}]^{-1}$, satisfies Eq. (7.2.21) and is positive definite as the Hessian has to be so at a minimum. In the Davidon-Fletcher-Powell method (Davidon 1959; Fletcher and Powell 1963), an initial symmetric and positive definite matrix is assumed for $\mathbf{B}_0$ and $\mathbf{B}_{i+1}$ is evaluated from $\mathbf{B}_i$ in the following way (see Rao 2009):

$$\mathbf{B}_{i+1} = \mathbf{B}_i + \mathbf{M}_i + \mathbf{N}_i \tag{7.2.22a}$$

where

$$\mathbf{M}_i = \lambda_i^* (\mathbf{S}_i \mathbf{S}_i^T)/(\mathbf{S}_i^T \mathbf{g}_i) \tag{7.2.22b}$$

$$\mathbf{N}_i = -\{(\mathbf{B}_i \mathbf{g}_i)\,(\mathbf{B}_i \mathbf{g}_i)^T\}/\left\{\mathbf{g}_i^T \mathbf{B}_i \mathbf{g}_i\right\} \tag{7.2.22c}$$

$$\mathbf{S}_i = -\mathbf{B}_i \nabla f_i \quad \text{and} \quad \mathbf{g}_i = \nabla f_{i+1} - \nabla f_i \tag{7.2.22d}$$

In the Broyden-Fletcher-Goldfarb-Shanno method (Broyden 1970; Fletcher 1970; Goldfarb 1970; Shanno 1970), it is the Hessian $\mathbf{H}_i$ that is updated iteratively rather than the inverse of the Hessian. In this case, $\mathbf{B}_{i+1}$ is obtained from $\mathbf{B}_i$ in the following way (Rao 2009):

$$\mathbf{B}_{i+1} = \mathbf{B}_i + \left(1 + \mathbf{g}_i^T \mathbf{B}_i \mathbf{g}_i / \mathbf{d}_i^T \mathbf{g}_i\right) \mathbf{d}_i \mathbf{d}_i^T / \mathbf{d}_i^T \mathbf{g}_i - \mathbf{d}_i \mathbf{g}_i^T \mathbf{B}_i / \mathbf{d}_i^T \mathbf{g}_i - \mathbf{B}_i \mathbf{g}_i \mathbf{d}_i^T / \mathbf{d}_i^T \mathbf{g}_i \tag{7.2.23a}$$

where

$$\mathbf{d}_i = \mathbf{X}_{i+1} - \mathbf{X}_i = \lambda_i^* \mathbf{S}_i. \tag{7.2.23b}$$

Of the two methods, the BFGS method is said to be superior because it is less influenced by errors in finding λ_i^*, and is considered to be the best unconstrained variable metric (quasi-Newton) method for objective functions that are not necessarily quadratic. Further discussion of these methods can be found in Rao (2009).

7.2.3 *Non-traditional Methods or Evolutionary Approaches*

A number of non-classical methods based on mimicking features of natural systems through stochastic simulations have been developed in recent years for multidimensional optimization problems. These methods have the advantage that they do not require evaluation of the derivatives of the objective function and use only its value. It is not even necessary for the objective function to be continuous. Further, design variables of all types—integers, real numbers, non-numerical options or a mixture of these—can be readily tackled. On the down side, due to their stochastic nature, they require a large sample population which requires large number of evaluations of the objective function. Some of the approaches that belong to this category of optimization methods are genetic algorithms, simulated annealing, particle swarming, ant colony optimization, etc. In each case, a specific feature of the approach nudges a sample population towards the optimal position. We discuss below the principles and concepts involved in these without going into the details. The interested reader is referred to specialized books and current literature for algorithms and further developments.

Genetic algorithm (GA) is the perhaps the most widely studied method among the non-traditional approaches to multidimensional optimization. Here, as the name implies, ideas from the field of genetics are used to direct the evolution of a set of initial population of feasible points towards the optimal solution. The guiding principle here is the "survival of the fittest" among a generation of feasible points through preferential breeding of the fitter ones among the population at the expense of the not-so-fit ones. This is done by a roulette-wheel type of selection among the existing points that will be used to breed the new population. The probability for selection for breeding is made proportional to its objective function value. When such a criterion is implemented on a limited set of initial population, evolution of successive generations will quickly gravitate towards the best among themselves, which may not be the most optimal. It may also degenerate into a set of equally good, but not sufficiently optimal, set of feasible points.

In order to prevent the occurrence of these possibilities, two new concepts, namely, crossover and mutation, are brought into play. Crossover enables random generation of new points in the neighbourhood of the current set of points with the hope and possibility that some of these may be better than the current best one. This is often done by expressing the value of the decision variable in the form of binary string, splitting each string into two halves (or sub-strings), and exchanging the halves (or sub-strings) of the same type (for example, the top half with the top half and the bottom half with the bottom half) randomly among the population. (It is left to the reader to show that this will generate a pair of new numbers which are in the neighbourhood, i.e., with values of design variables that are close to those, of the two parent numbers.) In practice, this need for binary number representation limits the accuracy with which the value of the decision variable can be evaluated. In order to induce the possibility of the finding a fitter point well away from the current set of point good points, the mutation operator is introduced through which the binary digits of a sub-string are flipped, i.e., the zeroes are converted to ones and the ones are converted to zeros, each with equal probability based on the length of the sub-string. The reader can verify that, in comparison with crossover, this can bring about a more radical change in the values of the design variables of the new pair of points. Mutation is typically done on a small percentage of the population. This could generate new points which are well away in the multidimensional space from the current set of fit points. This might also end up generating new points that are not so good, and thereby reduce the overall goodness of fit of the current generation of points. But there is also the possibility that points which are better than current, and perhaps closer to the true global optimum, may be found. Thus the GA approach contains the possibility of finding the global optimum if a sufficiently large number of generations is simulated. It does not get confined to a local minimum which could be the case with a gradient-based optimization process.

A GA simulation starts with M number of randomly chosen feasible points, where M is large. This constitutes a generation of population. The objective function for each of these points is evaluated. In the GA literature, the objective function is called the fitness function, and the objective is to maximize the fitness function. Using these objective function values, a discrete probability density function is created in such a way that the probability of selection for breeding of each point is proportional to its fitness. A new generation of M sample points is generated randomly as per this discrete probability function. While some bad ones may still be present, one would expect that the new generation would be fitter as the weaker ones have less probability of selection. In order to bring in variety, a crossover operation is carried out on a significant portion of the new population.

A large proportion of the population is randomly chosen and is divided into pairs. Each of the N attributes of each pair of the selected points is cut into two parts. One of the parts, randomly chosen, is exchanged with similar (first or second half) of the other member pair. This is done for all the selected pairs. Once this crossover operation is completed, a mutation operation is carried out on a small percentage, typically, a few percent of the population, in which values of the design variables are radically changed by flipping the values of the binary digits of sub-strings of freshly selected members of the population. With this, the identities of the M members of the new (second) generation are established. The fitness function is evaluated again for each of these, a fresh probability density function is created, and a new generation of sample points is created through rebirth (or reproduction), crossover and mutation. On the whole, with each generation, the overall fitness is expected to increase, and successive generations are created until a sufficiently high fitness value of the population is obtained. The genetic algorithm approach was formulated originally by Holland (1962, 1975) and is discussed extensively by Goldberg (1989), among others.

The simulated annealing method is modeled after the annealing process used in metallurgical and metal working industries to relieve residual stresses that are induced in the process of fabrication of a machine component. For example, the forging of a crankshaft used in an automobile introduces stresses within the material due to repeated hammering. Stress relieving is done by heating the component to a high target temperature and cooling it slowly in air or in controlled atmosphere to allow time for the desired phase transformations to occur, desired crystalline structure to emerge and grain boundaries to grow. In thermodynamics, phase transformations and crystalline structure and grain boundary formations are associated with free energy changes. The key to the formation of these stable, low energy structures lies in slow and controlled cooling so that the excess energy in the system is dissipated, and ordered structuring of the atoms and molecules emerges. In the simulated annealing method of optimization, the free energy is analogous to the objective function value, and the phase transformations or crystalline structure formations can be considered as local minima of the objective function. The rate of cooling is controlled by setting the value of a temperature-like variable. The method proceeds as follows. One chooses a random point $\mathbf{X}_0$ and finds the objective function value $f(\mathbf{X}_0)$ which is denoted as E_0 to indicate the connection to the energy of the system. Another point, $\mathbf{X}_1$, is again randomly selected, and its energy E_1 is determined by calculating the objective function value $f(\mathbf{X}_1)$. If $E_1 < E_0$, then X_1 is taken as the current point and we proceed further to choose another point with lower energy level. If $E_1 > E_0$, one would not normally reject it as the objective is to minimize the energy. However, Metropolis et al. (1953) suggested that the new value should be accepted with a probability given by the Boltzmann function

$$p(E_1|\, E_0) = \min\{1,\ \exp[-(E_1 - E_0)/kT]\} \tag{7.2.24}$$

Here, $p(E_1|E_0)$ is the probability that in a dynamically evolving system, the energy level at step 1 is E_1 given that the energy level at the previous step is E_0. In

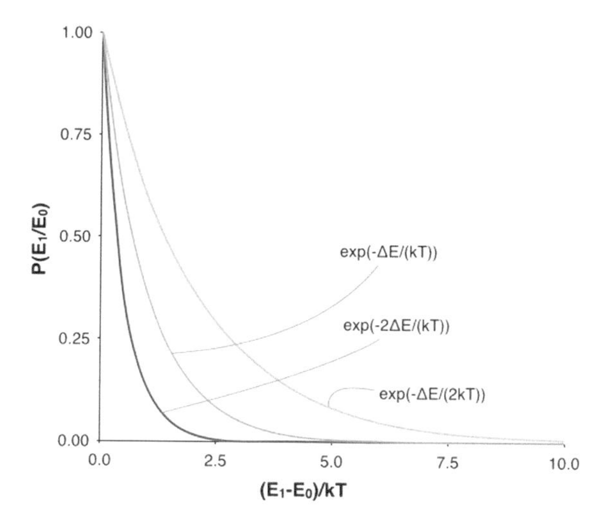

Fig. 7.4 Simulated annealing: Boltzmann function for different relative energy levels

analogy with a thermodynamic system tending towards equilibrium, k is the Boltzmann constant and T is a temperature-like parameter that can be used to control the rate of "cooling". As one can see from Fig. 7.4, as $(E_1 - E_0)$ increases, the probability of the new but higher value being accepted as the point from which to launch the next search decreases for a fixed temperature. At higher temperatures, the probability is less for the same difference in energy. Conversely, at lower temperatures, the probability of accepting the new point that has a significantly higher is higher. This can be used to control the evolution of the search in the following way. One would start the simulated annealing method with a fixed, high temperature, and let the solution evolve slowly, but not monotonically, towards a lower objective function value. After sufficient exploration at that temperature level, the temperature is reduced so as to increase the possibility of jumping into another local minimum, if there is one.

The advantage of this method over GA is that the search starts straightaway without having to wait for M function evaluations to be done. However, the method lacks the purposeful search towards better points, which is built into conventional gradient-based search methods as well as in the GA method through the survival of the fittest policy and preferential reproduction of the attributes of the better points. Algorithmic application of the simulated annealing method to the travelling salesman problem is discussed by Kirkpatrick et al. (1983) and Cerny (1985). Detailed discussion of the method is given by van Laarhoven and Aarts (1987).

Particle swarm optimization uses swarm intelligence to find the optimal solution. It takes inspiration from the movement of a swarm of insects, of a flock of birds or of a school of fish to guide the search towards the optimum point. The particle, which is an individual entity in these collections of insects/birds/fish, is endowed with its own intelligence in addition to some that comes from belonging to the swarm. The algorithm, originally proposed by Kennedy and Eberhart (1995), models the behaviour of a particle in the swarm using simple rules such as the following: the particle does not get too close to other particles nor does it leave too large a gap; it steers itself in the average direction of the swarm. It is assumed that once a member of the swarm sees "food", then all the members get to know of it instantly, and that all the members steer themselves towards it, although not instantly. The action of an individual member is also guided its past memory. The evolution of the particle swarm is modelled by the position and velocity of each "agent" in the swarm as a combination of these factors. For an ith particle in the swarm, these are calculated recursively as follows:

$$v^{n+1} = v^{n} + \varphi_1 \beta_1 (p_i - x_i) + \varphi_2 \beta_2 \left(p_g - x_i \right) \tag{7.2.25}$$

and

$$x^{n+1} = x^{n} + v^{n+1} \Delta t = x^{n} + v^{n+1} \tag{7.2.26}$$

Here v is the velocity and x is the position, φ_1 and φ_2 are the factors that determine the contribution from individual intelligence (φ_1) and group intelligence (φ_2), p_i and p_g are the previous best position of the individual i and the group, respectively, and β_1 and β_2 are scaled random numbers. The time step is taken as unity as it has no specific meaning in optimization. A number of improvements have been made over the years to avoid problems and pitfalls such as particle accelerating out of solution space or premature collapse on to a sub-optimal minimum, etc. The review papers of Banks et al. (2007, 2008) discuss several developments such as including inertia (Eberhart and Shi 2000) and constriction (Clerc and Kennedy 2002) parameters. The method is still under intense development and refinement, and considerable work is being conducted on parameter tuning and application to dynamic environments.

Ant colony optimization (Colorni et al. 1991; Dorigo et al. 1996; Dorigo and Stützle 2004) uses ideas from the social behaviour of ants in search of food. Ants are said to be nearly blind but are known to have the uncanny ability to establish the shortest route paths between their colonies and the food sources. This is achieved by marking the trail by pheromone, a chemical secretion with a high degree of specificity that is used by an ant to communicate with members of its own species and with insects of other species (Wilson 1962). An individual ant may move haphazardly until it comes across a pheromone trail, and may then use it to find the best way around obstacles and towards food sources etc. It may add to the trail its own findings and impressions by secreting its own pheromone in coded quantities by moving along the path back and forth. Ant colony optimization mimics this behaviour, and simulates the process of development of the pheromone trail and the subsequent establishment of an optimal path to the objective. For details of specific algorithms for optimization problems, the reader is referred to Dorigo et al. (1996) as a point of departure. The approach has been found to be useful in network routing problems.

We end this section by recapitulating briefly the main features of the evolutionary approaches to optimization. Compared to classical methods, they offer a number of advantages; these include general applicability across a wide range of problems, no requirement for prior knowledge of the problem space, ease of coding, scope for heuristic development and tuning of parameters. However, there is no single approach that works best for all problems, and some approaches are more suited under some conditions. Further, there is no guarantee of finding a globally optimal solution.

Apart from those cited in this section, a large number of heuristic evolutionary approaches mimicking a variety of imagined or real natural dynamic systems are being proposed. A survey of these is beyond the scope of the present book. In the next section, we discuss a couple of case studies illustrating how these concepts can be coupled with CFD to obtain optimal solutions for flow-related problems.

7.3 Case Studies of Shape Optimization

The application of optimization techniques for the design of fluid flow problems is no different in principle from those in other fields. However, when the influence of design variables on the equipment performance needs to be ascertained using CFD, then a major problem arises in applying optimization techniques. In most cases, the performance of a fluid flow equipment cannot be easily captured in simple analytical functions. When a CFD is simulation is needed to evaluate the objective function value, then the computational requirements for optimization can become very high. In the case of multi-parameter problems, which is often the case in industrial equipment, the determination of the gradient and the Hessian of the objective function space (see Sect. 7.2.2) can become computationally very expensive. In this section, we will consider, by way of illustration, two case studies in which optimization concepts are used to arrive at a solution to a fluid flow problem involving CFD. The first of these, namely, the case of the U-bend, deals with the minimization of the pressure drop as the flow goes through a 180° bend in the duct. The second case deals with the design of a T-junction in such a way that a desired flow split between the two outlets is achieved. The problem formulation for these cases is discussed in Sects. 7.3.1 and 7.3.2, respectively. A detailed description of the optimization technique is given in Sect. 7.3.3. The path to the optimal solution is discussed in Sect. 7.3.4.

7.3.1 *Formulation of the Optimization Problem: The Case of a U-Bend*

The case of shape optimization of a U-bend is a well-studied problem. The objective here is to find the optimal shape of the U-bend such that the pressure drop across a U-bend (180° bend) is minimum. This has recently been studied in the literature as a 26-parameter problem (Verstraete et al. 2013). For the sake of illustration, we consider a two-dimensional problem in which a duct of width W takes a U-bend and then goes straight, as shown schematically in Fig. 7.5a. The pressure drop in fluid flow through a bend arises from both wall friction and change in flow direction. Since both depend on flow velocity, the pressure drop can be reduced by increasing the cross-sectional area of the duct. However, the bend needs to be attached to a duct of width W at entrance and exit. Too sharp an increase or decrease in the cross-section can lead to flow separation and additional losses. At the same time, too gradual increase can lead to excessively long flow paths which increase wall friction. We therefore need to find the optimal change in the cross-section as the flow takes a U-turn.

This design problem can be formulated as an optimization problem in the following way (Srinivasan et al. 2017a). Consider Fig. 7.5a where the local radius of curvature and the width of the duct vary as a function of θ, the angle into the bend. If we plot the width as a function of radius of curvature, we obtain a closed curve since the entry duct and the exit duct have the same width (Fig. 7.5b). We can approximate this curve, which we expect to be smooth (because abrupt changes in cross-section are likely to lead to pressure losses), using an algebraic shape function which allows the curve to be represented using only a few parameters. One good choice for the profiles of fluid passages can be Bézier curves, which can be represented as (Sederberg 2012)

$$\mathrm{P(t)} = \sum_{\mathrm{i=0}}^{\mathrm{n}} \mathrm{B_i^n(t) P_i} \quad \mathrm{t} \in [0, 1] \tag{7.3.1}$$

where $\mathrm{P_i}$ are the control points, t is a parameter that varies from 0 to 1 as the curve goes from point $\mathrm{P_0}$ to point $\mathrm{P_n}$, and the basis function, $\mathrm{B_i^n}$ (t), is given by the Bernstein polynomials

$$\mathrm{B_i^n(t)} = \frac{n!}{\mathrm{(n-i)!i!}} \mathrm{(1-t)^{n-i}} t^{\mathrm{i}} \tag{7.3.2}$$

The points form an open control polygon (see Fig. 7.6a); the curve passes through the first control point ($\mathrm{P_0}$) and the last one ($\mathrm{P_n}$), and is tangent to the control polygon at those end points. Figure 7.6b shows four different cubic (n = 3) Bézier curves which show that a Bézier curve mimics thc control polygon and takes different shapes based on where the control points are relative to each other.

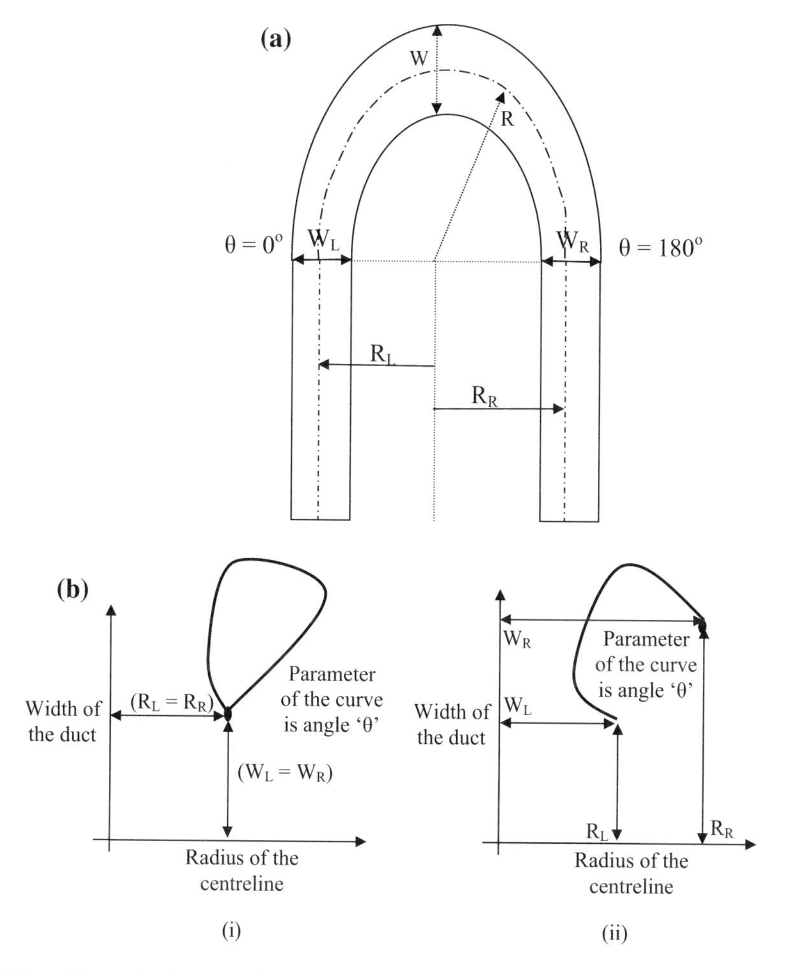

Fig. 7.5 **a** Schematic diagram of the geometry to be optimized, and **b** possible variation of the width versus radius of the centerline for the case of (i) equal and (ii) unequal duct widths (Srinivasan 2015)

For given x- and y-coordinates of the two control points, the shape of the U-bend is defined. The optimization problem then reduces to finding the x- and y-coordinates of the two control points that give a shape of the bend with minimal pressure drop.

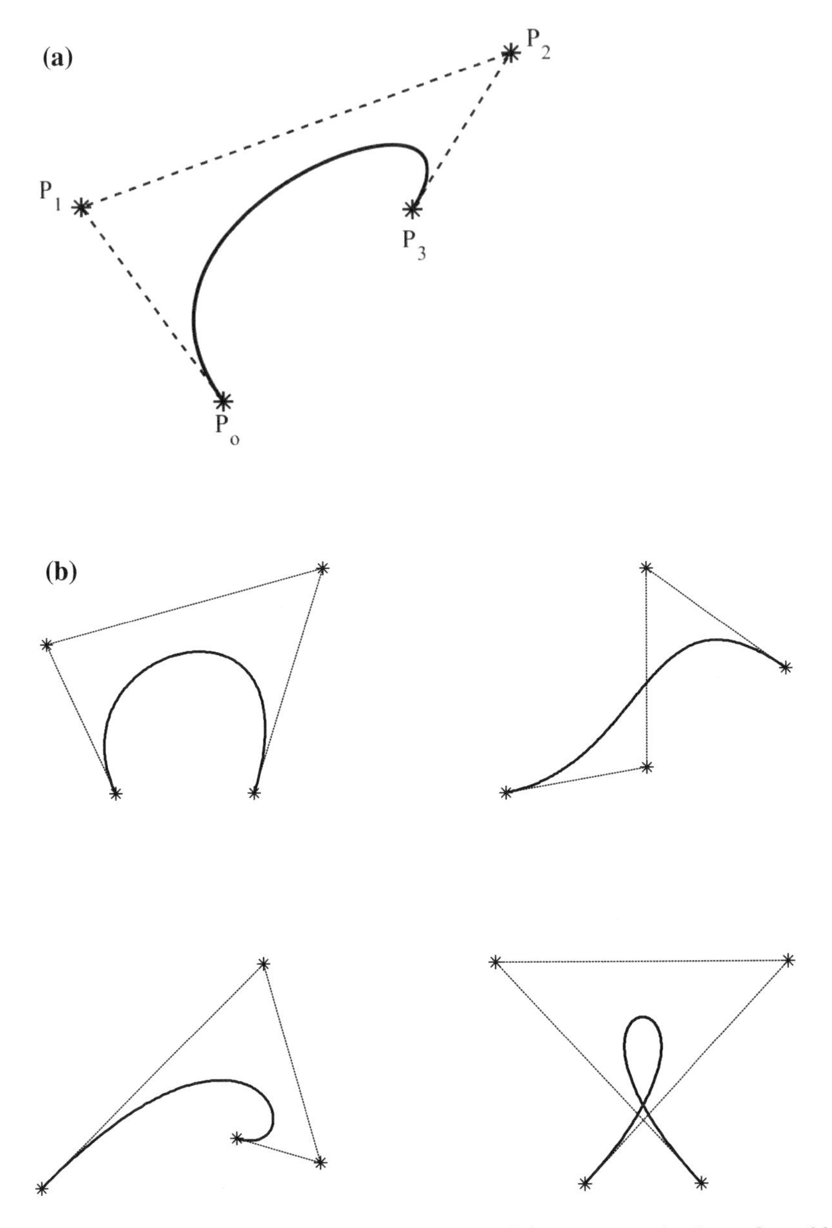

Fig. 7.6 **a** Cubic Bézier curve with four control points and the open control polygon formed by the control points P_0 to P_3. **b** Different shapes obtained by moving the control points (Srinivasan 2015)

7.3.2 *Formulation of the Optimization Problem: The Case of a T-Junction*

Before we move on to the solution, let us consider a very different problem, namely, the two-dimensional problem of flow through a dividing T-junction in which flow comes in through an inlet pipe and divides into a straight outlet and a branch outlet which is at right angles to it. This is shown schematically in Fig. 7.7. The objective here is to design the junction in such a way that the flow automatically, i.e., without the need for a flow regulating valve, splits as per a desired fraction. We wish to do this by contouring the junction so that the required amount of flow is diverted into the branch outlet. A possible shape of the junction is shown in Fig. 7.7; here, the junction is made up of three curved walls, which, between them, create the two path ways for the fluid to flow through. We wish to obtain the shape of the junction for the cases where the branch outlet has 30% of the total flow rate, 50 and 70% of the flow rate. We expect the shapes of the curves to be different for each case; we also expect that the pressure drop would be higher for the case where 70% of the incoming flow rate is to be persuaded to go through the branch outlet than for the case where branch flow rate is only 30%. A conventional CFD solution predicts the flow rates through the two outlets for a given shape of the junction; here, we wish to do the inverse problem, namely, predict the shape for a given flow split.

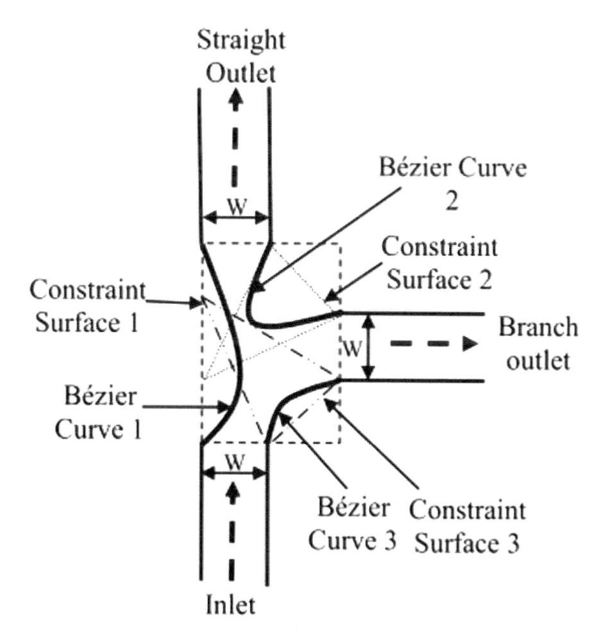

Fig. 7.7 Geometry for the T-junction problem. The domain in the dotted line is to be shape-optimized (Srinivasan 2015)

This is arguably a tougher problem and is still in the research domain. Here, we illustrate how this can be done using one method (without claiming it to be the only one or the best one) reported in Srinivasan et al. (2017b). Following the previous example, we make up the shape of the junction using three Bézier curves, one for each corner. For simplicity, we assume that a quadratic Bézier curve, i.e., one which has only one movable control point, would be sufficient in each case. Since three such curves are needed to make the junction, the (x, y) coordinates of the three control points will be our decision or design variables. In order to pose it as an optimization problem, we define the objective function as the root mean square value of the deviation in each outlet from the desired value. In quantitative terms, this can be written as

$$f = \left[\sqrt{\left(Q_{st} - \widetilde{Q}_{st} \right)^2 + \left(Q_{br} - \widetilde{Q}_{br} \right)^2} \right] / Q_{in} \tag{7.3.3}$$

where Q is the volumetric flow rate, the tilde indicates the desired flow rate, the subscripts *in*, *br* and *st* indicate the flow rates through the inlet, the branch outlet and the straight outlet, respectively. One can see that when the flow split matches with the desired value, then $f = 0$. The definition of f in Eq. (7.3.3) can be extended to a case where there are multiple outlets. Although we cannot prove it, we can reasonably be assured that by "pinching" either of the two outlets sufficiently, we can get any flow split. However, we have no guarantee that there is only one solution; indeed, it looks likely there can be multiple solutions if the flow split is the sole criterion. For example, fixing the shape of one Bézier curve, one is likely to be able to pinch the other junction sufficiently to enforce the required flow split. Finally, in order to get to one of these solutions fast, we would like to look for solution within reasonable bounds of the decision variables. Therefore, we need to put constraints on what values these can take. One condition that readily comes to mind is that the curves should not intersect because if they do so in two-dimensional flow, there is no way flow can go through the channel. With reference to Fig. 7.7, we can say that the control points would likely be lie in the dotted zone ("the control polygon") if they were not to intersect but allow, at the same time, for sufficient pinching. If necessary, one can see what pathways are produced by looking at various combinations of locations of the control point of the three curves. Thus, we have a case of constrained optimization in which the objective function is to be minimized subject to the six decision variables lying within their declared control polygons.

7.3.3 Search Method for the Optimal Solution

We seek to find (x_i, y_i) such that f is minimum and (x_i, y_i) lie within their respective limits of variability. Given the complicated nature of the flow (we need to solve the steady, two-dimensional form of the Navier-Stokes equations along with a turbulence model if the flow is turbulent), we cannot hope for an analytical solution and have to use CFD techniques. Since CFD solution requires the geometry (including the shapes of the curved surfaces), we need to start with guess shapes and improve on them iteratively and incrementally until we reach a satisfactory solution. There are several ways in which this can be done (see Sect. 7.2); here, we will use the Box complex method (Box 1965), which has been used recently in some problems of engineering optimization and which is also relatively easy to implement. The method is similar to the simplex method discussed in Sect. 7.2.1, and is slightly modified to account for constraints. It works as follows. If there are N decision variables, we start with a population of 2N solutions, i.e., 2N sets of initial guesses for the values of the decision variables. This constitutes the initial simplex. For each of set of values, we evaluate the objective function. In the N-dimensional space, among the 2N points, there will be one (perhaps more) which is the worst point, i.e., it has the highest value of the objective function. We seek to replace this point with another which has a lower value of objective function. Since we do not know how to choose a better point, we guess that the parameter values of the new one are likely to be in the direction of the better ones, i.e., those with lower objective function value. We therefore find the centroid of all the other (2N-1) points and reflect the worst point along the line joining itself and the centroid by a factor α, which Box (1965) recommended should be 1.3 (see Fig. 7.8a). That is, we move away from the current worst point by choosing a new point which is diametrically opposite (but at 30% more distance) with respect to the centroid. We evaluate the objective function at this new point. If it is better than the worst point, we replace the worst point with it. If not, we retract it towards the centroid, and see if it gives a better objective function. We repeat this process until we find a better point than the current worst point. Thus, replacing the current worst point with an improved point may require several function evaluations.

Figure 7.8 illustrates the Box method for a two-parameter problem. The contours of the objective function are shown schematically in terms of the two parameters, X_1 and X_2. For this two-parameter problem, our initial simplex contains four points P_1 to P_4. The colour contours show the magnitude of the objective function, with the lowest value lying in the smallest ellipse. As the objective function value of point P_1 is the highest, we wish to replace it. As shown in Fig. 7.8b, we find the centroid X_c of the triangle having the other three points as its vertices. Point P_1 is reflected about X_c along the line P_1X_c. Since the first reflection takes it to a point with higher value of the objective function, we retract it along the same line in steps to reach point P_5. Having successfully found a better point, we replace P_1 by P_5 in the simplex. At this stage, the worst point is P_2. We replace this by point P_6 by reflecting it along the line P_2X_c where X_c is the centroid of the triangle having P_3, P_4 and P_5 as its vertices. This process is shown in Fig. 7.8c; now the simplex consists of points P_3, P_4, P_5 and P_6. Of these, P_5 is the worst point and this is reflected about the centroid X_c of the triangle formed by the remaining three

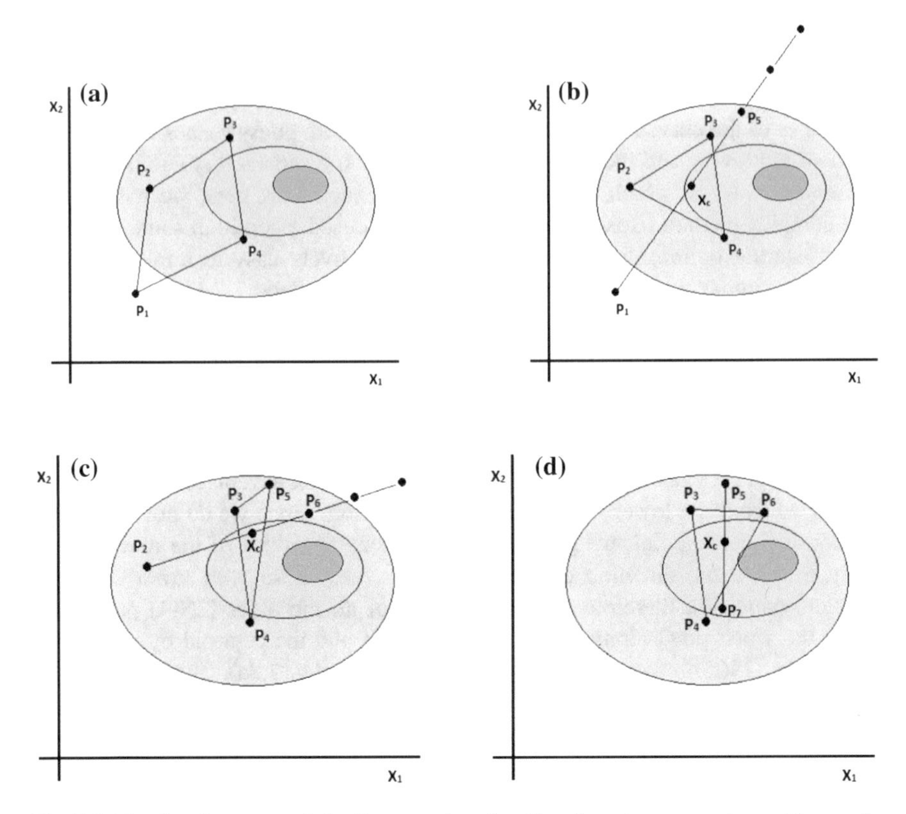

Fig. 7.8 The first four steps of the Box complex algorithm for a two-parameter problem: **a** the initial simplex consisting of points P_1 to P_4, **b** replacing point P_1 by P_5, **c** replacing point P_2 by P_6, and **d** replacing point P_5 by P_7. X_c is the centroid of the triangle formed by the three points which are being retained, and about which the point to be replaced is reflected and retracted if necessary. The colour contours represent the value of the objective function; the minimum lies in the pink region

vertices to arrive at a new point P_7. This process is illustrated in Fig. 7.8d. Comparing Fig. 7.8a, d, one can see that the new simplex consisting of points P_3, P_4, P_6 and P_7 has shrunk in size compared to the original simplex consisting of points P_1–P_4. The average objective function value of the simplex is lower and the best point of the simplex is closer to the true minimum than at the beginning. We continue this process until we reach sufficiently close to the minimum. The stopping criterion may be (a) a specified value of the objective function of the best point in the simplex, or (b) a given factor of reduction of the average objective function value, or (c) a given factor of shrinkage of the simplex. Some of these ideas are explored later.

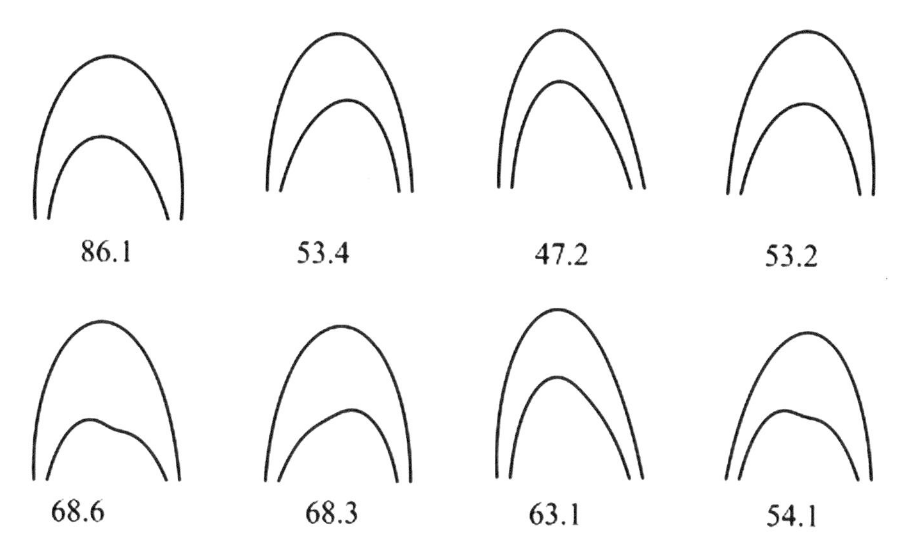

Fig. 7.9 Eight initial guesses of the bend geometry for the U-bend case. The numbers below each figure indicates the pressure drop (in Pa) obtained from CFD calculation of flow through the bend with extended inlet and outlet ducts (Srinivasan 2015)

7.3.4 Optimal Solutions

In the case of the U-bend (see Srinivasan et al. 2017a), there are two control points whose optimal (x, y) locations need to be determined. This makes it a four-parameter problem. The eight initial guesses of the bend geometry, which are obtained from feasible control point coordinates generated randomly, are shown in Fig. 7.9 along with the pressure drop obtained in each case from CFD simulations of each of bend geometry (with inlet and outlet ducting attached). This forms the initial simplex. We then use the Box method to generate successive values of the parameters; these are plotted in Fig. 7.10. The pressure drop across the bend decreases steadily as the number of iterations increases (Fig. 7.11). The situation at the end of 45 iterations is summarized in Fig. 7.12a which shows the final simplex. One can see that all the eight solutions have significantly reduced pressure drop ($\sim$41 Pa) compared with the average value of 62 Pa at the beginning or with respect to the base case of a semicircular bend of constant width (which has a pressure drop of 53 Pa). The best of the final simplex, the second shape with a pressure drop of 40.2 Pa in Fig. 7.12a, can be said to be the optimal for this case. The variation of the duct width with the radius of curvature for the optimal case, which was the focus of optimization, is shown in Fig. 7.12b. One can see a smooth variation with increasing width as the centerline radius increases followed by a similar (but not symmetric) decrease, which is consistent with the geometries shown in Fig. 7.12a. Thus, a gradual expansion of the duct cross-section followed

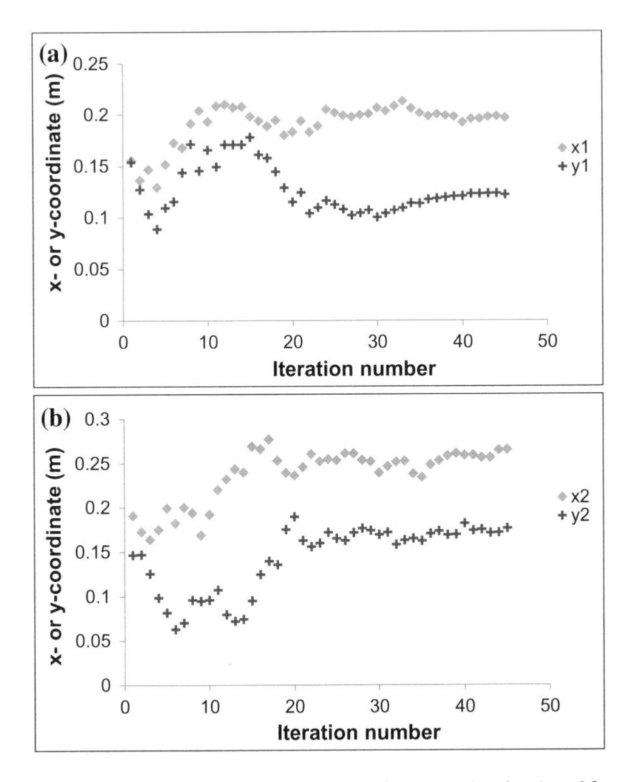

Fig. 7.10 Evolution of x- and y-coordinate positions of **a** control point 1 and **b** control point 2

by a gradual reduction back to the outlet duct size is the key to minimizing the pressure drop. This is intuitive but the set of calculations and the optimization strategy enable us to describe it quantitatively.

In the T-junction case, there are three control points, one for each Bézier curve; there are therefore six decision variables. Twelve initial guesses are generated randomly and these are used to launch the Box method. The progress of the search process is illustrated in Fig. 7.13 where the maximum, the average and the minimum values of the objective function of the simplex are plotted against the iteration number for the case where we seek a branch flow rate of 70%. One can see that, over the course of the search, all three reduce; while the maximum and the average value reduce with every iteration, the minimum value reduces only sporadically. This means a solution which is better than the current best solution is found only occasionally. After 125 iterations, we see a sufficiently low value of the objective function and we nearly achieve the desired target of 70% flow through the branch outlet. The shape of the T-junction and the velocity vectors are shown in Fig. 7.14.

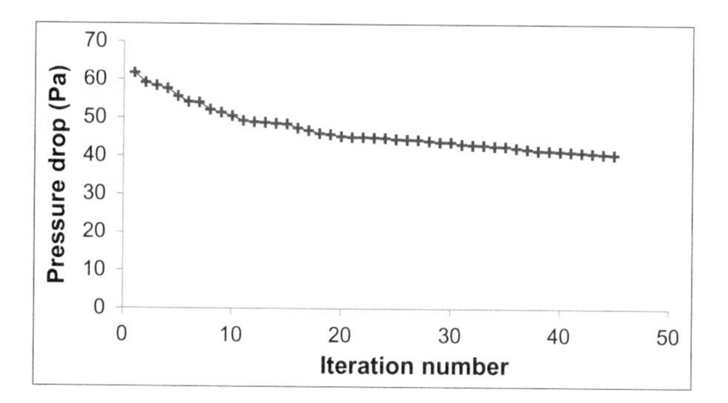

Fig. 7.11 Evolution of pressure drop as the iteration progresses

Figure 7.15 compares the predicted junction shapes and velocity vectors for the case of where the fractional branch flow rate is 0.29 and 0.59. One can see that the shapes are reasonable in the sense when more branch flow rate is required, the straight outlet is constricted more and the branch outlet opens up more. The shapes produced are thus intuitive; quantification of the design for a given set of flow conditions such as the duct sizes, the fluid, the constraints, etc. is what can be achieved by combining CFD with shape parameterization and a search method. Further examples of application and details of simulations can be found in Srinivasan (2015) and Srinivasan and Jayanti (2015) and Srinivasan et al. (2017a, b, c).

7.4 Issues in Shape Optimization

The case studies discussed in Sect. 7.3 are by no means reflective of the state of the art of optimization; however, they give a flavour of the issues involved in using CFD for shape design. The first of these is the computational time. One may note here that each evaluation of the objective function corresponds to one CFD simulation of flow through the U-bend or T-junction with given shapes of the curves. In the case of the U-bend, it has taken 45 steps (i.e., 45 CFD computations) to reach the approximately optimal solution; the T-junction case required 125 steps to reach the solution. This is fairly intensive computational effort by usual standards although it looks reasonable for a six-parameter optimization problem. Apart from the computational time, there are several other issues that need to be tackled. Although these arise specifically from the two case studies that we considered, many of the issues are generic to shape optimization. For ease of discussion, we can group them into issues arising from problem formulation and those arising from solving the optimization problem. These are discussed briefly below.

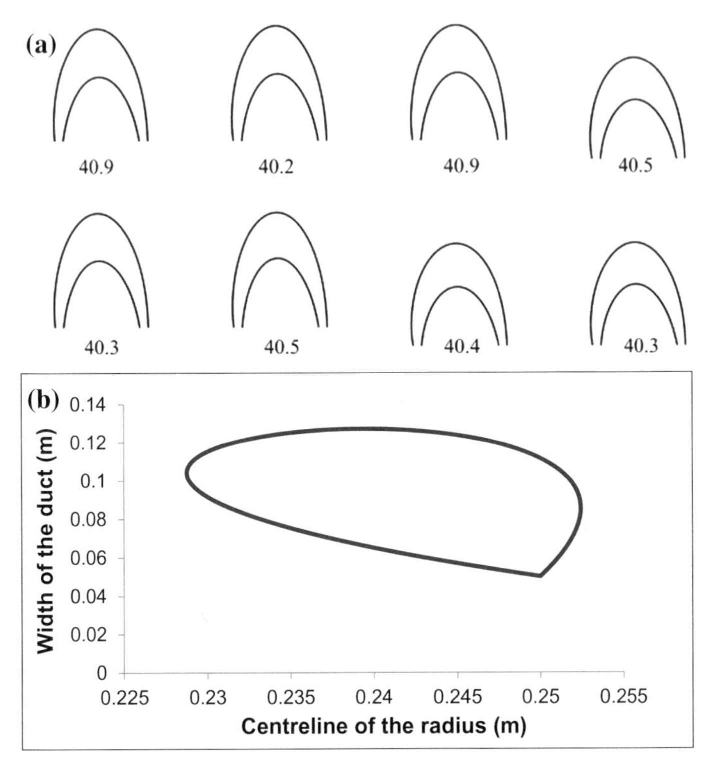

Fig. 7.12 **a** The eight bend geometries in the final simplex with the corresponding computed bend pressure drops, and **b** variation of centreline radius and duct width of the optimal U-bend (Srinivasan 2015)

7.4.1 Issues Arising Out of Problem Formulation

The formulation of the two problems as presented in Sect. 7.3.1 appears to be intuitively correct but there are hidden issues. Some of these are as follows:

- In case of single objective such as minimization of pressure drop in the first case and having a desired flow split in the second case, the formulation of the objective function is fairly straightforward. In practical cases, there are always caveats such as manufacturability of the duct in the first case and allowable pressure drop in the second case. Other criteria may also be of interest, including cost, compatibility with existing equipment, possibility of retrofitting, erosion, etc. The problem formulation in case of multiple criteria is not trivial. While some criteria, such as geometric compatibility with existing equipment, may be treated as constraints, others, such as a desired flow split at a junction at minimum pressure drop, may be more difficult to handle. While one can define an overall objective function combining the two, for example, sum of squares of

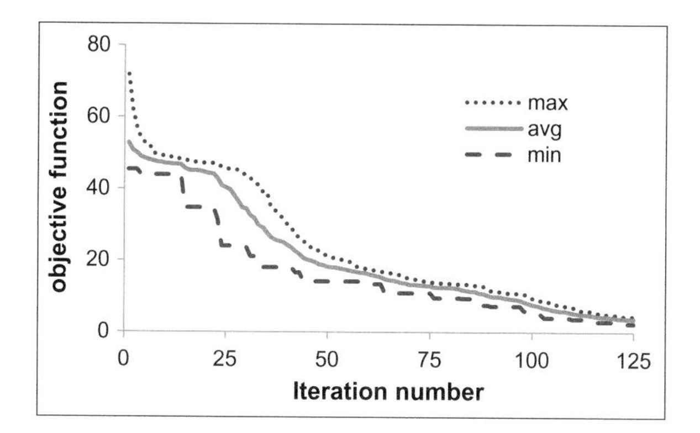

Fig. 7.13 Variation of the maximum, the average and the minimum value of the objective function of the simplex with iteration number for the T-junction with a branch flow rate fraction of 70%

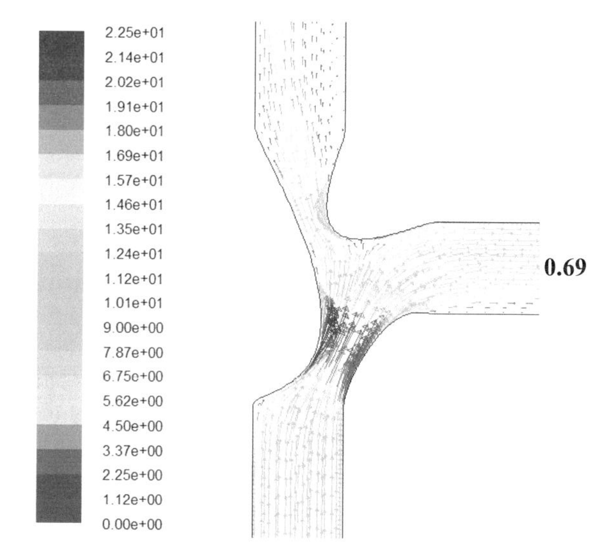

Fig. 7.14 Predicted junction shape and velocity vectors for a branch flow rate fraction of 0.69 (Srinivasan 2015)

flow maldistribution and pressure drop, the two cannot be added naïvely because they are dimensionally different. One can argue that both can be non-dimensionalized but a good value for non-dimensionalization may not be found. Also, the penalty for flow maldistribution may not be the same as the high pressure drop and the two therefore need to be assigned suitable weight factors so that they can be compared on par. For example, in a process plant, a 10% flow maldistribution may be intolerable whereas a 10% increase in pressure

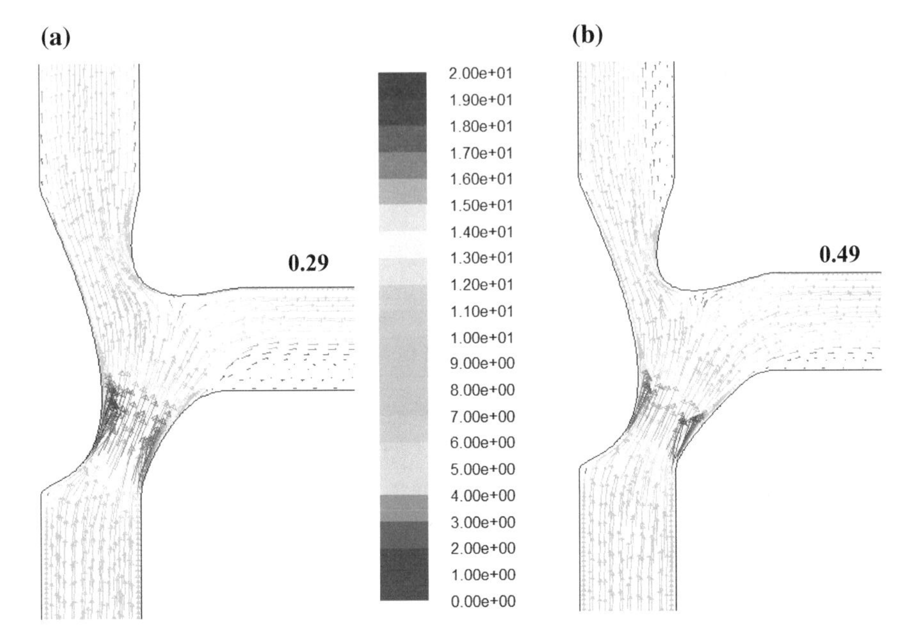

Fig. 7.15 Predicted junction shape and velocity vectors for a branch flow rate fraction of **a** 0.29 and **b** 0.49 (Srinivasan 2015). Note how the predicted junction has a narrower neck in the straight outlet section in case of (**b**) so as to force a larger fraction of the flow to go through the branch outlet

drop can be readily accommodated by adjusting the operating point of an induced draft fan. One may try to impose the pressure drop condition as a constraint but one may not know what would be a suitably low value.

- Parameterization of the shape of the flow domain is important to reduce the number of parameters as the computational effort for finding the optimal solution increases tremendously (though not at the same rate for all the optimization algorithms) as the number of parameters increases. Bézier curves have been used to define curved edges in the two examples considered. There a number of other options to parameterize shapes, e.g. Overhäuser curves, B-splines, NURBS (Samareh 1998). Even with Bézier curves, the number of control points to be used and the number of such curves to be used to define the flow domain may vary from case to case.
- Another important aspect of shape parameterization is fixing the parameter space. The parameter space should be wide enough to explore a number of possible shapes but not so wide as to give rise to negative flow domains. For example, the three Bézier curves in the T-junction case should not intersect but they need to get close enough to provide sufficient "throttling" of the flow into an outlet channel to provide flow control. In the T-junction example, the parametric space of the left Bézier curve 1 in the T-junction case (see Fig. 7.7)

has been fixed to vary such that it will not intersect with the other two curves. Two extreme shapes of the junction are shown in Fig. 7.16 (Srinivasan 2015). The shape shown in Fig. 7.16a will ensure very little flow rate through the straight outlet while that in Fig. 7.16b will lead to a high-pressure case of nearly equal flow through the two outlets.

- Yet another point of practical importance is the goodness of the choice of parameters. How does one know that the parameters chosen are the most effective in exercising control over the design? One could let the optimizer arrive at that intelligence by including many parameters but this would increase the load on the optimal search method. Yet another possibility would be to start with a wide range of parameters and weed out the insignificant ones as the search progresses. This requires the use of sophisticated data analytic tools, such as principal component analysis, adjoint methods, data reconciliation techniques, to be integrated into the optimization process.
- Yet another issue, which is relatively little explored, is the issue of inherent inaccuracies in the CFD simulations that are needed to evaluate the objective function, and the effect that this has on the quality of the optimal solution. In practical cases, the errors and approximations made during CFD solution of the governing equations, as well as those due to modeling of physical phenomena (see Chap. 6), will lead to approximations in the solution. Of these, those arising out of numerical simulations can be controlled to a large extent by a proper choice of robust algorithms and fine grid. The latter option makes the solution computationally very expensive, especially when numerous simulations need to

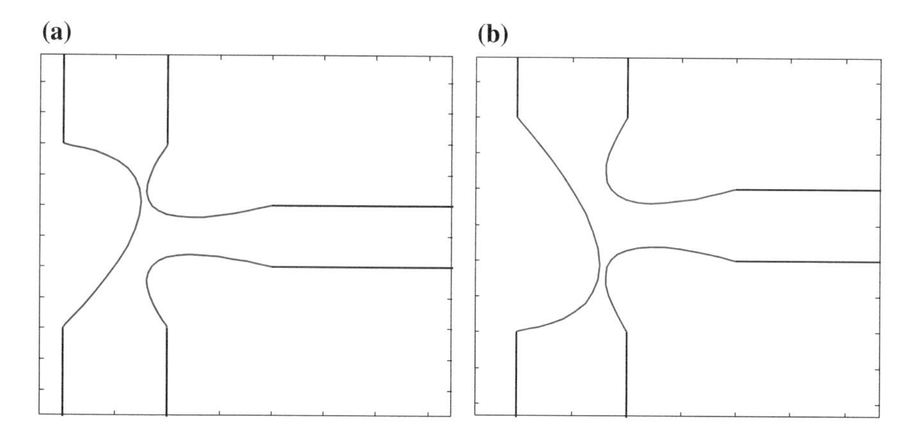

Fig. 7.16 Limiting the variation of the left Bezier curve so as not to intersect with the other curves. The extreme position in **a** will direct most of the flow through the side branch while the other extreme position of the left curve shown in **b** will give a high pressure drop solution with nearly equal flow rates in the two outlets

be conducted to find the optimal solution. One may start with a low fidelity solution and gradually increase the accuracy as the optimal solution is approached. The modeling approximations will still be there and one needs to interpret the optimality of the solution taking into account the uncertainty in the CFD predictions.

- Shape optimization of three-dimensional domains calls for parameterization of surfaces and not just curve segments. This will increase the complexity of parameterization and the number of parameters that need to be used. Also, the CFD simulations will be more expensive because these will be 3-d and not 2-d.

7.4.2 Issues Arising Out of the Search for Optimal Solution

A well-formulated optimization problem is a good beginning but many complications may still arise while searching for the optimal solution:

- First among these is the algorithm for minimization. While the Box's complex method discussed in Sect. 7.3.2 has proved to be successful, there are a number of other options for search methods. As discussed in Sect. 7.2, these fall into two principal categories: gradient-based methods such as the steepest descent method, the conjugate gradient method, the adjoint method, etc.; and direct search methods such as genetic/particle swarm/firefly algorithms, neural networks, response surface methods, topology optimization, etc. Direct search methods do not use any derivative information of the objective function; only the value of the objective function is used to guide the search process. Gradient-based methods need the derivative information to guide the search process. The interested reader is referred to current literature (e.g., Chen et al. 2015) on high fidelity global optimization for shape design problems.
- Each optimization method has its advantages and limitations. It may be said that no single optimization algorithm exists which will solve all the optimization problems equally efficiently. Some algorithms work well on some problems but perform poorly on other problems. Evolutionary algorithms such as genetic algorithms need a fairly large population of possible solutions to start the optimization process. They are also known to scale poorly with increasing number of parameters to be determined. Although gradient-based optimization algorithms are more efficient, they can be very sensitive to accuracy in determining the gradient. This can pose difficulties when nothing much is known on how the objective function varies within the parameter space.
- The progress of the search towards the optimal solution may not be smooth. For example, in the Box complex method, every attempted reflection or retraction does not result in a new point in the parameter space with a lower objective

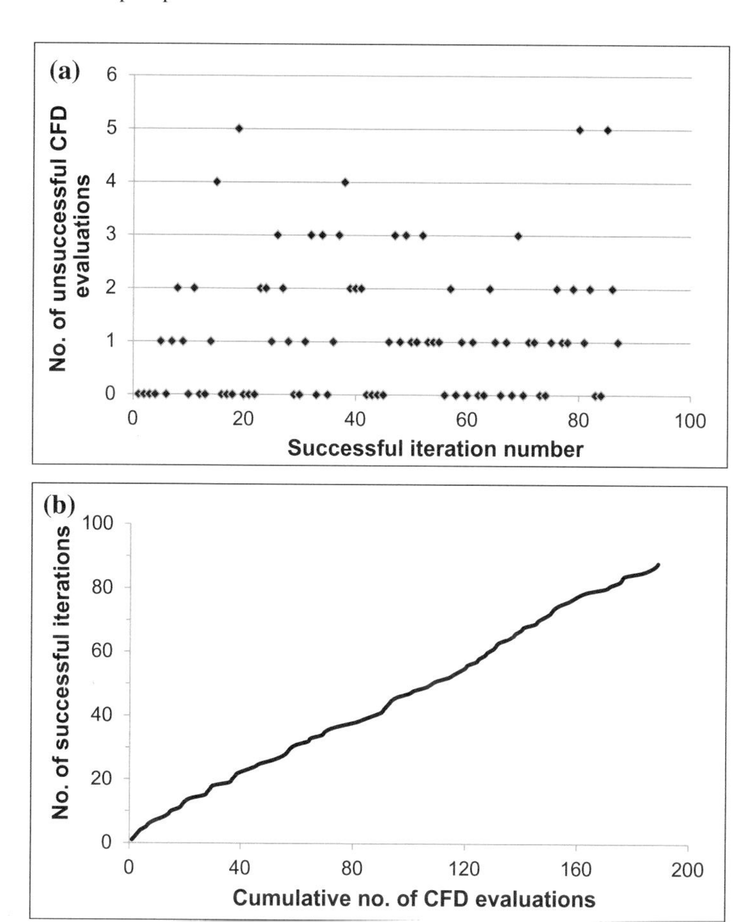

Fig. 7.17 **a** The number of wasted CFD computations per successful Box complex method for the U-bend optimization (Srinivasan 2015), and **b** the cumulative number of CFD computations versus the number of successful Box method iterations showing that less than half of the reflected/retracted points led to a solution with lower value of the objective function

function than the worst point. In such a case, a new point needs to be found with further refraction and the CFD solution needs to be determined at the new point. Sometimes, this process needs to be repeated a few times in order to locate a better point, as illustrated in Fig. 7.17a (Srinivasan 2015). As a result, the global (actual) number of steps (each step requires one CFD solution) can be much higher than the number of successful iterations (see Fig. 7.17b) for the U-bend optimization. Less than half the values predicted by the Box algorithm have actually led to improved solution. One can also see from the changing slope of the curve that it becomes increasingly more difficult to find a better solution as the approaches the optimal point.

- Search algorithms may breakdown or diverge or fail to reach a global optimum. For example, the Box complex is known to breakdown when the worst point is at the centroid. Gradient-based methods can fall into a limit cycle under certain conditions. The topology of the objective function in the parameter space can have a major influence on the rate of convergence or of success of search methods. A smoothly varying texture can be effectively exploited by gradient-based methods while fine texture can give them immense problems. The sensitivity of the CFD solution can also play an important role when the topology has many peaks and valleys which are of the size of the resolution of the CFD simulation. Finally, in dealing with CFD-based optimization—where one may not have a priori knowledge of the objective function space—, one may have to ascertain if an optimal solution exists at all within the parameter range while satisfying all the constraints. The choice of parameters and the imposition of constraints need to be carefully in such cases.

7.5 Summary

CFD-based shape design in an evolving field and is still in the research domain. It is possible in some cases to formulate the shape design problem as an optimization problem and arrive at optimal solutions. One may note that the final solution is only as good as the accuracy of the CFD simulation. If the grid is not fine enough and/or if the modeling (for example of turbulence) is not accurate enough, the optimal shape may not match with experimental results. In optimization studies in which CFD is used, the principal contributor to the cost of computation is the computational resources needed for evaluating the objective function through a CFD simulation. The efficiency of the search method in finding the optimal solution in as few iterations (or evaluations of the objective function) as possible is important. Also equally important is the robustness, i.e., its ability to find a solution without getting stuck or without becoming unstable. It can be said that there is no universally good search method that works for all cases. Multi-objective optimization problems (e.g. correct flow distribution while having to incur minimum pressure drop; simultaneous minimization of pressure drop and cost of fabrication; flow control with minimal erosion, etc.) are much more difficult to tackle. One of the approaches is to fix one objective and treat the others as constraints. Fixing reasonable values for the constraints then becomes a tricky problem. Fixing them too tightly may not yield a solution while too liberal constraints may end up having effectively no control! Yet another problem often faced in real world problems is to find the right set of decision variables, i.e., those that have an influence on the objective function. Having too few of them may not enable a satisfactory solution while having too many of them may make the optimization problem intractable. Finally, the question as to whether the solution offered by the optimizer is globally optimal cannot be answered unequivocally as a number of factors contribute to it and one has to examine critically the choice of parameters, the parametric space explored, the texture and topology of the objective function, the resolution of the exploration algorithm and the numerical and modeling accuracy of the CFD simulation. The interested reader is advised to read current literature on this fascinating subject.

Problems

1. Find the minima and maxima of the function

$$f_1(x) = 2x^4 + 4x^3 - 7x^2 - 8x + 10 \tag{Q7.1}$$

over the interval [−3, 3] using the conditions on $f_1'(x)$, $f_1''(x)$ and $f_1'''(x)$ for maxima or minima to occur.

2. A tent has a square base of side *a*, four equal sides of height *b* and a regular pyramid of height *h* on top. If the whole tent is made of canvas, then find the dimensions *a* and *h* such that the area of the canvas required is minimum for a fixed height *b* and a fixed volume *V* inside the tent.
3. Consider the spring-cart system shown in Fig. Q7.1. Cart A is connected to the left wall by a spring with a spring constant k and to Cart B by another spring with the same spring constant. Cart B is also connected to the left wall by a spring with a spring constant of 2k. In the initial positions of the two carts, the springs are unstretched. A horizontal force F is applied on Cart B. Find the displacements x_A* and x_B* of the carts in the new equilibrium position which is achieved under the condition of minimum potential energy (P) of the system. Assume that the springs are perfectly elastic and the carts are rigid and move frictionlessly. Note that for cart displacements of x_A and x_B, the potential energy of the system is the sum of the strain energies in the three springs minus the work done by the force F, i.e.,

$$P(x_A, x_B) = \frac{1}{2}kx_A^2 + \frac{1}{2}k(x_B - x_A)^2 + \frac{1}{2}\,2kx_B^2 - Fx_B \tag{Q7.2}$$

Use the gradient condition and the Hessian condition to find the minimum.

4. Minimize

$$f_2(x_1, x_2, x_3) = x_1^2 + x_2^2 - x_3^2 \tag{Q7.3}$$

subject to

$$x_1 + x_2 + x_3 \geq 5 \tag{Q7.4a}$$

$$x_2 x_3 \geq 2 \tag{Q7.4b}$$

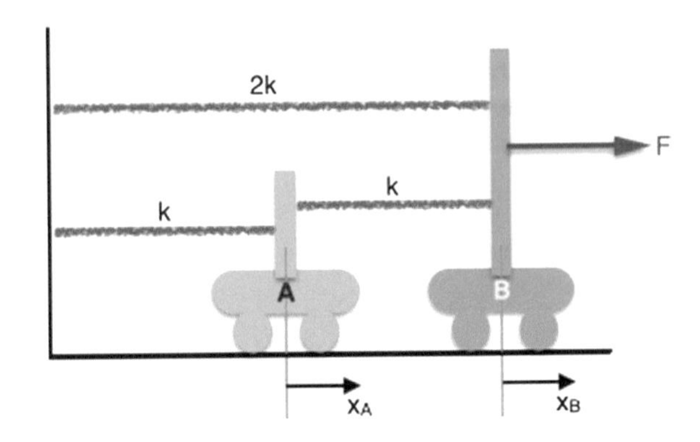

Fig. Q7.1 Geometry and specification of the spring-cart problem

$$x_1 \geq 0,\ x_2 \geq 0,\ x_3 \geq 2 \quad \text{(Q7.4c)}$$

5. Find the minima of $f_1(x)$ of Problem #1 using the bracketing method and the Golden Section method.
6. Repeat Problem #5 using the Newton method and the quasi-Newton method.
7. Plot and find the minimum graphically of the De Jong function in two dimensions given by

$$f_3(x,\ y) = x^2 + y^2, \quad -5 \leq x,\ y \leq 5 \quad \text{(Q7.5)}$$

   Find the minimum using the univariate search method.

8. Repeat Problem #5 using the conjugate directions method and the simplex method.
9. Plot and find the minimum graphically of the Rosenbrock's valley function, which is also known as the second function of De Jong:

$$f_4(x,\ y) = 100\left(y - x^2\right)^2 + \left(1 - x^2\right)^2 \quad -3 \leq x,\ y,\ \leq 3 \quad \text{(Q7.6)}$$

Find the minimum using the steepest descent method and the conjugate gradient method.

10. Repeat Problem #7 using the Newton method, the Marquardt method and the BFGS method.
11. Repeat Problem #7 using genetic algorithm method.
12. Consider the Himmelblau function given by

$$f_5(x_1,\ x_2) = \left(x_1^2 + y - 6\right)^2 + \left(x_1 + x_2^2 - 10\right)^2 \ -10 \leq x_1,\ x_2 \leq 10 \quad \text{(Q7.7)}$$

Locate its local minima and the global minimum by plotting the function in the given interval of the variables.

13. Use the conjugate gradient method and the BFGS method to find the minima of $f_5(x_1, x_2)$.
14. Use the Box method to find the minima of $f_5(x_1, x_2)$.
15. Use the genetic algorithm approach to find the global minimum of $f_5(x_1, x_2)$.

References

Acharya S, Baliga BR, Karki K, Murthy JY, Prakash C, Vanka SP (2007) Pressure-based finite-volume methods in computational fluid dynamics. J Heat Transfer, Trans ASME 129:407–424

Agrell J, Birgersson H, Boutonnet M (2002) Steam reforming of methanol over a Cu/ZnO/Al2O3 catalyst: A kinetic analysis and strategies for suppression of CO formation. J Power Sources 106:249–257

Ammara I, Masson C (2004) Development of a fully coupled control-volume finite element method for the incompressible Navier-Stokes equations. Int J Numer Methods Fluids 44: 621–644

Anderson WK, Thomas JL, van Leer B (1986) A comparison of finite volume flux vector splittings for the Euler equations. AIAA J 24:1453–1460

Avvari R, Jayanti S (2016) Flow apportionment algorithm for optimization of power plant ducting. Appl Thermal Engg 94:715–726

Axelsson O (1972) A generalized SSOR method. BIT Numer Methods 12(4):443–467

Axelsson O (1974) On preconditioning and convergence acceleration in sparse matrix problems. CERN Technical Report 74-10, Data Handling Division, Geneva, Switzerland

Axelsson O (1994) Iterative solution methods. Cambridge University Press, Cambridge, UK

Banks A, Vincent J, Anyakoha C (2007) A review of particle swarm optimization. Part I: background and development. Nat Comput 6:467–484

Banks A, Vincent J, Anyakoha C (2008) A review of particle swarm optimization. Part II: hybridisation, combinatorial, multicriteria and constrained optimization, and indicative applications. Nat Comput 7:109–124

Bayliss A, Turkel E (1982) Far field boundary conditions for compressible flows. J Comp Phys 48:182–199

Beam RM, Warming RF (1976) An implicit finite difference algorithm for hyperbolic systems in conservation law form. J Comp Phys 22:87–110

Beam RM, Warming RF (1978) An implicit factored scheme for the compressible Navier-Stokes equations. AIAA J 16:393–401

Beam RM, Warming RF (1982) Implicit numerical methods for the compressible Navier-Stokes and Euler equations. Von Karman Institute Lecture Series 1982-04, Rhode Saint Genese, Belgium

Benton ER, Platzman GW (1972) A table of solutions of the one-dimensional Burgers equation. Quart Appl Math 30:195–212

Bird RB, Armstrong RC, Hassager O (1987) Dynamics of polymeric liquids (vol I: Fluid Mechanics, 2nd ed). John Wiley, Hoboken

Bird RB, Stewart WE, Lightfoot EN (2002) Transport phenomena, 2nd edn. Wiley, New York

Birkhoff G (1972) The numerical solution of elliptic equations, CBMS-NSF Regional Conference Series in Applied Mathematics, SIAM

Blazek J (2001) Computational fluid dynamics: principles and applications. Elsevier, Amsterdam

S. Jayanti, *Computational Fluid Dynamics for Engineers and Scientists*,
https://doi.org/10.1007/978-94-024-1217-8

Book DL, Boris JP, Hain K (1975) Flux-corrected transport II: generalizations of the method. J Comp Phys 18:248–283
Boris JP, Book DL (1973) Flux corrected transport I:SHASTA, a fluid transport algorithm that works. J Comp Phys 11:38–69
Boris JP, Book DL (1976) Flux-corrected transport. III. Minimal-error FCT algorithms. J Comp Phys 20:397–431
Bolten M, Brandt A, Brannick J, Frommer A, Kahl K, Livshits I (2011) A bootstrap algebraic multilevel method for Markov chains. SIAM J Sci Comput 33(6):3425–3446
Bowyer A (1981) Computing dirichlet tessellations. Comput J 24:162–166
Box MJ (1965) A new method of constrained optimization and a comparison with other methods. Comput J 8:42–52
Brackbill JU, Kothe DB, Zemach C (1992) A continuum method for modeling surface tension. J Comput Phys 100:335–354
Brandt A (1977) Multilevel adaptive solutions to boundary value problems. Math Comput 31: 333–390
Briggs WL (1987) A multigrid tutorial. SIAM, Lancaster Press, Philadelphia, PA, USA
Briley WR, McDonald H (1975) Numerical prediction of incompressible separation bubbles. J Fluid Mech 69:631–656
Broyden CG (1967) Quasi-Newton methods and their application to function minimization. Math Comput 21:368
Burgers JM (1948) A mathematical model illustrating the theory of turbulence. Adv Appl Mech 1:171–199
Caretto LS, Curr RM, Spalding DB (1972a) Two numerical methods for three-dimensional boundary layers. Comput Methods Appl Mech Eng 1:39–57
Caretto LS, Gosman AD, Patankar SV, Spalding DB (1972b) Two calculation procedures for steady, three-dimensional flows with recirculation. In: Proceedings of Third International Conference Numerical Methods Fluid Dynamics, Paris; Lecture Notes Physics, vol 19. Springer, New York, pp 60–68
Cauchy A (1847) Methodes generals pour la resolution des systemes d'equations simultanees. CR Acad Sci Par 25:536–538
Černý V (1985) Thermodynamical approach to the traveling salesman problem: an efficient simulation algorithm. J Optim Theory Appl 45(1):41–51
Chakravarthy SR, Osher S (1985) A new class of high accuracy TVD schemes for hyperbolic conservation laws, AIAA Paper 85-0363, Reno, Nevada, USA
Chang JL, Kwak D (1984) On the method of pseudo compressibility for numerically solving incompressible flows, AIAA Paper 84-0252, AIAA 22nd Aerospace Sciences Meeting
Charney JG, Fjortoft KO, von Neumann J (1950) Numerical integration of the barotropic vorticity equation. Tellus 2:237–254
Chen ZJ, Przekwas AJ (2010) A coupled pressure-based computational method for incompressible/compressible flows. J Comp Phys 229:9150–9165
Chen X, Diez M, Kandasamy M, Zhang Z, Campana EF, Stern F (2015) High-fidelity global optimization of shape design by dimensionality reduction, metamodels and deterministic particle swarm. Eng Optim 47(4):473–494
Choi D, Merkle CL (1985) Application of time-iterative schemes to incompressible flow. AIAA J 23:1518–1524
Chorin AJ (1967) A numerical method for solving incompressible viscous flow problems. J Comp Phys 2:12–26
Chorin AJ (1968) Numerical solution of the Navier-Stokes equations. Math Comput. 22:745–762
Chung TJ (2002) Computational fluid dynamics. Cambridge University Press, Cambridge, UK
Ciarlet PG (1989) Introduction to numerical linear algebra and optimisation. Cambridge University Press, Cambridge, UK
Clerc M, Kennedy J (2002) The particle swarm: explosion, stability and convergence in a multi-dimensional complex space. IEEE Trans Evol Comput 6:58–73

Colorni A, Dorigo M, Maniezzo V (1991) Distributed Optimization by Ant Colonies. In: Proceedings of European conference on artificial life, Paris, France, pp 134–142
Concus P, Golub GH, O'Leary DP (1976) A generalized conjugate gradient method for the numerical solution of elliptic partial differential equations. In: Bunch JR, Rose DJ (eds) Sparse matrix computations. Academic Press, Orlando, FL, USA, pp 309–322
Coulson CA, Jeffrey A (1977) Waves: a mathematical approach to the common types of wave motion (2nd ed). Longman Group Limited, Harlow
Courant R, Friedrichs KO, Lewy H (1928) Uber die partiellen differenzgleichungen der mathematischen Physik. Mathematische Annalen 100:32–74
Courant R, Isaacson E, Reeves M (1952) On the solution of nonlinear hyperbolic differential equations by finite differences. Comm Pure Appl Math 5:243–255
Crank J, Nicolson P (1947) A practical method for numerical integration of solutions of partial differential equations of heat conduction type. Proc Cambridge Philos Soc 43:50–67
Darwish M, Sraj I, Moukalled F (2007) A coupled incompressible flow solver on structured grids. Numerical Heat Transfer, Part B: Fundamentals 52:353–371
Darwish M, Sraj I, Moukalled F (2009) A coupled finite volume solver for the solution of incompressible flows on unstructured grids. J Comp Phys 228:180–201
Davidon WC (1959) Variable metric method of minimization. ReportANL-5990, Argonne National Laboratory, Argonne, IL, USA
Davidson PA (2004) Turbulence: an introduction for scientists and engineers. Oxford Univ Press, Oxford, UK
Deb K (2004) Optimization for engineering design: algorithms and examples. Prentice Hall of India
Deen WH (1998) Analysis of transport phenomena. Oxford University Press, New York, USA
Delery J, Dussauge JP (2009) Some physical aspects of shock wave/ boundary layer interactions. Shock Waves 19:453–468
Demirdžic I, Lilek Z, Perić M (1993) A collocated finite volume method for predicting flows at all speeds. Int J Numer Methods Fluids 16:1029–1050
Dhir VK (2006) Mechanistic prediction of nucleate boiling heat transfer–Achievable or a hopeless task? J Heat Transfer 128:1–12
Dorigo M, Maniezzo V, Colorni A (1996) Ant system: optimization by a colony of cooperating agents. IEEE Trans Syst Man Cybern-Part B Cybern 26(1):29–41
Dorigo M, Stützle T (2004) Ant colony optimization. MIT Press, Cambridge, MA, USA
Drew DA, Lahey RT Jr (1979) Application of general constitutive principles to the derivation of multidimensional two-phase flow equations. Int J Multiph Flow 5:243–264
DuFort EC, Frankel SP (1953) Stability conditions in the numerical treatment of parabolic differential equations. Math Tables Other Aids Comput 7:135–152
Duraiswami R, Prosperetti A (1992) Orthogonal mapping in two dimensions. J Comp Phys 98:254–268
Eberhart RC, Shi Y (2000) Comparing inertia weights and constriction factors in particle swarm optimization. In: Proceedings of IEEE congress on evolutionary computation, San Diego, CA, pp 84–88
Enquist B, Osher S (1980) Stable and entropy satisfying approximations for transonic flow calculations. Math Comput 34:45–75
Enquist B, Osher S (1981) One-sided difference approximations for nonlinear conservation laws. Math Comput 36:321–352
Ferziger JH, Peric M (1999) Computational methods for fluid dynamics, 2nd edn. Springer, Berlin
Fletcher CAJ (1991) Computational techniques for fluid dynamics 1, 2nd edn. Springer-Verlag, Berlin
Fletcher R (1970) A new approach to variable metric algorithms. Computer Journal 13:317–322
Fletcher R, Powell MJD (1963) A rapidly convergent descent method for minimization. Computer Journal 6:163–168

Fletcher R, Reeves CM (1964) Function minimization by conjugate gradients. Computer Journal 7:149–154
Freund RW, Nachtigal MH (1991) QMR: A quasi-minimal residual method for non-Hermitian matrices. Numer Math 60:315–339
Frisch U (1995) Turbulence: the legacy of A.N. Kolmogorov. Cambridge University Press, Cambridge, UK
Fromm JE (1968) A method for reducing dispersion in convective difference schemes. J Comp Phys 3:176–189
Galpin PF, Van Doormall JP, Raithby GD (1985) Solution of the incompressible mass and momentum equations by application of a coupled equation line solver. Int J Numer Meth Fluids 5(7):615–625
Gerlach D, Alleborn N, Buwa V, Durst F (2007) Numerical simulation of periodic bubble formation at a submerged orifice with constant gas flow rate. Chem Eng Sci 62:2109–2125
Ghia U, Ghia KN, Shin CT (1982) High resolution incompressible flow using the Navier-Stokes equations and a multi-grid method. J Comp Phys 48:387–411
Gidaspow D, Jung J, Singh RK (2003) Hydrodynamics of fluidization using kinetic theory: An emerging paradigm 2002 Fluor-Daniel lecture, Powder Technol 148(Special Issue):123–141
Godunov SK (1959) Finite-difference method for numerical computation of discontinuous solutions of the equations of fluid dynamics. Mat Sbornik 47:271–306
Goldfarb D (1970) A family of variable metric methods derived by variational means. Math Comput 24:23–26
Goldberg DE (1989) Genetic algorithms in search, optimization and Machine Learning. Addison-Wesley, Reading, MA, USA
Golub GH, van Loan C (1990) Matrix computations. Johns Hopkins Univ. Press, Baltimore, USA
Gordon S, McBride BJ (1971) Computer program for calculation of complex chemical equilibrium compositions. Rocket Performance, Incident and Reflected Shocks, and Chapman-Jouguet Detonation, NASA SP-273
Gordon W, Hall C (1973) Construction of curvilinear coordinate systems and application to mesh generation. Int J Numer Methods Engg 7:461–477
Gresho PM (1991) Some current CFD issues relevant to the incompressible Navier-Stokes equations. Comput Methods Appl Mech Eng 87:201–252
Gresho PM, Gartling DK, Torczynski JR, Cliffe KA, Winters KH, Garratt TJ, Spence A, Goodrich JW (1993) Is the steady viscous incompressible two-dimensional flow over a backward-facing step at Re = 800 stable? Int J Numer Methods Fluids 17:501–541
Guin JA (1967) Modification of the complex method of constrained optimization. Comput J 10 (4):416–417
Hackbush W (1985) Multigrid methods and applications. Springer, Berlin, Germany
Hackbush W, Trottenberg U (1982) Multigrid methods. Lect Notes Math. 960 (Springer, New York)
Hadamard J (1952) Lectures on cauchy's problem in linear partial differential equations. Dover, New York
Hanby RF, Silvester DJ, Chew JW (1996) A comparison of coupled and segregated iterative solution techniques for incompressible swirling flow. Int J Num Method Fluids 22:353–373
Harlow FH, Welch JE (1965) Numerical calculation of time-dependent viscous incompressible flow of fluids with free surface. Phys Fluids 8:2182–2189
Harlow FH, Welch JE (1966) Numerical simulation of large-amplitude free-surface motions. Phys Fluids 9(5):842–851
Harten A (1983) High resolution schemes for hyperbolic conservation laws. J Comp Phys 49: 357–393
Harten A, Hyman JM (1983) Self-adjusting grid methods for one-dimensional hyperbolic conservation laws. J Comp Phys 50:235–269
Harten A (1984) On a class of high resolution total variation stable finite difference schemes, SIAM. J Numer Anal 21:1–23

Harten A, Engquist B, Osher S, Chakravathy SR (1987) Uniformly high order accurate essentially non-oscillatory schemes, III. J Comp Phys 71:231–303

Henson VE (2003) Multigrid methods for nonlinear problems: an overview. In: Proceedings of SPIE 5016, Computational Imaging, pp 36–48. doi:10.1117/12.499473

Hestenes MR, Stiefel E (1952) Methods of conjugate gradient s for solving linear systems. J Res Nat Bur Stand Sect B 49:409–436

Hildebrand FB (1956) Introduction to numerical analysis. McGraw-Hill, New York

Hirsch C (1988) Numerical computation of internal and external flows, volume 1: fundamentals of numerical discretization. John Wiley & Sons, Chichester, UK

Hirsch C (1990) Numerical computation of internal flows and external flows, vol 2: computational methods for inviscid and viscous flows. John Wiley, New York

Hirt CW, Nichols BD (1982) Volume of fluid method for dynamics of free boundaries. J Com Phys 39:201–225

Holland JH (1962) Outline for a logical theory of adaptive systems. J Assoc Comput Machinery 3:297–314

Holland JH (1975) Adaptation in natural and artificial systems. University of Michigan Press, Ann Arbour

Hottel HC, Sarofim AF (1967) Radiative heat transfer, 1st edn. McGraw-Hill, New York

Hugoniot H (1889) Sur la propagation du movement dans les corps et spécialement dans les gaz parfaits. J de l'Ecole Polytech 58:1–25

Issa RI, Lockwood FC (1977) On the prediction of two-dimensional supersonic viscous interaction near walls. AIAA J 15:182–188

Issa RI (1986) Solution of the implicitly discretized fluid flow equations by operator splitting. J Comput Phys 62:40–65

Issa RI, Gosman AD, Watkins AP (1986) The computation of compressible and incompressible recirculating flows by a non-iterative implicit scheme. J Comp Phys 62:66–82

Ivan L, Groth CP (2014) High-order solution-adaptive central essentially non-oscillatory (CENO) method for viscous flows. J Comp Phys 257:830–862

Ives DC (1982) Conformal grid generation, in numerical grid generation. In: JF Thompson (ed) Proceedings of a Symposium on the numerical grid generation of curvilinear coordinate Systems and Their Use in the Numerical Solution of Partial Differential Equations. Elsevier, New York, pp 107–130

Jang DS, Jetli R, Acharya S (1986) Comparison of the PISO, SIMPLER and SIMPLEC algorithms for the treatment of pressure-velocity coupling in steady flow problems. Numer Heat Transfer 10(3):209–228

Jayanti S, Wilkes NS, Clarke DS, Hewitt GF (1990) The prediction of flow over rough surfaces and its application to the interpretation of mechanisms of horizontal annular flow. Proc Roy Soc Lond A431:71–88

Jayanti S (2001) Hydrodynamics of jet mixing in vessels. Chem Eng Sci 56:193–210

Jones WP, Launder BE (1972) The prediction of laminarization with a two-equation model of turbulence. Int J Heat Mass Trans 15:301–314

Jones WP, Lindstedt RP (1988) Global reaction schemes for hydrocarbon combustion. Combustion and Flame 73(3):233–249

Karki KC, Patankar SV (1989) Pressure-based calculation procedure for viscous flows at all speeds in arbitrary configurations. AIAA J 27:1167–1174

Kennedy J, Eberhart R (1995) Particle swarm optimization. In: Proceedings of IEEE International Conference on Neural Networks IV, pp 1942–1948. doi:10.1109/ICNN.1995.488968

Kennedy J, Eberhart RC (2001) Swarm intelligence. Morgan Kaufmann. ISBN 1-55860-595-9

Kirkpatrick S, Gelatt CD Jr, Vecchi MP (1983) Optimization by simulated annealing. Science 220:671–680

Kim KH, Kim C, Rho O-H (2001) Methods for the accurate computations of hypersonic flows, Part I: ASUMPW+ scheme. J Comp Phys 174:38–80

Launder BE (1989) Second-moment closure and its use in modeling industrial turbulent flows. Int J Numer Meth Fluids 9(8):963–985
Lawn CJ (1987) Principles of combustion engineering for boilers. Academic Press, Cambridge
Lax PD (1954) Weak solutions of nonlinear hyperbolic equations and their numerical computation. Comm Pure Appl Math 7:159–193
Lax PD (1973) Hyperbolic systems of conservation laws and the mathematical theory of shock waves. SIAM, Philadelphia, PA, USA
Lax PD, Wendroff B (1960) Systems of conservations laws. Commun Pure Appl Math 13:217–237
Lax PD, Wendroff B (1964) Difference schemes for hyperbolic equations with high order of accuracy. Commun Pure Appl Math 17:381–397
Leister HJ, Peric M (1994) Vectorized strongly implicit solving procedure for a seven-diagonal coefficient matrix. Int J Num Methods Heat Fluid Flow 4(2):159–172
Leonard BP (1979) A stable and accurate convective modeling procedure based on quadratic upstream interpolation. Comp Meth Appl Mech Eng 19:59–98
Leonard BP, Niknafs HS (1991) Sharp monotonic resolution of discontinuities without clipping of narrow extrema. Comput Fluids 19:141–154
Lesieur M (1986) Turbulence in Fluids. Kluwer Academic Press
Levenspiel O (1972) Chemical reaction engineering. Wiley, Hoboken
Lighthill J (1978) Waves in fluids. Cambridge University Press, Cambridge, UK
Lilek Z (1995) Ein Finite-Volumen Verfahren zur Berechnung von inkompressiblen und kompressiblen Strömungen in komplexen Geometrien mit beweglichen Rändern und freien Oberflachen, Dissertation, University of Hamburg, Germany
Lilek Z, Muzaferija S, Perić M (1997) Efficiency and accuracy aspects of a full-multigrid SIMPLE algorithm for three-dimensional flows. Numer Heat Transf B 31(1):23–42
Liou M-S (1996) A sequel to AUSM: AUSM$^+$. J Comp Phys 129:364–382
Liou M-S (2006) A sequel to AUSM, Part II: ASUM$^+$-up for all speeds. J Comp Phys 214:137–170
Liou M-S, Steffen CJ Jr (1993) A new flux splitting scheme. J Comp Phys 107:23–39
Lo SH (1985) A new mesh generation scheme for arbitrary planar domains. Int J Numer Methods Eng 21:1403–1426
MacCormack RW (1969) The effect of viscosity in hypervelocity impact cratering, AIAA Paper 69-354. Cincinnati, Ohio
Magnussen BF, Hjertager BH (1977) On mathematical models of turbulent combustion with special emphasis on soot formation and combustion. In: 16th symposium (international) on combustion. The Combustion Institute, pp 719–729
Marion MJ (1982) Numerical analysis, a practical approach. Macmillian, New York, NY
Marquardt D (1963) An algorithm for least squares estimation of nonlinear parameters. SIAM J Appl Math 11:431–441
Mavriplis DJ (1995) Unstructured mesh generation and adaptivity. ICASE Report No 95-26, NASA Contractor Report 195069, NASA Langley, Hampton, VA
McDonald PW (1971) The computation of transonic flow through two-dimensional gas turbine cascade, ASME Paper 71-GT-89
McGuirk J, Page G (1991) Shock capturing using a pressure correction method, AIAA Paper No 89-0561, Reno, Nevada
Melaaen MC (1992) Calculation of fluid flows with staggered and nonstaggered curvilinear nonorthogonal grids. The theory, Numerical Heat Transfer, Part B: Fundamentals, 21, 1–19; A comparison 21:21–39
Merriam ML (1991) An efficient advancing-front algorithm for Delaunay triangulation. AIAA paper 91-0792
Metropolis N, Rosenbluth A, Rosenbluth M, Teller A, Teller E (1953) Equation of state calculations by fast computing machines. J Chem Phys 21:1087–1092
Meurant G (1999) Computer solution of large linear systems. North Holland Press
Mohammadi B, Pironneau O (2010) Applied shape optimization for fluids. Oxford University Press, New York

Mudde RF, van den Akker HEA (2001) 2D and 3D simulations of an internal airlift loop reactor on the basis of a two-fluid model. Chem Eng Science 56:6351–6358
Murthy JY, Mathur SR (1997) A pressure-based method for unstructured meshes. Numer Heat Trans Fundam 31(2):195–215
Navier CLMH (1823) Mémoir sure les lois du mouvement des fluides. Mem Acad Inst Sci Fr 6:389–440
Nelder JA, Mead R (1965) A simplex method for function minimization. Comput J 7:308
Nelson JM (1978) A triangulation algorithm for arbitrary planar domains. Appl Math Modeling 2 (151–159):1978
Nocedal J, Wright SJ (2006) Numerical optimization, 2nd edn. Springer, Berlin
Nowneswara Reddy C, Jayanti S (2013) A model for the prediction of safe heat flux from a downward-facing hot patch. Nucl Eng Des 265:45–52
O'Hayre R, Cha SW, Colella W, Prinz FB (2009) Principles of fuel cells, 2nd edn. Wiley, Hoboken
Osher S, Chakravarthy SR (1983) Upwind schemes and boundary conditions with applications to Euler equations in general coordinates. J Comp Phys 50:447–481
Osher S, Sethian JA (1988) Fronts propagating with curvature dependent speed: algorithms based on Hamilton-Jacobi formulations. J Comput Phys 79:12–49
Patankar SV (1980) Numerical Heat Transfer and Fluid Flow. Hemisphere, Washington D.C, USA
Patankar SV, Spalding DB (1972) A calculation procedure for heat, mass and momentum transfer in three-dimensional parabolic flows. Int J Heat Mass Transfer 15:1787–1806
Peaceman DW, Rachford HH (1955) The numerical solution of parabolic and elliptic differential equations. J Soc Ind Appl Math 3:28–41
Peters N (1984) Laminar diffusion flamelet models in non-premixed turbulent combustion. Prog Energy Combust Sci 10:319–339
Pope SB (2000) Turbulent flows. Cambridge University Press, Cambridge, UK
Powell MJD (1964) An efficient method for finding the minimum of a function of several variables without calculating derivatives. Comp J 7:303–307
Pozrikidis C (1998) Numerical computation in science and engineering. Oxford University Press, New York
Prandtl L (1925) Bericht uber Untersuchungen zur ausgebildeten Turbulenz. Z Agnew Math Mech 5(2):136–139 (English translation available as NACA TM 1231, 1949)
Rankine WJM (1870) On the thermodynamic theory of waves of finite longitudinal disturbance. Trans Roy Soc 160:277–288 London
Rao SS (2009) Engineering optimization. John Wiley & Sons, Hoboken, NJ, USA
Rasquin M, Deconinck H, Degrez G (2010) FlexMG: a new library of multigrid preconditioners for a spectral/finite element incompressible flow solver. Int J Numer Methods Engg 82 (12):1510–1536
Reid J (1971) On the method of conjugate gradients for the solution of large sparse systems of linear equations. Large sparse sets of linear equations. Academic Press, Orlando, FL, USA, pp 231–254
Renardy M (1997) Imposing no boundary condition at outflow: why does it work? Int J Numer Methods Fluids 24:413–417
Reynolds WC (1987) Fundamentals of turbulence for turbulence modeling and simulation. Lecture Notes for Von Karman Institute Agard. Report No 755
Rhie CM, Chow WL (1983) Numerical study of the turbulent flow past an air foil with trailing edge separation. AIAA J 21:1525–1532
Rhine JM, Tucker RT (1991) Modelling of gas-fired furnaces and boilers. McGraw-Hill, New York
Richtmyer RD, Morton KW (1967) Difference methods for initial-value problems, 2nd edn. Interscience Publishers, New York
Rizzi A, Eriksson LE (1985) Computation of inviscid incompressible flow with rotation. J Fluid Mech 153:275–312

Roe PL (1981) Approximate Riemann solvers, parameter vectors and difference schemes. J Comp Phys 43:357–372
Roe PL (1985) Some contributions to the modelling of discontinuous flows, Proc. 1983 AMS-SIAM Summer Seminar on Large Scale Computing in Fluid Mechanics. Lect Appl Math 22:163–195
Rubin SG, Khosla PK (1982) Polynomial interpolation method for viscous flow calculations. J Comp Phys 27:153–182
Rubin S, Tannehill J (1992) Parabolized/ reduced Navier-Stokes computational techniques. Annu Rev Fluid Mech 24:117–144
Saad Y, Schultz MH (1986) GMRES: a generalized residual algorithm for solving non-symmetric linear systems. SIAM J Sci Stat Comput 7:856–869
Sani RL, Gresho PM (1994) Resume and remarks on the open boundary conditions, Minisymposium. Int J Numer Methods Fluids 18:983–1008
Samareh JA (1999) A survey of shape parameterization techniques. NASA/CP-1999-209136: 333–343
Scardovelli R, Zaleski S (1999) Direct numerical simulation of free-surface and interfacial flow. Annu Rev Fluid Mech 31:567–603
Schneider GE, Zedan M (1981) A modified strongly implicit procedure for the numerical solution of field problems. Numer Heat Transfer 4:1–19
Schlichting H (1960) Boundary layer theory. McGraw-Hill, New York
Sederberg TW (2012) Computer aided geometric design. Lecture Notes, Brigham Young University. http://www.tom.cs.byu.edu/557/text/cagd.pd. Accessed 7 Jul 2013
Shanno DF (1970) Conditioning of quasi-Newton methods for function minization. Math Comput 24:647–656
Shyy W, Sun C-S (1993) Development of a pressure-correction/staggered-grid based multigrid solver for incompressible recirculating flows. Comput Fluids 22:51–76
Siegel R, Howell JR (2002) Thermal radiation heat transfer. Taylor & Francis, New York
Simcox S, Wilkes NS, Jones IP (1992) Computer simulation of the flows of hot gases from the fire at King's Cross underground station. Fire Saf J 18(1):49–73
Smoot LD, Smith PJ (1985) Coal gasification and combustion. Plenum Press, New York
Sod GA (1978) A survey of several finite difference methods for systems of nonlinear hyperbolic conservation laws: a review. J Comp Phys 27:1–31
Sonneveld P (1989) CGS, a fast Lanczos type solver for non-symmetric linear systems. SIAM J Sci Stat Comput 10:36–52
Spalding DB (1971) Mixing and chemical reaction in steady confined turbulent flames. In: Proceeding of 13th symposium (international) on combustion. The Combustion Institute, pp 649–657
Spendley W, Hext GR, Himsworth FR (1962) Sequential application of simplex designs in optimization and evolutionary operation. Technometrics 4:441
Srinivasan K (2015) Design of fluid flow ducting elements using shape functions, a search method and CFD. PhD Thesis, Department of Chemical Engineering, Indian Institute of Technology Madras, Chennai, India
Srinivasan K, Jayanti S (2015) An automated procedure for the optimal positioning of guide plates in a flow manifold using Box complex method. Appl Therm Eng 76:292–300
Srinivasan K, Balamurugan V, Jayanti S (2017) Optimization of U-bend using shape function, Search Method and CFD. ISME J Thermofluids 03(1):55–67
Srinivasan K, Balamurugan V, Jayanti S (2017) Flow control in t-junction using CFD based optimization. In: Saha AK, Das D, Srivastava R, Panigrahi PK, Muralidhar K (eds) Fluid mechanics and fluid power- contemporary research. Lecture Notes in Mechanical Engineering, Springer Nature, Berlin, pp 687–696
Srinivasan K, Balamurugan V, Jayanti S (2017) Shape optimization of flow split ducting elements using an improved Box complex method. Eng Optim 49:199–215

Steger JL (1978) Coefficient matrices for implicit finite-difference solution of inviscid fluid conservation law equations. Comput Methods Appl Mech Eng 13:175–188

Steger JL, Sorenson RL (1980) Use of hyperbolic differential equations to generate body-fitted coordinates, Numerical Grid Generation Techniques, NASA CP-2166

Steger JL, Warming RF (1981) Flux vector splitting of the inviscid gas dynamic equations with application to finite-difference methods. J Comp Phys 40:263–293

Stokes GG (1845) On the theories of the internal friction of fluids in motion, and of the equilibrium and motion of elastic solids. Trans Camb Phil Soc 8:287–319

Stone HL (1968) Iterative solution of implicit approximations of multidimensional partial differential equations. SIAM J Numer Anal 5:530–558

Stüben K (2001) A review of algebraic multigrid. J Comput Appl Math 128:281–309

Stüben K (2003) Algebraic multigrid (AMG): experiences and comparisons. Appl Math Comput 13:419–452

Suresan H, Jayanti S (2004) The case of an oscillating manometer with variable density and dissipation: experimental and numerical study. Nucl Eng Des 229:59–73

Sussman M, Smereka P, Osher S (1994) A level set approach for computing solutions to incompressible two-phase flow. J Comput Phys 114:146–159

Sweby PK (1984) High resolution schemes using flux limiter for hyperbolic conservation laws. SIAM J Num Anal 21:995–1011

Tannehill JC, Holst TL, Rakich JV (1975) Numerical computation of two-dimensional viscous blunt body flows with an impinging shock, AIAA Paper 75-154, Pasadena, California

Tannehill JC, Anderson DA, Pletcher RH (1997) Computational fluid mechanics and heat transfer, 2nd edn. Taylor and Francis, Philadelphia, PA

Temam R (1969) Sur l'approximation de la solution des equations de Navier-Stokes par la méthode des pas fractionnaires. (I): Arch Ration Mech Anal 32:135–153; (II): Arch Ration Mech Anal 33:377–385

Tennekes H, Lumley JL (1981) A first course in turbulence. MIT Press, USA

Thévenin D, Jániga G (eds) (2008) Optimization and computational fluid dynamics. Springer-Verlag, Berlin

Thomas JL, Salas MD (1986) Far field boundary conditions for transonic lifting solutions to the Euler equations. AIAA J 24:1074–1080

Thompson JF, Thames FC, Mastin CW (1974) Automatic numerical generation of body-fitted curvilinear coordinate system for field containing any number of arbitrary two-dimensional bodies. J Comp Phys 15:299–319

Thompson JF, Warsi ZUA, Mastin CW (1985) Numerical grid generation-foundations and applications. Elsevier, New York

Vanka SP (1986) Block-implicit multigrid solution of Navier-Stokes equations in primitive variables. J Comp Phys 65:138–158

van den Vorst HA, Sonneveld P (1990) CGSTAB, a more smoothly converging variant of CGS. Tech. Report 90–50. Delft University of Technology, The Netherlands

van der Vorst HA (1992) Bi-CGSTAB: A fast and smoothly converging variant of Bi-CG for the solution of non-symmetric linear systems. SIAM J Sci Stat Comput 13:631–644

van Doormal JP, Raithby GD (1984) Enhancement of the SIMPLE method for predicting incompressible fluid flows. Num heat Transfer 7:147–163

van Doormal JP, Raithby GD, McDonald BH (1987) The segregated approach to predicting viscous compressible fluid flows. J Turbomach 109:268–277

van Laarhoven PJ, Aarts EH (1987) Simulated annealing: theory and applications. Springer, Dordrecht

van Leer B (1977) Towards the ultimate conservative difference scheme III: upstream-centered finite-difference schemes for ideal compressible flow. J Comp Phys 23:263–275

van Leer B (1982) Flux vector splitting for the Euler equations. In: Proceedings of 8th International Conference on Numerical Methods in Fluid Dynamics. Lecture Notes Physics, vol 170. Springer, New York, pp 507–512

Verstraete T, Coletti F, Bulle J, Vanderwielen T, Arts T (2013) Optimization of a U-bend for minimal pressure loss in internal cooling channels – part I: numerical method. J Turbomach 135:051015-1–051015-10
Vinokur M (1974) Conservation equations of gas dynamics in curvilinear coordinate systems. J Comp Phys 14:105–125
Viskanta R, Menguc MP (1987) Radiation heat transfer in combustion systems. Prog Energy Comb Sci 13:97–160
von Neumann J, Richtmyer RD (1950) A method for the numerical calculation of hydrodynamic shocks. J Appl Phys 21:232–237
Warming RF, Hyett BJ (1974) The modified equation approach to the stability and accuracy analysis of finite-difference methods. J Comp Phys 14:159–179
Warming RF, Beam RM (1976) Upwind second-order difference schemes and applications in aerodynamic flows. AIAA J 14(9):1241–1249
Warnatz J, Maas U, Dibbe RW (1995) Combustion. Springer, Berlin, Germany
Warsi ZUA (2006) Fluid dynamics: theoretical and computational approaches, 3rd edn. Taylor and Francis, UK
Watson DF (1981) Computing the n-dimensional Delaunay tessellation with application to Voronoi polytopes. Comput J 24:167–172
Wesseling P (1990) Multigrid methods in computational fluid dynamics. Z Angew Math Mech 70: T337–T347
Wesseling P (2004) Principles of computational fluid dynamics. Springer, Berlin, Germany
Whitfield DL, Janus JM (1984) Three-dimensional unsteady Euler equations solution using flux vector splitting. AIAA Paper 84-1552
White FM (1986) Fluid mechanics. McGraw-Hill, New York
White (1991) Viscous fluid flow, 2nd edn. McGraw-Hill, New York
Wilcox DC (1988) Reassessment of the scale-determining equation for advanced turbulence models. AIAA J 26:1299–1310
Wilcox DC (1993) Turbulence modeling for CFD. DCW Industries, Inc., 5354 Palm Drive, La Canada, California 91011
Wilson EO (1962) Chemical communication among workers of the fire ant Solenopsis saevissima (Fr. Smith). 3. The experimental induction of social responses. Anim Behav 10:159–164
Woodward P, Colella P (1984) The numerical simulation of two-dimensional fluid flow with strong shocks. J Comp Phys 54:115–173
Yakhot V, Orzag SA (1986) Renormalization group analysis of turbulence: I. Basic theory. J Sci Comput 1(1):1–51
Young DM, Jea KC (1981) Generalized conjugate gradient acceleration of iterative methods, part II: the nonsymmetrizable case. Report #163, Centre for Numerical Analysis. University of Texas at Austin
Zedan M, Schneider GE (1983) A three-dimensional strongly implicit procedure for heat conduction. AIAA J 21(2):295–303

Index

S. Jayanti, *Computational Fluid Dynamics for Engineers and Scientists*,
https://doi.org/10.1007/978-94-024-1217-8

J

K

L

M

N

O

P

Q

R

Printed in the United States
By Bookmasters